ENGINEERING YOUR FUTURE

A Comprehensive Introduction to Engineering

ENGINEERING YOUR FUTURE

A Comprehensive Introduction to Engineering
SEVENTH EDITION

William C. Oakes, PhD
Purdue University

Les L. Leone, PhD
Michigan State University

Craig J. Gunn, MS
Michigan State University

Contributors

Frank M. Croft, Jr., PhD
Ohio State University

John B. Dilworth, PhD
Western Michigan University

Heidi A. Diefes, PhD
Purdue University

Ralph E. Flori, PhD
University of Missouri-Rolla

Marybeth Lima, PhD
Louisiana State University

Merle C. Potter, PhD
Michigan State University

Michael F. Young, MS
Michigan Technological University

Editor

John L. Gruender

New York Oxford
OXFORD UNIVERSITY PRESS

Oxford University Press, Inc., publishes works that further Oxford University's
objective of excellence in research, scholarship, and education.

Oxford   New York
Auckland   Cape Town   Dar es Salaam   Hong Kong   Karachi
Kuala Lumpur   Madrid   Melbourne   Mexico City   Nairobi
New Delhi   Shanghai   Taipei   Toronto

With offices in
Argentina   Austria   Brazil   Chile   Czech Republic   France   Greece
Guatemala   Hungary   Italy   Japan   Poland   Portugal   Singapore
South Korea   Switzerland   Thailand   Turkey   Ukraine   Vietnam

For titles covered by Section 112 of the US Higher Education Opportunity Act,
please visit www.oup.com/us/he for the latest information about pricing
and alternate formats.

Published by Oxford University Press, Inc.
198 Madison Avenue, New York, New York 10016
http://www.oup.com

Library of Congress Cataloging-in-Publication Data

Oakes, William C., Ph.D.
  Engineering your future: a comprehensive introduction to engineering / William C. Oakes,
Les L. Leone, Craig J. Gunn; contributors, Frank M. Croft, Jr. ... [et al.]; editor,
John L. Gruender. — 7th ed.
     p. cm.
  Includes bibliographical references and index.
  ISBN 978-0-19-979756-1
  1. Engineering—Vocational guidance. 2. Engineering. I. Leone, Les L.   II. Gunn,
Craig J.   III. Croft, Frank M.   IV. Gruender, John L.   V. Title.
 TA157.O223 2012
 620.0023—dc22   2011001715

9 8 7 6 5 4

Printed in the United States of America
on acid-free paper

Contents

Studying Engineering

Preface

You can't make an educated decision about what career to pursue without adequate information. *Engineering Your Future* endeavors to give you a broad introduction to the study and practice of engineering. In addition to presenting vital information, we've tried to make it interesting and easy to read as well.

You might find Chapter 3, Profiles of Engineers, to be of particular interest to you. The chapter includes information from real people—engineers practicing in the field. They discuss their jobs, their lives, and the things they wish they had known going into the profession. Chapter 2, Engineering Majors, also should be a tremendous help to you in determining what areas of engineering sound most appealing to you as you begin your education.

The rest of the book presents such things as a historical perspective of engineering; some thoughts about the future of the profession; some tips on how best to succeed in the classroom; advice on how to gain actual, hands-on experience; exposure to computer-aided design; and a nice introduction to several areas essential to the study and practice of engineering.

We have designed this book for modular use in a freshman engineering course that introduces students to the field of engineering. Such a course differs in content from university to university. Consequently, we have included many topics, too numerous to cover in one course. We anticipate that several of the topics will be selected for a particular course with the remaining topics available to you for outside reading and for future reference.

As you contemplate engineering, you should consider the dramatic impact engineers have had on our world. Note the eloquent words of American Association of Engineering Societies Chair Martha Sloan, a professor of electrical engineering at Michigan Technological University:

> In an age when technology helps turn fantasy and fiction into reality, engineers have played a pivotal role in developing the technologies that maintain our nation's economic, environmental and national security. They revolutionized medicine with pacemakers and MRI scanners. They changed the world with the development of television and the transistor, computers and the Internet. They introduced new concepts in transportation, power, satellite communications, earthquake-resistant buildings, and strain-resistant crops by applying scientific discoveries to human needs.
>
> Engineering is sometimes thought of as applied science, but engineering is far more. The essence of engineering is design and making things happen for the benefit of humanity.

Joseph Bordogna, President of IEEE, adds:

> *Engineering will be one of the most significant forces in designing continued economic development and success for humankind in a manner that will sustain both the planet and its growing population. Engineers will develop the new processes and products. They will create and manage new systems for civil infrastructure, manufacturing, communications, health care delivery, information management, environmental conservation and monitoring, and everything else that makes modern society function.*

We hope that you, too, will find the field of engineering to be attractive, meaningful, and exciting—one that promises to be both challenging and rewarding, and one that matches well with your skills and interests.

For the instructor's convenience, there is a companion website with ancillary materials (PowerPoint lectures and a test bank). This material may be found at

http://www.oup.com/us/companion.websites/9780199797561/

New to this Edition
- Updated Computer Tools chapter covers computer use from the history of the Internet through advanced engineering packages, including spreadsheets, mathematical programs, social media, and communicating via e-mail.
- Effective coverage of teamwork for engineering classes shows students how a diversity of skills is essential to their future careers.
- Updated green and bioengineering information addresses current issues including sustainability and advancing human health.
- Fresh new profiles of contemporary engineers and their current research give students an insider's glimpse into what fields they may be most suited for.
- Profiles of students who are currently majoring or considering majoring in engineering with candid discussions about what they find challenging and the positive experiences they are having in engineering.
- Up-to-date statistics for the job market provide critical data to help inform students' career decisions.
- Updated and timely financial aid information covers student loans, financial assistance, grants, and avoiding scams.

Acknowledgments
The authors are especially grateful to the reviewers whose opinion and comments directly influenced the development of this edition:

Keith Gardiner, Lehigh University
Chris Geiger, Florida Gulf Coast University
Yoon Kim, Virginia State University
Keith Level, Las Positas College
S.T. Mau, California State University at Northridge
John Nicklow, Southern Illinois University at Carbondale
Charles E. Pierce, University of South Carolina
Megan Piccus, Springfield Technical Community College
Ken Reid, Ohio Northern University
Gregory Wight, Norwich University
David Willis, University of Massachusetts at Lowell

—The Authors

ENGINEERING YOUR FUTURE

A Comprehensive Introduction to Engineering

Chapter 1

The History of Engineering

1.1 INTRODUCTION

Engineering involves a continuum, where every new innovation stands on the shoulders of those who have gone before—perhaps including yours some day. How do you get the necessary insight to participate? Look to the stories of history.

As we begin, you will notice that there were few engineering innovations in the early years. As time passed, innovations occurred more rapidly. Today, engineering discoveries are made almost daily. The speed with which things now change indicates the urgency of understanding the process of innovation. But if we don't maintain a big-picture awareness regarding our projects, we may be surprised by unintended consequences. History provides a great opportunity to observe the context of before, during, and after some of the greatest engineering problems ever faced—and of those we face today.

A main thing to realize is that history is not about memorizing names and dates. History is about people. The insights you'll gain from stories about how engineers developed everything from kitchen appliances to high-tech industrial equipment can be very motivating. And you'll really be able to relate to the history of bridges, for instance, after you've been assigned to build a model bridge in class yourself!

Engineers are professionals. Professionals are leaders. To lead you need to understand the origins of your profession. The stories of history give you the foundation you need, as you learn from the great innovators and see how they handled all aspects of problem-solving. Such broad learning can greatly help your career and aid your development as a leader in your profession.

Many types of professionals are required to master the history of their trade as part of their degrees. The best professionals keep studying throughout their careers. Reading papers, magazines, and journals is simply contemporary "historic" study. This habit builds on their foundational knowledge of history. Such a background gives professionals their best chance to know what it takes to move forward in their field—an amazing challenge, as the stories to come will show!

The study of history, of course, not only helps us create new futures, but it also helps us understand what good qualities from the past are worth emulation. Craftsmanship, integrity, and dedication are clearly evident in our forebears' engineering artifacts. And history is full of interesting educational stories, characters, and ingenious development. You can learn what it means to do quality work in a quality way, no matter what the level of your contribution to the profession.

Definition of Engineering

Even if you already have a general knowledge of what engineering involves, a look at the definition of the profession may give you more insight. ABET—The Accreditation Board for Engineering and Technology—defines engineering as:

> *The profession in which knowledge of the mathematical and natural sciences, gained by study, experience, and practice, is applied with judgment to develop ways to use, economically, the materials and forces of nature for the benefit of mankind.*

In simple terms, engineering is about using natural materials and forces for the good of mankind—a noble endeavor. This definition places three responsibilities on an engineer: (1) to develop judgment so that you can (2) help mankind in (3) thrifty ways. Looking at case *histories* and *historic* overviews might help us achieve the insight needed to fulfill those responsibilities.

An engineering professor once said that the purpose of an engineer is *"to interpret the development and activity of man."* Technical coursework teaches us skills, but history can teach us how to *interpret* scenarios, and to sort out the pros and cons of various options. History helps us forge bonds of fellowship that connect us to the past and inspire us to be our best for tomorrow. A solid knowledge of history transforms our schooling from training to true education.

1.2 GETTING STARTED

As we proceed, you should note that the engineer has always had a monumental impact on the human race at every stage of societal development. The few items mentioned here are only the tip of the iceberg when it comes to the contributions that engineers have made to the progress of humanity.

Prehistoric Culture

If you look back at the definition of engineering given by ABET, you will notice an important statement: "The profession in which knowledge of the mathematical and natural sciences . . . is applied. . . ." Individuals involved during prehistoric times in activities which we recognize today as engineering—problem-solving, tool-making, etc.—did not have a grasp of mathematical principles nor knowledge of natural science *as we know it today*. They designed and built needed items by trial and error and intuition. They built some spears that worked and some that failed, but in the end they perfected weapons that allowed them to bring down game animals and feed their families. Since written communication and transportation did not exist at that time, little information or innovation was exchanged with people from faraway places. Each group inched ahead on its own.

However, the innovators of yore would have made fine engineers today. Even in light of their limited skill, their carefully cultivated knowledge of their surroundings was more extensive than we typically can comprehend. Their skill in craftsmanship was often marvelous in its effectiveness, integrity, and intricacy. They passed on knowledge of all aspects of life, which they typically treated as a whole entity, with utmost seriousness to the next generation. This information was carefully memorized and kept accurate, evolving, and alive. Early man even tried to pass on vital information by way of coded cave paintings and etchings as an extra safeguard. Breakthroughs in transportation and exploration are being located ever earlier as we continue to make discoveries about various peoples traveling long before we thought they did—influencing others and bringing back knowledge. We can still see some-thing of the prehistoric approach in some of today's cultures, such as Native Americans, aborigines, and others.

However, despite the strong qualities of prehistoric man—the importance placed on respect for life and a sense of the sacred—their physical existence was harshly limited. The work of engineers can be seen in this light as a quest to expand outer capacity without sac-rificing inner integrity. The physical limitations of prehistoric man can perhaps be highlighted as follows. (How does engineering impact each of the areas listed?)

Physical limitations of prehistoric cultures:

- They had no written language.
- Their verbal language was very limited.
- They had no means of transportation.
- They had no separate concept of education or specialized methodology to discover new things.
- They lived by gathering food and trying to bring down game with primitive weapons.
- Improvement of the material aspects of life came about very slowly, with early, primi-tive engineering.

Our Computer Age

You, on the other hand, live in the information age. With such tools as the Internet, the answers to millions of questions are at your fingertips. We have traveled to the moon and our robots have crawled on Mars. Our satellites are exploring the ends of our known universe. Change is no longer questioned; it is expected. We excitedly await the next model in a series, knowing that as the current model is being sold it has already become obsolete. Speed, furi-ous activity, and the compulsion to never sit still are part of our everyday lives. Let's look at the times when constant change was not the norm. As we progress from those primitive times into the 20th century, you will observe that fury of engineering activity.

We will present a panoramic view of engineering by briefly stating some of the more inter-esting happenings during specific time periods. Notice the kinds of innovations that were introduced. Take a careful look at the relationships that many inventions had with each other. Think about the present, and the connectivity between all areas of engineering and the crit-ical importance of the computer. Innovations do not happen in a vacuum; they are interre-lated with the needs and circumstances of the world at the time.

Activity 1.1

Prepare a report that focuses on engineering in one of the following eras. Analyze the events that you consider to be engineering highlights and explain their importance to the progress of man.

a) Prehistoric man
b) Egypt and Mesopotamia
c) Greece and Rome
d) Europe in the Middle Ages
e) Europe in the Industrial Revolution
f) The 20th century

The assignment above might seem overwhelming—covering an enormous amount of time and information. Actually, the key developments might be simpler to identify than you think. In the pages that follow, however, we will only set the stage for your investigation of your profession. It will be your job to fill in the details for your particular discipline.

The Pace of History

The rate of innovation brings up some interesting points. As we move through the past 6000 years, you should realize that the rate at which we currently introduce innovations is far more rapid than in the past. It used to take years to accomplish tasks that today we perform in a very short time—tasks that we simply take for granted. Think about the last time that your computer was processing slower than you thought it should. Your words may be echoing now: "Come on! Come on! I don't have all day!" In the past, there were often decades without noticeable technological progress. Think of the amount of time that it takes to construct a building today with the equipment that has been developed by engineers. It is not uncommon to see a complete house-frame constructed in a single day. Look in your history books and read about the time it took to create some of the edifices in Europe. You can visit churches today that took as long as 200 years to construct. Would we ever stand for that today?

Now, shift gears and evaluate the *purpose* behind the monumental efforts of the "slow" past. Was their only goal "to get the job done"? Ask them! Look into their stories. There you'll find coherent, colorful explanations for the case of keeping all aspects of culture connected to the main goals of life. The proper connection to God, truth, justice, fate, reality, life, and ancestry was the goal of early science and of many cultures. Not much was allowed to interfere. Even so, the ancients accomplished fantastic feats with only a rudimentary knowledge of the principles you learned as a child. But do we know what else they knew? Perhaps their lack of physical speed was partly voluntary! Perhaps it was surpassed by strengths and insights in other areas. Perhaps today's prowess comes at the loss of other qualities. We need to study history so that we can avoid making Faustian bargains. (Hey! Faust is a character from literary history!) Archimedes, for instance, refused to release information that could be used to make more effective weapons; he knew it would be used for evil and not the pursuit of wisdom; only when his home city of Syracuse was no longer able to hold off Roman attackers did he release his inventions to the military.

The study of history confronts us with dilemmas. Neither side of a true dilemma ever goes away. The story of history is never over. Speed is relative, after all!

Quick Overview

Let's begin with a quick review of the history of engineering—six thousand years in a single paragraph . . .

> Our technological roots can be traced back to the seed gatherer/hunter. Prehistoric man survived by collecting seeds and killing what animals he could chase down. He endured a very lean physical existence. As he gradually improved his security through innovation, humanity increased in numbers, and it was important to find ways to feed and control the growing population. To support larger populations reliably, the methods and implements of farming and security were improved. Much later, after many smaller innovations, specialized industrial man stepped onto the scene, ready to bring the world productivity and material wealth. The pursuit of science prospered due to its usefulness and profitability.

Much abstract research was done with surplus funding. The microchip was developed. And today, the resulting Technological/Post-Industrial/Computer-Information Man is **you**—ready to use the vast information available in the present day to build the world of the future.

Most innovation would not have come to pass were it not for the work of engineers. The sections that follow will present a brief look at some of the highlights of those 6000 years. Spend some time poring over a few classic history texts to get the inside stories on the innovations that interest you the most. Though you might have interest in one particular field of engineering, you might find stories of innovation and discovery in other disciplines to be equally inspiring.

1.3 THE BEGINNINGS OF ENGINEERING

The Earliest Days

The foundations of engineering were laid with our ancestors' effort to survive and to improve their quality of life. From the beginning they looked around their environment and saw areas where life could be made easier and more stable. They found improved ways to hunt and fish. They discovered better methods for providing shelter for their families. Their main physical concern was day-to-day survival. As life became more complicated and small collections of families became larger communities, the need grew to look into new areas of concern: power struggles, acquisition of neighboring tribes' lands, religious observances. All of these involved work with tools. Engineering innovations were needed to further these interests. Of course, in those days projects weren't thought of as separate from the rest of life. In fact, individuals weren't generally thought of as being separate from their community. They didn't look at life from the point of view of specialties and individual interests. Every person was an engineer to an extent.

Modern aborigines still live today much as their ancestors did in prehistoric times. However, frequently even they take advantage of modern engineering in the form of tools, motors and medicine. The Amish can also fall into this category of being a roots-oriented culture. Such cultures tend to use tools only for physical necessities so that the significance of objects doesn't pollute their way of life. This struggle, as we know from contemporary "historic" media reports, has only been partly successful, and has caused conflict and misunderstanding on occasion.

Egypt and Mesopotamia

As cities grew and the need for addressing the demands of the new fledgling societies increased, a significant change took place. People who showed special aptitude in certain areas were identified and assigned to ever more specialized tasks. This labeling and grouping was a scientific breakthrough. It gave toolmakers the time and resources to dedicate themselves to building and innovation. This new social function created the first real engineers, and for the first time innovation flourished rapidly.

Between 4000 and 2000 B.C., Egypt and Mesopotamia were the focal points for engineering activity. Stone tools were developed to help man in his quest for food. Copper and bronze axes were perfected through smelting. These developments were not only aimed at hunting. The development of the plow was allowing man to become a farmer so that he could reside in one place and leave the nomadic life. Mesopotamia also made its mark on engineering by

Figure 1.1 The Step Pyramid of Sakkara.

giving birth to the wheel, the sailing boat, and methods of writing. Engineering skills that were applied to the development of everyday items immediately improved life as they knew it. We will never be able to understand completely the vast importance of the Greeks, Romans, and Egyptians in the life of the engineer.

During the construction of the pyramids (c.2700–2500 B.C.) the number of engineers required was immense. They had to make sure that everything fit correctly, that stones were properly transported long distances, and that the tombs would be secure against robbery. Imhotep (chief engineer to King Zoser) was building the Step Pyramid at Sakkara (pictured in Fig. 1.1) in Egypt about 2700 B.C. The more elaborate Great Pyramid of Khufu (pictured in Fig. 1.2) would come about 200 years later. The story of the construction of the pyramids is one that any engineer would appreciate, so consider doing some research on the pyramids on your own. Or perhaps some day you'll have a chance to take a trip to Egypt to see them for yourself. By investigating the construction of the pyramids, you will receive a clear and fascinating education about the need for designing, building, and testing with any engineering project. These early engineers, using simple tools, performed with great acuity, insight, and technical rigor, tasks that even today give us a sense of pride in their achievements.

The Great Pyramid of Khufu (pictured in Fig. 1.2) is the largest masonry structure ever built. Its base measures 756 feet on each side. The 480-foot structure was constructed of over 2.3 million limestone blocks with a total weight of over 58,000,000 tons. Casing blocks of fine limestone were attached to all four sides. These casing stones, some weighing as much as 15 tons, have been removed over the centuries for a wide variety of other uses. It is hard for us to imagine the engineering expertise needed to quarry and move these base and casing stones, and then piece them together so that they would form the pyramid and its covering.

Here are additional details about this pyramid given by Roland Turner and Steven Goulden in *Great Engineers and Pioneers in Technology, Volume 1: From Antiquity through the Industrial Revolution*:

Buried within the pyramid are passageways leading to a number of funeral chambers, only one of which was actually used to house Khufu's remains.

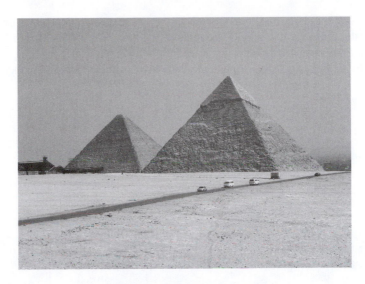

Figure 1.2 The Great Pyramid of Khufu.

The granite-lined King's Chamber, measuring 17 by 34 feet, is roofed with nine slabs of granite which weigh 50 tons each. To relieve the weight on this roof, located 300 feet below the apex of the pyramid, the builder stacked five hollow chambers at short intervals above it. Four of the "relieving chambers" are roofed with granite lintels, while the topmost has a corbelled roof. Although somewhat rough and ready in design and execution, the system effectively distributes the massive overlying weight to the sturdy walls of the King's Chamber.

Sheer precision marks every other aspect of the pyramid's construction. The four sides of the base are practically identical in length—the error is a matter of inches—and the angles are equally accurate. Direct measurement from corner to corner must have been difficult, since the pyramid was built on the site of a rocky knoll (now completely enclosed in the structure). Moreover, it is an open question how the builder managed to align the pyramid almost exactly north-south. Still, many of the techniques used for raising the pyramid can be deduced.

After the base and every successive course was in place, it was leveled by flooding the surface with Nile water, no doubt retained by mud banks, and then marking reference points of equal depth to guide the final dressing. Complications were caused by the use of blocks of different heights in the same course.

The above excerpt mentions a few of the fascinating details of the monumental job undertaken to construct a pyramid with primitive tools and only human labor.

1.4 AN OVERVIEW OF ANCIENT ENGINEERING

The following sections will give you a feel for what was going on from 2000 B.C. to the present. In this section we will review one specific engineering feat of each of the cultures of ancient Greece, Rome, and China. These overviews are meant to demonstrate the effort that went into engineering activities of the past. They represent only a small portion of the many developments of the time. It is important for you to look closely at what was being accomplished and the impact it had upon the people of the time. We will mention a few of those accomplishments, but it is important that you consider on your own what good came from the activities of those early engineers. How did their innovations set the groundwork for what was to come?

Figure 1.3 The Parthenon in Athens.

Engineering the Temples of Greece

The Parthenon, shown in Fig. 1.3, was constructed by Iktinos in Athens in 447 B.C. and was completed by 438 B.C. The temple as we know it was to be built on the foundation of a previous temple. The materials that were used came from the salvaged remains of the previous temple. The Parthenon was designed to house a statue of Athena, which was to be carved by Phidias and stand almost forty feet tall. The temple was to make the statue seem proportioned relative to the space within which it was to be housed. Iktinos performed the task that he was assigned, and the temple exists today as a monument to engineering capability.

> Structural work on the Parthenon enlarged the existing limestone platform of the old temple to a width of 160 feet and a length of 360 feet. The building itself, constructed entirely of marble, measured 101 feet by 228 feet; it was the largest such temple on the Greek mainland. Around the body of the building Iktinos built a colonnade, customary in Greek temple architecture. The bases of the columns were 6 feet in diameter and were spaced 14 feet apart. Subtle harmonies were thus established, for these distances were all in the ratio of 4:9. Moreover, the combined height of the columns and entablatures (lintels) bore the same ratio to the width of the building.

Remember that this was the year 438 B.C. If we asked a contractor today about the skill needed to build this structure with tools that were available 2500 years ago, we'd get quite a reply.

Figure 1.4 Roman Aqueduct.

The Roman Roads and Aqueducts

Construction of the first great Roman Road, the Appian Way, began around 312 B.C. It connected Rome and Capua, a distance of 142 miles. The Appian Way eventually stretched to Brundisium at the very southernmost point in Italy, and covered 360 miles. With this start, the Roman engineers continued building roads until almost A.D. 200. Twenty-nine major roads eventually connected Rome to the rest of the empire. By A.D. 200 construction ceased, and repair and maintenance were the only work done on these roads. Aqueducts were part of this construction. One such aqueduct is shown in Fig. 1.4.

For those interested in civil engineering, the Roman roads followed elaborate principles of construction. A bedding of sand, 4 to 6 inches thick, or sometimes mortar one inch thick, was spread upon the foundation. The first course of large flat stones cemented together with lime mortar were placed upon this bedding of sand. If lime was not available, the stones (none smaller than a man's hand) were cemented together with clay. The largest were placed along the edge to form a retaining wall. This course varied from 10 inches thick on good ground to 24 inches on bad ground. A layer of concrete about 9 inches deep was placed on top of this, followed by a layer of rich gravel or sand concrete. The roadway would generally be 12 inches thick at the sides of the road and 18 inches in the middle, thus creating a crown which caused runoff. While this third course was still wet, the fourth or final course was laid. This was made of carefully cut hard stones. Upon completion these roads would be from 2 feet to 5 feet thick, quite a feat for hand labor.

It is interesting to note that after the fall of Rome, road building was no longer practiced by anyone in the world. It would be many hundreds of years before those who specialized in road building again took on the monumental task of linking the peoples of the world.

Figure 1.5 The Great Wall of China.

The Great Wall of China

In 220 B.C. of the Ch'in Dynasty, Meng T'ien, a military general, led his troops along the borders of China. His primary role was that of a commander of troops charged with the task of repelling the nomadic hordes of Mongolians who occasionally surged across the Chinese border. The Ch'in emperor, Shih Huang Ti, commissioned him to begin the building of what would become known as the Great Wall of China; see Fig. 1.5.

The emperor himself conceived the idea to link all the fortresses that guarded the northern borders of China. The general and the emperor functioned as engineers, even though this was not their profession. They solved a particular problem by applying the knowledge they possessed in order to make life better for their people. The ancient wall is estimated to have been 3,080 miles in length, while the modern wall runs about 1,700 miles. The original wall is believed to have passed Ninghsia, continuing north of a river and then running east through the southern steppes of Mongolia at a line north of the present Great Wall. It is believed to have reached the sea near the Shan-hal-huan River. After serving as a buffer against the nomadic hordes for six centuries, the wall was allowed to deteriorate until the sixth and seventh centuries A.D., when it underwent major reconstruction under the Wei, Ch'i and Sui dynasties. Although the vast structure had lost military significance by the time of China's last dynasty, the Ch'ing, it never lost its significance as a "wonder of the world" and as a feat of massive engineering undertaking.

1.5 TRAVELING THROUGH THE AGES

At the same time that the previous monuments were being built, there were a number of other engineering feats under way. Let's look at a few of those other undertakings.

As you read through the following, we encourage you to investigate on your own any of the engineering accomplishments which grab your attention from this list. You should get an overall sense of both the pace and the focus of development over the centuries from these lists.

Note that as we enter the modern age, the scope of invention appears to narrow, with much of the activity relating to computers. What impact does this have on the notion of "ever-faster historic development"?

What is the role of the inventor in history? Sometimes the name is important, other times not. At times during certain eras, innovations were being made simultaneously by a number of people. So no individual really stands out. Perhaps the developments were more collective during such times. Perhaps the players involved were often racing each other to the patent office. At other times, with certain inventions, a single person made a significant breakthrough on his own.

When is it the person and when is it the times? Here are a few clues. When it's the person, his peers might think he's a nut, his work might even be outlawed or ignored. When it's the times, there are frequently many innovators doing similar work in close proximity or even in cooperation. Sometimes even when a name stands out, you still get the impression that his effort was more communal than singular. These are concepts to consider as you watch time and innovations flow past in these lists and in your studies.

1200 B.C. – A.D. 1

- The quality of wrought iron is improved.
- Swords are mass produced.
- Siege towers are perfected.
- The Greeks develop manufacturing.
- Archimedes introduces mathematics in Greece.
- Concrete is used for the arched bridges, roads, and aqueducts in Rome.

Activities 1.2

a) Investigate the nature of manufacturing in these early times.

b) Investigate warfare as it was first waged. What was the engineer's role in the design of war-related equipment? What did these engineers build during peace-time?

c) Concrete was being used in Rome before A.D. 1. Trace the history of concrete from its early use to its prominence in building today.

d) Metals have always been an important part of the history of engineering. Investigate the progression in the use of metals from copper and bronze to iron and steel. What effect did the use of these metals have on the societies in which they were used?

A.D. 1 –1000

- The Chinese further develop the study of mathematics.
- Gunpowder is perfected.
- Cotton and silk are manufactured.

1000–1400

- There is growth in the silk and glass industries.
- Leonardo Fibonacci (1170–1240), medieval mathematician, writes the first Western text on algebra.

Activity 1.3

What influence does the introduction of the first Western algebra text have on the society of the time? Why would engineers be interested in this text?

1400–1700

- Georgius Agricola's *De re metallica*, a treatise on mining and metallurgy, is published posthumously.
- Federigo Giambelli constructs the first time bomb for use against Spanish forces besieging Antwerp, Belgium.
- The first water closet (toilet) is invented in England.
- Galileo begins constructing a series of telescopes, with which he observes the rotation of the sun and other phenomena supporting the Copernican heliocentric theory.
- Using dikes and windmills, Jan Adriaasz Leeghwater completes drainage of the Beemstermeer, the largest project of its kind in Holland (17,000 acres).
- Otto von Guericke, mayor of Magdeburg, first demonstrates the existence of a vacuum.
- Christian Huygens begins work on the design of a pendulum-driven clock.
- Robert Hooke develops the balance spring to power watches.
- Charles II charters the Royal Society, England's first organization devoted to experimental science.
- Isaac Newton constructs the first reflecting telescope.
- Work is completed on the Languedoc Canal, the largest engineering project of its kind in Europe.
- Thomas Savery patents his "miner's friend," the first practical steam pump.
- The agriculture, mining, textile and glassmaking industries are expanded.
- The concept of the scientific method of invention and inquiry is originated.
- The humanities and science are first thought to be two distinctly separate entities.
- Robert Boyle finds that gas pressure varies inversely with volume (Boyle's Law).
- Leibniz makes a calculating machine to multiply and divide.

1700–1800

- The Leyden jar stores a large charge of electricity.
- The Industrial Revolution begins.

- James Watt makes the first rotary engine.
- The instrument-maker Benjamin Huntsman develops the crucible process for manufacturing steel, improving quality and sharply reducing cost.
- Louis XV of France establishes the Ecole des Ponts et Chausses, the world's first civil engineering school.
- John Smeaton completes construction of the Eddystone lighthouse.
- James Brindley completes construction of the Bridgewater Canal, beginning a canal boom in Britain.
- James Watt patents his first steam engine.
- The spinning jenny and water frame, the first successful spinning machines, are patented by James Hargreaves and Richard Arkwright, respectively.
- Jesse Ramsden invents the first screw-cutting lathe, permitting the mass production of standardized screws.
- The Society of Engineers, Britain's first professional engineering association, is formed in London.
- David Bushnell designs the first human-carrying submarine.
- John Wilkinson installs a steam engine to power machinery at his foundry in Shropshire, the first factory use of the steam engine.
- Abraham Darby III constructs the world's first cast iron bridge over the Severn River near Coalbrookdale.
- Claude Jouffroy d'Abbans powers a steamboat upstream for the first time.
- Joseph-Michel and Jacques-Etienne Montgolfier construct the first passenger-carrying hot air balloon.
- Henry Cort patents the puddling furnace for the production of wrought iron.
- Joseph Bramah designs his patent lock, which remains unpicked for 67 years.
- British civil engineer John Rennie completes the first building made entirely of cast iron.

Activities 1.4

a) The Industrial Revolution changed the whole landscape of the world. How did the engineer fit into this revolution? What were some of the major contributions?

b) Research and compare the first rotary engine to the rotary engines of today.

1800–1825

- Automation is first used in France.
- The first railroad locomotive is unveiled.
- Jean Fourier, French mathematician, states that a complex wave is the sum of several simple waves.
- Robert Fulton begins the first regular steamboat service with the *Clermont* on the Hudson River in the U.S.
- Chemical symbols as they are used today are developed.
- The safety lamp for protecting miners from explosions is first used.
- The single wire telegraph line is developed.
- Photography is born.

- Electromagnetism is studied.
- The thermocouple is invented.
- Aluminum is prepared.
- Andre Ampere shows the effect of electric current in motors.
- Sadi Carnot finds that only a fraction of the heat produced by burning fuel in an engine is converted into motion. This forms the basis of modern thermodynamics.

1825–1875

- Rubber is vulcanized by Charles Goodyear in the United States.
- The first iron-hulled steamer powered by a screw propeller crosses the Atlantic.
- The rotary printing press comes into service.
- Reinforced concrete is used.
- saac Singer invents the sewing machine.
- George Boole develops symbolic logic.
- The first synthetic plastic material—celluloid—is created by Alexander Parkes.
- Henry Bessemer originates the process to mass-produce steel cheaply.
- The first oil well is drilled near Titusville, Pennsylvania.
- The typewriter is perfected.
- The Challenge Expedition (1871–1876) forms the basis for future oceanographic study.

1875–1900

- The telephone is patented in the United States by Alexander Graham Bell.
- The phonograph is invented by Thomas Edison.
- The incandescent lightbulb also is invented by Edison.
- The steam turbine appears.
- The gasoline engine is invented by Gottlieb Daimler.
- The automobile is introduced by Karl Benz.

1900–1925

- The Wright brothers complete the first sustained flight.
- Detroit becomes the center of the auto industry.
- Stainless steel is introduced in Germany.

- Tractors with diesel engines are produced by Ford Motor Company.
- The first commercial airplane service between London and Paris commences.
- Diesel locomotives appear.

1925–1950

- Modern sound recordings are introduced.
- John Logie Baird invents a high-speed mechanical scanning system, which leads to the development of television.
- The Volkswagen Beetle goes into production.
- The first nuclear bombs are used.
- The transistor is invented.

1950–1975

- Computers first enter the commercial market.
- Computers are in common use by 1960.
- The first artificial satellite—Sputnik 1, USSR—goes into space.
- Explorer I, the first U.S. satellite, follows.
- The laser is introduced.
- Manned space flight begins.
- The first communication satellite—Telstar—goes into space.
- Integrated circuits are introduced.
- The first manned moon landing occurs.

1975–1990

- Supersonic transport from U.S. to Europe begins.
- Cosmonauts orbit the earth for a record 180 days.
- The Columbia space shuttle is reused for space travel.
- The first artificial human heart is implanted.

1990–Today

- Robots walk on Mars.
- Computer processor speed is dramatically improved.
- The Channel Tunnel (the "Chunnel") between England and France is completed.
- World's new tallest building opens in Kuala Lumpur, Malaysia (1,483 feet).
- Global Positioning Satellite (GPS) technology is declassified, resulting in hundreds of safety, weather and consumer applications.

You are the potential innovator of tomorrow. Research the progress made over the past decade in engineering on your own. Consider possible engineering innovations to which you might one day contribute.

1.6 A CASE STUDY OF TWO HISTORIC ENGINEERS

The important thing to realize about history is that it's about people. Much of it is based on biography, in fact. To fully understand the story of an invention, you need to investigate the people involved. An invention and its inventor are inextricably woven together.

As you study the details of history, you'll likely find yourself most interested as you get to know more about the people involved and the challenges they were up against. Hopefully, the two studies that follow will inspire you to further investigation.

Leonardo da Vinci

Leonardo da Vinci, shown in Fig. 1.6, had an uncanny ability to envision mechanized innovations that would one day see common usage—especially in the field of weaponry (helicopters, tanks, artillery, see Figs. 1.7–1.11). He seemed to know almost everything that was knowable in his time. He was a true Renaissance man and harnessed his immense genius to make improvements to most every aspect of the lives of his contemporaries, and to greatly impact the lives of future generations. Da Vinci's handicap, however, was his inability to read Latin, which prevented him from learning from the common scientific writings of the day, which focused on the works of Aristotle and other Greeks and their relation to the Bible. Instead, da Vinci was forced to make his assessments solely from his observations of the world around him. He was not interested in the thoughts of the ancients; he simply wanted to use his engineering skills to improve his environment.

Figure 1.6 Leonardo da Vinci (born April 15, 1452, Vinci, Italy, died May 2, 1519).

He did indeed bring his genius to bear on an enormously wide range of subjects, but little of his work had any relationship to that of his contemporaries. He designed, he built, and he tested. But various essential aspects of development were always lacking from his innovations. He could do many great things, but he couldn't do it all. He lacked a community of engineering peers with whom he could integrate his efforts. Being so far ahead of his time relative to the development of the field of engineering in his day, it was centuries before many of his innovations came to fruition.

Da Vinci's life as an engineer was one of total immersion in the intellectual activities of his time. He created frescoes, he tried his hand at sculpture, he was an architect, and he engineered hundreds of useful devices and many others that never came to life. He was a respected member of the community and was consistently in the employment of members of the aristocracy.

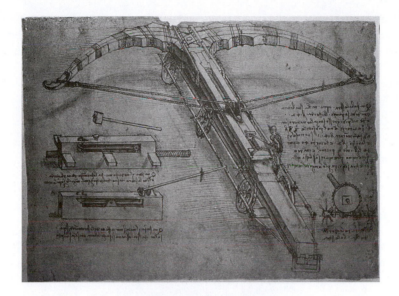

Figure 1.7 Weapon design.

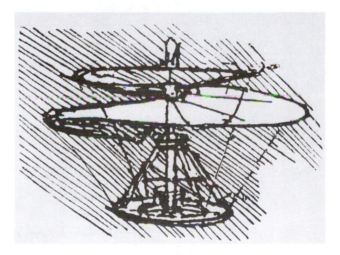

Figure 1.8 A flying machine.

To obtain his first commission as an engineer he wrote a letter claiming that he could: construct movable bridges; remove water from the moats of fortresses under siege; destroy any fortification not built of stone; make mortars, dart-throwers, flame-throwers and cannon capable of firing stones and making smoke; design ships and weapons for war at sea; dig tunnels without making any noise; make armored wagons to break up the enemy in advance of the infantry; design buildings; sculpt in any medium; and paint.

With that resume, he was hired with the title of Painter and Engineer to the Duke. Did he do all the things that he said he could? We don't know. But we do know that, among other things, he worked diligently perfecting the canal system around Milan.

DaVinci's chief court function during the period he was employed by the Duke was to produce spectacular shows for the entertainment of the aristocracy who came to court.

He produced musical events, designed floats, and delighted the court with processions that dazzled the eyes and flying devices that wowed the audience. The interesting thing is that during this time he wrote over 5000 pages of notes, detailing every conceivable kind of invention. This truly was a great engineer at work.

There is so much you will encounter if you investigate Leonardo da Vinci. It is interesting to note that he seemingly examined every aspect of his world and left behind a significant record from which we can learn.

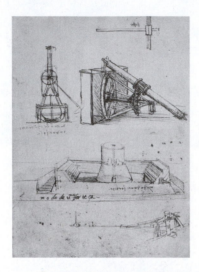

Figure 1.9 Cannon design.

Figure 1.10 Machine guns.

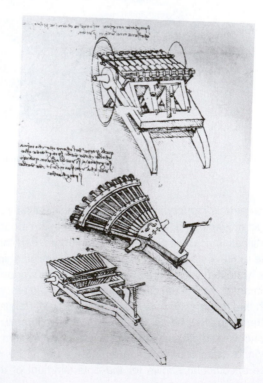

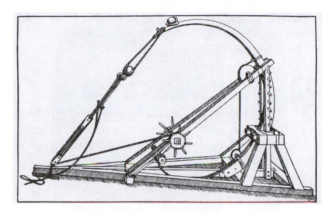

Figure 1.11 Stone throwers.

Gutenberg and His Printing Press

Another important innovator of this era developed one of the most history-altering inventions of all time—the printing press. In 1455, Johannes Gutenberg, pictures in Fig. 1.12, printed the first book, a Bible. He lived before Leonardo da Vinci and wasn't a generalist like da Vinci, but his invention changed society forever. With the invention of the printing press, man was able to use, appreciate, and disseminate information as never before.

The development of printing came at a time when there was a growing need for the ability to spread information. There had been a long phase of general societal introversion from which mankind was prepared to emerge. Gutenberg's printing press was the spark that ignited the flame of widespread communication.

Figure 1.12   Johannes Gutenberg, creator of the first mass-producing printing press (born 1394–1398, Mainz, Germany, died Feb. 3, 1468).

There were already printing presses when Gutenberg introduced his. As early as the 11th century, the Chinese had developed a set of movable type from a baked mixture of clay and glue. The type pieces were stuck onto a plate where impressions could be taken by pressing paper onto the stationary type. Since the type was glued to the plate, the plate could be heated and the type removed and re-situated. The process was not used in any form of mass production, but it did form a basis for the future invention of the Gutenberg

Figure 1.13 The printing press.

printing press. In Europe there were attempts at creating presses that would produce cheap playing cards and frivolous items, but it took Gutenberg to see the practical need for a press that could provide the common man with reading material. The time was right; the inventor was prepared. History is made by the engineer who delivers the major innovation society is waiting for—whether society has been eagerly awaiting the breakthrough or is caught pleasantly by surprise.

Johannes Gutenberg was born into a noble family of the city of Mainz, Germany. His early training is reputed to have been in goldsmithing. By 1428 he had moved to Strasbourg. It was there that he began to formulate the idea for his printing press. His first experiments with movable type stemmed from his notions of incorporating the techniques of metalworking—such as casting, punch-cutting, and stamping—for the mass production of books. Since all European books at this time were hand-written by scribes with elaborate script, Gutenberg decided to reproduce this writing style with a font of over 300 characters, far larger than the fonts of today. To make this possible, he invented the variable-width mold, and perfected a rugged blend of lead, antimony, and tin used by type foundries up to the present century.

Activities 1.5

a) Study some of the inventions of Leonardo da Vinci. Draw conclusions on how these early works influenced later inventions or innovations.
b) Research and build one of da Vinci's inventions, or a model of one, from his plans.
c) Where does Boyle's Law fit in your future studies?
d) Do you agree that the humanities and science should be separated? Why or why not?
e) Explain why the printing press revolutionized the world and the life of the engineer.

1.7 THE HISTORY OF THE DISCIPLINES

As you investigate the many areas of engineering throughout this textbook, you will discover that there are a wide variety of career paths from which to choose. As a prelude to the in-depth look at engineering fields which we provide in Chapter 2, this section takes a brief look at the historic backgrounds of the following disciplines:

- Aerospace Engineering
- Agricultural Engineering
- Chemical Engineering
- Civil Engineering
- Computer Engineering
- Electrical Engineering
- Industrial Engineering
- Mechanical Engineering

From the core areas listed, many additional engineering specialties have evolved over time, for a total today of over 30 different fields in engineering.

The information here touches only lightly on the backgrounds of some of these fields. We recommend that you research in greater detail the history of disciplines of interest to you. An in-depth description of a number of these fields is found in the next chapter.

The individual disciplines of engineering were actually named only a short time ago relative to the history of the world. What we present here is a short history of the major disciplines dating from after the time that they were individually recognized.

Aerospace Engineering

Aerospace engineering is concerned with engineering applications in the areas of *aeronautics* (the science of air flight) and *astronautics* (the science of space flight). Aerospace engineering deals with flight of every kind: balloon flight, sailplanes, propeller- and jet-powered aircraft, missiles, rockets, satellites, and advanced interplanetary concepts such as ion-propulsion rockets and solar-wind vehicles. It is the field of future interplanetary travel. The challenge is to produce vehicles that can traverse the long distances of space in ever-shorter periods of time. New propulsion systems will require that engineers venture into areas never before imagined. By reading histories and biographies, you get a feel for the changes that aerospace pioneers brought about, which will help you to understand the mind-set needed to make great leaps forward yourself.

Activity 1.6

Where do you think the next breakthrough will come for the aerospace engineer? How should this affect the approach we take in the education of the next generation of aerospace engineers? Do you think your generation will walk on Mars?

Agricultural Engineering

At the turn of the 20th century, a large majority of the working population was engaged in agriculture. The mass entry into industrial employment would come a few decades later. This movement to the factories of America lowered the number of workers in agriculture to under five percent. This drastic change occurred for a variety of reasons, the principal reason being the integration of technology and engineering into agriculture, which allowed modern farmers to feed approximately ten times as many families per farmer as their ancestors did a hundred years prior.

As you investigate the many facets of agricultural engineering, you will discover that America leads the world in agricultural technology. The world depends upon the United States to feed those in areas of famine, to supply agricultural implements to countries that do not have the resources to perfect such implements, and to continue to perfect technologies to feed more and more of the world's growing population. The agricultural engineer has one of the largest responsibilities in the engineering community. The modern world cannot survive without the efficiencies of a mass-produced food supply.

Agricultural engineering focuses on the following areas:

- soil and water
- structures and environment
- electrical power and processing
- food engineering
- power and machinery

Activities 1.7

a) Compare the growth of industry with the number of people leaving the farms of America. What innovations in farming helped to offset this exodus? How were we able to continue to feed the population with such a change?

b) Inspect an early piece of farm machinery. Find a current model of that machine. How are they different? What has not changed in their design?

c) Farm machinery can be highly dangerous if not used properly. Investigate the dangers posed by farm machinery and how the operators are protected from those dangers.

d) Follow a piece of farm machinery through its many changes from the 17th century to the present day.

e) Investigate where agriculture has gone in the late 20th century. What new aspects has it undertaken?

Chemical Engineering

Chemical Engineering is one of the newer disciplines among engineering professions. A closely related field, Chemistry, has been studied for centuries in its many forms from alchemy to molecular structure, but chemical engineering began its rise to prominence shortly after World War I. The number of activities involving chemical processing forced engineers and chemists to combine their efforts, bonding the two disciplines together. The chemical engineer needed to understand the way in which chemicals moved and how they could be joined together. This necessity lead the chemical engineer to begin experimenting with the ways in which mechanisms were used to separate and combine chemicals. Experimenting with mass transfer, fluid flow, and heat transfer proved beneficial. The chemical engineering field is relatively new and is a critical engineering profession.

Chemical engineering applies chemistry to industrial processes which change the composition or properties of an original substance for useful purposes. Chemical engineering is involved in the manufacture of drugs, cements, paints, lubricants, pesticides, fertilizers, cosmetics, foods; in oil refining, combustion, extraction of metals from ores; and in the production of ceramics, brick, and glass. The petrochemical industry is a sizable market for the chemical engineer. Chemical engineers can apply their skills in food engineering, process dynamics and control, environmental control, electrochemical engineering, polymer science technology, unit operations, and plant design and economics.

Activities 1.8

a) Pick a product that you believe has had to be engineered by a chemical engineer. Explain the process that started with the raw materials and ended with the product you identified. Now look into the recent history of each step of this process. Where have the latest improvements occurred? Try to identify the oldest technique used today in the process. Any book covering the background of this discipline and its products should include such information.

b) Look at the history of the study of chemistry and try to pinpoint those times when the application of chemistry involves the chemical engineer.

c) What inventions and innovations did the first chemical engineers develop?

d) Contact a chemical engineering society and collect information on their history—especially about their early members.

Civil Engineering

The 17th and 18th centuries gave birth to most of the major modern engineering disciplines in existence today. But civil engineering is the oldest. Civil engineering traces its roots to early eighteenth century France, though the first man definitely to call himself a "civil engineer" was a keen Englishman, John Smeaton, in 1761, [Kirby 1956, p. xvi]. Smeaton was a builder of lighthouses who helped distinguish the profession of civil engineering from architecture and military engineering. The surveying of property, the building of roads and canals to move goods and people, and the building of bridges to allow safe passage over raging waters all fall within the parameters of what we now call civil engineering. Construction was the primary focus of the civil engineer while the military engineer focused on destruction. In America, construction of the earliest railroads began in 1827 and demonstrated the dedication of those early engineers in opening up the West to expansion.

Today, civil engineering focuses primarily on structural issues such as rapid transit systems, bridges, highway systems, skyscrapers, recreational facilities, houses, industrial plants, dams, nuclear power plants, boats, shipping facilities, railroad lines, tunnels, harbors, offshore oil and gas facilities, pipelines, and canals. Civil engineers are heavily involved in improving the movement of populations through their many physical systems. Mass transit development allows the civil engineer to protect society and the environment, reducing traffic and toxic emissions from individual vehicles. Highway design provides motorists with safe motorways and access to recreation and work. The civil engineer is always at the center of discussion when it comes to transport and buildings.

Activities 1.9

a) Trace the evolution of the first accounts of civil engineering in history. How has it changed? What new areas have entered the field and why did they become part of civil engineering?
b) Research a significant American civil engineering project and explain its importance in the history of the United States.
c) Investigate one of the following and describe in simple terms how it was constructed: a railroad, a bridge, a canal, or a sewer.
d) Research and discuss the connection between mining engineering and civil engineering.

Computer Engineering

The role that computers play in engineering design has reached a level where no major business can exist comfortably without access to computer technology. Large, medium, and small organizations alike all depend on computers to perform inventories, create billing, record sales, and order new stock. The storage of data alone makes the computer indispensable. The once-mammoth computers of the past have been downsized to the size of your fist, and computing speed today was unthinkable just years ago. Even homeowners seem to be vying with their neighbors to see who can purchase the fastest computer. The world has become almost completely computer-centered, and the computer's role should only increase.

Activity 1.10

It's time to use your imagination: what will computer engineers of the future be required to do in order to both lead and to keep up with innovations? What new things will computers do?

Electrical Engineering

The amount of historical perspective in this discipline is vast. "The records of magnetic effects date back to remotest antiquity . . . The whole of Electrical Engineering is based on magnetic and electrical phenomenon." [Dunsheath 1962, p. 21] In the late 1700s, experiments concerning electricity and its properties began. By the early 1800s, Volta began his studies of electric conduction through a liquid. Electric current and its movement became the topic of the day. As the need for electricity to power the many devices that were being fabricated by the mechanical engineers grew, the profession of electrical engineering prospered. Direct current, alternating current, the electric telegraph—all became areas to be investigated by the early electrical engineer. The profession started with magnetism and electrical phenomenon, but it has grown more recently in new directions as new requirements have been imposed by modern society.

Electrical engineering is the largest branch of engineering, employing over 400,000 engineers. Among the major specialty areas in electrical engineering are electronics and solid-state circuitry, communication systems, computers and automatic control, instrumentation and measurements, power generation and transmission, and industrial applications.

Activities 1.11

a) Investigate early electrical phenomenon and how it was perceived by early man.
b) Take one pioneer in electrical engineering and show how his/her experiments aided the growth of electrical engineering.
c) What are some of the highpoints in the history of electrical engineering?

Industrial Engineering

Industrial engineering is a growing branch of the engineering family because of the awareness that the application of engineering principles and techniques can help create better working conditions. Industrial engineers must design, install, and improve systems which integrate people, materials, and equipment to provide efficient production of goods. They must coordinate their understanding of the physical and social sciences with the activities of workers to design areas in which the workers will produce the best results. The daily lives of the people with whom industrial engineers work are closely tied to the designs that they create.

Activity 1.12

Observe an area where people work. How would you as an industrial engineer improve the working conditions and productivity of these workers?

Mechanical Engineering

As coke replaced charcoal in the blast furnaces of England in the early 1700s, the beginnings of the modern mechanical engineering profession dawned. The introduction of coke allowed for larger blast furnaces and a greater ability to use iron. Higher-quality wrought iron could also be produced. These improvements laid the foundation for the Industrial Revolution, during which the production of great quantities of steel was made possible. As these materials became more readily available, the mechanical engineer began to design improved lathes and milling and boring machines. The mechanical engineer realized that a wide variety of devices needed to be created to work with the quantities of iron and steel being produced. As the pace of manufacturing increased, the numbers of mechanical engineers also showed a marked increase. The profession would continue to grow in its size and importance. Steam power became vitally important during the industrial revolution. The mechanical engineer was needed to create the tools to harness the power of steam. From the steam engine to the automobile, from the automobile to the airplane, from the airplane to the space shuttle, the mechanical engineer has been and always will be in great demand for the development of new devices for the betterment of man.

Automobiles, engines, heating and air-conditioning systems, gas and steam turbines, air and space vehicles, trains, ships, servomechanisms, transmission mechanisms, radiators, mechatronics, and pumps are a few of the systems and devices requiring mechanical engineering knowledge. Mechanical engineering deals with power, its generation, and its application. Power affects the rate of change or "motion" of something. This can be change of temperature or change of motion due to outside stimulus. Mechanical engineering is the broadest-based discipline in engineering. The breadth of study required to be a mechanical engineer allows one to diversify into many of the other engineering areas. The major specialty areas of mechanical engineering are: applied mechanics; control; design; engines and power plants; energy; fluids; lubrication; heating, ventilation, and air-conditioning (HVAC); materials, pressure vessels and piping; and transportation and aerospace.

Activities 1.13

a) Look around you. How many things can you find that have been influenced by the work of a mechanical engineer?
b) Investigate the early lathes and explain the improvements made in today's models.
c) Explore in greater depth the effect the iron industry had upon the Industrial Revolution.
d) Who are some of the important names in the early days of the Industrial Revolution and what did they do for the revolution?
e) Where would mechanical engineering be without the Industrial Revolution?
f) Pick a period in the time between 1700–1999. Look closely at the contributions made by mechanical engineers.

Bioengineering

Bioengineering is the application of biotechnology. It is also referred to as biological engineering or biosystems engineering, and involves the application of engineering principles to living things. The U.S. National Institute of Health defines bioengineering as integrating "physical, chemical, or mathematical sciences and engineering principles for the study of biology, medicine, behavior, or health."

Bioengineers can work in a variety of fields, including medicine and medical device development, drug manufacturing, agriculture, and environmental science. They can work in industry, hospitals, academic institutions, and government agencies, among others. Biomechanical engineers, for instance, help design equipment for surgeons to perform arthroscopic procedures. Other bioengineers work to design better exercise equipment or therapeutic devices. Some work with nanotechnology to repair cell damage or control gene function. Still others build artificial limbs or organs.

Green Engineering

Green engineering is a general term used to refer to the application of engineering principles in an effort to conserve natural resources or to minimize adverse impact on the environment. Efficiency, renewability, and durability are some of the key aspects of green design.

Examples of green engineering include efforts by architects and engineers to develop structures that consume a minimum of energy, both in their construction and in their ongoing operation. This may involve the use of solar energy or other design elements. LEED (Leadership in Energy and Environmental Design) certification pertains to standards for construction related to energy conservation and can involve criteria related to material selection, determination of location of manufacture, water use efficiency, indoor air quality, and even landscaping and irrigation.

Green engineering principles are also applied to automobile manufacture, including hybrid vehicles and flex fuels. Other careers involve solar energy generation, thermal engineering, and providing clean water supplies.

REFERENCES

A History of Technology, Volume II, Ed. Charles Singer, New York, Oxford University Press, 1956.

American Society of Civil Engineers, *The Civil Engineer: His Origins*, New York, ASCE, 1970.

Burghardt, M. David, *Introduction to Engineering*, 2nd Ed., New York, HarperCollins, 1995.

Burstall, A., *A History of Mechanical Engineering*, London, Faber, 1963.

De Camp, L. Sprague, *The Ancient Engineers*, Cambridge, The MIT Press, 1970.

Dunsheath, P., *A History of Electrical Engineering*, London, Faber, 1962.

Gray, R. B., *The Agricultural Tractor 1855–1950*, St. Joseph, MI, American Society of Agricultural Engineers, 1975.

Great Engineers and Pioneers in Technology, Volume 1: From Antiquity through the Industrial Revolution, Eds. Roland Turner and Steven L. Goulden, New York, St. Martin's Press, 1981.

Greaves, W. F., and J. H. Carpenter, *A Short History of Electrical Engineering*, London, Longmans, Green and Co Ltd, 1969.

Kirby, R., et al., *Engineering in History*, New York, McGraw-Hill, 1956.

Miller, J. A., *Master Builders of Sixty Centuries*, Freeport, New York, Books for Libraries Press, 1972.

Red, W. Edward, *Engineering—The Career and The Profession*, Monterey, California, Brooks/Cole Engineering Division, 1982.

Rosenberg, S. H., *Rural America A Century Ago*, St. Joseph, MI, American Society of Agricultural Engineers, 1976.

EXERCISES AND ACTIVITIES

1.1 The history of engineering is long and varied. It contains many interesting inventions and refinements. Select one of these inventions and discuss the details of its creation. For example, you might explain how the first printing presses came into being and what previous inventions were used to create the new device.

1.2 Build a simple model of one of the inventions mentioned in this chapter—a bridge, aqueduct, or submarine, for instance. Explain the difficulties of building these devices during the time they were invented.

1.3 Explain how easy it would be to create some inventions of the past using our present-day knowledge and capability.

1.4 Engineering history is filled with great individuals who have advanced the study and practice of engineering. Investigate an area of engineering that is interesting to you and write a detailed report on an individual who made significant contributions in that area.

1.5 Explain what it would have been like to have been an engineer during any particular historical era.

1.6 Compare the lives of any two engineers from the past. Are there similarities in their experiences, projects, and education?

1.7 What kind of education were engineers of old able to obtain?

1.8 What period of engineering history interests you most? Why? Explain why this period is so important in the history of engineering.

1.9 If you had to explain to a 7-year-old child why engineering is important to society, what information from the history of engineering would you relate? Why?

Chapter 2

Engineering Majors

2.1 INTRODUCTION

Engineers produce things that impact us every day. They invent, design, develop, manufacture, test, sell, and service products and services that improve the lives of people. The Accreditation Board for Engineering and Technology (ABET), which is the national board that establishes accreditation standards for all engineering programs, defines engineering as follows [Landis]:

> *Engineering is the profession in which a knowledge of the mathematical and natural sciences, gained by study, experience, and practice, is applied with judgment to develop ways to utilize, economically, the materials and forces of nature for the benefit of mankind.*

Frequently, students early in their educational careers find it difficult to understand exactly what engineers do, and often more to the point, where they fit best in the vast array of career opportunities available to engineers.

Common reasons for a student to be interested in engineering include:

1. Proficiency in math and science
2. Suggested by a high school counselor
3. Has a relative who is an engineer
4. Heard it's a field with tremendous job opportunity
5. Read that it has high starting salaries

While these can be valid reasons, they don't imply a firm understanding of engineering. What is really important is that a student embarking upon a degree program, and ultimately a career, understands what that career entails and the options it presents. We all have our own strengths and talents. Finding places to use those strengths and talents is the key to a rewarding career.

The purpose of this chapter is to provide information about some of the fields of engineering in order to help you decide if this is an area that you might enjoy. We'll explore the role of engineers, engineering job functions, and the various engineering disciplines.

The Engineer and the Scientist

To better understand what engineers do, let's contrast the roles of engineers with those of the closely related field of the scientist. Many students approach both fields for similar reasons: they were good at math and science in high school. While this is a prerequisite for both fields, it is not a sufficient discriminator to determine which is the right career for a given individual.

The main difference between the engineer and the scientist is in the object of each one's work. The scientist searches for answers to technological questions to obtain a knowledge of why a phenomenon occurs. The engineer also searches for answers to technological questions, but always with an application in mind.

Theodore Von Karman, one of the pioneers of America's aerospace industry, said, "Scientists explore what is; engineers create what has not been." [Paul Wright, *Introduction to Engineering*].

In general, science is about discovering things or acquiring new knowledge. Scientists are always asking, "Why?" They are interested in advancing the knowledge base that we have in a specific area. The answers they seek may be of an abstract nature, such as understanding the beginning of the universe, or more practical, such as the reaction of a virus to a new drug.

The engineer also asks, "Why?" but it is because of a problem which is preventing a product or service from being produced. The engineer is always thinking about the application when asking why. The engineer becomes concerned with issues such as the demand for a product, the cost of producing the product, the impact on society, and the environment of the product.

Scientists and engineers work in many of the same fields and industries but have different roles. Here are some examples:

- Scientists study the planets in our solar system to understand them; engineers study the planets so they can design a spacecraft to operate in the environment of that planet.
- Scientists study atomic structure to understand the nature of matter; engineers study the atomic structure in order to build smaller and faster microprocessors.
- Scientists study the human neurological system to understand the progression of neurological diseases; engineers study the human neurological system to design artificial limbs.
- Scientists create new chemical compounds in a laboratory; engineers create processes to mass-produce new chemical compounds for consumers.
- Scientists study the movement of tectonic plates to understand and predict earthquakes; engineers study the movement of tectonic plates to design safer buildings.

The Engineer and the Engineering Technologist

Another profession closely related to engineering is engineering technology. Engineering technology and engineering have similarities, yet there are differences; they have different career opportunities. ABET, which accredits engineering technology programs as well as engineering programs, defines engineering technology as follows:

> *Engineering technology is that part of the technological field which requires the application of scientific and engineering knowledge and methods combined with technical skills in support of engineering activities; it lies in the occupational*

spectrum between the craftsman and engineering at the end of the spectrum closest to the engineer.

Technologists work with existing technology to produce goods for society. Technology students spend time in their curricula working with actual machines and equipment that are used in the jobs they will accept after graduation. By doing this, technologists are equipped to be productive in their occupation from the first day of work.

Both engineers and technologists apply technology for the betterment of society. The main difference between the two fields is that the engineer is able to create new technology through research, design, and development. Rather than being trained to use specific machines or processes, engineering students study additional mathematics and engineering science subjects. This equips engineers to use these tools to advance the state of the art in their field and move technology forward.

There are areas where engineers and engineering technologists perform very similar jobs. For example, in manufacturing settings, engineers and technologists are employed as supervisors of assembly line workers. Also, in technical service fields both are hired to work as technical support personnel supporting equipment purchased by customers. However, most opportunities are different for engineering and engineering technology graduates.

- The technologist identifies the computer networking equipment necessary for a business to meet its needs and oversees the installation of that equipment; the engineer designs new computer boards to transmit data faster.
- The technologist develops a procedure to manufacture a shaft for an aircraft engine using a newly developed welding technique; the engineer develops the new welding machine.
- The technologist analyzes a production line and identifies new robotic equipment to improve production; the engineer develops a computer simulation of the process to analyze the impact of the proposed equipment.
- The technologist identifies the equipment necessary to assemble a new CD player; the engineer designs the new CD player.
- The technologist identifies the proper building materials and oversees the construction of a new building; the engineer determines the proper support structures, taking into account the local soil, proposed usage, earthquake risks and other design requirements.

What Do Engineers Do?

Engineering is an exciting field with a vast range of career opportunities. In trying to illustrate the wide range of possibilities, Professors Jane Daniels and Richard Grace from Purdue University constructed the cubic model of Figure 2.1. One edge of the cube represents the engineering disciplines that most students identify as the potential majors. A second edge of the cube represents the different job functions an engineer can have within a specific engineering discipline. The third edge of the cube represents industrial sectors where engineers work. A specific engineering position, such as a mechanical engineering design position in the transportation sector, is the intersection of these three axes. As one can see from the cube, there is a vast number of possible engineering positions.

In the following sections of this chapter, the engineering functions and majors are described. The remaining axis, the industrial sectors, are dependent on the companies and governmental agencies that employ engineers.

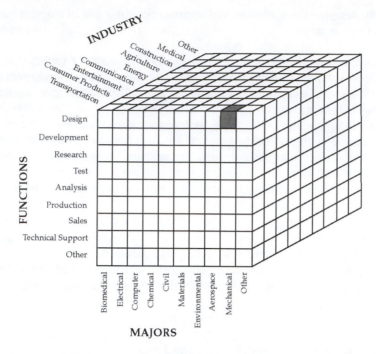

Figure 2.1 Engineering positions. [Grace and Daniels]

To obtain more information about the various industrial sectors:

*   Explore your school's placement center
*   Visit job fairs
*   Attend seminars on campus sponsored by various companies
*   Search the Internet (visit websites describing career opportunities)
*   Talk to faculty familiar with a certain industry
*   "Shadow" a practicing engineer
*   Work as an intern or co-op engineer
*   Take an engineering elective course

Most engineering curricula include common courses within an engineering major for the first two years of study. It is not until the junior or senior year that students take technical electives that can be industry specific. Students are encouraged to explore various career opportunities so they can make better decisions when required to select their junior and senior level electives.

Example 2.1

Let's consider a mechanical engineer (an "ME") who performs a design function to illustrate how a specific job in one industrial sector can vary from that in another sector.

*   **Aerospace**—*Design of an aircraft engine fan blade*: Detailed computer analyses and on-engine testing are required for certification by the FAA. Reliability, efficiency, cost,

and weight are all design constraints. The design engineer must push current design barriers to optimize the design constraints, potentially making tradeoffs between efficiency, cost, and weight.

- **Biomedical**—*Design of an artificial leg and foot prosthesis giving additional mobility and control to the patient*: Computer modeling is used to model the structure of the prosthesis and the natural movement of a leg and foot. Collaboration with medical personnel to understand the needs of the patient is a critical part of the design process. Reliability, durability, and functionality are the key design constraints.
- **Power**—*Design of a heat recovery system in a power plant, increasing the plant productivity*: Computer analyses are performed as part of the design process. Cost, efficiency, and reliability are the main design constraints. Key mechanical components can be designed with large factors of safety since weight is not a design concern.
- **Consumer products**—*Design of a pump for toothpaste for easier dispensing*: Much of the development work might be done with prototypes. Cost is a main design consideration. Consumer appeal is another consideration and necessitates extensive consumer testing as part of the development process.
- **Computer**—*Design of a new ink-jet nozzle with resolution approaching laser printer quality*: Computer analyses are performed to ensure that the ink application is properly modeled. Functionality, reliability, and cost are key design concerns.

2.2 ENGINEERING FUNCTIONS

Within engineering there are basic classifications of jobs that are common across the various engineering disciplines. What follows are brief descriptions of these different engineering job functions. A few examples are provided for each function. It is important to realize that all the fields of engineering have roles in each of the main functions described here.

Research

The role of the engineering researcher is the closest to that of a scientist of all the engineering functions. Research engineers explore fundamental principles of chemistry, physics, biology, and mathematics in order to overcome barriers preventing advancement in their field. Engineering researchers differ from scientists in that they are interested in the application of a breakthrough, whereas scientists are concerned with the knowledge that accompanies a breakthrough.

Research engineers conduct investigations to extend knowledge using various means. One of the means is conducting experiments. Research engineers may be involved in the design and implementation of experiments and the interpreting of the results. Typically, the research engineer does not perform the actual experiment. Technicians are usually called upon for the actual testing. Large-scale experiments may involve the coordination of additional supporting personnel including other engineers, scientists, technologists, technicians, and craftspeople.

Research is also conducted using the computer. Computational techniques are developed to calculate solutions to complex problems without having to conduct costly and time-consuming experiments. Computational research requires the creation of mathematical models to simulate the natural occurring phenomena under study. Research engineers also might develop the computational techniques to perform the complex calculations in a timely and cost-effective fashion.

Most research engineers work for some type of research center. A research center might be a university, a government laboratory such as NASA, or an industrial research center. In most research positions an advanced degree is required, and often a Ph.D. is needed. If research appeals to you, a great way to explore it is by doing an undergraduate research project with an engineering professor. This will allow you to observe the operation of a laboratory first-hand and find out how well you enjoy being part of a research team.

Development

Development engineers bridge the gap between laboratory research and full-scale production. The development function is often coupled with research in so-called R&D (research and development) divisions. Development engineers take the knowledge acquired by the researchers and apply it to a specific product or application. The researcher may prove something is possible in a laboratory setting; the development engineer shows that it will work on a large, production-size scale and under actual conditions encountered in the field. This is done in pilot manufacturing plants or by using prototypes.

Development engineers are continuously looking for ways to incorporate the findings of researchers into prototypes to test their feasibility for use in tomorrow's products. Often, an idea proven in a laboratory needs to be significantly altered before it can be introduced on a mass production scale. It is the role of development engineers to identify these areas and work with the design engineers to correct them before full-scale production begins.

An example of a development process is the building of concept cars within the automotive industry. These are unique cars that incorporate advanced design concepts and technology. The cars are then used as a test case to see if the design ideas and technology actually perform as predicted. The concept cars are put through exhaustive tests to determine how well the new ideas enhance a vehicle's performance. Each year, new technology is introduced into production automobiles that was first proven in development groups using concept vehicles.

Testing

Test engineers are responsible for designing and implementing tests to verify the integrity, reliability, and quality of products before they are introduced to the public. The test engineer devises ways to simulate the conditions a product will be subjected to during its life. Test engineers work closely with development engineers in evaluating prototypes and pilot facilities. Data from these initial development tests are used to decide whether full production versions will be made or if significant changes are needed before a full-scale release. Test engineers work with design engineers to identify the changes in the product to ensure its integrity.

A challenge that engineers face is simulating the conditions a product will face during its life span, and doing so in a timely, cost-effective manner. Often the conditions the product will face are difficult to simulate in a laboratory. A constant problem for the test engineer is simulating the aging of a product. An example of such a testing challenge is the testing of a pacemaker for regulating a patient's heart which is designed to last several decades. An effective test of this type cannot take 20 years or the product will be obsolete before it is introduced. The test engineer must also simulate conditions within the human body without exposing people to unnecessary risks.

Other challenges facing test engineers involve acquiring accurate and reliable data. The test engineer must produce data that show the product is functioning properly or identify areas of concern. Test engineers develop data acquisition and instrumentation methods to achieve this. Techniques such as radio telemetry may be used to transmit data from the inside of a device being tested. The measurement techniques must not interfere with the operation of the device, presenting a tremendous challenge for small, compact products. Test engineers must cooperate with design engineers to determine how the device being tested can be fitted with instrumentation yet still meet its design intent.

Test engineers must have a wide range of technical and problem-solving skills. They must also be able to work in teams involving a wide range of people. They work with design and development engineers, technicians, and craftspeople, as well as management.

Example 2.2

Test engineers must understand the important parameters of their tests. The development of a certain European high-speed train provides an example of the potential consequences which may result when test engineers fail to understand these parameters. A test was needed to show that the windshield on the locomotive could withstand the high-velocity impacts of birds or other objects it might encounter. This is also a common design constraint encountered in airplane design. The train test engineers borrowed a "chicken gun" from an aerospace firm for this test. A chicken gun is a mechanism used to propel birds at a target, simulating in-flight impact. With modern laws governing cruelty to animals, the birds are humanely killed and frozen until the test.

On the day of the test, the test engineers aimed the gun at the locomotive windshield, inserted the bird and fired. The bird not only shattered the windshield but put a hole through the engineer's seat.

The design engineers could not understand what went wrong. They double-checked their calculations and determined that the windshield should have held. The problem became clear after the test engineers reviewed their procedure with the engineers from the aerospace firm from whom they had borrowed the equipment. The aerospace test engineers asked how long they had let the bird thaw.

The response was, "Thaw the bird?"

There is a significant difference in impact force between a frozen eight-pound bird and a thawed bird. The test was successfully completed later with a properly thawed bird.

Design

The design function is what many people think of when they think of engineering, and this is where the largest number of engineers are employed. The design engineer is responsible for providing the detailed specifications of the products society uses.

Rather than being responsible for an entire product, most design engineers are responsible for a component or part of the product. The individual parts are then assembled into a product such as a computer, automobile, or airplane. Design engineers produce detailed dimensions and specifications of the part to ensure that the component fits properly with adjoining pieces. They use modern computer design tools and are often supported by technicians trained in computer drafting software.

The form of the part is also a consideration for the design engineer. Design engineers use their knowledge of scientific and mathematical laws, coupled with experience, to generate a shape to meet the specifications of the part. Often, there is a wide range of possibilities and considerations. In some fields, these considerations are ones that can be calculated. In others, such as in consumer products, the reaction of a potential customer to a shape may be as important as how well the product works.

The design engineer also must verify that the part meets the reliability and safety standards established for the product. The design engineer verifies the integrity of the product. This often requires coordination with analysis engineers to simulate complex products and field conditions, and with test engineers to gather data on the integrity of the product. The design engineer is responsible for making corrections to the design based on the results of the tests performed.

In today's world of ever-increasing competition, the design engineer must also involve manufacturing engineers in the design process. Cost is a critical factor in the design process and may be the difference between a successful product and one that fails. Communication with manufacturing engineers is therefore critical. Often, simple design changes can radically change a part's cost and affect the ease with which the part is made.

Design engineers also work with existing products. Their role includes redesigning parts to reduce manufacturing costs and time. They also work on redesigning products that have not lived up to expected lives or have suffered failure in the field. They also modify products for new uses. This usually requires additional analysis, minor redesigns, and significant communication between departments.

Analysis

Analysis is an engineering function performed in conjunction with design, development and research. Analysis engineers use mathematical models and computational tools to provide the necessary information to design, development, or research engineers to help them perform their function.

Analysis engineers typically are specialists in a technology area important to the products or services being produced. Technical areas might include heat transfer, fluid flow, vibrations, dynamics, system modeling, and acoustics, among others. They work with computer models of products to make these assessments. Analysis engineers often possess an advanced degree and are experienced in their area of expertise.

In order to produce the information required of them, they must validate their computer programs or mathematical models. To do so may require comparing test data to their predictions. This requires coordination with test engineers in order to design an appropriate test and to record the relevant data.

An example of the role of analysis is the prediction of temperatures in an aircraft engine. Material selection, component life estimates, and design decisions are based in large part on the temperature the parts attain and the duration of those temperatures. Heat transfer analyses are used to determine these temperatures. Engine test results are used to validate the temperature predictions. The permissible time between engine overhauls can depend on these temperatures. The design engineers then use these results to ensure reliable aircraft propulsion systems.

Systems

Systems engineers work with the overall design, development, manufacture, and operation of a complete system or product. Design engineers are involved in the design of individual components, but systems engineers are responsible for the integration of the components and systems into a functioning product.

Systems engineers are responsible for ensuring that the components interface properly and work as a complete unit. Systems engineers are also responsible for identifying the overall design requirements. This may involve working with customers or marketing personnel to accurately determine market needs. From a technical standpoint, systems engineers are responsible for meeting the overall design requirements.

Systems engineering is a field that most engineers enter only after becoming proficient in an area important to the systems, such as component design or development. Some graduate work often is required prior to taking on these assignments. However, there are some schools where an undergraduate degree in systems engineering is offered.

Manufacturing and Construction

Manufacturing engineers turn the specifications of the design engineer into a tangible reality. They develop processes to make the products we use every day. They work with diverse teams of individuals, from technicians on the assembly lines to management, in order to maintain the integrity and efficiency of the manufacturing process.

It is the responsibility of manufacturing engineers to develop the processes for taking raw materials and changing them into the finished pieces that the design engineers detailed. They utilize state-of-the-art machines and processes to accomplish this. As technology advances, new processes often must be developed for manufacturing the products.

The repeatability or quality of manufacturing processes is an area of increasing concern to modern manufacturing engineers. These engineers use statistical methods to determine the precision of a process. This is important, since a lack of precision in the manufacturing process may result in inferior parts that cannot be used or that may not meet the customer's needs. Manufacturing engineers are very concerned about the quality of the products they produce. High quality manufacturing means lower costs since more parts are usable, resulting in less waste. Ensuring quality means having the right processes in place, understanding the processes and working with the people involved to make sure the processes are being maintained at peak efficiency.

Manufacturing engineers also keep track of the equipment in a plant. They schedule and track required maintenance to keep the production line moving. They also must track the inventories of raw materials, partially finished parts, and completely finished parts. Excessive inventories tie up substantial amounts of cash that could be used in other parts of the company. The manufacturing engineer is also responsible for maintaining a safe and reliable workplace, including the safety of the workers at the facility and the environmental impact of the processes.

Manufacturing engineers must be able to work with diverse teams of people including design engineers, tradesmen, and management. Current "just in time" manufacturing practices reduce needed inventories in factories but require manufacturing engineers at one facility to coordinate their operation with manufacturing engineers at other facilities. They also must coordinate the work of the line workers who operate the manufacturing equipment. It is imperative that manufacturing engineers maintain a constructive relationship with their company's trade workers.

Manufacturing engineers play a critical role in modern design practices. Since manufacturing costs are such an important component in the success of a product, the design process must take into account manufacturing concerns. Manufacturing engineers identify high-cost or high-risk operations in the design phase of a product. When problems are identified, they work with the design engineers to generate alternatives.

In the production of large items such as buildings, dams, and roads, the production engineer is called a construction engineer rather than a manufacturing engineer. However, the role of the construction engineer is very similar to that of the manufacturing engineer. The main difference is that the construction engineer's production facility is typically outdoors while the manufacturing engineer's is inside a factory. The functions of construction engineers are the same as mentioned above when "assembly line" and "factory" are replaced with terms like "job site," reflecting the construction of a building, dam, or other large-scale project.

Operations and Maintenance

After a new production facility is brought on-line, it must be maintained. The operations engineer oversees the ongoing performance of the facility. Operations engineers must have a wide range of expertise dealing with the mechanical and electrical issues involved with maintaining a production line. They must be able to interact with manufacturing engineers, line workers and technicians who service the equipment. They must coordinate the service schedule of the technicians to ensure efficient service of the machinery, minimizing its downtime impact on production.

Maintenance and operations engineers also work in non-manufacturing roles. Airlines have staffs of maintenance engineers who schedule and oversee safety inspections and repairs. These engineers must have expertise in sophisticated inspection techniques to identify all possible problems.

Large medical facilities and other service sector businesses require operations or maintenance engineers to oversee the operation of their equipment. It is obviously critical that emergency medical equipment be maintained in peak working order.

Technical Support

A technical support engineer serves as the link between customer and product and assists with installation and setup. For large industrial purchases, technical support may be included in the purchase price. The engineer may visit the installation site and oversee a successful start-up. For example, a new power station for irrigation might require that the technical support engineer supervises the installation and helps the customer solve problems to get the product operational. To be effective, the engineer must have good interpersonal and problem-solving skills as well as solid technical training.

The technical support engineer may also trouble-shoot problems with a product. Serving on a computer company's help line is one example. Diagnosing design flaws found in the field once the product is in use is another example.

Technical support engineers do not have to have in-depth knowledge of each aspect of the product. However, they must know how to tap into such knowledge at their company.

Modern technical support is being used as an added service. Technical support engineers work with customers to operate and manage their own company's equipment as well as others. For example, a medical equipment manufacturer might sell its services to a hospital to manage and operate its highly sophisticated equipment. The manufacturer's engineers would not only maintain the equipment, but also help the hospital use its facilities in the most efficient way.

Customer Support

Customer support functions are similar to those of technical support as a link between the manufacturer and the customer. However, customer support personnel also are involved in the business aspect of the customer relationship. Engineers are often used for this function because of their technical knowledge and problem-solving ability. Typically, these positions require experience with the products and customers, and also require some business training.

The customer support person works with technical support engineers to ensure proper customer satisfaction. Customer support is also concerned with actual or perceived value for customers. Are they getting what they paid for? Is it cost-effective to continue current practices? Customer support persons are involved in warranty issues, contractual agreements, and the value of trade-in or credits for existing equipment. They work very closely with the technical support engineers and also with management personnel.

Sales

Engineers are valuable members of the sales force in numerous companies. These engineers must have interpersonal skills conducive to effective selling. Sales engineers bring many assets to their positions.

Engineers have the technical background to answer customer questions and concerns. They are trained to identify which products are right for the customer and how they can be applied. Sales engineers can also identify other applications or other products that might benefit the customer once they become familiar with their customer's needs.

In some sales forces, engineers are utilized because the customers are engineers themselves and have engineering-related questions. When airplane manufacturers market their aircraft to airlines, they send engineers. The airlines have engineers who have technical concerns overseeing the maintenance and operation of the aircraft. Sales engineers have the technical background to answer these questions.

As technology continues to advance, more and more products become technically sophisticated. This produces an ever-increasing demand for sales engineers.

Consulting

Consulting engineers are either self-employed or they work for a firm that does not provide goods or services directly to consumers. Such firms provide technical expertise to organizations that do. Many large companies do not have technical experts on staff in all areas of operation. Instead, they use consultants to handle issues in those technical areas.

For example, a manufacturing facility in which a cooling tower is used in part of the operation might have engineers who are well versed in the manufacturing processes but not in cooling tower design. The manufacturer would hire a consulting firm to design the cooling tower and related systems. Such a consultant might oversee the installation of such a system or simply provide a report with recommendations. After the system is in place or the report is delivered, the consultant would move on to another project with another company.

Consulting engineers also might be asked to evaluate the effectiveness of an organization. In such a situation, a team of consultants might work with a customer and provide suggestions and guidelines for improving the company's processes. These might be design methods, manufacturing operations, or even business practices. While some consulting firms provide only engineering-related expertise, other firms provide both engineering and business support and require the consulting engineers to work on business-related issues as well as technical issues.

Consulting engineers interact with a wide range of companies on a broad scope of projects, and come from all engineering disciplines. Often a consultant needs to be registered as a professional engineer in the state where he or she does business.

Management

In many instances, engineers work themselves into project management positions, and eventually into full-time management. National surveys show that more than half of all engineers will be involved in some type of management responsibilities—supervisory or administrative—before their career is over. Engineers are chosen for their technical ability, their problem-solving ability and their leadership skills.

Engineers may manage other engineers or support personnel, or they may rise to oversee the business aspects of a corporation. Often, prior to being promoted to this level of management, engineers acquire some business or management training. Some companies provide this training or offer incentives for employees to take management courses in the evening on their own time.

Other Fields

Some engineering graduates enter fields other than engineering, such as law, education, medicine, and business. Patent law is one area in which an engineering or science degree is almost essential. In patent law, lawyers research, write, and file patent applications. Patent lawyers must have the technical background to understand what an invention does so that they can properly describe the invention and legally protect it for the inventor.

Another area of law that has become popular for engineering graduates is corporate liability law. Corporations are faced with decisions every day over whether to introduce a new product, and must weigh potential risks. Lawyers with technical backgrounds have the ability to weigh technical as well as legal risks. Such lawyers are also used in litigation for liability issues. Understanding what an expert witness is saying in a suit can be a tremendous advantage in a courtroom and can enable a lawyer to effectively cross-examine the expert witness. Often, when a corporation is sued over product liability, the lawyers defending the corporation have technical backgrounds.

Engineers are involved in several aspects of education. The one students are most familiar with is engineering professors. College professors usually have their Ph.D.s. Engineers with master's degrees can teach at community colleges and in some engineering technology programs. Engineering graduates also teach in high schools, middle schools, and elementary schools. To do this full-time usually requires additional training in educational methods, but the engineering background is a great start. Thousands of engineers are involved in part-time educational projects where they go to classes as guest speakers to show students what "real engineers" do. The beginning of this chapter described one such engineer. The American Society for Engineering Education (ASEE) is a great resource if you are interested in engineering education.

Engineers also find careers in medicine, business, on Wall Street, and in many other professions. In modern society, with its rapid expansion of technology, the combination of problem-solving ability and technical knowledge makes engineers very valuable and extremely versatile.

Example 2.3

A dean of a New York engineering school passed along a story about her faculty, which was concerned that most of that year's graduates had not accepted traditional engineering positions. The largest employers were firms that hire stockbrokers. The school's engineering graduates were prized for their ability to look at data and make rational conclusions from the data. In other words, they were very effective problem solvers. They also had the ability to model the data mathematically. This combination made the engineering graduates a perfect fit for Wall Street. This is not an isolated story. Many engineering graduates find themselves in careers unrelated to their engineering education. All engineers are trained problem solvers and in today's technically advanced society, such skills are highly regarded.

2.3 ENGINEERING MAJORS

The following section is a partial listing of the various engineering disciplines in which engineering students can earn degrees. The list includes all of the most common majors. Some smaller institutions may offer programs that are not mentioned in this chapter. Be aware that some disciplines are referred to by different names at various institutions. Also, some programs may be offered as a subset of other disciplines. For instance, aeronautical and industrial engineering is sometimes combined with mechanical engineering. Environmental engineering might be offered as part of civil or chemical engineering.

These descriptions are not meant to be comprehensive. To describe a field completely, such as electrical engineering, would take an entire book by itself. These descriptions are meant to give you an overview of the fields and to answer some basic questions about the differences among the fields. It is meant to be a starting point. When selecting a major, a student might investigate several sources, including people actually working in the field.

An engineering student should keep in mind that the list of engineering fields is fluid. Areas such as aerospace, genetic, computer, and nuclear engineering did not even exist 50 years ago. Yet men and women developed the technology to create these fields. In your lifetime, other new fields will be created. The objective is to gain the solid background and tools to handle future challenges. It is also important to find a field you enjoy.

Example 2.4

"Show me the money!" As academic advisors, we see students regularly who are interested in engineering but not sure which discipline to pursue. Students are tempted to decide which field to enter by asking which offers the greatest salary. There are two issues a student should consider when looking at salaries of engineers.

First, all engineers make a good starting salary—well over the average American household income. However, a high salary would not make up for a job that is hated. The salary spread between the engineering disciplines is not large, varying only by about 10% from the average. The career you embark on after graduation from college will span some forty years—a long time to spend in a discipline you don't really enjoy.

Second, if you consider money a critical factor, you should consider earning potential, not starting salary. Earning potential in engineering is dependent, to a large extent, on job performance. Engineers who do well financially are those who excel professionally. Again, it is very rare for someone to excel in a field that he or she does not enjoy.

Aerospace Engineering

Aerospace engineering was previously referred to as aeronautical and astronautical engineering. Technically, aerospace engineering involves flight within the Earth's atmosphere and had its birth when two bicycle repairmen from Dayton, Ohio, made the first flight at Kitty Hawk, North Carolina, in 1903. Since that time, aerospace engineers have designed and produced aircraft that have broken the sound barrier, achieved near-orbit altitudes, and become the standard mode of transportation for long journeys. Aerospace engineering involves flight in space which began on October 4, 1957, when the Soviet Union launched Sputnik into orbit. The United States achieved one of its proudest moments in the field of aerospace on July 20, 1969, when Neil Armstrong set foot on the moon's surface—the first person ever to do so. In order to describe what the broad field of aerospace engineering has become, it can be separated into the categories of aerodynamics, propulsion, structures, controls, orbital mechanics, and life sciences.

Aerodynamics is the study of the flow of air over a streamlined surface or body. The aerodynamicist is concerned with the lift that is produced by a body such as the wing of an aircraft. Similarly, there is a resistance or drag force when a body moves through a fluid. An engineer looks to optimize lift and minimize drag. While mechanical, civil, and chemical engineers also study flow over bodies, the aerospace engineer is interested especially in high-speed air flows. When the speed of the aircraft approaches or exceeds the speed of sound, the modeling of the airflow becomes much more complex.

Air-breathing propulsion systems that power airplanes and helicopters include propellers, and gas turbine jets, as well as ramjets and scram-jets. Propulsion systems used to launch or operate spacecraft include liquid and solid rocket engines. **Propulsion engineers** are continually developing more efficient, quieter, and cleaner-burning conventional engines as well as new engine technologies. New engine concepts include wave rotors, electric propulsion, and nuclear-powered craft.

The structural support of an aircraft or spacecraft is critical. It must be both lightweight and durable. These conditions are often mutually exclusive and provide challenges for structural design engineers. **Structural engineers** utilize new alloys, composites, and other new materials to meet design requirements for more efficient and more maneuverable aircraft and spacecraft.

The control schemes for aircraft and spacecraft have evolved rapidly. The new commercial airplanes are completely digitally controlled. Aerospace engineers work with electrical engineers to design **control systems** and subsystems used to operate the craft. They must be able to understand the electrical and computational aspects of the control schemes as well as the physical systems that are being controlled.

Aerospace engineers are also interested in **orbital mechanics**. They calculate where to place a satellite to operate as a communication satellite or as a global positioning system (GPS). They might also determine how to use the gravity fields of the near-Earth planets to help propel a satellite to the outskirts of the solar system.

An aerospace engineer must be aware of human limitations and the effects of their craft on the human body. It would serve no useful purpose to design a fighter plane that was so maneuverable it incapacitated the pilot with a high G-force. Designing a plane and a system which keeps the pilot conscious is an obvious necessity. Understanding the physiological and psychological effects of lengthy exposure to weightlessness is important when designing a spacecraft or space station.

The American Institute of Aeronautics and Astronautics (AIAA) is one of the most prominent aerospace professional societies. It is composed of the following seven technical groups, which include 66 technical committees:

- Engineering and Technology Management
- Aircraft Technology Integration and Operations
- Propulsion and Energy
- Space and Missile Systems
- Aerospace Sciences
- Information and Logistics Systems
- Structures, Design, and Testing

Agricultural Engineering

Agricultural engineering traces its roots back thousands of years to the time when people began to examine ways to produce food and food products more efficiently. Today, the role of the agricultural engineer is critical to our ability to feed the ever-expanding population of the world. The production and processing of agricultural products is the primary concern of agricultural engineers. Areas within agricultural engineering include power machinery, food processing, soils and water, structures, electrical technologies, and bioengineering.

The mechanical equipment used on modern farms is highly sophisticated and specialized. **Harvesting equipment** not only removes the crop from the field but also begins to process it and provides information to the farmers on the quality of the harvest. This requires highly complicated mechanical and electrical systems that are designed and developed by the agricultural engineer.

Often, once the food is harvested it must be processed before it reaches the marketplace. Many different technologies are involved in the efficient and safe processing and delivery of food products. **Food process engineers** are concerned with providing healthier products to consumers who increasingly rely on processed food products. This often requires the agricultural engineer to develop new processes. Another modern-day concern is increased food safety. Agricultural engineers design and develop means by which food is produced free of contamination such as the irradiation techniques used to kill potentially harmful bacteria in food.

The effective management of **soil and water resources** is a key aspect of productive agriculture. Agricultural engineers design and develop means to address effective land use, proper drainage, and erosion control. This includes such activities as designing and implementing an irrigation system for crop production and designing a terracing system to prevent erosion in a hilly region.

Agricultural **structures** are used to house harvested crops, livestock, and their feed. Agricultural engineers design structures including barns, silos, dryers, and processing centers. They look for ways to minimize waste or losses, optimize yields, and protect the environment.

Electrical and information technology development is another area important to the agriculture community due to the fact that farms are typically located in isolated regions. Agricultural engineers design systems that meet the needs of the rural communities.

Bioengineering has rapidly evolved into a field with wide uses in health products as well as agricultural products. Agricultural engineers are working to harness these rapidly developing technologies to further improve the quality and quantity of the agricultural products necessary to continue to feed the world's population.

The American Society of Agricultural Engineers (ASAE) is one of the most prominent professional societies for agricultural engineers. It has eight technical divisions which address areas of agricultural engineering:

- Food Processing
- Information and Electrical Technologies
- Power and Machinery
- Structures and Environmental

- Soil and Water
- Forest
- Bioengineering
- Aqua Culture

Architectural Engineering

In ancient times, major building projects required a master builder. The master builder was responsible for the design and appearance of the structure, for selecting the appropriate materials, and for personally overseeing construction. With the advent of steel construction in the 19th century, it became necessary for master builders to consult with specialists in steel construction in order to complete their designs. As projects became more complex, other specialists were needed in such areas as mechanical and electrical systems. Modern architectural engineers facilitate the coordination between the creativity of the architect and the technical competencies of a variety of technology specialists.

Architectural engineers are well-grounded in the engineering fundamentals of structural, mechanical, and electrical systems, and have at least a foundational understanding of aesthetic design. They know how to design a building and how the various technical systems are interwoven within that design. Their strengths are both creative and pragmatic.

The National Society of Architectural Engineers, which is now the Architectural Engineering Institute, defined architectural engineering as:

> "... the profession in which a knowledge of mathematics and natural sciences, gained by study, experience, and practice, is applied with judgment to the development of ways to use, economically and safely, the materials and forces of nature in the engineering design and construction of buildings and their environmental systems." [Belcher]

There are four main divisions within architectural engineering: structural; electrical and lighting; mechanical systems; and construction engineering and management.

Structural is primarily concerned with the integrity of the structure of buildings. Determining the integrity of a building's structure involves the analysis of loads and forces resulting from normal usage and operation as well as from earthquakes, wind forces, snow loads, or wave impacts. The structural engineer takes the information from the analysis, and designs the structural elements that will support the building while at the same time meeting the aesthetic and functional needs involved.

The area of **electrical and lighting** systems is concerned with the distribution of utilities and power throughout the building. This requires knowledge of electricity and power distribution as well as lighting concerns. Lighting restrictions may require energy-efficient lighting systems that are sufficient to meet the functional requirements related to use of the building. The architectural engineer also must insure that the lighting will complement the architectural designs of the building and the rooms within that structure.

Mechanical systems control the climate of a building, which includes cooling and heating the air in rooms as well as controlling humidity and air quality. Mechanical systems are also used to distribute water through plumbing systems. Other mechanical systems include transportation by way of elevators and escalators within a building. These systems must be integrated to complement the architectural features of the building.

The fourth area is **construction engineering and management**, which combine the technical requirements with the given financial and legal requirements to meet project deadlines. The architectural engineer is responsible for implementing the design in a way which assures the quality of the construction and meets the cost and schedule of the project. Modern computer tools and project management skills are implemented to manage such complex projects.

Other areas which are emerging for architectural engineers include energy management, computerized controls, and new building materials, including plastics and composites and acoustics. Many programs have added explicit areas of emphasis in the application of alternative energy, energy efficiency, and sustainability into architectural engineering programs. Designing structures to reduce energy use, material consumption, and environmental impact in the design, construction, and operation phases of buildings is a growing area of interest and a topic of a great deal of research within the field.

Architectural engineers are employed by consulting firms, contractors, and government agencies. They may work on complex high-rise office buildings, factories, stadiums, research labs, or educational facilities. They also may work on renovating historic structures or developing affordable low-income housing. As the construction industry continues to grow and as projects continue to become more complex, the outlook for architectural engineers is bright.

There are less than 20 ABET-accredited programs in architectural engineering in the United States. Some of these are four-year programs, while others provide the combined engineering foundation and architectural insights in a five-year program.

The Architectural Engineering Institute (AEI) was formed in 1998 with the merger of the National Society of Architectural Engineers and the American Society of Civil Engineers Architectural Engineering Division. AEI was created to be the home for professionals in the building industry. It is organized into divisions, which include:

- Commercial Buildings
- Industrial Buildings
- Residential Buildings
- Institutional Buildings
- Military Facilities
- Program Management
- Mitigation of the Effects of Terrorism

- Building Systems
- Education
- Architectural Systems
- Fully Integrated and Automated Project Process
- Glass as an Engineered Material
- Sick/Healthy Buildings
- Designing for Facilities for the Aging

Reference: Belcher, M. C., *Magill's Survey of Science: Applied Science*, Salem Press, Pasadena, CA, 1993.

Biological Engineering

Biological engineering is one of the newer fields of engineering that spans the fields of biology and engineering. Traditionally, engineering has applied concepts from mathematics and the physical sciences to design, analyze, and manufacture inanimate products and processes to address human and environmental needs. Biological engineering, also called bioengineering, or biotechnological engineering, integrates the study of living systems into engineering, opening opportunities to adapt and even create living products. A search of the Accreditation Board of Engineering and Technology's website (www.abet.org, accessed November 2010) for accredited biological engineering shows a small but growing number of accredited programs. One example is at the Massachusetts Institute of Technology, whose website defines this new discipline as follows:

> The goal of this biological engineering discipline is to advance fundamental understanding of how biological systems operate and to develop effective biology-based technologies for applications across a wide spectrum of societal needs including breakthroughs in diagnosis, treatment, and prevention of disease, in design of novel materials, devices, and processes, and in enhancing environmental health. web.mit.edu/be/index.shtml (accessed November 2010)

This definition shows the breadth and opportunities for this emerging field. Biological engineering is being practiced as separate programs as well as integrated with more traditional engineering majors at many schools. Some schools, for example, include biological engineering as a complement to their agricultural engineering programs to differentiate the parts of their curriculum and research programs that address new opportunities in biological areas that have been opened via modern technology. Another example is the inclusion of bioengineering with biomedical engineering programs. Similar to other fields of engineering that have grown out of the more traditional disciplines, biological engineering is evolving rapidly. At this point in the discipline's development, this means that the term "biological engineering" may mean slightly different things at different institutions. Students are encouraged to research individual programs to understand what this exciting and emerging field means at their respective institution.

Biomedical Engineering

Biomedical engineering is one of the newer fields of engineering, first recognized as a discipline in the 1940s. Its origins, however, date back to the first artificial limbs made of wood or other materials. Captain Ahab in "Moby Dick" probably didn't think of the person who made his wooden leg as a biomedical engineer, yet he was a predecessor for biomedical engineers who design modern prosthetics.

Biomedical engineering is a very broad field that overlaps with several other engineering disciplines. In some institutions, it may be a specialization within another discipline. Biomedical engineering applies the fundamentals of engineering to meet the needs of the medical community. Because of the wide range of skills needed in the fields of engineering and medicine, biomedical engineering often requires graduate work. The broad field of biomedical engineering encompasses the three basic categories of bioengineering, medical, and clincal.

Bioengineering is the application of engineering principles to biological systems. This can be seen in the production of food or in genetic manipulation to produce a disease-resistant strain of a plant or animal. Bioengineers work with geneticists to produce bioengineered products. New medical treatments are produced using genetically altered bacteria to produce human proteins needed to cure diseases. Bioengineers may work with geneticists in producing these new products in the mass quantities needed by consumers.

The medical aspect of biomedical engineering involves the design and development of devices to solve medical challenges. This includes designing mechanical devices such as a prosthesis which gives individuals the desired mobility. It also involves the development of the chemical processes necessary to make an artificial kidney function and the electrical challenges in designing a new pacemaker.

Medical engineers develop instrumentation for medical uses including non-intrusive surgical instruments. Much of the trauma of surgery results from the incisions made to gain access to the area of concern. Procedures allowing a surgeon to make a small incision, such as orthoscopic surgery, have greatly reduced risk and recovery time for patients.

Rehabilitation is another area in which biomedical engineers work. It involves designing and developing devices for the physically impaired. Such devices can expand the capabilities of impaired individuals, thereby improving their quality of life or shortening recovery times.

Clinical engineering involves the development of systems to serve hospitals and clinics. Such systems exhaust anesthetic gases from an operating room without impairing the surgical team. Air lines in ventilating systems must be decontaminated to prevent microorganisms from spreading throughout the hospital. Rehabilitation centers must be designed to meet the needs of patients and staff.

Chemical Engineering

Chemical engineering differs from most of the other fields of engineering in its emphasis on chemistry and the chemical nature of products and processes. Chemical engineers take what chemists do in a laboratory and, applying fundamental engineering, chemistry, and physics principles, design and develop processes to mass-produce products for use in our society. These products include detergents, paints, plastics, fertilizers, petroleum products, food products, pharmaceuticals, electronic circuit boards, and many others.

The most common employment of chemical engineers is in the design, development, and operation of large-scale chemical production facilities. In this area, the design function involves the **design of the processes** needed to safely and reliably produce the final product. This may involve controlling and using chemical reactions, separation processes, or heat and mass transfer. While the chemist might develop a new compound in the laboratory, the chemical engineer would develop a new process to make the compound in a pilot plant, which is a small-scale version of a full-size production facility. An example would be the design of a process to produce a lower-saturated-fat product with the same nutritional value yet still affordable. Another example would be the development of a process to produce a higher-strength plastic used for automobile bumpers.

With respect to **energy sources,** chemical engineers are involved in the development of processes to extract and refine crude oil and natural gas. Petroleum engineering grew out of chemical engineering and is described in a separate section. Chemical engineers are also involved in alternative fuel development and production.

The processing of petroleum into plastics by chemical engineers has created a host of consumer products that are used every day. Chemical engineers are involved in adapting these products to meet new needs. GE Plastics, for instance, operates a research house in Massachusetts made entirely from plastic. It is used as a research facility to develop new building materials that are cheaper and more beneficial to the consumer.

Many of the chemicals and their byproducts used in industry can be dangerous to people and/or the environment. Chemical engineers must develop processes that minimize harmful waste. They work with both new and traditional processes to treat hazardous by-products and reduce harmful emissions.

Chemical engineers are also very active in the bio-products arena. This includes the **pharmaceutical industry**, where chemical engineers design processes to manufacture new lines of affordable drugs. Geneticists have developed the means to artificially produce human proteins that can be used successfully to treat many diseases. They use genetically altered bacteria to produce these proteins. It is the job of the chemical engineer to take the process from the laboratory and apply it to a larger-scale production process to produce the quantities needed by society.

Chemical engineers are also involved in **bio-processes** such as dialysis, where chemical engineers work along with other biomedical engineers to develop new ways to treat people. Chemical engineers might be involved in the development of an artificial kidney. They could also be involved in new ways to deliver medicines, such as through skin implants or patches.

Chemical engineers have become very active in the production of **circuit boards**, such as those used in computers. To manufacture the very small circuits required in modern electronic devices, material must be removed and deposited very precisely along the path of the circuit. This is done using chemical techniques developed by chemical engineers. With the demand for smaller and faster electronics, challenges continue for chemical engineers to develop the processes to make them possible.

Chemical engineers are also involved in the **modeling of systems**. Computer models of manufacturing processes are made so that modifications can be examined without having to build expensive new facilities. Managing large processes involving facilities which could be spread around the globe requires sophisticated scheduling capabilities. Chemical engineers who developed these capabilities for chemical plants have found that other large industries with scheduling challenges, such as airlines, can utilize these same tools.

The American Institute of Chemical Engineering (AIChE) is one of the most prominent professional organizations for chemical engineers. It is organized into 13 technical divisions that represent the diverse areas of the chemical engineering field:

- Catalysis and Reaction Engineering
- Computing and Systems Technology
- Engineering and Construction Contracting
- Environmental
- Food, Pharmaceutical, and Bioengineering
- Forest Products
- Fuels and Petrochemicals

- Heat Transfer and Energy Conversion
- Management
- Materials Engineering and Sciences
- Nuclear Engineering
- Safety and Health
- Separations

Civil Engineering

Ancient examples of early **civil engineering** can be seen in the pyramids of Egypt, the Roman roads, bridges and aqueducts of Europe, and the Great Wall in China. These all were designed and built under the direction of the predecessors to today's civil engineers. Modern civil engineering continues to face the challenges of meeting the needs of society. The broad field of civil engineering includes these categories: structural, environmental, transportation, water resources, surveying, urban planning, and construction engineering.

A humorous aside perhaps illustrates one aspect of civil engineering. . . . An Air Force general with an engineering background was asked what the difference is between civil and aerospace engineers. "That's easy," he responded. "The aerospace engineers build modern, state-of-the-art weapon systems and the civil engineers build the targets."

Structural engineers are the most common type of civil engineer. They are primarily concerned with the integrity of the structure of buildings, bridges, dams, and highways. Structural engineers evaluate the loads and forces to which a structure will be subjected. They analyze the structural design in regard to earthquakes, wind forces, snow loads, or wave impacts, depending on the area in which the building will be constructed.

A related field of structural engineering is **architectural engineering**, which is concerned with the form, function and appearance of a structure. The architectural engineer works alongside the architect to ensure the structural integrity of a building. The architectural engineer combines analytical ability with the concerns of the architect.

Civil engineers in the **environmental** area may be concerned with the proper disposal of wastes—residential and industrial. They may design and adapt landfills and waste treatment facilities to meet community needs. Industrial waste often presents a greater challenge because it may contain heavy metals or other toxins which require special disposal procedures. Environmental engineering has come to encompass much more and is detailed in a later section of this chapter.

Transportation engineers are concerned with the design and construction of highways, railroads, and mass transit systems. They are also involved in the optimization and operation of the systems. An example of this is **traffic engineering**. Civil engineers develop the tools to measure the need for traffic control devices such as signal lights and to optimize these devices to allow proper traffic flow. This can become very complex in cities where the road systems were designed and built long ago but the areas around those roads have changed significantly with time.

Civil engineers also work with water resources as they construct and maintain dams, aqueducts, canals, and reservoirs. **Water resource engineers** are charged with providing safe and reliable supplies of water for communities. This includes the design and operation of purification systems and testing procedures. As communities continue to grow, so do the challenges for civil engineers to produce safe and reliable water supplies.

Before any large construction project can be started, the construction site and surrounding area must be mapped or surveyed. **Surveyors** locate property lines and establish alignment and proper placement of engineering projects. Modern surveyors use satellite technology as well as aerial and terrestrial photogrammetry. They also rely on computer processing of photographic data.

A city is much more than just a collection of buildings and roads. It is a home and working place for people. As such, the needs of the inhabitants must be taken into account in the design of the city's infrastructure. **Urban planning engineers** are involved in this process. They incorporate the components of a city (buildings, roads, schools, airports, etc.) to meet the overall needs of the population. These needs include adequate housing, efficient transportation, and open spaces. The urban planning engineer is always looking toward the future to fix problems before they reach a critical stage.

Another civil engineering area is **construction engineering**. In some institutions, this may even be a separate program. Construction engineers are concerned with the management and operation of construction projects. They are also interested in the improvement of construction methods and materials. Construction engineers design and develop building techniques and building materials that are safer, more reliable, cost-effective, and environmentally friendly. There are many technical challenges that construction engineers face and will continue to face in the coming decades.

An example of one such construction challenge is the rebuilding of a city's infrastructure, such as its sewers. While the construction of a sewer may not seem glamorous or state-of-the-art, consider the difficulty of rebuilding a sewer under an existing city. A real problem facing many cities is that their infrastructure was built decades or centuries ago, but it has a finite life. As these infrastructures near the end of their expected lives, the construction engineer must refurbish or reconstruct these systems without totally disrupting the city.

The American Society of Civil Engineers (ASCE) is one of the most prominent professional organizations for civil engineers. It is organized into the following 16 technical divisions covering the breadth of civil engineering:

- Aerospace
- Air Transport
- Architectural Engineering
- Construction Division
- Energy
- Engineering Mechanics
- Environmental Engineering
- Geomatics
- Highway
- Materials Engineering
- Pipeline
- Urban Planning and Development
- Urban Transportation
- Water Resources Engineering
- Water Resources Planning and Management
- Waterways, Ports, Coastal, and Ocean Engineering

Computer Engineering

Much of what computer and electrical engineers do overlaps. Many **computer engineering** programs are part of electrical engineering or computer science programs. However, computer technology and development have progressed so rapidly that specialization is required in this field, and thus computer engineering is treated separately. Given the wide range of computer applications and their continued growth, computer engineering has a very exciting future.

Computer engineering is similar to computer science, yet distinct. Both fields are extensively involved with the design and development of software. The main difference is that the computer scientist focuses primarily on the software and its optimization. The computer engineer, by contrast, focuses primarily on computer hardware—the machine itself. Software written by the computer engineer is often designed to control or to interface more efficiently with the hardware of the computer and its components.

Computer engineers are involved in the design and development of operating systems, compilers, and other software that requires efficient interfacing with the components of the computer. They also work to improve computer performance by optimizing the software and hardware in applications such as computer graphics.

Computer engineers work on the design of computer architecture. Designing faster and more efficient computing systems is a tremendous challenge. As faster computers are developed, applications arise for even faster machines. The continual quest for faster and smaller microprocessors involves overcoming barriers introduced by the speed of light and by circuitry so small that the molecular properties of components become important.

Computer engineers develop and design electronics to interface with computers. These include modems, Ethernet connections, and other means of data transmission. Computer engineers created the devices that made the Internet possible.

In addition to having computers communicate with each other, the computer engineer is also interested in having computers work together. This may involve increasing the communication speed of computer networks or using tightly coupled multiple processors to improve the computing speed.

Security is becoming a bigger concern as more and more information is transferred using computers. Computer engineers are developing new means of commercial and personal security to protect the integrity of electronic communications.

Artificial intelligence, voice recognition systems, and touch screens are examples of other technologies with which computer engineers are involved. The computers of the future will look and operate much differently than they do now, and it is the computer engineer who will make this happen.

Electrical Engineering

Considering the wide range of electronic devices people use every day, it is not surprising that **electrical engineering** has become the most populated of the engineering disciplines. Electrical engineers have a wide range of career opportunities in almost all of the industrial sectors. In order to provide a brief discussion of such a broad field, we will divide electrical engineering into eight areas: computers, communications, circuits and solid state devices, control, instrumentation, signal processing, bioengineering, and power.

Engineers specializing in **computer** technology are in such high demand that numerous institutions offer a separate major for computer engineering. Please refer back to the earlier section which described computer engineering separately.

Electrical engineers are responsible for the explosion in **communication** technologies. Satellites provide nearly instantaneous global communication. Global positioning systems (GPS) allow anyone with the required handheld unit to pinpoint precisely where they are located anywhere in the world. Fiber optics and lasers are rapidly improving the reliability and speed with which information can be exchanged. Wireless communication allows people to communicate anywhere, with anyone. Future breakthroughs in this field will have a tremendous impact on how we live in tomorrow's society.

Electrical engineers also design and develop electronic **circuits**. Circuit design has changed rapidly with the advent of microelectronics. As circuits continue to shrink, new barriers such as molecular size emerge. As the limits of current technology are approached, there will be incentives to develop new and faster ways to accomplish the same tasks.

Almost all modern machines and systems are digitally controlled. **Digital controls** allow for safer and more efficient operation. Electronic systems monitor processes and make corrections faster and more effectively than human operators. This improves reliability, efficiency, and safety. The electrical engineer is involved in the design, development, and operation of these control systems. The engineer must determine the kind of control required, what parameters to monitor, the speed of the correction, and many other factors. Control systems are used in chemical plants, power plants, automotive engines, airplanes, and a variety of other applications.

For a control system to operate correctly, it must be able to measure the important parameters of whatever it is controlling. Doing so requires accurate **instrumentation**, another area for the electrical engineer. Electrical engineers who work in this area develop electrical devices to measure quantities such as pressure, temperature, flow rate, speed, heart rate, and blood pressure. Often, the electrical devices convert the measured quantity to an electrical signal that can be read by a control system or a computer. Instrumentation engineers also design systems to transmit the measured information to a recording device, using telemetry. For example, such systems are needed for transmitting a satellite's measurements to the recording computers back on earth as it orbits a distant planet.

Signal processing is another area where electrical engineers are needed. In many instances the electrical signals coming from instrumentation or other sources must be conditioned before the information can be used. Signals may need to be electronically filtered, amplified, or modified. An example is the active noise control system on a stethoscope which allows a paramedic to listen to a patient's heart while in a helicopter. The active noise control system can block out the sound of the helicopter so the paramedic can make a quick, accurate assessment of the patient. Signal processing also comes into play in areas such as voice recognition for computers.

Electrical engineers also work in biomedical, or **bioengineering,** applications, as described earlier. Electrical engineers work with medical personnel to design and develop devices used in the diagnosis and treatment of patients. Examples include non-intrusive techniques for detecting tumors through magnetic resonance imaging (MRI) or through computerized axial tomography (CAT) scans. Other examples include pacemakers, cardiac monitors, and controllable prosthetic devices.

The generation, transmission, and distribution of electric **power** is perhaps the most traditional aspect of electrical engineering. Electrical engineers work closely with mechanical engineers in the production of electrical power. They also oversee the distribution of power through electrical networks and must ensure reliable supplies of electricity to our communities. With modern society's dependence on electricity, interruptions in the flow of electricity can be catastrophic. Many of today's power-related challenges revolve around reliability and cost of delivery. Electrical engineers also work with materials engineers to incorporate superconductivity and other technology in more efficient power transmission.

The Institute of Electrical and Electronics Engineers (IEEE) is the largest and most prominent professional organization for electrical engineers. It is organized into 37 technical divisions, which indicates the breadth of the field of electrical engineering. The technical divisions include:

- Aerospace and Electronic Systems
- Antennas and Propagation
- Broadcast Technology
- Circuits and Systems
- Communications
- Components Packaging and Manufacturing Technology
- Computer
- Consumer Electronics
- Control Systems
- Dielectrics and Electrical Insulation
- Education
- Electromagnetic Compatibility
- Electron Devices

- Microwave Theory and Techniques
- Neural Networks
- Nuclear and Plasma Sciences
- Oceanic Engineering
- Power Electronics
- Power Engineering
- Professional Communication
- Reliability
- Robotics and Automation
- Signal Processing
- Social Implications of Technology
- Solid-State Circuits
- Geoscience and Remote Sensing
- Industrial Electronics

- Engineering in Medicine and Biology
- Engineering Management
- Instrumentation and Measurement
- Lasers and Electro-Optics
- Magnetics

- Industrial Applications
- Information Theory
- Systems, Man, and Cybernetics
- Ultrasonics, Ferroelectrics, and Frequency Control
- Vehicular Technology

Environmental Engineering

Environmental engineering is a field which has evolved to improve and protect the environment while maintaining the rapid pace of industrial activity. This challenging task has three parts to it: disposal, remediation, and prevention.

Disposal is similar to that covered under civil engineering. Environmental engineers are concerned with disposal and processing of both industrial and residential waste. Landfills and waste treatment facilities are designed for residential waste concerns. The heavy metals and other toxins found in industrial wastes require special disposal procedures. The environmental engineer develops the techniques to properly dispose of such waste.

Remediation involves the cleaning up of a contaminated site. Such a site may contain waste which was improperly disposed of, requiring the ground and/or water to be removed or decontaminated. The environmental engineer develops the means to remove the contamination and return the area to a usable form.

Prevention is an area that environmental engineers are becoming more involved in. Environmental engineers work with manufacturing engineers to design processes that reduce or eliminate harmful waste. One example is in the cleaning of machined parts. Coolant is sprayed on metal parts as they are cut to extend the life of the cutting tools. The oily fluid clings to the parts and has to be removed before the parts are assembled. An extremely toxic substance had been used in the past to clean the parts because there was not a suitable alternative. Environmental engineers discovered that oil from orange peels works just as well and is perfectly safe (even edible). Since this new degreasing fluid did not need to be disposed of in any special way, manufacturing costs were reduced, and the environment of the workers improved.

Prevention does not just mean avoiding disasters. It also means reducing the environmental impact or "footprint" of products and processes. Environmental engineers may be engaged in improving the sustainability of processes, including manufacturing and construction as well as the design and operation of new products. The environmental engineers study the environmental impact of the processes or products; identify alternatives; and work to implement these alternatives to reduce the impact on our environment. Reducing the environmental impact of the products we use every day is vital to protecting the global environment as the world's population continues to grow.

Environmental engineers must be well grounded in engineering fundamentals and current on environmental regulations. Within their companies, they are the experts on compliance with the ever-changing environmental laws. An environmental engineer must be able to understand the regulations and know how to apply them to the various processes they encounter.

Industrial Engineering

Industrial engineering is described by the Institute of Industrial Engineers as the design, improvement, and installation of integrated systems of people, material, and energy. Industrial engineering is an interdisciplinary field that involves the integration of technology, mathematical models, and management practices. Traditional industrial engineering is done on a factory floor. However, the skills of an industrial engineer are transferable to a host of other applications. As a result, industrial engineers find themselves working within a wide variety of industries. Four of the main areas of emphasis for industrial engineers are production, manufacturing, human factors, and operations research.

The **production** area includes functions such as plant layout, material handling, scheduling, and quality and reliability control. An industrial engineer would examine the entire process involved in making a product, and optimize it by reducing cost and production time, and by increasing quality and reliability. In addition to factory layout, industrial engineers apply their expertise in other ways. For example, with an amusement park an industrial engineer would analyze the flow of people through the park to reduce bottlenecks and provide a pleasant experience for the patrons.

Manufacturing differs from production in that it addresses the components of the production process. While production concerns are on a global scale, manufacturing concerns address the individual production station. The actual material processing, such as machining, is optimized by the industrial engineer.

The **human factors area** involves the placement of people into the production system. An industrial engineer in this area studies the interfaces between people and machines in the system. The machines may include production machinery, computers, or even office chairs and desks. The industrial engineer considers ergonomics in finding ways to improve the interfaces. He or she looks for ways to improve productivity while providing a safe environment for workers.

Operations research is concerned with the optimization of systems. This involves mathematically modeling systems to identify ways to improve them. Project management techniques such as critical path identification fall under operations research. Often, computer simulations are required to either model the system or to study the effects of changes to the system. These systems may be manufacturing systems or other organizations. The optimizing of sales territories for a pharmaceutical sales force provides a non-manufacturing example.

The Institute for Industrial Engineering (IIE) is one of the most prominent professional organizations for industrial engineers. It is organized into three societies, 10 technical divisions and 8 interest groups. The following technical divisions show the breadth of industrial engineering:

- Aerospace and Defense
- Energy, Environment, and Plant Engineering
- Engineering Economy
- Facilities Planning and Design
- Financial Services
- Logistics Transportation and Distribution
- Manufacturing
- Operations Research
- Quality Control and Engineering Reliability
- Utilities

Marine and Ocean Engineering

Nearly 80 percent of the Earth's surface is covered with water. Engineers concerned with the exploration of the oceans, the transportation of products over water, and the utilization of resources in the world's oceans, seas, and lakes are involved in **marine and ocean engineering**.

Marine engineers focus on the design, development, and operation of ships and boats. They work together with **naval architects** in this capacity. Naval architects are concerned with the overall design of the ship. They focus on the shape of the hull in order to provide the appropriate hydrodynamic characteristics. They are also concerned with the usefulness of the vessel for its intended purpose, and with the design of the ship's subsystems including ventilation, water, and sanitary systems to allow the crew to work efficiently.

The **marine engineer** is primarily concerned with the subsystems of the ship which allow the ship to serve its purpose. These include the propulsion, steering, and navigation systems. The marine engineer might analyze the ship for vibrations or stability in the water. The ship's electrical power distribution and air-conditioning fall under the responsibility of marine engineers. They also might be involved in the analysis and design of the cargo handling systems of the ship.

The responsibilities of an **ocean engineer** involve the design, development, and operation of vehicles and devices other than boats or ships. These include submersible vehicles used in the exploration of the oceans and in the obtaining of resources from the ocean depths. He or she might be involved in the design of underwater pipelines or cables, offshore drilling platforms, and offshore harbor facilities.

Ocean engineers also are involved with the interaction of the oceans and things with which oceans come in contact. They study wave action on beaches, docks, buoys, moorings, and harbors. Ocean engineers design ways to reduce erosion while protecting the marine environment. They study ways to protect and maintain marine areas which are critical to our food supply. Ocean engineers become involved with pollution control and treatment in the sea and alternative sources of energy from the ocean.

One of the professional societies to which these engineers may be involved is the Society of Naval Architects and Marine Engineers. The society is subdivided into the following nine technical and research committees:

- Hull Structure
- Hydrodynamics
- Ship's Machinery
- Ship Technical Operations
- Offshore

- Ship Production
- Ship Design
- Ship Repair and Conversion
- Small Craft

Materials Engineering

The origins of **materials engineering** can be traced to around 3000 B.C. when people began to produce bronze for use in creating superior hunting tools. Since that time, many new materials have been developed to meet the needs of society. Materials engineers develop these new materials and the processes to create them. The materials may be metals or nonmetals: ceramics, plastics, and composites. Materials engineers are generally concerned with four areas of materials: structure, properties, processes, and performance.

Materials engineers study the **structure** and composition of materials on a scale ranging from the microscopic to the macroscopic. They are interested in the molecular bonding and chemical composition of materials. The materials engineer is also concerned with the effect of grain size and structure on the material properties.

The **properties** in question might include strength, crack growth rates, hardness, and durability. Numerous technological advances are impeded by a lack of materials possessing the properties required by the design engineers. Materials engineers seek to develop materials to meet these demands.

A given material may have very different properties depending on how the material is **processed.** Steel is a good example. Cooling can affect its properties drastically. Steel that is allowed to cool in air will have different properties than steel that is cooled through immersion in a liquid. The composition of a material also can affect its properties. Materials such as metallic alloys contain trace elements that must be evenly distributed throughout the alloy to achieve the desired properties. If the trace elements are not well distributed or form clumps in the metal, the material will have very different properties than the desired alloy. This could cause a part made with the alloy to fail prematurely. Materials engineers design processes and testing procedures to ensure that the material has the desired properties.

The materials engineer also works to ensure that a material meets the **performance** needs of its application by designing testing procedures to ensure that these requirements are met. Both destructive and nondestructive testing techniques are used to serve this process.

Materials engineers develop new materials, improve traditional materials, and produce materials reliably and economically through synthesis and processing. Subspecialties of materials engineering, such as metallurgy and ceramics engineering, focus on classes of materials with similar properties.

Metallurgy involves the extraction of metals from naturally occurring ore for the development of alloys for engineering purposes. The metallurgical engineer is concerned with the composition, properties and performance of an alloy. Detailed investigation of a component failure often identifies design flaws in the system. The materials engineer can provide useful information regarding the condition of materials to the design engineer.

Ceramics is another area of materials engineering. In ceramic engineering, the naturally occurring materials of interest are clay and silicates, rather than an ore. These non-metallic minerals are employed in the production of materials that are used in a wide range of applications, including the aerospace, computer, and electronic industries.

Other areas of materials engineering focus on polymers, plastics, and composites. **Composites** are composed of different kinds of materials which are synthesized to create a new material to meet some specific demands. Materials engineers are also involved in **biomedical** applications. Examples include the development of artificial tissue for skin grafts, or bone replacement materials for artificial joints.

One of the professional societies to which materials engineers may belong is the Minerals, Metals, and Materials Society. It is organized into the following five technical divisions:

- Electronic, Magnetic, and Photonic Materials
- Extraction and Processing
- Light Metals
- Materials Processing and Manufacturing
- Structural Materials

Mechanical Engineering

Mechanical engineering is one of the largest and broadest of the engineering disciplines. It is second only to electrical engineering in the number of engineers employed in the field. Fundamentally, mechanical engineering is concerned with machines and mechanical devices. Mechanical engineers are involved in the design, development, production, control, operation, and service of these devices. Mechanical engineering is composed of two main divisions: design and controls, and thermal sciences.

The **design** function is the most common function of mechanical engineering. It involves the detailed layout and assembly of the components of machines and devices. Mechanical engineers are concerned about the strength of parts and the stresses the parts will need to endure. They work closely with materials engineers to ensure that correct materials are chosen. Mechanical engineers must also ensure that the parts fit together by specifying detailed dimensions.

Another aspect of the design function is the design process itself. Mechanical engineers develop computational tools to aid the design engineer in optimizing a design. These tools speed the design process by automating time-intensive analyses.

Mechanical engineers are also interested in controlling the mechanical devices they design. **Control** of mechanical devices can involve mechanical or hydraulic controls. However, most modern control systems incorporate digital control schemes. The mechanical engineer models controls for the system and programs or designs the control algorithm.

The noise generated from mechanical devices is often a concern, so mechanical engineers are often involved in acoustics—the study of noise. The mechanical engineer works to minimize unwanted noise by identifying the source and designing ways to minimize it without sacrificing a machine's performance.

In the **thermal sciences**, mechanical engineers study the flow of fluids and the flow of energy between systems. Mechanical engineers deal with liquids, gases, and two-phase flows, which are combinations of liquids and non-liquids. Mechanical engineers might be concerned about how much power is required to supply water through piping systems in buildings. They might also be concerned with aerodynamic drag on automobiles.

The flow of energy due to a temperature difference is called **heat transfer**, another thermal science area in which mechanical engineers are involved. They predict and study the temperature of components in environments of operation. Modern personal computers have microprocessors that require cooling. Mechanical engineers design the cooling devices to allow the electronics to function properly.

Mechanical engineers design and develop **engines**. An engine is a device that produces mechanical work. Examples include internal combustion engines used in automobiles and gas turbine engines used in airplanes. Mechanical engineers are involved in the design of the mechanical components of the engines as well as the overall cycles and efficiencies of these devices.

Performance and efficiency are also concerns for mechanical engineers involved in the production of power in large **power generation systems.** Steam turbines, boilers, water pumps, and condensers are often used to generate electricity. Mechanical engineers design these mechanical components needed to produce the power that operates the generators. Mechanical engineers also are involved in alternative energy sources including solar and hydroelectric power, and alternative fuel engines and fuel cells.

Another area in the thermal sciences is **heating, ventilating, and air-conditioning** (HVAC). Mechanical engineers are involved in the climate control of buildings, which includes cooling and heating the air in buildings as well as controlling humidity. In doing so, they work closely with civil engineers in designing buildings to optimize the efficiency of these systems.

Mechanical engineers are involved in the **manufacturing processes** of many different industries. They design and develop the machines used in these processes, and develop more efficient processes. Often, this involves automating time-consuming or expensive procedures within a manufacturing process. Mechanical engineers also are involved in the development and use of **robotics** and other automated processes.

In the area of **biomedical engineering**, mechanical engineers help develop artificial limbs and joints that provide mobility to physically impaired individuals. They also develop mechanical devices used to aid in the diagnosis and treatment of patients.

The American Society of Mechanical Engineering (ASME) is one of the most prominent professional societies for mechanical engineers. It is divided into 35 technical divisions, indicating the diversity of this field. These divisions are:

- Advanced Energy Systems
- Aerospace Engineering
- Applied Mechanics
- Basic Engineering Technical Group
- Bioengineering
- Design Engineering
- Dynamic Systems and Control
- Electrical and Electronic Packaging
- FACT
- Fluids Engineering
- Fluids Power Systems and Technology Systems

- Heat Transfer
- Information Storage/Processing
- Internal Combustion Engine
- Gas Turbine
- Manufacturing Engineering
- Materials
- Materials Handling Engineering
- Noise Control and Acoustics
- Non-destructive Evaluation Engineering
- Nuclear Engineering
- Ocean Engineering
- Offshore Mechanics / Arctic Engineering
- Petroleum
- Plant Engineering and Maintenance
- Power
- Pressure Vessels and Piping
- Process Industries
- Rail Transportation
- Safety Engineering and Risk Analysis
- Solar Energy
- Solid Waste Processing
- Technology and Society
- Textile Engineering
- Tribology

Mining Engineering

Modern society requires a vast amount of products made from raw materials such as minerals. The continued production of these raw materials helps to keep society functioning. **Mining engineers** are responsible for maintaining the flow of these raw materials by discovering, removing, and processing minerals into the products society requires.

Discovering the ore involves exploration in conjunction with geologists and geophysicists. The engineers combine the utilization of seismic, satellite, and other technological data, utilizing a knowledge of rocks and soils. The exploration may focus on land areas, the ocean floor, or even below the ocean floor. In the future, mining engineers may also explore asteroids, which are rich in mineral deposits.

Once mineral deposits are identified, they may be **removed**. One way minerals are removed is by way of mining tunnels. The engineers design and maintain the tunnels and the required support systems including ventilation and drainage. Other times, minerals are removed from open pit mines. Again, the engineers analyze the removal site and design the procedure for removing the material. The engineer also develops a plan for returning the site to a natural state. Mining engineers use boring, tunneling, and blasting techniques to create a mine. Regardless of the removal technique, the environmental impact of the mining operation is taken into account and minimized.

The mining engineer is also involved in the **processing** of the raw minerals into usable forms. Purifying and separating minerals involves chemical and mechanical processes. While mining engineers may not be involved in producing a finished product that consumers recognize, they must understand the form their customers can use and design processes to transform the raw materials into usable forms.

Mining engineers also become involved in the **design** of the specialized equipment required for use in the mining industry. The design of automated equipment capable of performing the most dangerous mining jobs helps to increase safety and productivity of a mining operation. Since mines typically are established in remote areas, mining engineers are involved in the **transportation** of minerals to the processing facility.

The expertise of mining engineers is not used exclusively by the mining industry. The same boring technology used in developing mines is used to create subway systems and railroad tunnels, such as the one under the English Channel.

Nuclear Engineering

Nuclear engineers are concerned primarily with the use and control of energy from nuclear sources. This involves electricity production, propulsion systems, waste disposal, and radiation applications.

The production of **electricity** from nuclear energy is one of the most visible applications of nuclear engineering. Nuclear engineers focus on the design, development, and operation of nuclear power facilities. This involves using current fission technology as well as the development of fusion, which would allow sea water to be used as fuel. Nuclear energy offers an environmentally friendly alternative to fossil fuels. A current barrier to production of nuclear facilities is the high cost of construction. This barrier provides a challenge for design engineers to overcome. Research is currently being performed on the viability of smaller, more efficient nuclear reactors.

Nuclear power is also used in **propulsion systems.** It provides a power source for ships and submarines, allowing them to go years without refueling. It is also used as a power source for satellites. Nuclear-powered engines are being examined as an alternative to conventional fossil-fueled engines, making interplanetary travel possible.

One of the main drawbacks to nuclear power is the production of **radioactive waste**. This also creates opportunities for nuclear engineers to develop safe and reliable means to dispose of spent fuel. Nuclear engineers develop ways to reprocess the waste into less hazardous forms.

Another area in which nuclear engineers are involved is the use of **radiation** for medical or agricultural purposes. Radiation therapy has proven effective in treating cancers, and radioactive isotopes are also used in diagnosing diseases. Irradiating foods can eliminate harmful bacteria and help ensure a safer food supply.

Due to the complex nature of nuclear reactions, nuclear engineers are at the forefront of advanced computing methods. High-performance computing techniques, such as parallel processing, constitute research areas vital to nuclear engineering.

The American Nuclear Society (ANS) is one of the professional societies to which nuclear engineers belong. It is divided into these 16 technical divisions:

- Biology and Medicine
- Decommissioning, Decontamination, and Reutilization
- Education and Training
- Environmental Sciences
- Fuel Cycle and Waste Management
- Fusion Energy
- Human Factors
- Isotopes and Radiation
- Materials Science and Technology
- Mathematics and Computations
- Nuclear Criticality Safety
- Nuclear Operations
- Nuclear Installations Safety
- Power
- Radiation Protection and Shielding
- Reactor Physics

Petroleum Engineering

Petroleum and petroleum products are essential components in today's society. **Petroleum engineers** maintain the flow of petroleum in a safe and reliable manner. They are involved in the exploration for crude oil deposits, the removal of oil, and the transporting and refining of oil.

Petroleum engineers work with geologists and geophysicists to identify potential oil and gas reserves. They combine satellite information, seismic techniques, and geological information to **locate deposits of gas or oil**. Once a deposit has been identified, it can be removed. The petroleum engineer designs, develops, and operates the needed drilling equipment and facilities. Such facilities may be located on land or on offshore platforms. The engineer is interested in removing the oil or gas in a safe and reliable manner—safe for the people involved as well as for the environment. **Removal of oil** is done in stages, with the first stage being the easiest and using conventional means. Oil deposits are often located in sand. A significant amount of oil remains coating the sand after the initial oil removal. Recovery of this additional reserve requires the use of secondary and tertiary extraction techniques utilizing water, steam, or chemical means.

Transporting the oil or gas to a processing facility is another challenge for the petroleum engineer. At times this requires the design of a heated pipeline such as the one in Alaska to carry oil hundreds of miles over frozen tundra. In other instances this requires transporting oil in double-hulled tankers from an offshore platform near a wildlife refuge. Such situations necessitate extra precautions to ensure that the wildlife is not endangered.

Once the oil or gas arrives at the processing facility, it must be **refined** into usable products. The petroleum engineer designs, develops, and operates the equipment to chemically process the gas or oil into such end products. Petroleum is made into various grades of gasoline, diesel fuel, aircraft fuel, home heating oil, motor oils, and a host of consumer products from lubricants to plastics.

Other Fields

The most common engineering majors have been described in this chapter. However, there are other specialized engineering programs at some institutions. Here is a partial listing of some of these other programs:

- Automotive Engineering
- Acoustical Engineering
- Applied Mathematics
- Bioengineering
- Engineering Science
- Engineering Management
- Excavation Engineering
- Fire Engineering
- Forest Engineering
- General Engineering
- Genetic Engineering

- Geological Engineering
- Inventive Design
- Manufacturing Engineering
- Packaging Engineering
- Pharmaceutical Engineering
- Plastics Engineering
- Power Engineering
- Systems Engineering
- Theatre Engineering
- Transportation Engineering
- Welding Engineering

2.4 EMERGING FIELDS

The fields of engineering have been, and will continue to be, dynamic. As new technologies emerge, new definitions are needed to classify disciplines. The boundaries will continue to shift and new areas will emerge. The explosion of technological advances means that there is a good chance you will work in a field that is not currently defined. Technological advances are also blurring the traditional delineations between fields. Areas not traditionally linked are coming together and providing cross-disciplinary opportunities for engineers. A good example of this is in the development of smart buildings which sense the onset of an earthquake, adapt their structures to survive the shaking, and then return to normal status afterward. Research is being conducted on these technologies that bridge computer engineering with structural (civil) engineering to produce such adaptable buildings.

The incredible breakthroughs in biology have opened many new possibilities and will continue to impact most of the fields of engineering. Historically, biology has not been as integrated with engineering as has physics and chemistry, but that is rapidly changing and will have enormous impact on the future of engineering. The ability to modify genetic codes has implications in a wide range of engineering applications including the production of pharmaceuticals that are customized for individual patients, alternative energy sources, and environmental reclamation.

Nanotechnology is an area that is receiving a great deal of attention and resources, and is blurring the boundaries of the fields of engineering. Nanotechnology is an emerging field in which new materials and tiny structures are built atom-by-atom, or molecule-by-molecule, instead of the more conventional approach of sculpting parts from pre-existing materials. The possibilities for nano-applications include:

- the creation of entirely new materials with superior strength, electrical conductivity, resistance to heat, and other properties
- microscopic machines for a variety of uses, including probes that could aid diagnostics and repair
- a new class of ultra-small, super-powerful computers and other electronic devices, including spacecraft
- a technology in which biology and electronics are merged, creating "gene chips" that instantly detect food-borne contamination, dangerous substances in the blood, or chemical warfare agents in the air
- the development of "molecular electronics" and devices that "self assemble," similar to the growth of complex organic structures in living organisms

Nanotechnology requires specialized laboratory and production facilities that provide further challenges and opportunities for future engineers.

The explosion of information technology with the Internet and wireless communication has produced a melding of disciplines to form new fields in information science and technology. Information management and transfer is an important and emerging issue in all disciplines of engineering and has opened opportunities for engineers who want to bridge the gaps between the traditional fields and information technologies.

Sustainability is another important concept that spans many engineering disciplines. Sustainable development is a term that was defined by a report to the United Nations, known as the Brundtland Report, in 1987 as: "development that meets the needs of the present without compromising the ability of future generations to meet their own needs" (*Our Common Future, 1987*). Designing products and processes that contribute to sustainable development is important for all engineering disciplines and is becoming a larger part of undergraduate and graduate education as well as professional development programs. As the global population continues to rise, so too do the pressures on the global environment. Sustainability will have to play a large and significant role in product development, manufacturing, and construction if we are to meet these challenges.

Undoubtedly, more new fields will open and be discovered as technology continues to advance. As the boundaries of the genetic code and molecular-level device are crossed, new frontiers will open. As an engineer, you will have the exciting opportunity to be part of the discovery and definition of these emerging fields which will have tremendous impact on society's future.

2.5 CLOSING THOUGHTS

The information presented in this chapter is meant to provide a starting point on the road to choosing a career. There may have been aspects of one or more of these engineering fields that appealed to you, and that's great. The goal, however, is not to persuade you that engineering is for everyone; it is not. The goal is to provide information to help you decide if engineering would be an enjoyable career for *you*.

As a student, it is important to choose a career path that will be both enjoyable and rewarding. For many, an engineering degree is a gateway to just such a career. Each person has a unique set of talents, abilities, and gifts that are well matched for a particular career. In general, people find more rewarding careers in occupations where their gifts and talents are well used. Does engineering match your talents, abilities, and interests?

Right now, choosing a career may seem overwhelming. But at this point, you don't have to. What you are embarking on is an education that will provide the base for such decisions. Think about where a degree in engineering could lead. One of the exciting aspects of an engineering education is that it opens up a wide range of jobs after leaving college. Most students will have several different careers before they retire, and the important objective in college is obtaining a solid background that will allow you to move into areas that you enjoy later in life.

As you try to make the right choice for you, seek out additional information from faculty, career centers, professional societies, placement services, and industrial representatives. Ask a lot of questions. Consider what you would enjoy studying for four years in college. What kind of entry-level job would you be able to get with a specific degree? What doors would such a degree open for you later in life? Remember, your decision is unique to you. No one can make it for you.

2.6 ENGINEERING AND TECHNICAL ORGANIZATIONS

The following is a list of many of the engineering technical societies available to engineers. They can be a tremendous source of information for you. Many have student branches which allow you to meet both engineering students and practicing engineers in the same field. Join one or more of these organizations during your freshman or sophomore year.

American Association for the Advancement of Science
1200 New York Avenue, NW
Washington, DC 20005
(202) 326-6400
www.aaas.org

American Association of Engineering Societies
1111 19th Street, NW
Suite 403
Washington, DC 20036
(202) 296-2237
www.aaes.org

American Ceramic Society
735 Ceramic Place
Westerville, OH 43081-8720
(614) 890-4700
www.acers.org

American Chemical Society
1155 16th Street, NW
Room 1209
Washington, DC 20036-1807
(202) 872-4600
www.acs.org

American Concrete Institute
38800 Country Club Drive
Farmington Hills, MI 48331
(248) 848-3700
www.aci-int.org

American Congress on Surveying and Mapping
5410 Grosvenor Lane
Suite 100
Bethesda, MD 20814-2122
(301) 493-0200

American Consulting Engineers Council
1015 15th St., NW, Suite 802
Washington, DC 20005
(202) 347-7474
www.acec.org

American Gas Association
1515 Wilson Blvd.
Arlington, VA 22209
(703) 841-8400
www.aga.com

American Indian Science and Engineering Society
5661 Airport Blvd.
Boulder, CO 80301-2339
(303) 939-0023
www.colorado.edu/AISES

American Institute of Aeronautics and Astronautics
1801 Alexander Bell Drive, Suite 500
Reston, VA 20191-4344
(800) NEW-AIAA or (703) 264-7500
www.aiaa.org

American Institute of Chemical Engineers
345 East 47th Street
New York, NY 10017-2395
(212) 705-7000 or (800) 242 4363
www.aiche.org

American Institute of Mining, Metallurgical and Petroleum Engineers
345 East 47th Street
New York, NY 10017
(212) 705-7695

American Nuclear Society
555 North Kensington Avenue
La Grange Park, IL 60526
(708) 352-6611
www.ans.org

American Oil Chemists' Society
1608 Broadmoor Drive
Champaign, IL 61821-5930
(217) 359-2344
www.aocs.org

American Railway Engineering Association
8201 Corporate Drive
Landover, MD
(301) 459-3200

American Society of Agricultural Engineers
2950 Niles Road
St. Joseph, MI 49085-9659
(616) 429-0300
www.asae.org

American Society of Civil Engineers
1801 Alexander Bell Drive
Reston, VA 20191-4400
(800) 548-2723 or (703) 295-6000
www.asce.org

American Society of Naval Engineers
1452 Duke Street
Alexandria, VA 22314
(703) 836-6727
www.jhuapl.edu/ASNE

American Society for Engineering Education
1818 N St., NW, Suite 600
Washington, DC 20036-2479
(202) 331-3500
www.asee.org

American Society for Heating, Refrigeration and Air Conditioning Engineers
1791 Tulie Circle NE
Atlanta, GA 30329
(404) 636-8400
www.ashrae.org

American Society of Mechanical Engineers
345 East 47th Street
New York, NY 10017-2392
(212) 705-7722
www.asme.org

American Society of Plumbing Engineers
3617 Thousand Oaks Blvd.
Suite 210
Westlake Village, CA 91362-3649
(805) 495-7120
www.aspe.org

American Society of Nondestructive Testing
1711 Arlingate Lane
P.O. Box 28518
Columbus, OH 43228-0518
(614) 274-6003
www.asnt.org

American Society for Quality
611 East Wisconsin Avenue
Milwaukee, WI 53202
(414) 272-8575
www.asq.org

American Water Works Association
6666 West Quincy Avenue
Denver, CO 80235
(303) 443-9353
www.aws.org

The Architectural Engineering Institute
1801 Alexander Bell Drive, 1st Floor
Reston, VA 20191-4400
(703) 295-6370 Fax (703) 295-6132
http://www.aeinstitute.org

Association for the Advancement of Cost Engineering International
209 Prairie Avenue
Suite 100
Morgantown, WV 26505
(304) 296-8444
www.aacei.org

Board of Certified Safety Professionals
208 Burwash Avenue
Savoy, IL 61874-9571
(217) 359-9263
www.bcsp.com

Construction Specifications Institute
601 Madison Street
Alexandria, VA 22314-1791
(703) 684-0300
www.csinet.org

Information Technology Association of America
1616 N. Fort Myer Drive, Suite 1300
Arlington, VA 22209
(703) 522-5055
www.itaa.org

Institute of Electrical and Electronics Engineers
1828 L Street NW, Suite 1202
Washington, DC 20036
(202) 785-0017
www.ieee.org

Institute of Industrial Engineers
25 Technology Park
Norcross, GA 30092
(770) 449-0461
www.iienet.org

Iron and Steel Society
410 Commonwealth Drive
Warrendale, PA 15086-7512
(412) 776-1535
www.issource.org

Laser Institute of America
12424 Research Parkway, Suite 125
Orlando, FL 32826
(407) 380-1553
www.laserinstitute.org

Mathematical Association of America
1529 18th Street, NW
Washington, DC 20036
(202) 387-5200
www.maa.org

The Minerals, Metals and Materials Society
420 Commonwealth Drive
Warrendale, PA 15086
(412) 776-9000
www.tms.org

NACE International
1440 South Creek Drive
Houston, TX 77084-4906
(281) 492-0535
www.nace.org

National Academy of Engineering
2101 Constitution Avenue, NW
Washington, DC 20418
(202) 334-3200

**National Action Council for
    Minorities in Engineering, Inc.**
The Empire State Building
350 Fifth Avenue, Suite 2212
New York, NY 10118-2299
(212) 279-2626
www.naof cme.org

The National Association of Minority Engineering Program Administrators, Inc.
1133 West Morse Blvd., Suite 201
Winter Park, FL 32789
(407) 647-8839
(407) 629-2502 Fax

National Association of Power Engineers
1 Springfield Street
Chicopee, MA 01013
(413) 592-6273
www.powerengineers.com

National Conference of Standards Laboratories International
2995 Wilderness Place, Ste. 101
Boulder, CO 80301-5404
(303) 440-3339
www.ncsli.org

National Institute of Standards and Technology
Publications and Programs Inquiries
Public and Business Affairs
Gaithersburg, MD 20899
(301) 975-3058
www.nist.gov

National Science Foundation
4201 Wilson Blvd.
Arlington, VA 22230
(703) 306-1234
www.nsf.gov

National Society of Black Engineers
1454 Duke Street
Alexandria, VA 22314
(703) 549-2207
www.nsbe.org

National Society of Professional Engineers
1420 King Street
Alexandria, VA 22314
(888) 285-6773
www.nspe.org

Society of Allied Weight Engineers
5530 Aztec Drive
La Mesa, CA 91942
(619) 465-1367

Society of American Military Engineers
607 Prince Street
Alexandria, VA 22314
(703) 549-3800 or (800) 336-3097
www.same.org

Society of Automotive Engineers
400 Commonwealth Drive
Warrendale, PA 15096
(412) 776-4841
www.sae.org

Society of Fire Protection Engineers
7315 Wisconsin Avenue
Suite 1225W
Bethesda, MD 20814
(301) 718-2910

Society of Hispanic Professional Engineers
5400 East Olympic Blvd.
Suite 210
Los Angeles, CA 90022
(213) 725-3970
www.engr.umd.edu/organizations/shpe

Society of Manufacturing Engineers
One SME Drive
P.O. Box 930
Dearborn, MI 48121-0930
(313) 271-1500
www.sme.org

Society for Mining, Metallurgy Exploration, Inc.
8307 Shaffer Parkway
Littleton, CO 80127
(303) 973-9550
www.smenet.org

Society of Naval Architects and Marine Engineers
601 Pavonia Avenue
Jersey City, NJ 07306
(800) 798-2188
www.sname.org

Society of Petroleum Engineers
P.O. Box 833836
Richardson, TX 75083-3836

Society of Plastics Engineers
14 Fairfield Drive
Brookfield, CT 06804-0403
(203) 775-0471
www.4spe.org

Society of Women Engineers
120 Wall Street
11th Floor
New York, NY 10005
(212) 509-9577
www.swe.org

SPIE-International Society for Optical Engineering
P.O. Box 10
Bellingham, WA 98227-0010
(360) 676-3290
www.spie.org

Tau Beta Pi
508 Dougherty Engineering Hall
P.O. Box 2697
Knoxville, TN 37901-2697
(423) 546-4578
www.tbp.org

Women in Engineering Initiative
University of Washington
101 Wilson Annex
P.O. Box 352135
Seattle, WA 98195-2135
(206) 543-4810
www.engr.washington.edu/~wieweb

REFERENCES

Burghardt, M.D., *Introduction to the Engineering Profession,* 2nd Edition, Harper Collins College Publishers, 1995.

Garcia, J., *Majoring {Engineering},* The Noonday Press, New York, New York, 1995.

Grace, R., and J. Daniels, *Guide to 150 Popular College Majors,* College Entrance Examination Board, New York, New York, 1992, pp. 175-178.

Irwin, J.D., *On Becoming an Engineer,* IEEE Press, New York, 1997.

Kemper, J.D., *Engineers and Their Profession,* 4th Edition, Oxford University Press, New York, 1990.

Landis, R., *Studying Engineering, A Road Map to a Rewarding Career*, Discovery Press, Burbank, California, 1995.

Smith, R.J., B.R. Butler, W.K. LeBold, *Engineering as a Career,* 4th Edition, McGraw-Hill Book Company, New York, 1983.

Wright, P.H., *Introduction to Engineering,* 2nd Edition, John Wiley and Sons, Inc., New York, 1994.

World Commission on Environment and Development, *Our Common Future*, Oxford University Press, 1987.

EXERCISES AND ACTIVITIES

2.1   Contact a practicing engineer in a field of engineering that interests you. Write a brief report on his or her activities and compare them to the description of that field of engineering as described in this chapter.

2.2   For a field of engineering that interests you, make a list of potential employers that hire graduates from your campus, and list the cities in which they are located.

2.3   Visit a job fair on your campus and briefly interview a company representative. Prepare a brief report on what that company does, what the engineer you spoke to does, and the type of engineers they are looking to hire.

2.4   Make a list of companies that hire co-op and/or intern students from your campus. Write a brief report on what these companies are looking for and their locations.

2.5   Select a company that employs engineers in a discipline that interests you, and visit their web page. (You can search for them using Yahoo or Alta Vista, or you can call them and ask for their Web address.) Prepare a brief report on what the company does, hiring prospects, engineering jobs in that organization, and where the company is located.

2.6   Contact a person with an engineering degree who is not currently employed in a traditional engineering capacity. Write a one-page paper on how that person uses his or her engineering background in their job.

2.7   Write a one-page paper on an engineering field that will likely emerge during your lifetime (a field that does not currently exist). Consider what background an engineering student should obtain in preparation for this emerging field.

2.8   Draft a sample letter requesting a co-op or intern position.

2.9   Identify a modern technological problem. Write a brief paper on the role of engineers and technology in solving this problem.

2.10 Select an engineering discipline that interests you and a particular job function within that discipline. Write a brief paper contrasting the different experiences an engineer in this discipline would encounter in each of three different industries.

2.11 Make a list of your own strengths and talents. Write a brief report on how these strengths are well matched with a specific engineering discipline.

2.12 Pick an engineering discipline that interests you. List and briefly describe the technical and design electives available in that discipline for undergraduates.

2.13 Select a consumer product you are familiar with (a stereo, clock radio, automobile, food product, etc.). List and briefly describe the role of all the engineering disciplines involved in producing the product.

2.14 Select an engineering discipline that interests you. Write a brief paper on how the global marketplace has altered this discipline.

2.15 Write a brief paper listing two similarities and two differences for each of the following engineering functions:

a) research and development
b) development and design
c) design and manufacturing
d) manufacturing and operations
e) sales and customer support
f) management and consulting

2.16 Find out how many engineering programs your school offers. How many students graduate each year in each discipline?

2.17 Which of the job functions described in this chapter is most appealing to you? Write a brief paper discussing why it is appealing.

2.18 Write a paper about an engineer who made a significant technical contribution to society.

2.19 Report on the requirements for becoming a registered professional engineer in your state. Also report on how registration would be beneficial to your engineering career.

2.20 Make a list of general responsibilities and obligations you would have as an engineer.

2.21 Write a brief paper on the importance of ethical conduct as an engineer.

2.22 Select one of the industrial sectors within engineering and list five companies that do business in that sector. Briefly describe job opportunities in each.

2.23 Prepare a report on how the following items work and what engineering disciplines are involved in their design and manufacture:
a) CD player
b) CAT scan machine
c) Computer disk drive
d) Dialysis machine
e) Flat TV screen

2.24 Answer the following questions and look for themes or commonality in your answers. Comment on how engineering might fit into these themes.

a) If I could snap my fingers and do whatever I wanted, knowing that I wouldn't fail, what would I do?

b) At the end of my life, I'd love to be able to look back and know that I had done something about _____.

c) What would my friends say I'm really interested in or passionate about?

d) What conversation could keep me talking late into the night?

e) What were the top five positive experiences I've had in my life, and why were they meaningful to me?

2.25 Write a short paper describing your dream job, regardless of pay or geographical location.

2.26 Write a short paper describing how engineering might fit into your answer to 2.25.

2.27 List the characteristics a job must have for you to be excited to go to work every morning. How does engineering fit with those characteristics?

2.28 List the top ten reasons why a student should study engineering.

2.29 List the ten inappropriate reasons for a student to choose to study engineering.

2.30 Select an industrial sector and describe what you suppose a typical day is like for:
a) Sales engineer
b) Research engineer
c) Test engineer
d) Manufacturing engineer
e) Design engineer

2.31 Make a list of five non-engineering careers an engineering graduate could have and describe each one briefly.

2.32 Select one of the professional societies that was listed at the end of this chapter. Prepare a report on what the organization does and how it could benefit you as a practicing engineer.

2.33 Select one of the professional societies that was listed at the end of this chapter. Identify one of its technical divisions and prepare a report on that division's activities.

2.34 Identify one of the student branches of an engineering professional society on your campus and prepare a report on the benefits of involvement with that organization for an engineering student.

2.35 Contact one of the professional societies listed in this chapter and have them send you information. Prepare a brief presentation summarizing the material.

2.36 Write a letter to a 9th-grade class explaining the exciting opportunities in your chosen major within engineering. Include a short discussion of the classes they should be taking to be successful in that same major.

2.37 Write a letter to a 9th-grade class describing the opportunities a bachelor's degree in engineering provides in today's society.

2.38 Write a letter to your parents detailing why you are going to major in the field you have chosen.

2.39 Write a one-page paper describing how your chosen field of engineering will be different in 25 years.

2.40 Select one field of engineering and write a one-page paper on how the advances in biology have influenced that field.

2.41 Select one field of engineering and write a one-page paper on how the advances in nanotechnology have influenced that field.

2.42 Select one field of engineering and write a one-page paper on how the advances in information technology have influenced that field.

2.43 Select two fields of engineering and describe problems that span these two disciplines.

2.44 Write a brief paper on an emerging area within engineering. Relate the area to your chosen engineering major.

2.45 Research the current spending priorities of the U.S. government in the area of technical research. How will these priorities impact your chosen major?

2.46 For each grouping of engineering disciplines, describe applications or problems that span the disciplines.
    a. Aerospace, Materials and Civil
    b. Mechanical, Agricultural and Computer
    c. Biological and Environmental
    d. Industrial, Chemical and Electrical
    e. Biomedical and Nuclear
    f. Agricultural and Aerospace
    g. Materials and Biomedical
    h. Civil and Computer

2.47 Select on professional organization and find out how they handle new and emerging technologies within their society (where do they put them and where can people working in emerging areas find colleagues?).

2.48 Interview a practicing engineering and a faculty member from the same discipline about their field. Compare and contrast their views of the discipline.

2.49 Select two engineering majors and compare and contrast the opportunities available between the two.

2.50 Identify how your skills as an engineer can be used within your local community—either as full-time work or as a volunteer. Share your findings with the class by preparing a short oral presentation.

Chapter 3

Profiles of Engineers

3.1 INTRODUCTION

This chapter contains a collection of profiles of engineering graduates to let you read first-hand accounts of what it is really like to be an engineer. Each engineer wrote his or her own profile.

Our intent is to provide you with a glimpse of the diversity of the engineering workforce. Engineers are people just like you, and had varying reasons for pursuing engineering as a career. Engineering graduates also take very diverse career paths.

In order to capture a flavor of this diversity, each engineer was asked to address three areas:

1. Why or how they became an engineer
2. Their current professional activities
3. Their life outside of work

This is not a comprehensive survey of the engineering workforce. To truly represent the breadth of engineering careers and the people in those careers would take several volumes, not just one chapter. This is meant to be only a beginning. It is also arranged simply to show you the wide range of engineering careers that are possible. To keep the wide-ranging feel, we've avoided ranking the profiles by subject, and instead present them alphabetically. This makes it easier to see the common bonds across all the disciplines.

We recommend that you follow up and seek out other practicing engineers or engineering students, to get more detailed information about the specific career path you might follow. As with the information we have collected in other chapters of this text, our goal is to assist you in finding that career path which is right for you. You are the one who must ultimately decide this for yourself.

The following Table 3.1 summarizes those who provided profiles.

TABLE 3.1 Summary of Profiled Engineers

Name	BS Degree	Graduate Degree	Current Job Title
Sue Abreu	BS/E/BioMed	MD	Lt. Col and Medical Director
Moyosola Ajaja	BS/EE		Software Engineer
Patrick Rivera Antony	BS/Aero		Project Manager
Artagnan Ayala	BS/Aero		Combustion Engineer
Sandra Begay-Campbell	BS/CE	MS/CE	Executive Director of AISES
Raymond C. Barrera	BS/EE	MS/Software	Computer Engineer
Linda Blevins	BS/ME	MS/ME & Ph.D.	Mechanical Engineer with NIST
Peter Bosscher	BS/CE	MS/CE	Professor
Timothy Bruns	BS/EE		Software Manager
Jerry Burris	BS/EE	MBA	General Manager
Bethany Fabin	BS/ABE		Design Engineer
Bob Feldmann	BS/EE	MS/Comp, MBA	Director, Tactical Aircraft Systems
Steven Fredrickson	BS/EE	Ph.D.	Project Manager, NASA
Myron Gramelspacher	BS/ME		Manufacturing Manager
Karen Jamison	BS/IE	MBA	Operations Manager
Beverly Johnson	BS/ME	MS/EM & MBA	Supervisor
James Lammie	BS/CE	MS/CE	Member of Board of Directors
Wendy Lawrence	BS/OE	MS/OE	NASA Astronaut
Ryan Maibach	BS/CEM		Project Engineer
Mary Maley	BS/AgE		Product Manager
Jeanne Mordarski	BS/IE	MBA	Sales Manager
Mark Pashan	BS/EE	MS/EE & MBA	Director, Hardware Operations
Douglas Pyles	BS/E/Mgt	MS/CEM	Engineering Consultant
Emery Sanford	BS/ME		Product Design Engineer, Apple Computer
Patrick Shook	BS/ME	MS/ME	Senior Engineer
Nana Tzeng	BS/ME	MS/ME	Design Engineer
Patrice Vanderbeck	BS/ME		Electronics systems Engineer
Jack Welch	BS/ChE	MS/ChE & Ph.D.	Chief Executive Officer (retired)
Shawn Williams	BS/EE		Product General Manager
Adel Zakaria	BS/ME	MS/IE & Ph.D.	Sr. VP Engr & Manufacturing

Profile of a Biomedical Engineer:
Sue H. Abreu, Ft. Bragg, North Carolina

Occupation

Lieutenant Colonel, Medical Corps, United States Army
Medical Director, Quality Assurance, Womack Army Medical Center

Education

IDE (BSE, Biomedical Engineering), 1978
MD, Uniformed Services University of the Health Sciences, 1982

Studying Engineering

As I started college, I was planning to be a teacher. Because of taking an elective class in athletic training, I developed an interest in sports medicine. I ended up taking most of my classes in aeronautical engineering so I could study the lightweight structures and materials that could be used to design artificial limbs or protective equipment for sports. Late in college, I decided to go to medical school and ended up graduating from college with an interdisciplinary engineering degree.

Career Life

After medical school, I specialized in nuclear medicine. In nuclear medicine we use small amounts of radioactive compounds to see how things work inside of people. By using special cameras that detect radiation and computers that help gather the data, we can watch how various organs function. We can do three-dimensional studies and quantify results. In nuclear medicine, I am a consultant to other physicians: I help them decide what tests might be helpful and discuss the meaning of the results of the nuclear medicine procedures we do for their patients.

I ended up in a field I never had heard of when I started college, but I found it as I kept exploring areas that intrigued me. I tried new classes and looked for opportunities that interested me, even if they didn't fit the paths most students followed. As a result, I found a specialty I enjoy, and I'm now doing a great deal of teaching within my specialty of nuclear medicine and in my current work in quality assurance.

So, be sure to follow your dreams—if you can take something you love doing and find a way to earn a living doing it, you will end up much happier than if you set money or prestige as your goals.

Life Outside of Work

Outside of work I enjoy skydiving. I volunteer as the team doctor for the U.S. Parachute Team and have traveled all over the world with them. I currently live in a large house on six acres in the country, not far from an airport with a parachuting center. I share the house with an airline pilot and an artist—both expert skydivers—who help make it a great place to live. Although I was married, I chose not to have children; but the dog and cats help keep us company here.

Profile of a Computer Engineer:
Moyosola O. Ajaja, Chandler, Arizona

Occupation

Software Engineer at Intel Corporation

Education

BS, Computer and Electrical Engineering, 1997

Studying Engineering

 I came into engineering the easy way—by excelling in math and physics in high school. Deciding to enroll at Purdue and pursue a dual degree in computer and electrical engineering was a little more complicated. I wanted to learn more about computers, and I wanted to seek my fortune in a distant land. I picked the U.S. and justified the 7000-mile journey from Lagos, Nigeria, where my family lived, to West Lafayette, Indiana, where Purdue is located, with the phrase "dual degree." (I understand that degree option is no longer available. Fellow adventurers will have to justify their journeys with a different explanation.)

During my first year at Purdue I set two goals for myself: first, find a scholarship to fund my education, and second, gain useful work experience. I applied for dozens of scholarships. I was partial to those offered by engineering firms that provided internships, since internships for first-year engineering students were very scarce. In addition, I attended every resume or interview preparation workshop offered during that year. My efforts paid off. I was invited to join the cooperative education program with a summer placement with Intel Corporation, and later I was awarded an Intel Foundation scholarship which paid my tuition.

Career Life

Today, I work as a software engineer with Intel in Arizona. I develop hardware emulation units and validation test suites for new processors. What that means in plain English: I take descriptions of hardware features and functions and write software programs that behave like the hardware should. This is cheaper than fabricating silicon devices each time a change is made during design. As an Intel engineer with access to the latest and greatest technologies, I am constantly challenged to learn new things to remain at the leading edge of computer technology.

Life Outside of Work

I have tried to maintain a balance between my work and my non-work activities. My weekend mornings are spent running with my dog or hiking up Camelback Mountain in Phoenix. The evenings are spent in classes like dog training, theology, or photography. My real passion is traveling, and through engineering school, internships, and my current assignment, I have met people who helped fuel my interest in increasingly diverse destinations.

I've discovered that engineering is a *discipline*, not just a major. The distinction here is that a discipline involves the development of the faculties through instruction and exercise, while a major is simply a field of study, an area of mental focus, or a concentration. For me this means the qualities of an engineer should be apparent in all I do. The guide I use is the Code of Ethics approved by the IEEE, which is presented in the Ethics chapter.

Profile of an Aerospace Engineer: Patrick Rivera Antony

Occupation:

Project Manager, Boeing Space Beach

Education:

BS, Aerospace Engineering

Studying Engineering

When Apollo 11 landed on the moon I was eight years old. I still remember to this day how excited I was and exactly where I was and how I felt when I saw Neil Armstrong step off the ladder of the lunar module and become the first man to walk on the moon. Math and science was always my favorite subject in school, so it gave me a good background to get started in my college studies.

Along with space travel, rockets and airplanes have always fascinated me. The whole concept of being able to fly and be free is amazing. I also knew that if I wanted to become an astronaut, I would have to learn to fly. So I got my pilot's license and found that the fundamental principles of flying are exactly what you learn in the courses taken for Aerospace Engineering.

Now, I would probably put more emphasis on my non-engineering classes. To be a good engineer now, you have to be able to communicate effectively, work in a team environment, and know different aspects of the business. Being good technically is not enough anymore. I would take more English, Business, and Social Science classes to make me a more well rounded individual and more valuable employee.

Career Life

Currently I am the Project Manager of the Performance Management and Continuous Improvement team for Boeing Space Beach, HB/SB Host Site Engineering. My responsibilities include developing and maintaining the management system used for organizational oversight, planning and operational assessment. I coordinate company policy deployment of business goals and strategies from the Vision Support Plan to the Engineering functional organization. Additionally, I interface with Program Management and Technical Managers to define performance requirements and develop innovative process and organizational solutions.

In my current management role, I spend a lot of time in planning meetings, so every day is different. I also do not often have an opportunity to use my engineering education, but as a leader of an organization, my engineering background allows me to communicate competently on technical matters related to the business.

Life Outside of Work

Most of my free time I try to spend with my wife and two children. But when I am not with them, I am studying. I am currently pursuing a Doctorate degree in Organizational Psychology with an emphasis in Change Management. With what little free time I have after that, I enjoy golf, hiking, fishing, and biking.

Profile of an Aerospace Engineer:
Artagnan Ayala, Gilbert, Arizona

Occupation

Combustion Engineer, Diversity Organizations Manager, Honeywell

Education

BS, Aeronautical and Astronautical Engineering, 1995
MS (in progress), Mechanical & Aerospace Engineering

Studying Engineering

I have been interested in space since I was very young; I always wanted to be an Astronaut. This was my motivation to become an engineer. I figured that, if one day I was to climb on a rocket and go to outer space, I'd better know how it works. I am still working towards that goal.

My education has definitely met my expectations. I have applied what I learned in my job, and some of it to life in general. Engineering is not only a field you go into, but also a way of thinking. You are taught to solve problems, which can be applied to everything you do.

If I could start over, I would interact more with professors, take better notes, and learn more about statistics.

Career Life

Currently I am a Combustion Engineer, a Six Sigma Plus Black Belt, and the Diversity Organizations Manager for the Society of Hispanic Professional Engineers at Honeywell Engines & Systems.

As a combustion engineer I design and develop combustion systems that are installed in Auxiliary Power Units and Industrial Power Generators. I have finished the development on one system, designed a technology demonstrator, and am currently designing a premixed fuel delivery system.

As a Six Sigma Plus Black Belt I apply statistical tools to improve all sorts of processes, from combustion system development and manufacturing.

As Diversity Organizations Manager, I am responsible for the company's contact and participation with the Society.

What I like the most about what I do is the diversity of my responsibilities. I get to apply my engineering skills every day, and I get to learn more skills. This has allowed me to receive my Black Belt certification, and participate in the Honeywell Quest for Excellence, a company event where teams with outstanding results present their work in a competition to win the Premier Achievement Award, the biggest team honor. I am particularly proud of my participation as presenter in 2 events, and making it to the finals in one of them.

Life Outside of Work

My wife Laura, a Graphic Designer, and I recently expanded our family with the arrival of our first daughter, Deanna Isabella. I devote most of my free time to my family and some to my studies.

Before we had a baby, I participated in a volleyball league, went dancing at clubs and concerts on weekends, and traveled outside of Arizona.

Profile of a Civil Engineer:
Sandra Begay-Campbell, Boulder, Colorado

Occupation

AISES Executive Director

Education

BSCE, 1987; MS, Structural Engineering, 1991

Studying Engineering

I am a Navajo and the executive director of the American Indian Science and Engineering Society (AISES), which is a non-profit organization whose mission is to increase the number of American Indian scientists and engineers. I am the third executive director in the Society's twenty-year history and the first woman to serve in this position. I manage the Society's operations and educational programs. For more AISES information, check out www.aises.org.

In 1987, I received a BSCE degree from the University of New Mexico. I worked at Lawrence Livermore National Laboratories before I earned a MS, Structural Engineering degree from Stanford University. I also worked at Los Alamos National Laboratory and Sandia National Laboratories before accepting my current leadership position. Within AISES, I served as a college chapter officer, a national student representative, and board of directors member. I was the first woman AISES board of directors Chairperson.

In the sixth grade, I was very interested in architecture, but I knew I was not an artist. I also enjoyed math and solving problems so I looked into the engineering profession. I attended a "minority introduction to engineering" program as a high school junior and I discovered that civil engineers worked on a variety of interesting public projects, which included work with architects. This program solidified my decision to become an engineer.

Career Life

One of the earliest challenges I faced was in continuing my structural engineering studies following the 1989 San Francisco Bay-Area earthquake. I was a first quarter graduate student at Stanford when the earthquake hit. Through prayer and reflection, I understood my unique role as an American Indian engineer. I must use my best knowledge to design structures for earthquake resistance, but my cultural heritage taught me the wisdom that engineers ultimately cannot control Nature and that we have to accept the consequences from natural phenomena.

Life Outside of Work

Life outside of work is difficult to describe at this point in time. With the re-building of the AISES organization and relocation of the offices, I don't have much time for outside activities. I have also been commuting between Boulder, Colorado, and Albuquerque, New Mexico. In brief, my hobbies are watching college basketball, watching movies, and working on my home's backyard. My husband and I have two dogs and a cat.

Profile of a Computer Engineer:
Raymond C. Barrera, Gaithersburg, MA

Occupation

Computer Engineer, Advanced Concepts and Engineering Division, Space and Warfare Systems Command Systems Center, San Diego

Education

BS, Electrical and Computer Engineering 1989
MS, Software Engineering 1999

Studying Engineering

I was very fortunate during high school to work for an archaeologist and her husband who were great mentors. To me archaeology is like detective work— finding bits of information here and there and putting them together to form the big picture. Dr. Bernice McAllister taught me the scientific methodology an archaeologist needs to base sound conclusions on evidence. I think I would be happy had I become an archaeologist, but I really enjoy building things. My dad's training as an electronic technician had gotten me interested in electronics when I was very young. That, with some encouragement from Dr. McAllister's husband, Capt. James McAllister, USN (ret), helped convince me to select Electrical Engineering as my specialty.

Career Life

I work at a research, development, test, and evaluation laboratory for the US Navy. I am involved in testing and system engineering of command and control systems. Command and control systems are used by tactical commanders for decision making and direction. I began working here in 1989 so I was here during Desert Storm. Perhaps even more important than the technical work is the ability to communicate. Not very many engineers work alone. A former Navy Admiral, Grace Hopper (who is said to have coined the computer term "bug"), used a length of wire to describe a nanosecond to programmers. It was about a foot long, the distance that electricity could travel in one billionth of a second. But then she showed a microsecond—a coil of wire almost a thousand feet long. She was trying to convince programmers not to waste even a microsecond. Often the most difficult engineering challenge is to share an idea with others in oral and written presentations, but that is the only way these ideas can come to life.

Life Outside of Work

My wife Martha and I spend most of our time outside of work with our new daughter Laura. I do have some flexibility on my work schedule so I can spend more time with her. I've been able to select job assignments that don't require too much travel. Since this is a research laboratory there are always new things to do. In the over ten years I've been here, no two have been the same. In this command alone there are engineers working with supercomputers, lasers, networking, marine mammals, 3-D displays, simulators, and sensors.

Profile of a Mechanical Engineer:
Linda G. Blevins, Gaithersburg, Maryland

Occupation

Mechanical Engineer, National Institute of Standards and Technology

Education

BSME, 1989; MSME, 1992; PhD, 1996

Studying Engineering

During high school I discovered that I enjoyed mathematics. I learned about engineering when I participated in a six-week summer honors program at the University of Alabama before my senior year in high school. I took college calculus that summer, and I was hooked. I chose to study mechanical engineering because the course subjects are diverse and the industrial demand for mechanical engineers remains steady. As a co-op at Eastman Chemical Co., I worked on engineering problems in power and chemical plants. The concepts that I learned in classes came to life during the alternate semesters that I worked, and the money I earned helped pay for school. After earning a BS degree from the University of Alabama, I obtained an MS degree from Virginia  Tech, and a PhD degree from Purdue University. I never would have set or achieved these goals without encouragement and advice from faculty members. Because these mentors played such valuable roles in my life, I would advise college students to get to know their professors well. These personal investments will be rewarding for years to come.

Career Life

I am a mechanical engineer in the Building and Fire Research Laboratory at the National Institute of Standards and Technology (NIST), a national research laboratory operated by the U.S. Department of Commerce, located in Gaithersburg, Maryland. Our goals are to study the ways that fires ignite, spread, and extinguish so that our nation can minimize the loss of lives and property to fires. My primary job function is to improve the accuracy of measurements made during fire research. A few things routinely measured are toxic gas concentration, temperature, and heat intensity. I spend my time developing laser-based instrumentation, devising computer (math) models of instrument behavior, designing laboratory equipment, tinkering with electronics, publishing papers, writing and reviewing research proposals, and presenting talks at conferences. In addition, I work on a project funded by the National Aeronautics and Space Administration (NASA) to study fires in space. Working in a research laboratory ensures that I am constantly learning and growing, and I realize every day how lucky I am to be here. My job is exciting, fun, and rewarding.

Life Outside of Work

During my free time, I enjoy hiking, rollerblading, and reading. I participate in a weekly bowling league and I manage a softball team each summer. I also volunteer as a member of the Mechanical Engineering Advisory Board at the University of Alabama. This allows me to travel home to Alabama (and visit my family) several times a year. Finally, I volunteer regularly to educate children and community members about the excitement of engineering.

Profile of a Civil Engineer:
Peter J. Bosscher, Madison, Wisconsin

Occupation

Professor of Civil & Environmental, and Geological Engineering at University of Wisconsin-Madison

Education

B.S. Calvin College, 1975
B.S. Civil Engineering, University of Michigan, 1976 (Magna cum laude)
M.S. Civil Engineering, University of Michigan, 1977
Ph.D. Civil Engineering, University of Michigan, 1981

Studying Engineering

My first BS degree involved an excellent mix of both the technical and non-technical aspects of engineering. Much of this education considered how one's engineering needs to be and can be a people-serving profession. My further degrees mainly contributed to my technical background and in forming my teaching and research skills. Hands-on experiences in one's education can be very formative; being a teaching assistant for me was one of these.

Career Life

In order to get tenure at a major research university, one must demonstrate a strong record of teaching, research, and service. These areas of work have provided deep satisfaction in turning the light on for students in their education, permitting me to pursue interesting problems to fairly deep levels, and giving me connections with the broader engineering community through conferences. More recently I have found great joy in working on engineering for the developing world where small contributions are amplified by the great need.

Life Outside of Work

My life is given needed rhythm by the daily support of family, weekly by my church, each semester by my students, and yearly by the seasons. This has been critical to my success and my sense of success. Each of these groups hones and sharpens me, urging me to do my best work and to contribute fully to the good of my communities. Life is about relationships... something I didn't understand when young.

Profile of an Electrical Engineer:
Timothy J. Bruns, St. Louis, Missouri

Occupation

Software Manager at Boeing Co.

Education

BSEE, 1983

Studying Engineering

I became interested in electronics at a young age by building electronic kits from companies like Radio Shack and Heathkit. As a
teenager, I became very active in local citizen's band (CB) radio
groups. It was an easy decision for me to pursue a degree in engineering. The technology has changed so much since I graduated,
and I have needed to stay current with the latest technology and
to find ways to apply it to my line of work. If you are just starting out
in engineering, I encourage you to apply yourself and do your very
best in all your classes. When I arrived at Purdue I felt as if I was

the least prepared of any of my classmates, but I worked hard and did very well. Some of
the better-prepared students did not apply themselves from the beginning and suffered as
a result. One thing I would have done differently is to get to know my professors and teaching assistants better. In large universities and organizations it is easy to get lost in the crowd,
and I wish that I had formed better friendships and relationships with my instructors.

Career Life

I am the software manager for a team of 15 developers that is creating a Windows NT application. This application uses the latest technologies such as MFC, COM and ActiveX. A typical day is spent reviewing the technical work of the team, along with reviewing schedules and
making estimates for future work. I often meet with customers of our product and suppliers of
our software development tools. Since our program is just getting started, I have been spending a lot of time interviewing people who would like to join our team. It is difficult to say how I
apply my engineering training directly to my current job. I know that my engineering degree
has given me the ability to plan and organize the work of our team, and to solve the many
problems that come up. The thing I like best about my job is the wide variety of assignments
I have had in my 15 years with Boeing. Working in a large company gives me the ability
to have several "mini-careers," all while working for the same company. A significant accomplishment that I have made while working at Boeing is the introduction of new tools and
technology into the software development process. One tool that we have introduced automatically produces source code from a graphical representation. This tool enables us to
bypass much of the labor-intensive and error-prone aspects of software design.

Life Outside of Work

In the engineering field, particularly in electrical and computer engineering, you will find that
the technology changes very rapidly. In my case, I stay abreast of the latest technologies by
enrolling in evening computer classes through the local universities. I enjoy home "engineering" projects such as designing a new deck. My wife, Donna, and I keep very busy raising our two sons, Garrett and Gavin.

Profile of an Electrical Engineer:
Jerry W. Burris, Louisville, Kentucky

Occupation

General Manager of Refrigeration Programs for General Electric Appliances

Education

BSEE, 1985; MBA, 1994

Studying Engineering

I have always had a curiosity about how things work (especially electronic devices). My parents recognized this at a very early age. They encouraged me to think about becoming an engineer. I was the child in the family who was always asked to fix the TV or electronic games. This continued through high school, where I excelled in math and science.

Purdue University was a natural choice for me, not only for its reputation for engineering excellence, but also due to the added bonus of having Marion Blalock and her Minority Engineering Program. This program has served as a recruiting magnet for Purdue and also has served as a mechanism for helping retain and matriculate students of color at Purdue.

While at Purdue, I was active in many extracurricular activities including leadership roles with NSBE and Kappa Alpha Psi fraternity. My early involvement in academics and extracurricular activities led to a full scholarship, which I received from PPG during my freshman year. Summer internships with PPG and IBM were invaluable in terms of giving me insight into what career path I wanted to pursue (design, manufacturing, or sales/marketing).

Career Life

I chose the technical sales and marketing route with General Electric's Technical Leadership Program. This premier program gave me advantages over direct hires in terms of exposure and training. After working six years I earned an MBA from Northwestern's Kellogg School of Management; I focused on global business, teamwork, and marketing.

My career has taken me from a role as a sales engineer, calling on industrial and OEM customers, to branch manager, with profit and loss responsibility, leading a team of nine people; to general manager of Refrigeration Programs at GE Appliances, where I now manage a $2 billion refrigeration product line.

I have been blessed with a lovely wife, who is also a Purdue and Northwestern graduate. We have two active children—Jarret, who is 7, and Ashlee, 4. We are managing dual careers at GE. This comes with significant challenges. However, GE has been very supportive of both of our careers.

Life Outside of Work

Life can not be all about work! You have to strive for balance. I have sought to keep God first in my life. I enjoy coaching my children in soccer, baseball and basketball, and I try to stay active with my own personal sporting activities. My favorite activities are listening to jazz music, traveling to exotic locations, managing our investment portfolio, and improving my golf game.

Profile of an Agricultural Engineer:
Bethany A. Elkin Fabin, Waterloo, Iowa

Occupation

Design Engineer, 8000 Chassis Design Team – John Deere Waterloo Works

Education

BS, Agricultural and Biological engineering

Studying Engineering

When I began to explore career options, I was told that an engi-  neering degree was the ticket to achieving success in a variety of fields. I investigated the Agricultural and Biological Engineering program at Penn State and discovered therein the opportunity to examine many aspects of engineering and agriculture under one discipline. I found my niche. This major provided the chance to "sample" many engineering topics and thus make knowledgeable decisions on what areas I wanted to pursue in future jobs. My Business Management minor also afforded many opportunities, and I would recommend that every engineer take at least a few business classes. I would also recommend getting involved in professional societies whenever possible. They provide many networking opportunities and a good preview of the job market. If I were to start my schooling over, I would take more of the hands-on classes. Also, I cannot begin to convey the importance of an internship or some kind of related work experience. Having the opportunity to work for a variety of companies in a variety of positions has helped me greatly in my career.

Career Life

In my current position as a chassis design engineer for John Deere, I work with others to design parts for tractor frames, coordinate homologation and standard reviews for update programs, and coordinate projects with supporting teams. In the latter role, I develop general specifications to ensure that we meet customer requirements and implement verification processes.

In a typical day of work, I spend a couple of hours working on Pro/E software designing and modeling parts. I also spend time working with suppliers and purchasing personnel to get parts quoted and ordered. In addition, I spend some time in our shop checking on prototype builds or test procedures, and some time in meetings working with different groups to keep people informed. The thing I like best about my job is the freedom I have to work on a variety of projects. It's nice to work for a company that has developed a strong name for itself and works diligently to stand behind their products.

Life Outside of Work

Outside of work, I welcome every opportunity to travel with my husband and play host to out-of-state friends, relatives, and foreign exchange students. MBA classes are taking up much of my time off the job currently, but in my free time I find I enjoy music, sports, rowing, training my dog, remodeling my house, and gardening. My membership in the local chapter of the American Society of Agricultural Engineers also keeps me busy with meetings and seminars.

Profile of an Electrical Engineer:
Bob Feldmann, St. Louis, Missouri

Occupation

Director, Tactical Aircraft Mission Systems, The Boeing Company

Education

BSEE, 1976; MS, Computer Science, 1980; MBA, 1999

Studying Engineering

My interest in engineering evolved naturally from my lifelong interest in science. Mathematics, while not my life's ambition, was interesting and satisfying. High school offered me the opportunity to enjoy learning about science, and I knew that I wanted to explore it even more in college.

In the mid-1970s, computers were not commonplace except in colleges, and I was hooked with my first FORTRAN programming class as a freshman. From that point on, I wanted to learn more about the hardware and software that made computers work. I oriented my electrical engineering curriculum toward digital electronics and used every elective I could to take software or software theory courses. While in college, I began a four-quarter stint as a cooperative engineering student at McDonnell Douglas in St. Louis. As a co-op, I was able to design software in the flight simulators (McDonnell was a world leader in simulation) and to work on a research design team for advanced flight control systems. Each semester when I went back to school, I would be at the library when Aviation Week magazine arrived, and I would read it cover to cover. When I graduated, I started my career as a software designer. My first day on the job, I was told that I would be responsible for the design of the software that controls the automatic carrier landing system on the F/A-18 aircraft. Ever since that first day, I have never been disappointed with the technical issues that have challenged me.

Career Life

Today I am leading a team of over 800 engineers in the design and production of Mission Systems (also known as avionics) for the F-15, the F/A-18, the AV-8B, and the T-45 aircraft. My role as team leader for the organization is to ensure that the various product teams are providing outstanding value to our customers with the quality of our designs. I no longer write software, but I interact with the technical teams, coaching and guiding them through the difficult challenges of today's technically-exploding world. My proudest recent accomplishment was leading a team of engineers in the design and flight test of a reconnaissance system for the F/A-18. That system will provide the United States with its first manned tactical reconnaissance capability since the mid-1980s.

Life Outside of Work

My life outside the office centers around outdoor activities and my family. My wife and I have three sons, all of whom play soccer and baseball (I have coached each one at various times). I really enjoy golf, bike riding, and other outdoor activities. My wife and I receive great pleasure from watching our sons grow up.

Profile of a Computer Engineer:
Steven E. Fredrickson, Houston, Texas

Occupation

Project manager of the Autonomous Extravehicular Robotic Camera for NASA; Electronics Engineer, NASA Johnson Space Center, 1995 – present

Education

BS, Computer and EE 1992; PhD, Engineering Science, 1995

Studying Engineering

To prepare for a leadership role in the emerging information society, I studied electrical and computer engineering as an undergraduate. At Purdue I supplemented engineering studies with non-engineering courses and extracurricular activities, and sought experiences to develop practical business skills. One highlight was the Cooperative Education Program. Three "Co-op" tours at NASA introduced me to software design, robotic control systems, and neural networks. This early work experience intensified my interest in advanced study of electrical engineering and robotics. To simultaneously satisfy my desires to engage in advanced academic research and to gain personal international experience, I pursued an engineering doctorate program in the Robotics Research Group at Oxford University.

I am extremely pleased with the universities I attended and the fields of study I completed to prepare for my current career. I would offer three recommendations to anyone pursuing an engineering path: 1) participate in Co-op or similar programs, 2) develop effective oral and written communications skills, 3) explore opportunities to study abroad.

Career Life

When I returned to NASA as a robotics research engineer, I transitioned from specialized research in artificial neural networks to broadly focused applied engineering. As project manager of the Autonomous Extravehicular Robotic Camera (AERCam) project, I have led a multidisciplinary team of engineers in development of a free-flyer robotic camera to provide "bird's eye" views of the Space Shuttle or International Space Station. Despite this deliberate transition to a project leadership role, it has been imperative for me to maintain my core technical skills. To ensure continued technical proficiency, I participate in several training courses and technical conferences every year.

Life Outside of Work

As much as I enjoy working at NASA, I believe it is essential to maintain outside interests. For me, that starts by spending time with my wife, Becky. Since Becky is pursuing a joint engineering and medical career, it can be demanding at times. The key for us has been to develop outside activities that we can enjoy together. Currently these include teaching Sunday school, participating in Bible study, attending concerts and plays, jogging, lifting weights, climbing at an indoor rock gym, and traveling. In addition, we allow each other time to pursue individual interests, which for me include reading, aviation, and golf.

Profile of a Mechanical Engineer:
Myron D. Gramelspacher, Hartland, WI

Occupation

Manufacturing Manager, General Electric

Education

BSME, 1989

Studying Engineering

I started at Purdue University in August 1985 in the engineering program. My interests in math and science were what really drove me to initially pursue opportunities in the field of engineering. My initial focus was in civil engineering, since I liked the concept of being able to work on roads, bridges, and outdoor structures. By learning more about the various engineering disciplines through seminars during my freshman year, I changed my mind and decided to pursue mechanical engineering. I felt that a degree in mechanical engineering would allow me more versatility and options in the workplace. Looking back on my college days, I wish I had taken courses in both business and foreign language to supplement my technical background.

Career Life

I graduated from Purdue University in 1989 with a degree in mechanical engineering, and started with General Electric (GE) as part of the Manufacturing Management Program. This program provided me with an opportunity to have six-month rotational assignments in two GE businesses. My first year was with GE Transportation Systems in Grove City, Pennsylvania, followed by a year with GE Aircraft Engines in Cincinnati, Ohio. In 1991, I transferred to the GE Medical Systems division in Milwaukee, Wisconsin. Since that time, I have held various positions in the Sourcing group, including supplier quality engineer, buyer, and team leader of the mechanical sourcing department. I also had the opportunity to live in Paris, France, for a year, heading up an Eastern European initiative. During that time, my efforts focused on the identification and qualification of suppliers in Eastern Europe. This position required that I travel throughout Europe, making it possible for me to experience different cultures and surroundings. This was a truly challenging and rewarding experience, both for my wife and me.

I currently hold the position of a Black Belt in GE's Six Sigma quality program. I utilize the Six Sigma tools and methodology to drive both process and product improvements that reduce costs, and ultimately impact our customers. The analytical skills and systematic problem-solving techniques that I gained through my undergraduate engineering courses have greatly contributed to the many opportunities and successes I have had in my professional career.

Life Outside of Work

I now am attending Marquette University, working toward my MBA. An MBA will complement my technical background and enable me to strengthen my overall business knowledge. Outside work, I enjoy making landscaping improvements around the house and tackling various wood-working activities. My wife, Kim, and I enjoy traveling in our spare time. I also enjoy playing golf, tennis, and softball.

Profile of an Industrial Engineer:
Karen Jamison, Dayton, Ohio

Occupation

Operations Manager, Jamison Metal Supply, Inc.

Education

BSIE, 1988; MBA, 2000

Studying Engineering

I didn't grow up knowing I wanted to be an engineer, but luckily my high school guidance counselor recognized my science and math abilities and encouraged me to try engineering. I firmly believe that engineering is a wonderful career in and of itself, and that it can be an excellent stepping stone for any other career you may wish to pursue in the future.

I chose industrial engineering because I am highly interested in improving the processes people use to do their work. Industrial engineering provides both technical challenges and the opportunity to work with all kinds of people.

If you are just starting to think about engineering or are trying to choose a specific discipline, talk to as many practicing engineers and professors as you can. Become involved in organizations on campus that will let you interact with other engineering students and practicing engineers.

I also highly recommend the co-op program. I had over two years of work experience when I graduated, and I knew what types of work I would enjoy. It is definitely to your advantage during interviews to know what type of job will best suit you, and to be able to speak intelligently on that subject.

Finally, remember that grades aren't everything but that your education is invaluable. If I were to do one thing differently, I would study to truly learn and understand the content instead of with the goal of getting a good grade in the class.

Career Life

Until last year, I was a consultant focusing on process improvement and business process re-engineering. Now I am learning to run Jamison Metal Supply, which is a business my parents founded 25 years ago. My job includes anything and everything that needs to be done. My primary responsibilities are overseeing operations to ensure quality products and timely deliveries, ordering steel for inventory and special orders, and pricing the material we sell.

I use my engineering training in all kinds of ways. I am working on updating our physical inventory system to better utilize warehouse floor space; I schedule customer orders to meet promised delivery times; and I am updating our computer system. Most importantly, engineering has taught me how to approach solving a problem and how to manage my time.

Life Outside of Work

My time outside of work is concentrated on completing my MBA degree, but I do find time for having fun as well. One of my favorite hobbies is crewing for a hot air balloon. I also teach a sign language class at the University of Dayton, and am vice president of the Purdue Club of Greater Dayton, Ohio. I think engineering is a very flexible field that allows individuals to prioritize their lives any way they wish.

Profile of a Mechanical Engineer:
Beverly D. Johnson, Waterloo, Iowa

Occupation

Supervisor in Wheel Operations at John Deere Waterloo Works

Education

BSME; MS, Engineering Management

Studying Engineering

My education includes a BS in Mechanical Engineering from the United States Military Academy, an MS in Engineering Management from the University of Missouri, Rolla, and my current study in the Executive Master's Degree Program at Northwestern University, Evanston, IL.

I think engineering is a very rewarding career because you can see the results of your effort every day. Engineering offers opportunities to create, build, design, and sometimes even destroy. Also, the analytical tools you develop in your engineering coursework make studying other subjects easier, and they are applicable to everyday life.

I truly enjoy my career in engineering. It is a dynamic career field that has taken me to many different jobs and many different places. I have done everything from constructing buildings and roads in Germany and the Hawaiian Islands to my current work as a supervisor in the wheel operations for the John Deere Waterloo Works.

Career Life

I have been with the John Deere Waterloo Works for two years, working in various engineering assignments such as quality engineering, project management, and process redesign. My current assignment as a supervisor in Wheel Operations is focused in production. I am responsible for the assembly processes pertaining to the tires and wheels for the 7000 and 8000 series tractors. I am also responsible for the daily supervision of the wage department personnel. Although my job is sometimes hectic, it is also very rewarding as I watch what our department is able to accomplish every day.

Prior to joining John Deere I spent nine years as a military officer in the U.S. Army Corps of Engineers. My primary responsibilities included the construction of buildings and roads, and the development and training of other engineers. My work with the military allowed me to live in, and travel throughout, Europe and the Pacific Islands.

Life Outside of Work

Although I chose engineering over journalism, my favorite pastimes are reading and writing. I also exercise regularly and compete in sports. I volunteer my time to the Boys and Girls Club of Waterloo, the American Red Cross, and a local university. However, my most important responsibility, and the most enjoyable, is the time I commit to the care and development of my two children, Colbert, 6, and Randy, 3.

Profile of a Civil Engineer: James L. Lammie, New York

Occupation

Board of Directors, Parsons Brinckerhoff Inc.

Education

BS, Civil Engineering, 1953; MS, Civil Engineering, 1957

Career Life

When I grew up, my father worked in a steel mill in Pittsburgh, the City of Bridges. I was fascinated with the many different bridges and what could be done with steel. I knew that I wanted to build things. I was fortunate to win an appointment to West Point, which was founded as the first engineering school in the U.S.

After graduation, I spent 21 years in the Army Corps of Engineers working on a wide variety of military and civil engineering projects all over the world. After retiring from the Army I knew I wanted to be a Project Manager on big projects, so I joined Parsons Brinckerhoff, Inc. and spent seven years as a consultant Project Manager for design and construction on the Metropolitan Atlanta Rapid Transit project (MARTA), the most rewarding period of my professional career. Today, my grandchildren ride what I helped build—a most rewarding feeling.

After MARTA, I had the pleasure of serving as the CEO of Parsons Brinckerhoff, Inc., the largest transportation design firm in the U.S., for the next fourteen years. Today, as a member of the Board of Directors of our employee-owned firm, I am still involved in some of our mega projects: the Central Artery Highway project in Boston, the new Taiwan High-Speed Rail system, the Bay Area Rapid Transit extension to the San Francisco Airport, and many others. The high point of my job is getting involved in critical project decisions and being able to "kick the tires" of work under construction.

Life Outside of Work

Thanks to my varied career in engineering, construction and management, I am also able to participate in a variety of outside activities: the Transportation Research Board (TRB), the Institute for Civil Infrastructure Systems (ICIS), the Engineering Advisory Board at Purdue University, and the National Academy of Engineering (NAE). I also teach and lecture on Project Management, Leadership, and Engineering Ethics. During my career, the high points have been presenting proposals and winning major jobs, election to the National Academy of Engineering, and receiving an Honorary Doctorate at Purdue University.

I always enjoyed participating in a variety of sports, until my knees gave out. The most personally rewarding aspect of my life over the years has been the companionship of my wife, three children (all in the medical profession, thanks to my wife's nursing career), and my eight grandchildren (with three going to colleges close by, permitting frequent visits).

Profile of an Astronaut:
Wendy B. Lawrence (Captain, USN), NASA Astronaut

Personal Data

Born July 2, 1959, in Jacksonville, Florida.

Education

Graduated from Fort Hunt High School, Alexandria, Virginia, in 1977; received a bachelor of science degree in ocean engineering from U.S. Naval Academy in 1981; a master of science degree in ocean engineering from Massachusetts Institute of Technology (MIT) and the Woods Hole Oceanographic Institution (WHOI) in 1988.

Organizations

Phi Kappa Phi; Association of Naval Aviation; Women Military Aviators; Naval Helicopter Association.

Special Honors

Awarded the Defense Superior Service Medal, the Defense Meritorious Service Medal, the NASA Space Flight Medal, the Navy Commendation Medal and the Navy Achievement Medal. Recipient of the National Navy League's Captain Winifred Collins Award for inspirational leadership (1986).

Experience

Lawrence graduated from the United States Naval Academy in 1981. A distinguished flight school graduate, she was designated as a naval aviator in July 1982. Lawrence has more than 1500 hours flight time in six different types of helicopters and has made more than 800 shipboard landings. While stationed at Helicopter Combat Support Squadron SIX (HC-6), she was one of the first two female helicopter pilots to make a long deployment to the Indian Ocean as part of a carrier battle group. After completion of a master's degree program at MIT and WHOI in 1988, she was assigned to Helicopter Anti-Submarine Squadron Light THIRTY (HSL-30) as officer-in-charge of Detachment ALFA. In October 1990, Lawrence reported to the U.S. Naval Academy where she served as a physics instructor and the novice women's crew coach.

NASA Experience

Selected by NASA in March 1992, Lawrence reported to the Johnson Space Center in August 1992. She completed one year of training and is qualified for flight assignment as a mission specialist. To date, her technical assignments within the Astronaut Office have included: flight software verification in the Shuttle Avionics Integration Laboratory (SAIL), Astronaut Office Assistant Training Officer, Director of Operations for NASA at the Gagarin Cosmonaut Training Center in Star City, Russia, and Astronaut Office representative for Space Station training and crew support.

A veteran of four space flights, Lawrence has logged over 1225 hours in space.

Profile of an Electrical Engineer:
Ryan Maibach, Farmington, Michigan

Occupation

Project Engineer at Barton Malow Company

Education

BS–CEM (Construction Engr. and Management), 1996

Studying Engineering

I think I have construction in my blood. My family has been involved in construction for four generations, and I was always surrounded by it while growing up. I can remember walking along beams with dad at a very young age, much to the irritation of my mother. So when it came time to choose a career, construction seemed to be the obvious answer.

Knowing what I wanted to study before I entered college was very nice, but it did not make my freshman year any easier. The first year of engineering school requires that you gut-out the tough prerequisites. During my college days, two extra-curricular activities were particularly rewarding to me. The first was my involvement in student organizations, which helped me to develop my leadership skills. The second was taking advantage of summer job internships, which provided practical experience by bringing to life the theoretical classroom education. Looking back, I should have taken more technology-related classes. I have found that technological skills are highly sought after in the job market.

Career Life

After graduation, I went to work for Barton Malow Company, a national design and construction services firm, and was made a field engineer on a basketball arena construction project. Presently, I am a project engineer working on a hospital expansion project. In my position I have a great deal of flexibility in terms of how I use my time throughout the day. I have worked on design development, purchasing contracts, scheduling, and subcontractor supervision.

Construction has been just as exciting as I had hoped. Every day when I leave my project, I can see what new accomplishments have been made—always, more steel has been hung or walls erected since I arrived. An interesting aspect of construction work is learning about our clients' businesses. It helps to have an understanding of our their industries in order for projects to be successful. For example, in order to make the present hospital expansion successful, our project team must understand how the hospital functions, overall. Having the opportunity to work on a variety of projects allows people in the construction industry to learn a great deal about various industries throughout their career.

Life Outside of Work

The spirit of camaraderie common to a construction site fosters friendships with co-workers that often continues outside of the work place. I have participated in a variety of community activities with my co-workers, ranging from speaking to school groups to refurbishing old homes for the elderly. Individuals looking for a challenging career which offers a lifetime of learning experience will find the construction industry very rewarding.

Profile of an Agricultural Engineer: Mary E. Maley, Battle Creek, Michigan

Occupation

Product Manager,
Kellogg Company

Education

BS, Agricultural Engineering (food engineering)

Studying Engineering

Math and science were always my favorite subjects, with the best part being the story problems where the concepts were applied. The idea of using scientific principles to solve a problem is what led me to choose engineering as a major. I would get to learn some more about math and chemistry as well as do something useful with that knowledge. That happened at college and continues to happen in my job.

Career Life

Today at Kellogg Company, the work I do is varied from day to day. My role is to make sure our manufacturing facilities have all the information they need, at the right time, to bring new products to market. That means coordinating the work from many different departments and gaining a consensus on the critical tasks to meet the timeline. You might ask, "What does that have to do with engineering?" Primarily, I bring together a myriad of details into one final outcome, just as all engineers do in combining the known facts to reach a solution. I just get to add some more unknowns and assumptions, such as dealing with people and changing requirements. The biggest challenge is getting the project accomplished to meet the needs of the consumer (that's you) before any of our competitors do.

Since Kellogg Company is a global company, my work affects the entire world. These days I work on projects for North America, Mexico, and Southeast Asia. I have had the opportunity to learn about other cultures and adapt our food products to fit their lifestyles. With manufacturing being located outside the U.S. as well, I encounter the varying work procedures and government regulations of each country. It makes my job challenging and enjoyable.

Life Outside of Work

Certainly, working at Kellogg's is not all that I do. My job is just one part of life. I find I need outlets for creative activities and for making contributions for the betterment of our world. Through sailboat racing I find a time of total concentration and a chance to apply aerodynamic principles. This also provides a fun way to have some competition. On the creative side, I participate in the handbell choir at my church. For me, music is a way to use my whole brain in the interpretation of notes into an emotional song of praise.

Perhaps more important is how I can give back to the community and the world in which I live. To promote cultural activities in my town, I serve on the board of directors of the Art Center. Through the Presbyterian Church, I lead the Benevolence Committee in disbursing funds and educating the congregation about the outreach in our community, the nation, and the world.

Profile of an Industrial Engineer:
Jeanne Mordarski, Albuquerque, NM

Occupation

Sales Manager, LightPath Technologies, Inc.

Education

BSIE; MBA

Studying Engineering

The engineering workload at Purdue was quite a shock to me. My first two years were a struggle and I was afraid to get involved in extracurricular activities. By my junior year, I became more concerned that I was missing out on the "college experience" than I was about my grades. I became an active member in several campus organizations—the best decision I ever made. I was forced to balance my studies and personal life. I broadened my network of friends, developed leadership skills, and learned to manage my time more effectively. As a bonus, my grades improved tremendously.

My emphasis within the IE curriculum was on Production and Manufacturing Systems. I accepted a production supervisor position with Corning Inc. after graduation. This role put me in the middle of the action, and taught me to think on my feet and make sound decisions. For eight years I worked at Corning in various engineering and manufacturing capacities. During this time, I was able to land a one-year tenure in Japan implementing Process Management Systems at our facility in Shizuoka.

For three years I took evening classes working toward an MBA from Syracuse University. In my course work, I realized how much I enjoyed the business side of things. Upon completion of my degree, I accepted a sales manager position at Corning in the telecommunications market.

Career Life

I recently left Corning to work as a sales manager for a start-up company, LightPath Technologies, Inc. Working for a large company directly from college gave me invaluable experience. The structure enabled me to work more effectively. However, as I progressed through the ranks at Corning, I realized that this same structure was limiting my ability to contribute because of the many management layers. In my current role at LightPath, we are introducing new products to the telecommunications market.

Life Outside of Work

I have an eclectic mix of interests outside of work. I truly enjoy exercise and the outdoors. On weekends, you'll find me skiing, camping, hiking, rock climbing, or biking. I love international travel and scuba diving, and take every opportunity I can to participate in both. I have recently taken up Latin social dance and kickboxing. I am also involved with the Purdue Alumni Association in Albuquerque. I like to keep busy and have worked very hard to strike a balance between my career and personal life. I seldom work more than 40 hours a week. I made it a goal to be more productive during work hours to minimize overtime and unnecessary stress. It usually works.

Profile of an Electrical Engineer:
Mark Allen Pashan, Red Bank, New Jersey

Occupation

Director of Hardware Development, Lucent Technologies

Education

BS; MS; MBA

Studying Engineering

When I was in high school trying to decide which career to pursue, I had a number of criteria: I wanted a job that I'd look forward to each day, that offered continuous learning, and that offered a reasonable level of financial stability. Engineering satisfied those criteria for me. I enjoyed math and science (the foundations of engineering) in high school, but engineering is more than number crunching. The field of engineering rewards creativity, the ability to find a better way to solve a problem. If I had to do it over again, I'd still choose engineering, but I'd also have bought more shares of Wal-Mart, Lucent, and Yahoo when they were first offered.

Career Life

In my career, I have advanced through a number of levels of technical management, and currently have about 130 engineers reporting to me. My job is no longer at the level of designing integrated circuits. I guide my team's progress on a number of new product development activities. I work to make sure we have the right people working on the right things at the right time. I set priorities among the competing needs of the business, and evaluate new business opportunities. To do my job, I use a combination of business and technical judgment: what are the future customer needs, what are the available and soon-to-be-available technologies, what are my competitors doing and what may they do next, who can do the work and work well together, and can we get the work done in time and at a reasonable cost. The end results are new products introduced into the marketplace that turn a profit for the business. That goal can only be achieved through others. A good part of my job is getting my teams to achieve more than they thought possible.

This is the best time in history to be an engineer. There are more available alternatives than ever—from startup companies to large established firms, from full-time to part-time work hours. There are more opportunities for continuing education and there is the potential for significant financial reward for those willing to take a risk.

My organization is spread across three states and I have customers and suppliers all over the world. My job requires travel and long hours, and I couldn't do my job and have a family without the support of my wife, Reem. But we do it together and the kids are a joy (even when they don't always obey). I enjoy a number of activities outside of work such as basketball, traveling, and dining out.

Profile of an Engineer: Douglas E. Pyles

Occupation

Engineering Consultant, Ingersoll-Rand

Education

BS, Interdisciplinary Engineering

Studying Engineering

I enjoyed science in high school, especially physics. I would describe myself during that time as a loner, a trait that became a weakness for me when I went to college. If I were to give advice to a student just beginning to study engineering I would say the following: 'Don't live in a shell, become active in one or two student organizations, make friends in your classes, get help right away and when you need it, don't be afraid!' If I had it to do over again I would try to follow this advice.

Career Life

I work for Ingersoll-Rand, at the Tool & Hoist Division Business Development Center. My title is Engineering Consultant, which is the first level on Ingersoll-Rand's technical career ladder. The career ladder is for people who, after progressing through the ranks, want to continue pursuing areas of technical expertise rather than managing people.

My primary responsibility at this time is supporting New Product Development Teams in the areas of mechanisms, design, testing, and manufacturing. I am personally responsible for insuring that portions of a system meet all performance, cost, manufacturability, and life targets. During the course of a project we travel extensively in a team-wide effort to learn the needs of the customer. I also participate with and help guide the team as we develop customer specifications for the final product.

Life Outside of Work

As I write this, I am also running an analysis on a Unix workstation that sits next to my PC. In the past, we have had to rely on experience and a process of building and testing to perfect our products. Now we have the advantage of computer analysis. I hold several patents for designs that were developed in this way. I enjoy being able to answer complex questions and make decisions about designs before making prototypes. I also really enjoy working on projects that have the potential for increasing market share and profits for our division. This type of analysis is only one of the many tools we use to develop high-quality products for an ever more demanding and changing marketplace.

When I am not slaying dragons at work, I am likely to be with my wife and three kids. I still maintain that anyone with children has no spare time, and has to schedule all their fun. On Wednesday nights, I play music with a different group of people. We have several guitars, a bass and a keyboard, as well as dedicated vocalists. We are responsible for leading music at church most every Sunday morning. I play the twelve-string guitar. When I was in school, I never thought I would ever find myself up in front of 300 people singing and playing music! It shows me that whatever stage of life I am in I must always look for opportunities to reach and to grow.

Profile of a Mechanical Engineer:
Emery Sanford, San Francisco, California

Occupation

Product Design Engineer, Apple Computer

Education

BSME, University of California-Berkeley, 2004

Studying Engineering

I took two years of engineering class in high school. When I was young, I enjoyed taking apart my dad's old typewriters and tape players to see the mechanisms inside, but I didn't know I wanted to be an engineer until I took those classes in high school. My class participated in a Supermileage vehicle competition, where teams from different high schools across the country are given the same small gasoline engine with the task of building a vehicle to achieve the highest mile per gallon rating. The Supermileage project exposed me to all aspects of the product design process, from the earliest sketches, to late nights in the shop trying to build a carbon fiber chassis, to finally driving the vehicle around the track. It is no coincidence that I found a career where I get to do those exact same tasks, only now the products I work on are used on a much larger, worldwide scale.

Career Life

As a product design engineer, my team is responsible for the mechanical design of consumer electronic products. That includes brainstorming product architectures, determing the size of the product (exactly how thin can you make that iPod?), detailed design of all the products' parts, followed by refining the design so it looks, feels, and functions as intended, all while fulfilling reliability requirements to ensure it can stand up to customer use over its lifetime. We work closely with other engineering teams, including electrical engineers, industrial designers, and manufacturing experts, to create innovative designs that people love to use. Many of the products I helped design have reached the market, available for sale all over the world.

Life Outside of Work

Outside of work, I try to enjoy the outdoors as much as possible, whether it be running on land or surfing in the ocean. I enjoy racing sailboats on San Francisco Bay during the summer and snowboarding during the winter months. I am also taking a Chinese class.

Profile of a Mechanical Engineer:
Patrick J. Shook, Columbus, Indiana

Occupation

Senior Engineer, New Product Development, Cummins Engine Company

Education

BSME, 1992; MSME, 1994

Studying Engineering

Mr. Myers, my high school chemistry teacher, had a discussion with me one day about Purdue's co-operative education program. He could see my interest in math and science and pointed me toward a field which I knew very little about—engineering. I investigated, and with high expectations, made the decision to attend Purdue to study and to prepare for what seemed to be a very interesting career.

By the end of the first semester during my freshman year, I had decided to pursue a mechanical engineering degree. This was after many discussions with junior and senior engineering students as well as with my father and a few professors. I had grown up in a family which owned a general contracting business (house construction, remodeling, etc.) and the broad variety of topics of study within mechanical engineering seemed to fit my desires. I also signed up to become a co-op student in order to obtain valuable work experience as well as to help pay for my education.

After graduating with my BSME, I entered into a research assistantship at Purdue for an intense (but extremely rewarding) two years on the way to obtaining my MSME. The most exciting task given to me by Prof. Fleeter was to build and operate a helicopter engine compressor test stand.

After graduating with my MSME, I hired on at Cummins Engine Company and worked for four years as a mechanical development engineer. This time was filled with designing abuse tests for semi-truck engines and determining how to improve the components that wore out during those tests. Without a doubt, learning how to work with people is easily 50 percent of my job. Since I have been at Cummins, I have learned that being clear with people concerning the goals of a plan is extremely important. As in a football huddle, everyone on the "field" needs to know what the "play" is and how to execute it.

Since July 1998, I have been working in a new position which focuses on engine cycle simulation. This has been primarily computer work and has re-sharpened my skills in fluid mechanics, thermodynamics, and heat transfer. The variety in my job has been enjoyable: from defining customer requirements for specific components to maximizing work processes within the structure of a large company.

Life Outside of Work

Outside of work, my wife and I do our best to serve the Lord and our church. In the past, we have both taken and taught classes on what the Bible says about marriage. We wanted to build on a good foundation and have enjoyed our marriage more and more with each year.

Profile of a Mechanical Engineer:
Nana Tzeng, Seattle, Washington

Occupation

Design Engineer, The Boeing Company

Education

BSME, 1997; MSME, 1998

Studying Engineering

What I enjoyed most about being an engineering student was making stuff—in other words designing and fabricating mechanical parts. While I was at Purdue, I participated in the Solar Racing Club. As one of the few mechanical engineers on a team dominated by electrical engineers, I helped improve the braking system, performed computer aided stress analysis on the chassis, learned to weld, machined rotors and other parts, and got to be driver of the solar car. The experience was not only rewarding, it also helped me relate what I read about in textbooks with applying that knowledge. College offers many extracurricular opportunities and I would encourage any engineering student to become involved in hands-on activities and research projects.

Career Life

My career really is rocket science! I currently work in the Instrumentation Development and Design group at Rocketdyne, the division of The Boeing Company that designs and develops rocket engines. The team I work with is responsible for all the sensors and electrical components on the Space Shuttle Main Engine. My latest project is the redesign of the spark ignition system. This involves the design of components and tooling, creating and updating of drawings, and working with the manufacturing team to improve the fabrication process. I also help the members of my team analyze sensor data from hot fire tests and space shuttle flights. Because the nature of my work is highly technical, I regularly use the knowledge and skill I gained as an engineering student. Now that I am familiar with the complexity of rocket engines and the detailed work that goes into building one, it's even more amazing when I see everything come together during launch.

Life Outside of Work

Ever since I finished school, I have been able to develop other interests and hobbies, some of which are golf, photography and snowboarding. I also often enjoy hiking, camping, rollerblading, shopping, concerts, clubs, etc. The best advice I have to offer to students of any discipline is to keep an open mind and take advantage of your opportunities.

Profile of a Mechanical Engineer:
Patrice Vanderbeck, Cedar Falls, Iowa

Occupation

Electronics systems engineer for the 6000 and 7000 series John Deere tractors

Education

BSME, 1982

Studying Engineering

I like to understand how things work. Math was my favorite subject in high school, followed closely by the sciences. Engineering seemed like the appropriate career for me, based on my interests. My college advisor recommended I study mechanical engineering after assessing my capabilities and interests. It sure was the right direction. The engineering curriculum can be difficult, but an engineering degree gives you many options. You can go into design, management, marketing, sales, law, manufacturing, or research, to name just a few areas. If you decide you want a change, it makes it easier for you to move on in a new direction. The engineering degree will open doors for you.

Given the opportunity, I would have changed two things about the path I took. First, I would have developed better study skills in high school, or earlier in college. I had to do some backtracking because of this. Also, I would have worked at a company that manufactures a product (like tractors) directly out of college instead of starting at a consulting firm. I learned it was important for me to work where a product is manufactured.

Career Life

I have worked at three companies since graduating in 1982. I am currently the electronics systems engineer for the 6000 and 7000 series John Deere tractors. I make sure that the different engineering teams within the electronics and vehicle groups are communicating with each other. My job combines design, program management, negotiating, and communicating. I deal with current tractor-related issues and the designs for new tractor programs at the systems level. I love my job because it is never boring. I am always learning. I work with many talented, dedicated people. When I was in a previous position at John Deere, I had design responsibility for a device for left-hand control of the forward and reverse movement of a tractor. It had 38 subassemblies. Between the supplier and me, we designed these subassemblies into a very small package. It took a lot of development effort to get the assembly to work perfectly within the entire system of the tractor. In the end, the device was well received by our customers for its function and reliability.

Life Outside of Work

I enjoy biking, skiing, weaving, entertaining, and reading. I belong to an investment club and a reading club. I am married to another engineer. We have a vacation home on the Mississippi River where we do a lot of boating and entertaining. We love to travel. My husband and I do volunteer work with Habitat for Humanity and with our church. Our engineering jobs allow us to live a comfortable life and to enjoy fun things like travel and boating.

Profile of a Chemical Engineer:
Jack Welch, Fairfield, Connecticut

Occupation

CEO of General Electric (Retired, 2002)

Education

BSChE; MSChE; PhD

Career Life

The man called "CEO of the century" by the editor-in-chief of *Time* magazine is an engineer. Jack Welch, who led General Electric's transformation over the past two decades into a global technology and services giant, started with the company as an engineer in Pittsfield, Mass. He had earned his BS ChE from the U of Mass in 1957, and followed that with an MS and PhD from the U of Illinois in 1958 and 1960.

In high school he had captained the hockey and golf teams and earned the distinction of being voted "Most Talkative and Noisiest Boy" by his classmates. "No one in my family had ever gone to college, but I had that ambition," Welch recalls. "Of course, believe me, my mother had that ambition for both of us."

"Life is a series of experiences, a series of steps if you will," he continues. "Every time you're reaffirmed, every time someone tells you you're Okay, you can go on to the next step, the next challenge. Well, my teachers in the Engineering Department told me I was Okay. In fact, they told me I was really good. A couple of them practically adopted me, and told me I had what it took to go on to graduate school. I had never even thought of graduate school. But they really believed in me."

After earning his PhD, Welch returned to Massachusetts and GE's Chemicals Division for his first job as a development specialist. It was on that first job that he demonstrated many of the leadership traits that characterize him to this day.

"I was an entrepreneur in a small business outside the mainstream of GE—the plastics business. My technician and I were partners working on the same thing. We had two people, then four people, then eight people, then 12. Today, GE Plastics is a $6 billion business. But it started that way. Everyone's involved. Everyone knows. Everyone's got a piece of the action. The organization's flat. All these things are from when I was 26 years old."

Welch's rapid rise in GE continued, and in 1981 he became the eighth chairman and CEO of the company that was founded in 1892. Although he recently retired, the organization he led was named "Most Admired" by *Fortune* magazine and "Most Respected" by the *Financial Times*.

Yet he described his job running a company with 1998 revenues of approximately $100 billion as "not rocket science." Instead he saw his key role as allocating both human and financial resources in a way that will continue GE's growth. "My job is allocating capital, human and financial, and transferring the best practices. That's all. It's transferring ideas, putting the right people in the right jobs and giving them the resources to win," he says.

Welch, now the father of four and recently a grandfather for the fourth time, continues on the golf course his winning ways that began in high school. He's twice won his club championship and has even bested well-known pros in friendly play.

Profile of an Electrical Engineer:
Shawn D. Williams, Twinsburg, Ohio

Occupation

Product General Manager at GE Appliances

Education

BSEE, 1985

Studying Engineering

The opportunity to further my math and science abilities attracted me to engineering. I really enjoyed calculus, chemistry, and physics in high school, so I attended Marion Blalock's Target Cities Luncheon in Chicago. At the luncheon, Purdue engineering students spoke about engineering, and their comments opened my eyes to engineering as a possible career.

As a freshman in engineering at Purdue, I queried several juniors and seniors in various engineering disciplines on their course work. I concluded that electrical engineering would allow me to leverage my foundation in mathematics and physics. In addition, computers were becoming more popular, and I concluded that EE would provide insights into computers. My advice to students is that you remain persistent and disciplined in pursuing an engineering degree.

Career Life

Currently, I am a regional sales manager for General Electric Industrial Systems. My primary job responsibilities include delivering top-line sales growth on an annual sales volume of $130 million. I lead a team of 70 employees (mostly engineers) throughout six states in the Midwest as they execute business strategies in their local trading areas.

In the course of a day, I may do such things as interview for an open sales engineer position, review a trading area strategy with a general manager, expedite a delivery with a factory for a distributor principal, and provide coaching to a new sales engineer on a project. The best aspect of my job is the variety of strategic and tactical tasks in which I engage. My engineering degree provided the solid foundation for me to understand the technical nature of the products, but the ability to handle multiple tasks is key in executive management.

Life Outside of Work

Obtaining an engineering degree has assisted me in providing a lifestyle for my family that I never thought possible while growing up. It is imperative for me to balance work and family life. My wife and my son and I spend weekends visiting the zoo, seeing friends, playing board games, and attending church.

On a personal note, I am the primary provider for my family, so keeping in shape physically, mentally, spiritually, and intellectually are keys to a successful life. I accomplish this by running 3 to 4 miles three times a week; consistently working on lowering my golf handicap of 13 by playing golf with customers and friends weekly; and incessantly learning about life through Bible reading, management books, and by mentoring and coaching employees, students and family members. I can honestly say that without my engineering experiences, the life I have now would still be just a dream!

Profile of a Mechanical Engineer:
Dr. Adel A. Zakaria, Waterloo, Iowa

Occupation

Senior VP, Engineering and Manufacturing, Worldwide Agricultural Equipment Division

Education

BSME, Egypt, 1967; MSIE, 1971; PhD, Industrial Engineering, 1973

Studying Engineering

Growing up in a developing country, I viewed engineering as an instrument for making genuine progress. It has offered me an opportunity to be part of moving things forward in our generation. In particular, I was interested in how things are made. Manufacturing offered a unique way to work with three domains: things, people, and systems. In today's highly technical society, engineering will offer a student an ideal foundation for several careers. A co-op or summer job experience can be very helpful in getting early exposure and making one's study a lot more interesting and meaningful.

Career Life

Within a $7 billion Worldwide Agricultural Equipment Division, I guide the work of 20,000 employees who engineer, manufacture, and provide product support to our customers.

A typical day for me in the last year might include:

- visiting with key customers and their servicing dealer in Nebraska
- chairing a Worldwide Combine Product Council in Germany
- visiting a new joint venture site in India to review progress of the factory construction
- reviewing the results of a business improvement team

I enjoy the breadth of my job, working with a variety of people, the disciplines they represent, and the constant challenge of leading change. Not only have I been able to apply engineering training, but I have had to constantly augment it by learning and expanding my knowledge in new, evolving areas.

Significant accomplishments in my career include the early development of cellular manufacturing, the development of computer aided design and manufacturing tools (CAD/CAM), guiding the development and introduction of our company's first worldwide tractor product platform, and the breakthrough in forging a new win/win labor strategy with our unions.

Life Outside of Work

My hobbies outside of work include racquetball and photography. Our family (including our two daughters) has traveled together in over 30 countries. We enjoy learning about the world's people and cultures. We also have camped in most of the U.S. and Canadian national parks. At various times I have been active in a number of national engineering societies. I also have participated in a number of community volunteer groups, including a hospital board, United Way, and the Boys & Girls Club.

Student Profiles

James Michael Campbell

Describe your experiences in college.

My experiences in college have been pretty much what I expected them to be. College is the next step after high school in an individual's education, and that's what my classes are. Math has picked up where I left off, and I have even repeated some material in my other classes. Coming from a high school like West really prepares you for some of the topics that are taught in college. The biggest changes from high school are the lack of personal attention, and the whole "do it yourself" style of teaching. The major method of learning is reading the book on your own, and the lectures help to clarify some of the concepts (hopefully). For the first time in my life, I am in control of how much I learn.

If applicable, describe your work experiences in engineering.

I haven't had an engineering-related job yet in college, but for the two summers after my sophomore and junior years of high school I worked a part-time job helping out on an experiment that was being run by the nuclear physics department at the UW-Madison. From what I understood, the experiment involved creating high-energy plasma, and controlling them with magnetic fields. The goal was to possibly learn something about fusion. I didn't understand much of the science behind the experiment, but it was a good opportunity for me to see how projects like that are run, and the kind of work that goes into research projects.

What do you do for fun outside of work?

Some things I do for fun include swimming, water polo, pick-up basketball games, watching sports on TV with friends, playing video games, watching movies, hanging out with friends and doing whatever.

Natalia Duque

Where did you attend college or are you currently going to college?

I am a short girl born and raised in Bogotá, Colombia. Right now, I am a senior majoring in Industrial and Systems Engineering and continuing in school next year to obtain a Master's degree in the same field.

If applicable, describe your work experiences in engineering.

As a student intern, I worked at General Motors, in one of its powertrain manufacturing facilities. This was an exceptional experience because I got to apply some of the concepts learned in classes, interact with professional engineers, understand the different fields within engineering and position myself in the day-to-day of an engineer for a couple of months.

What do you do for fun outside of work?

I love dancing especially salsa! I teach salsa lessons at school through SHPE (Society of Hispanic Professional Engineers). I am the vice president of this organization at the University of Wisconsin-Madison chapter. I really enjoy being active in the organization and putting a lot of energy and effort into it; not only because it has been very beneficial as an engineering student, but there I also met my best friends in school and found a sense of family. I also love swimming and clowning! (I was a professional clown for 4 years while living in Miami, FL.)

Sarah Freidel

Where did you attend college or are you currently going to college?

I am currently a sophomore at University of Wisconsin-Milwaukee. I am seeking a major in architecture and a minor in structural engineering.

Describe your experiences in college.

I have found my structural engineering minor as an ideal complement to my architecture major. I am interested in math and science, primarily physics, and found academic success in these areas. I considered that given my proficiency in math and the fact that I enjoy it, I might as well incorporate it into my education. I also enjoy design and problem solving and aesthetic concerns that come with it. I found architecture as a field of study that merges math and design. As an architect I hope to gain proficiency in design, but then proceed one step further to take on the challenge of figuring out how a building functions structurally. Through my studies in engineering I am learning the principles I need to analyze how buildings are put together. This deeper understanding of tectonic systems strengthens my architectural design.

What do you do for fun outside of work?

I enjoy knitting and baking. I make a killer chocolate zucchini cake. I played soccer for twelve years, first with the Madison 56ers and then on the women's team here at UWM. While I no longer play, I continue to stay involved with the team and the soccer community. I also love to read—some of my favorites include *Atlas Shrugged* by Ayn Rand and *In the Fall* by Jeffery Lent.

Colin Quinn

Where did you attend college or are you currently going to college?

I am currently a junior studying mechanical engineering at University of Wisconsin-Madison.

Describe your experiences in college.

College has shown me the importance of hands-on experience. I have many incredibly smart people around me who ace their classes but have trouble getting jobs after graduating. Getting involved in projects is just as important, if not more important, than your grades. I am a member of the UW Formula SAE team. We build a Formula 1 style race car around a 600cc motorcycle engine. We design and fabricate all the parts, do all the engine programming, all the CAD work, and all the suspension simulations. We compete against over 130 international teams in Detroit every spring. I have learned more in this project than in all my classes combined, plus I get to drive the fastest car I will ever drive.

If applicable, describe your work experiences in engineering.

I am currently on co-op at Wolf Appliance, a subsidiary of Sub-Zero, in Madison, Wisconsin. I work with the design engineers on design and research and development projects. I have programmed and developed a steamer and fryer unit, modified drawings and models on Solidworks, done prototyping work, and done tear-downs on competitor's units. It has been a great learning experience and introduction into the professional world. I strongly recommend co-oping to anyone who wants a good job directly out of college.

What do you do for fun outside of work?

I play soccer, go snowboarding and wakeboarding, play cards with friends, and race the formula car on the weekend.

Jeff Schneider

Describe your experiences in college.

I graduated in May 2005 from the University of Wisconsin-Madison with a degree in Geological Engineering and Geology/Geophysics. I am now a graduate student at Madison in the Geological Engineering program. As an undergraduate I had the opportunity to be a teaching assistant for both an introduction to engineering course designed for freshman, and an intermediate-level civil engineering course. I also had the opportunity to work for a graduate student and participate in research as an undergraduate.

Perhaps the most unique experience outside the classroom was participating in an organization known as Engineers Without Borders. The organization provides sustainable engineering solutions to disadvantaged communities around the world. I served as Vice President of the UW-Madison chapter in 2005 and as the President in 2006. Each of these experiences helped me pursue my interest in research and allowed me to continue my education in graduate school.

If applicable, describe your work experiences in engineering.

I spent two summers working for an environmental consulting firm. I split my time between working in the field and in the office. While I used many of the technical skills learned from school, the ability to write and having communication skills were more important.

What do you do for fun outside of work?

When I manage to pull some free time together I enjoy playing guitar, sailing, snowboarding, running, reading and writing. Finding time for these activities is essential for staying focused during school and work.

Megan Sharrow

Describe your experiences in college.

College has been a whirlwind of experiences for me. As my father cautioned me when I first left home for school, I have "not let school get in the way of a good education." As a result, I went on study abroad, I co-oped, and I joined student groups on campus. I was less concerned with a 4.0 and more concerned with getting a broad-based, hands-on knowledge of the engineering profession. Subsequently, I have found that when applying for jobs or when interviewing for jobs that the experiences I had outside of the classroom were just as important, if not more important, to employers than my GPA.

If applicable, describe your work experiences in engineering.

I have had the pleasure of working for American Champion Aircraft, Inc; Orbital Technologies Corporation; and the UW-Madison College of Engineering. These jobs required me to do everything from 3-D modeling to prototype fabrication to giving "shoot-from-the-hip" presentations to visiting officials from NASA. While my work experiences have been diverse, I have found a common thread through them all: communication. Without communication, progress simply does not happen. However, with a finely tuned ability to communicate, one is likely to work one's way through a task, a project, or a career successfully and efficiently.

What do you do for fun outside of work?

Outside of work my time is largely dedicated to martial arts. I began participating in Jeet Kune Do a year ago and have supplemented this training with work in the areas of Kali, Brazilian Jiu Jitsu, and Muay Thai. I also enjoy long walks in the woods, surfing, and dancing in my bare feet.

Joanna Storm

Describe your experiences in college.

During my freshman year, I was able to meet and connect with a lot of great students living in my same dorm and I was fortunate to meet and make more genuine friends throughout my next 4 years. I got involved in student organizations right away by joining the *Wisconsin Engineer* magazine staff, and had the pleasure of working with them for 2 years. During my junior year, I got involved with LeaderShape, first as a participant and then as a hired on-site coordinator. Finally, I became a member of Alpha Pi Mu (a professional Industrial Engineering society) and the Iron Cross Society here at UW-Madison during my senior year. As a grad student in the Professional Master's Degree program (within IE), I am continuing my work with the Iron Cross Society and currently serving as Secretary for the Human Factors & Ergonomics Society (HFES).

If applicable, describe your work experiences in engineering.

During my undergraduate career, I was privileged to obtain 2 summer internships with the National Aeronautics and Space Administration (NASA) Headquarters in Washington D.C. I first worked in the Office of Space Flight (May 2004–July 2004) exploring the re-engineering process of large systems to factor in lessons learned from the Challenger and Columbia tragedies, along with the responses to these events from multiple NASA space centers. This work included interviewing NASA engineers and upper-management officials from around the country. In addition, I studied the applicability of the STAMP/Archetypes Analysis tools (as postulated by Dr. Leveson et al. of MIT) toward the large systems engineering issues facing NASA. During my second summer internship with NASA HQ, I worked in the Office of Space Mission Operations (May 2005–August 2005) reviewing, monitoring, and compiling the "pressing issues" for NASA's Return To Flight (RTF) project, given by the Independent Technical Authority (iTA), into a more user-friendly format for middle and upper management. I also edited and rewrote large sections of the NASA NPR document that describes how to plan and execute large-scale program projects in accordance with NASA guidelines. For both internship experiences, I was paired with a fellow engineering student from the University of Southern Alabama (USA), and at the end of each summer we provided the Chief Operations Engineer (our immediate supervisor) with recommendations for permanent and positive changes in all aspects of the organization, helping improve the work environment (along with related tools and/or documents) for staff and visitors, and helping improve system safety throughout the organization.

What do you do for fun outside of work?

I enjoy spending time with my friends and family, listening to music, watching movies, and playing games and solving logic puzzles. Sudoku rocks!... I even go out with co-workers (fellow self-proclaimed "engin-nerds") after work on Fridays for "Super Sudoku Challenges."

Mark Zahi

Describe your experiences in college.

When I first started college, I had almost three more years of engineering experience than other incoming freshmen. More importantly, I was already sure about my career choice. Other students who lacked previous experience in engineering were not used to the type of curriculum and many of my classmates either dropped out or switched to an easier degree program. College is intense in every aspect. Not only is the bar set higher for grades and school work, but also the level of social interaction. There are hundreds of clubs and organizations to join, thousands of students, teachers, and professors, each with their own unique experiences. As an engineering student, I still had a relatively heavy course load when compared to other easier majors. I am also part of several student organizations such as Formula SAE and ASME. Add in work and time going out to bars, the only thing I can complain about is the lack of sleep.

If applicable, describe your work experiences in engineering.

I just finished a co-op with Wolf Appliance, working for 40 hours a week for an entire semester. The co-op program is sponsored by ECS (engineering career services). Almost 90% of all engineering students go on co-op at some point or another. This co-op provided valuable real-life engineering experience, experiencing engineering in the real world before graduation. During this term I actually used information I learned in high school—drafting and 3D modeling (specifically SolidWorks).

What do you do for fun outside of work?

Outside of work I am the Formula SAE Frame and Body Group Leader, as well as the Chief Financial Officer. Formula SAE is a group of 40 or so engineering students who come together and build a full race car to compete in the most prestigious Society of Automotive Engineers competition. This event is attended by almost 140 teams from around the world, featuring top schools such as the University of Wisconsin, MIT, Stanford, University of Michigan, etc. I am responsible for designing and manufacturing the frame from 4130 steel and the body from carbon fiber—both skills I learned in high school.

EXERCISES AND ACTIVITIES

3.1 Select five of the engineers profiled in this chapter. Prepare a one-page summary of their current responsibilities.

3.2 Select five other engineers profiled in this chapter. What were some of the common factors that got them interested in engineering or helped them succeed in school?

3.3 Of the engineers profiled in this chapter, describe three who have unique, non-engineering jobs.

3.4 Make a list summarizing the current jobs and responsibilities of five of the engineers in this chapter.

3.5 What types of career paths are available to engineers? Make a list of seven different career paths taken by engineers in this chapter.

3.6 Select a historical engineering figure. Prepare a one-page profile on this person using the format used in this chapter.

3.7 Interview a practicing engineer and prepare a profile on him or her, similar to the format used in this chapter.

3.8 Select five of the engineers profiled in this chapter and list some of the challenges they had to overcome during their education and/or career.

3.9 Prepare a one-page paper summarizing some of the challenges engineers face today.

3.10 Select one of the engineers profiled in this chapter and prepare an oral presentation on that engineer's company or organization.

3.11   Which career path presented in this chapter sounds most appealing to you?

3.12   Which of the engineers' stories can you relate to the most? Why?

3.13   What are your goals for your life outside of work? In this chapter, did you read about anyone with similar goals and/or interests? Explain.

3.14   Imagine yourself five years from now. Write your profile using the same format as presented in this chapter.

3.15   Imagine yourself ten years from now. Write your profile from that perspective.

3.16   Imagine yourself 25 years from now. Write your profile from that perspective.

3.17   What are your thoughts about the success and fulfillment found by engineers who obtained an advanced degree outside the field of engineering?

3.18   Based on the profiles and your own experience, what are the advantages and disadvantages of getting a graduate degree in engineering?

3.19   Prepare a matrix of the bachelor's degrees the engineers in this chapter received and their current positions. How many work in jobs traditionally associated with bachelor's degrees?

3.20   Select your preferred major and prepare a list of potential career paths it could lead to. Provide a brief explanation for each option.

3.21   Today's engineering workforce is truly diverse, with both men and women from all ethnic backgrounds working together. This was not always the case. Select a historical figure who was a woman or minority engineer and prepare a one-page profile which discusses the difficulties she or he had to overcome.

3.22   Research a company you would be interested in working for and list the engineers in their upper-level management structure. Identify the highest level of management currently held by an engineer in that company.

3.23   Prepare a one-page paper on how some of the engineers profiled have addressed family responsibilities amidst an engineering career.

3.24   A stereotype of engineers is that they are boring loners who only care about numbers and technology. Based on the personal-interest sections of the profiles, prepare a report refuting this stereotype.

Chapter 4

A Statistical Profile
of the Engineering Profession

4.1 STATISTICAL OVERVIEW

Many students who choose a major in engineering know little about the profession or the individuals who work in the field. This chapter will answer such questions as: How many people study engineering? What are their majors? What is the job market for engineers like? How much do engineers earn? How many women and minorities are studying engineering? How many practicing engineers are there in the U.S.?

The picture of the engineering profession presented in this chapter will assist you in better understanding the field. This information has been gathered from a variety of surveys and reports that are published on a regular basis by various organizations. As with many types of data, some are very current and others not as current. We have endeavored to provide commentary that explains the information presented.

4.2 COLLEGE ENROLLMENT TRENDS
OF ENGINEERING STUDENTS

As observed in Figure 4.1, the number of first-year students enrolled in engineering programs has fluctuated greatly during the period from 1955 to 2008, according to the 2008 Engineering Workforce Commission's annual survey of enrollment patterns. During the 1950s and 1960s the number of first-year students pursuing bachelor's degrees in engineering programs ranged between 60,000 and 80,000. During the 1970s there was a significant drop in student interest, and freshman enrollment fell to a low of 43,000. This was followed by the largest long-term climb in student enrollment, which eventually peaked in the early 1980s at about 118,000 students. From that point until 1996 there was a steady decline in student enrollment, with a low of about 85,000 first-year students. This was followed in 1997 to 2002 with a modest increase in student enrollment each year, which was projected to last for several years. From 2003-2006 there was a decline in first year students enrolling in engineering programs, even though total undergraduate engineering enrollment remained high. 2007 and 2008 show a modest increase in first year students to 111,000.

Why has enrollment in engineering (and higher education in general) fluctuated so greatly over the last few decades? The primary factor influencing enrollment is the number of high school graduates in a given year. Of course, those numbers are directly correlated to the number of births 18 years previously. Given these facts and the information in Figure 4.2, it can be determined how the birth rates from 1960 to 2007 have influenced, and will influence, the enrollment of college-bound students. Likewise, Figure 4.3 provides data concerning high school graduations from 1988 to 2019. This shows that high school graduations started to climb in 1992 and increased in 1999 through 2005. Projections indicate that there should be modest increases in 2006 through 2009 and then a slight decline followed by a rise in graduations.

Obviously, there are other factors which influence a student's choice of a major, including the general economy and the relative popularity of various fields in any given year.

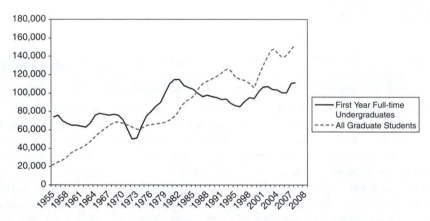

Figure 4.1 Engineering enrollments: selected indicators, 1955–2008

Sources: Through 1978, Table 23, "Engineering Enrollments and Degrees" in the Placement of Engineering and Technology Graduates, 1979 (New York: Engineering Manpower Commission); for 1979–1998, Engineering and Technology Enrollments, Fall 1979–1998, Engineering Workforce Commission of the American Association of Engineering Societies, 1979–2008, Elkridge, MD.

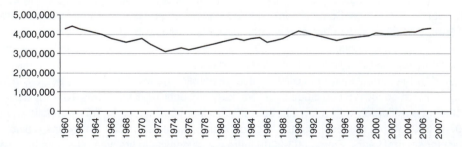

Figure 4.2 Births in the U.S., 1960–2007

Source: U.S. Department of Health and Human Services, Centers for Disease Control and Prevention National Center for Health Statistics, 2007

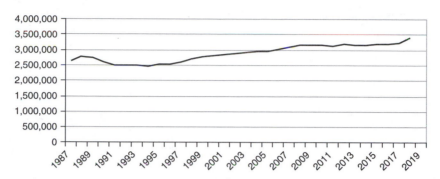

Figure 4.3 High school graduations, 1987–2019

Source: US Department of Education, National Center for Education Statistics, 2009. Projected data are for the years 2005-2019.

It also should be noted that the total number of students pursuing graduate engineering degrees increased significantly from the late 1970s to 1992. Since then, there was a relatively steady decline until 1997, when enrollment started increasing again, with 2004 numbers reaching an all-time high of over 145,000. 2005 and 2006 numbers show a decline to about 140,000, but 2008 figures rose to over 153,000 graduate students.

4.3 COLLEGE MAJORS OF RECENT ENGINEERING STUDENTS

Table 4.1 provides specific 2008 enrollment data for undergraduate and graduate students in several broad engineering fields. This information shows that over 440,000 students were majoring in one of the undergraduate engineering disciplines, with another 153,000 pursuing graduate engineering degrees. The largest field of study at all levels is the electrical and computer engineering disciplines, followed by mechanical and aerospace. It is interesting to note that over 88,800 students are in "other engineering disciplines" (which includes the many smaller, specialized fields of engineering) and over 14,755 are in pre-engineering programs.

4.4 DEGREES IN ENGINEERING

When examining the number of degrees awarded to engineering graduates during the years from 1983–2009, (see fig., page 115) there is a close correlation between enrollment and degrees. Since 1986, when over 78,000 engineering graduates were produced, to 1997 when approximately 65,000 degrees were awarded, there was a steady decline (with the exception of 1993) until 1995, and then a slight increase until 1997. However, there were declines in 1998 and 1999, and then a gradual increase between 2000–2006, to a total of 76,000. 2007, however, shows a slight decline in degrees awarded, followed by an increase in 2008 and a decline in 2009. Table 4.2 documents the number of bachelor's degrees awarded by field of study for the years, 1998 to 2007. It is interesting to note that the greatest number were earned by students in mechanical engineering, which in 2006 surpassed electrical engineering and the closely related fields of computer engineering or computer science (which at many schools are in the same department, or are closely aligned with electrical engineering programs).

TABLE 4.1 Fall, 2008 Engineering Enrollments by Broad Disciplinary Groups

	All Engineering Disciplines	Electrical and Computer	Mechanical and Aerospace	Civil and Environ-Mental	Chemical and Petroleum	Industrial, Management, & Manufacturing	All Other Disciplines	Pre-Engineer-ing
Full-Time Undergrads								
First Year	111,006	25,083	23,696	12,459	7,410	1,784	32,538	8,036
Second Year	92,364	22,320	22,753	12,409	7,598	2,871	20,087	4,326
Third Year	86,894	23,816	23,187	13,199	7,499	3,545	14,429	1,222
Fourth Year	112,878	33,065	29,918	18,434	9,923	5,114	16,719	335
Fifth Year	8,203	2,029	2,261	1,339	799	530	1,237	8
Full-Time Under-grad	411,345	106,313	101,815	57,840	33,229	13,844	85,010	13,927
Part-Time Undergrad	31,607	12,557	7,042	4,907	1,475	986	3,812	828
Total All Under-grad	442,952	118,870	108,857	62,747	34,704	14,830	88,822	14,755
Candidates for M.S.								
Full Time	60,169	25,567	9,637	6,808	2,107	6,000	10,047	3
Part Time	35,544	13,541	4,818	3,649	728	4,653	8,129	26
Total All M.S.	95,713	39,108	14,455	10,457	2,835	10,653	18,176	29
Candidates for Ph.D.								
Full Time	50,261	18,360	7,725	4,436	5,047	1,922	12,771	----
Part Time	7,132	2,809	991	836	360	548	1,588	----
Total All Ph.D.	57,393	21,169	8,716	5,272	5,407	2,470	14,359	----
Full-Time Grads	110,430	43,927	17,362	11,244	7,154	7,922	22,821	----
Part-Time Grads	42,676	16,350	5,809	4,485	1,088	5,201	9,743	----
Total All Grads	153,106	60,277	23,171	15,729	8,242	13,123	32,564	----

Source: Engineering Workforce Commission of the American Association of Engineering Societies

Engineering Degrees, 1983–2009 by Degree Level

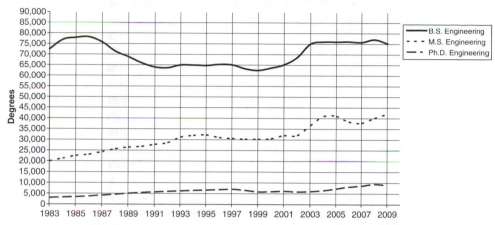

Source: 1983-2009: Engineering Workforce Commission of the American Association of Engineering Societies

4.5 JOB PLACEMENT TRENDS

From 1999 to mid-2001, the job market for graduating engineers was excellent. In fact experts agree that it was probably the hottest job market in over 30 years. According to surveys of employers conducted by the Collegiate Employment Research Institute at Michigan State University, overall hiring then contracted nearly 50% in 2002–2003. However, the survey of over 2,500 employers for 2009–2010 shows a continuing poor economy and a tighter labor market for new college graduates. If this trend continues, opportunities for engineers will still be available, but will require students to devote increased efforts to find the right opportunity. It should be noted that employment opportunities and starting salaries for engineers have traditionally surpassed those of most other majors on college campuses, and this trend continues.

The Bureau of Labor Statistics, in its 2008 report, shows that 2.5 million people work as engineers and other closely related professions, which is approaching a record high for the profession. Figure 4.4 illustrates the recent trends in engineering employment over the last decade. These trends involve some modest fluctuation, perhaps due to corporate downsizing, and some engineering job losses. However, the overall picture is one of fairly stable employment.

Figure 4.5 illustrates the relatively low unemployment in the engineering profession. In 1994 the unemployment rate of engineers was about 3.5%. Since that time, there has been a steady decrease, until 2001 when the unemployment rate was around 1.7%. Since 2001, engineering unemployment climbed over 3.0% in 2003, and then steadily dropped to 1.3% in 2007. However, 2008 numbers show engineering unemployment at 3.3% and climbed to 6% in mid-2009. While the 2009 engineering unemployment rate was the highest since 1994, the unemployment rate of engineers continues to be well below that of the general population.

4.6 SALARIES OF ENGINEERS

Salaries of engineering graduates consistently have been among the highest for all college graduates over the past several decades. To some extent, this has been due to the shortage of engineering candidates to fill available jobs. However, even in "down" years and leaner economic periods, engineers tend to do better in the job market than do their counterparts from other majors.

TABLE 4.2 Bachelor's Degree in Engineering, by Discipline, 2000–2009

Curriculum	2000	2001	2002	2003	2004	2005	2006	2007	2008	2009
Aerospace	1,274	1,542	1,773	2,024	2,298	2,395	2,681	2,763	2,897	2,989
Agricultural	624	583	322	330	305	302	343	351	355	383
Biomedical	1,172	1,215	1,679	1,962	2,360	2,638	3,028	3,055	3,478	3,787
Chemical	6,044	5,757	5,657	5,342	4,939	4,621	4,590	4,607	4,915	5,151
Civil	8,750	8,219	8,185	8,595	8,477	8,857	9,432	9,875	10,862	11,274
Computer	9,816	11,595	13,703	16,607	17,101	16,019	14,282	13,171	11,471	10,370
Electrical	12,643	12,929	13,031	14,177	14,495	14,742	14,329	13,783	13,113	11,978
Engr. Science	949	964	923	1,018	1,072	1,019	1,015	1,097	1,104	1,104
Industrial	3,133	2,982	3,252	3,313	3,225	3,220	3,079	3,150	3,064	3,077
Materials & Metallurgical	901	842	803	869	775	815	919	924	1,027	985
Mechanical	12,989	12,968	13,343	14,114	14,321	14,835	15,698	16,172	17,865	17,016
Mining	373	357	305	276	206	224	243	345	345	341
Nuclear	130	100	144	123	195	239	327	324	400	394
Petroleum	250	268	259	330	269	298	381	468	541	665
Other Engineering Disciplines	4,587	4,874	5,269	5,951	5,965	5,779	5,756	5,738	5,670	5,806
TOTAL	63,635	65,195	68,648	75,031	76,003	76,003	76,103	75,823	77,107	75,320

Source: Engineering Workforce Commission of the American Association of Engineering Societies, 2009

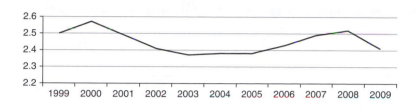

**Figure 4.4 Numbers of Employed Engineers and Other
Closely Related Occupations (in millions)**

Source: Bureau of Labor Statistics

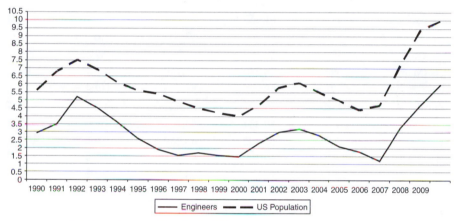

**Figure 4.5 U.S. General Population Unemployment Compared to Engineering
Unemployment: Recent Trends—1990–2009**

Source: Bureau of Labor Statistics

As reflected in Table 4.3, recent data indicate that average starting salaries for 2010 graduates in selected engineering fields have varied significantly. Starting salary offers in many disciplines have increased while a few have dropped slightly. Chemical engineers and petroleum engineering B.S. graduates are at the top of the salary list. This trend may reflect the shortage of trained graduates in those fields.

Table 4.4 provides starting salary data for 2008 graduates by engineering curriculum and some of the types of employers that hire students in those fields.

One question often raised by engineering students concerns the long-term earning potential for engineers. In other words, does the high starting salary hold up over the length of one's engineering career? Figure 4.6 presents 2009 median salaries for all engineers based on the number of years since their bachelor's degree. It is apparent that an engineer's relative earnings hold up well for many years following graduation.

One significant difference in salaries is shown in Figures 4.7 and 4.8. These data relate the 2009 salary difference between engineers who are placed into supervisory positions and those who remain in non-supervisory roles. Over the course of their career, supervisors tend to earn significantly higher salaries than those in non-supervisory positions.

TABLE 4.3 Starting Salary Offers in Engineering, Reported 2008

Curriculum	2010 Average	2009 Average	2008 Average	% change since 2009	% change since 2008
Bachelor's					
Aerospace	$58,208	$55,943	$55,815	4.0%	4.1%
Agricultural	53,535	54,690	55,265	-2.1%	-3.2%
Architectural	50,263	47,581	49,406	5.6%	1.7%
Bioengineering	53,416	53,417	54,013	0.0%	-1.1%
Chemical	64,889	65,675	63,773	-1.2%	1.7%
Civil	51,321	52,287	51,780	-1.8%	-0.9%
Computer Engr.	60,396	60,844	60,280	-0.7%	0.2%
Computer Sci.	60,473	61,467	61,110	-1.6%	-1.1%
Electrical	59,512	60,509	57,603	-1.6%	3.2%
Environmental	55,491	50,109	51,222	10.7%	7.7%
Industrial/Manuf	57,396	58,230	57,740	-1.4%	-0.6%
Materials	58,171	58,076	58,748	0.2%	-1.0%
Mechanical	58,110	59,222	57,024	-1.9%	1.9%
Metallurgical	59,037	59,837	54,727	-1.3%	7.3%
Mining	63,207	62,802	58,389	0.6%	7.6%
Nuclear	57,417	60,209	58,531	-4.6%	-1.9%
Petroleum	77,278	85,417	75,621	-9.5%	2.1%
Master's					
Aerospace	$66,072	$68,061	$70,309	-2.9%	-6.4%
Chemical	90,333	70,484	66,338	28.2%	26.6%
Civil	57,225	53,311	56,300	7.3%	1.6%
Computer Engr.	69,389	72,771	75,712	-4.6%	-9.1%
Computer Sci.	69,753	68,627	73,826	1.6%	-5.8%
Electrical	67,844	70,921	72,814	-4.3%	-7.3%
Industrial/Manuf	63,271	67,591	66,387	-6.4%	-4.9%
Mechanical	67,234	66,961	65,121	0.4%	3.1%
Doctoral					
Aerospace	$64,875	$90,505	$66,041	-28.3%	-1.8%
Chemical	82,488	85,250	82,419	-3.2%	0.1%
Civil	58,964	60,351	60,981	-2.3%	-3.4%
Computer Sci.	69,112	84,080	87,216	-17.8%	-26.2%
Electrical	81,188	89,715	85,045	-9.5%	-4.8%
Mechanical	73,036	75,186	72,068	-2.9%	1.3%

TABLE 4.4 Average Starting Salary by Curriculum in 2008 and Some Common Types of Employers for Each Field

* Curriculum Type of Employer[1]	FALL 2010
• Aerospace/Aeronautical Engineering	
Aerospace Products & Parts	*
Federal Government	*
Consulting Services	*
Engineering Services	*
• Agriculture Engineering	
N/A	
• Architectural Engineering	
Building,Developing, General Contracting	*
Architectural Services	*
Engineering Services	*
• Bioengineering/Biomedical Engineering	
N/A	
• Chemical Engineering	
Chemicals(Basic)	$66,693
Food,Beverage, Tobacco	63,295
Household & Personal Care Products	*
Petroleum & Coal Products	74,973
Pharmaceuticals & Medicine Mfg.	62,517
Consulting Services	60,107
Engineering Services	60,187
Utilities	*
• Civil Engineering	
Building, Developing, General Contracting	54,108
Federal Government	47,030
State & Local Government	43,470
Consulting Services	47,845
Engineering Services	50,387
Transportation Services	*
Utilities	58,763
• Computer Engineering	
Aerospace Products & Parts	58,074
Computer & Electronic Products Mfg.	*
Electrical Equip/Appliance/Component Mfg.	*
Computer Systems Design/Consulting/Programming	56,023
Engineering Services	*
• Computer Engineering	
Aerospace	61,923
Computer & Electronic Products Mfg.	63,736
Electrical Equip/Appliance/Component Mfg.	*
Machinery Mfg.	*
Federal Government	*
Communications (Broadcasting/Telecomm.)	61,250
Computer Systems Design/Consulting/Programming	63,453
Consulting Services	*
Engineering Services	57,938
Utilities	*

(continued)

TABLE 4.4 Average Starting Salary by Curriculum in 2008 and Some Common Types of Employers for Each Field *(continued)*

* Curriculum Type of Employer[1]	FALL 2010
• Electrical/Electronics Engineering	
Aerospace Products & Parts	$59,565
Computer & Electronic Products Mfg.	55,931
Electrical Equip/Appliance/Component Mfg.	60,395
Machinery Mfg.	*
Federal Government	56,098
Communications (Broadcasting/Telecomm.)	57,667
Computer Systems Design/Consulting/Programming	*
Consulting Services	*
Engineering Services	58,429
Utilities	60,692
• Environmental Engineering	
N/A	
• Industrial/Manufacturing Engineering	
Aerospace Products & Parts	*
Electrical Equip/Appliance/Component Mfg.	*
Food,Beverage, Tobacco	*
Machinery Mfg.	*
Transportation Equipment Mfg.	*
Consulting Services	61,714
Engineering Services	56,628
Financial Services	61,923
• Materials Engineering	
N/A	
• Mechanical Engineering	
Aerospace Products & Parts	57,983
Building,Developing, General Contracting	59,015
Chemicals (Basic)	*
Computer & Electronic Products Mfg.	*
Electrical Equip/Appliance/Component Mfg.	*
Household & Personal Care Products	*
Machinery Mfg.	56,301
Medical Equipment & Supplies Mfg.	*
Mining	*
Petroleum & Coal Products	70,163
Transportation Equipment Mfg.	53,160
Federal Government	52,475
Consulting Services	58,297
Engineering Services	57,563
Transportation Services	*
Utilities	55,582

[1] Employers who reported 10 offers or more
* Not provided for fewer than 10 offers

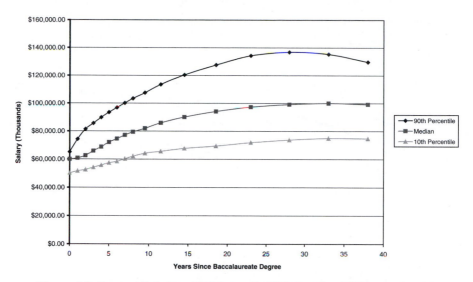

Figure 4.6 Current Salaries of Engineers in All Industries (All Degree Levels)

Source: "Engineers' Salaries: Special Industry Report, 2009"
Engineering Workforce Commission of the American Association of Engineering Societies, Inc.

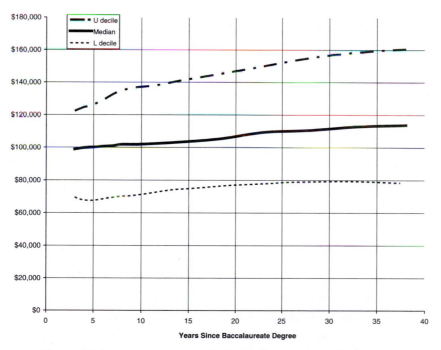

Figure 4.7 Salary Curves for Engineering Supervisors, All Degree Levels

Source: "Engineers Salaries: Special Industry Report, 2009"
Engineering Workforce Commission of the American Association of Engineering Societies, Inc.

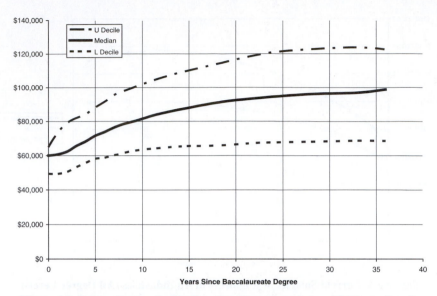

Figure 4.8 Salary Curves for Engineering Non-supervisors, All Degree Levels

Source: "Engineers' Salaries: Special Industry Report, 2009"
Engineering Workforce Commission of the American Association of Engineering Societies, Inc.

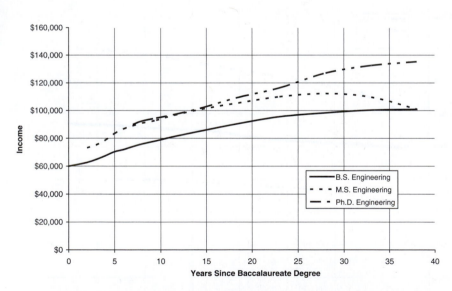

**Figure 4.9 Current Median Income of All Engineers in All Industries by
Highest Degree Earned and Length of Experience**

Source: "Engineers' Salaries: Special Industry Report, 2009"
Engineering Workforce Commission of the American Association of Engineering Societies, Inc.

TABLE 4.5 Income by Region

All Engineers—All Degree Levels

Graduates of 1998–99 versus Graduates of 1988–1991

Region	Graduates of 1998–99	Graduates of 1988–1991
New England (CT, ME, MA, NH, RI, VT)	86,918	99,027
Middle Atlantic (NJ, NY, PA)	77,364	86,154
East North Central (IL, IN, MI, OH, WI)	84,621	99,612
West North Central (IA, KS, MN, MO, NE, ND, SD)	73,593	97,963
South Atlantic (DE, DC, FL, GA, MD, NC, SC, VA, WV)	*75,937	*87,137
East South Central (AL, KY, MS, TN)	*74,718	*86,671
West South Central (AR, LA, OK, TX)	*92,882	*109,732
Mountain (AZ, CO, ID, MT, NM, NV, UT, WY)	80,816	95,696
Pacific (AK, CA, HI, OR, WA)	80,148	93,477

* 2008 data

Source: 2009 Engineers' Salaries: Special Industry Report
Engineering Workforce Commission of the American Association of Engineering Societies, Inc.

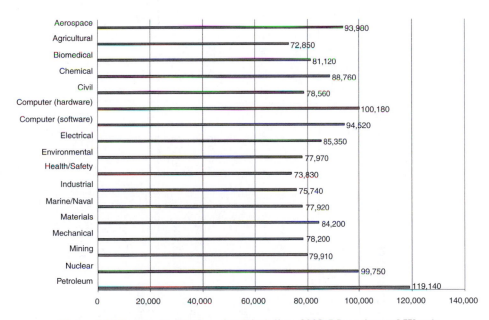

Figure 4.10 Income by Engineering Discipline, 2008 (Mean Annual Wage)

Source: Bureau of Labor Statistics, Division of Occupational Employment Statistics, 2008

Figure 4.9 shows how an engineer's income is influenced by the level of education achieved. As might be expected, those with master's and doctoral degrees have higher compensation levels. Length of experience in the engineering field is also an income determinant. Figure 4.9 demonstrates a consistent, regular growth in an engineer's income level as experience is gained. Generally, the average increase is at about 4.6% per year of experience.

Table 4.5 demonstrates 2009 median salaries by region of employment. This compares graduates of 1988–1991 with those of 1998–1999. For 10 year out graduates, salaries are

TABLE 4.6 Average Bachelor's Degrees Starting Salaries by Curriculum and Job Function

Curriculum	Research & Development	Testing	Design or Construction	Manufacturing	Sales	Consulting	Management Trainee
Aerospace	*	*	$58,914	*	*	*	*
Chemical	$66,717	*	*	$65,583	*	$56,000	*
Civil	*	*	53,796	*	*	*	*
Computer Engineering	*	*	*	*	*	*	*
Computer Science	*	*	*	*	*	66,333	*
Electrical	*	$60,765	58,175	*	$56,688	*	*
Industrial/Manufacturing	*	*	*	56,304	*	60,885	*
Mechanical	53,701	56,002	53,306	54,204	55,389	63,214	$62,278

*Less than 5 offers reported

TABLE 4.7 Income by Industry Sector
All Engineers—All Degree Levels
Graduates of 1998–99 versus Graduates of 1988–1991

Industry Sector	Graduates of 1998–99	Graduates of 1988–1991
All Manufacturing Industries	86,252	100,196
Metal Manufacturing Industries	NA	83,649
Metal Products Manufacturing	69,542	78,347
Machinery Manufacturing	72,390	83,786
Transportation Equipment Manufacturing	87,212	101,976
Motor Vehicle Manufacturing	87,535	102,275
All Non-Manufacturing	72,243	85,774
Transportation and Warehousing	69,640	79,995
Architectural, Engineering, and Related Services	83,676	101,303
Construction	79,824	93,539
Mining	83,532	102,577
Electric Power Generation	77,969	92,172
Research and Development Services	81,040	95,914
Professional, Scientific, and Technical Services	*79,483	*97,050
Public Administration	68,523	82,742
Utilities	77,645	93,875

*2008 Data

Source: 2009 Engineers' Salaries: Special Industry Report
    Engineering Workforce Commission of the American Association of Engineering Societies, Inc.

highest in the West South Central Region, followed by those in New England. The lowest reported salaries are for engineers in the West North Central. (Geographic strongpoints have shifted dramatically and repeatedly in recent years.)

For 20 year out graduates, salaries are highest in the West South Central Region. Lowest salaries are in the Middle Atlantic Region.

Figure 4.10 provides the 2008 median annual income by major branch of engineering for all working engineers. Petroleum engineers and computer hardware engineers show the highest income levels, with agricultural engineers at the lowest levels.

When viewing the average starting salaries of engineers by discipline and job function in Table 4.6, it is surprising to observe that there is great variance within discipline and function. The highest reported starting salaries are for computer engineers in the consulting function. The lowest are for civil engineers in the consulting function.

It is also interesting to examine median annual income by industry or service. Table 4.7 compares the 2009 median salaries of graduates from 1988–1991 with those of 1998–1999. The data show those involved in certain manufacturing areas are at the top of the salary list. The lowest from these classes are involved in public administration.

4.7 THE DIVERSITY OF THE PROFESSION

For many years, engineering was a profession dominated by white males. In recent years, this has been changing as more women and minority students have found engineering to be an excellent career choice. Unfortunately, the rate of change in both enrollment and degrees awarded to these students has been slower than expected. This is especially discouraging at a time when population statistics show a significant increase in the number of college-bound women and minority students.

Table 4.8 provides information concerning degrees awarded to women and minority groups over a recent 10-year period. While the numbers have been increasing, the growth rate has been slower than expected. Since recent demographic data show that the number of minority students attending college is growing, we would expect to see a larger percentage of these students earning engineering degrees. So far, this is not the case—especially at the graduate level, where increases have been steady, but modest.

TABLE 4.8 Degrees in Engineering, by Level, and Type of Student, 2000–2009

Level and Type	2000	2001	2002	2003	2004	2005	2006	2007	2008	2009
Bachelor's Degrees	63,635	65,195	68,648	75,031	76,003	76,003	76,103	75,823	77,107	75,320
Women	13,140	13,195	14,102	15,114	15,282	14,868	14,654	14,101	13,865	13,432
African Americans	3,150	3,182	3,358	3,429	3,699	3,756	3,673	3,735	3,470	3,392
Hispanic Americans	4,124	4,152	4,298	4,652	4,813	4,890	4,957	5,133	5,486	5,488
Native Americans	347	275	315	388	362	378	456	426	427	328
Asian Americans	7,529	8,340	8,669	9,705	9,941	10,033	9,719	9,466	9,143	8,743
Foreign Nationals	5,048	4,839	4,859	5,560	5,768	5,644	5,354	5,152	4,787	4,519
Master's Degrees	30,453	32,008	31,983	36,611	40,953	41,087	38,451	37,805	40,122	41,967
Women	6,431	7,026	7,067	8,024	8,842	9,212	8,731	8,393	9,237	9,596
African Americans	778	826	867	916	947	1,072	1,009	1,078	1,116	1,137
Hispanic Americans	851	815	817	927	1,143	1,194	1,185	1,292	1,318	1,393
Native Americans	96	70	72	89	91	123	128	81	78	94
Asian Americans	2,613	2,776	2,829	3,172	3,934	3,994	3,990	3,995	3,863	3,629
Foreign Nationals	11,680	13,698	13,615	16,831	18,447	17,536	15,441	14,604	16,408	18,426
Doctoral Degrees	5,929	6,141	5,863	6,027	6,504	7,276	8,116	8,614	9,449	8,907
Women	937	1,039	1,024	1,040	1,136	1,322	1,592	1,689	1,976	1,896
African Americans	96	103	94	97	102	111	121	122	127	139
Hispanic Americans	85	84	112	107	97	107	99	120	152	163
Native Americans	11	10	8	13	7	15	11	15	15	16
Asian Americans	397	403	401	337	397	391	496	432	508	582
Foreign Nationals	2,961	3,235	3,217	3,441	3,766	4,405	5,048	5,180	5,564	4,869

Source: 2000–2009: Engineering Workforce Commission of the American Association of Engineering Societies

TABLE 4.9 Women and Minorities in the Engineering Pipeline, 2008
Percentage of All Full-Time Undergraduates

	First-Year Full-Time B.S. Students	All Full-Time B.S. Students
Women	17.42%	17.76%
African Americans	6.61	5.40
Hispanic Americans	9.26	8.86
Native Americans	0.69	0.59
Asian Americans	10.00	10.97

Source: Engineering Workforce Commission of the American
Association of Engineering Societies

Table 4.9 shows enrollment data for women and minorities in all undergraduate engineering programs. The percentage of women in engineering is over 17%, while the percentage of under-represented minorities is about 15%. These figures have dropped during the past few years, despite increased recruitment efforts by colleges and universities.

4.8 DISTRIBUTION OF ENGINEERS BY FIELD OF STUDY

According to the Bureau of Labor Statistics (see Figure 4.11), over 290,000 of all practicing engineers are employed in the electrical engineering discipline. This is followed by about 261,000 civil engineers, and 233,000 mechanical engineers. It is interesting to note that

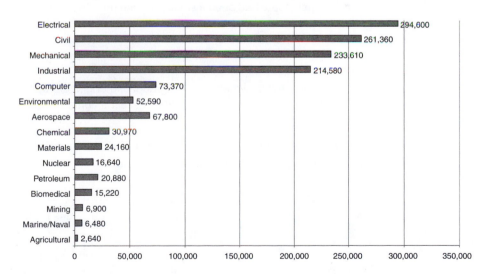

Figure 4.11 Practicing Engineers in the U.S., 2008

Source: Bureau of Labor Statistics, 2008

"other engineers" (which includes the smaller, specialized fields of engineering) include about 250,000 engineers. Comparing this information to previous data on enrollment and degrees, one would expect these numbers to remain somewhat steady.

4.9 ENGINEERING EMPLOYMENT BY TYPE OF EMPLOYER

According to the Bureau of Labor Statistics, in 2008 the largest sector of engineering employment, almost 40%, was in manufacturing industries. These include electrical and electronic equipment, industrial machinery, transportation equipment, and instruments.

In 2008, over 480,000 engineering jobs were in service industries which includes engineering and architectural services, research and testing services, and business services in which firms designed construction projects or did other engineering work on a contractual basis. Engineers were also working in communications, utilities, and computer industries.

Government agencies at the federal, state, and local levels employed approximately 192,000 engineers in 2008. The largest number were employed by the federal government in the Departments of Defense, Transportation, Agriculture, Interior, Energy, and the National Aeronautics and Space Administration (NASA). The majority of state and local workers are usually involved in local transportation, highway, and public works projects.

In 2000, almost 48,000 engineers were self-employed, typically as consultants.

4.10 PERCENT OF STUDENTS UNEMPLOYED OR IN GRADUATE SCHOOL

How does engineering unemployment compare to the number of students who go on to graduate school? As indicated in Table 4.10 and referred to earlier in Figure 4.1, the total number of students pursuing graduate engineering degrees increased significantly between the late 1970s and 1992. Since then, there had been a steady decline until 2000. During the period 2001–2008 enrollment in graduate engineering programs has soared to 153,000. The recent trend is similar to the early period covered by Table 4.10 when jobs for bachelor-level graduates were not as plentiful, and many are opting for graduate school, which may explain the rapid rise in graduate school enrollments.

Table 4.10 also shows the lower unemployment rate for graduating engineers between 1990 and 2008. The figures for 1992 and 1994 and 2002–2008 reflect a tighter job market

TABLE 4.10 Selected Disciplines' Engineering Graduate School Full and Part-time Enrollment Trends and Percent of Unemployed Engineers

	1990	1992	1994	1998	2000	2002	2004	2005	2007	2008
Aerospace	3,563	3,935	3,550	2,780	3,042	3,171	3,750	3,758	3,990	4,291
Chemical	6,657	6,926	7,292	6,683	6,714	6,963	7,282	6,878	6,949	7,268
Civil	11,625	13,900	14,164	11,009	11,151	12,546	12,942	12,561	12,787	13,430
Electrical	32,095	32,635	29,524	25,302	27,764	33,851	34,127	32,691	35,306	35,778
Industrial	9,713	11,552	11,401	9,428	5,234	6,330	5,789	5,340	5,677	6,302
Mechanical	17,115	18,566	17,310	14,032	15,201	17,085	17,945	17,554	18,349	18,880
TOTAL	80,768	87,514	83,241	69,234	69,136	79,946	81,835	78,782	83,058	85,949
% Unemployed Engineers	2.9%	5.2%	4.3%	1.7%	1.5%	3.0%	2.3%	2.1%	3.3%	3.1%

Sources: "Engineering Enrollments and Degrees," Engineering Manpower Commission of the American Association of Engineering Societies, Inc. 1990, 1992, 1994, and Engineering Workforce Commission of the American Association of Engineering Societies, Inc., 1998, 2000, 2002, 2004, 2005, 2007, 2008

than that of recent years. It should also be noted that, in general, the unemployment rate for new engineering graduates is far lower than for any other major.

The engineering job market continues to be tight; one might expect that the unemployment rate for engineers may increase, and the number going to graduate school will also continue to rise.

4.11 A WORD FROM EMPLOYERS

As discussed earlier, the current job market for graduating engineers is challenging. However, there are important issues that employers want students to consider. Even though this study is over 10 years old, the message is still important for today's engineering college student. In a 1999 survey of 450 employers done by the Collegiate Employment Research Institute at Michigan State University, recruiters were asked to provide commentary on any special concerns they had about new engineering graduates.

One interesting conclusion to the study found that "Employers want the *total package* when they hire their next engineering graduates. Not satisfied with academically well-prepared graduates, employers want individuals who possess and can demonstrate excellent communication and interpersonal skills, teamwork, leadership, and computer/technical proficiency. A willingness to learn quickly and continuously, to problem solve effectively, and to use their common sense is also desired. New employees must be hard-working, take initiative, and be able to handle multiple tasks." In addition, employers are looking for students with new emerging skills and aptitudes such as the ability to understand e-commerce, computer capabilities that include programming skills, and the ability to adapt to constant change. In the new global economy, increasing competition requires a strategy to respond quickly.

Therefore, it is important to realize that just earning an engineering degree is no longer sufficient to prepare students for the fast-paced, technologically changing global economy. Students must devote time in college developing those critical competencies that employers are seeking in order to be well prepared for the challenges that await them.

EXERCISES AND ACTIVITIES

4.1 If an administrator used only the data from 1975 to 1980 to predict future engineering enrollment, how many first-year, full-time students would have been expected in 2008? What would the percentage error have been?

4.2 What is the percentage increase in projected high school graduates in 2015 compared to those in 2005?

4.3 Using the data of Figure 4.4, estimate the number of employed engineers in 2015.

4.4 Using the starting salary data from Table 4.3, and assuming an annual increase at the current cost of living (currently about 2%), calculate what you could expect to receive as a starting salary in your chosen field of study if you graduate three years from now.

4.5 Using the information from Figure 4.9, estimate the engineering salary after 30 years assuming a) a bachelor's degree; b) a master's degree; and c) a doctorate degree.

4.6 Using the information from Figures 4.7 and 4.8, estimate the median salary after 30 years assuming the engineer is a) a supervisor and b) a non-supervisor.

4.7 If you were a supervisor in the upper decile rather than a non-supervisor in the upper decile, what would be your percentage increase in salary 20 years after graduation?

4.8 Calculate the percentage increase of the highest-paid bachelor's engineer in 2007 compared to the lowest-paid bachelor's engineer in 2007.

4.9 Suggest at least one reason for the relatively low income in the West North Central Region of Table 4.5.

4.10 Suggest at least one reason for the relatively low income of government or public administration workers compared to those in the private sector.

4.11 Project the percentage of women and minorities in engineering in the year 2010 using the data of Table 4.8. Do not include foreign nationals.

	'98	'99	'00	'01	'02	'03	'04	'05	'06	'07
% Minority Bachelor's	22.9	23.7	23.8	24.5	24.2	24.2	24.8	25.1	24.7	24.7
% Women Bachelor's	18.6	19.8	20.6	21.2	20.5	20.1	20.1	19.6	19.3	18.6

4.12 Estimate the percentage of engineers that are chemical engineers using the data of Figure 4.11.

4.13 Based on the current economic situation, do you expect the employment demand for graduating engineers to increase or decrease? Explain the basis for your answer.

4.14 If we experience a significant economic recovery, what do you think will happen to enrollment in graduate engineering programs?

4.15 Review the information presented in Figures 4.2 and 4.3 and Table 4.8. Make a prediction about the future enrollment of women and minorities in engineering. Explain your answer.

4.16 What factors do you think influence how employers determine hiring needs in any given year?

4.17 Attend a campus career fair. Talk with at least 3 recruiters from different firms. Make a list and discuss those factors which employers are stressing in their hiring decisions.

4.18 Based on the current economic situation, do you expect the salaries of engineers to continue to increase at the recent pace? Why or why not? Using other tables in this chapter, explain the basis for your answer.

4.19 Review the enrollment data in this chapter. What do you think high schools, colleges, and the engineering profession can do to increase the number of students choosing to pursue an engineering career?

4.20 What factors do you think contribute to the fact that entry-level engineering salaries are usually higher than any other major on campus?

Chapter 5

Global and International Engineering

5.1 INTRODUCTION

Many students who choose to study engineering make this decision primarily because of their aptitude in mathematics and science. As a result, most enter college thinking that they will no longer be required to take courses in the liberal arts and humanities. Many are especially pleased to avoid further foreign language study. Once they arrive on campus, they learn that some study in the liberal arts is still required. But the fact remains that few, if any, engineering students are required to complete a foreign language requirement. For beginning students, the concept of using and applying their background in an environment requiring use of a foreign language seems terribly remote to them. Most envision their career as remaining fairly stable, with one employer, in one location for many years, perhaps even a lifetime. The opportunity to "go global" with their engineering career probably has never been considered.

Perhaps no other factor has emerged more dramatically in shaping an engineering career than the rapid globalization of the engineering profession. This globalization evolved initially from the post-World War II movement toward international markets, sustained by large multinational corporations providing goods and services to many regions of the world. In more recent years the globalization of the engineering profession has increased rapidly due to corporate mergers and acquisitions and the availability of high-speed, low-cost communication systems, and cheap labor costs in many nations.

In the "old world" marketplace, countries competed by trading goods and services that were primarily developed and produced using capital and labor from within their own borders. However, as global restrictions eased and trade barriers changed, companies found they could maximize profits by investing capital resources in a variety of countries. Coupled with new policies that enable the movement of labor across political boundaries, many organizations have added incentives to "internationalize" their business. The current market climate encourages firms to maneuver factories, jobs, research facilities, distribution centers, and even corporate management operations in order to maximize market share, develop new products, access needed raw materials and less costly skilled and unskilled labor, while taking advantage of incentives that lower costs and maximize profits.

Since most of the products and services that are generated by these corporations are developed, designed and manufactured by engineers, these engineers now find themselves in a unique position as critical players in this rapidly expanding global economy. Today's engineers must be better educated and properly trained to meet the challenges of this role. They must have an understanding of the world community around them, world cultures, and the world marketplace. It has become apparent that strong technical skills by themselves are often insufficient. Unfortunately, in many cases engineers are the problem solvers, yet are weak in the cultural and language skills required in today's global marketplace.

This chapter will focus on several factors that drive the global economy. It also will explore many of the international opportunities available to engineers, and discuss the preparation needed for an international engineering career.

5.2 THE EVOLVING GLOBAL MARKETPLACE

Many factors have been instrumental in the rapid acceleration of the global marketplace. Some of these follow.

Changing World Maps and Political Alliances

One need only examine a map from 1990 and compare it to a current map to see the many changes that have occurred in that short period of time. New countries have been created by the break-up of the Former Soviet Union. Organizational and governmental changes have occurred in Eastern Europe (the reunification of Germany, the breakup of Czechoslovakia, etc.). New nationalistic tendencies have taken root in the Far and Middle East, Asia, and Africa. The effects of September 11, 2001, and world terrorism and the war in Iraq have also changed the world's political and economic climate. Each of these situations has had both positive and negative impacts on the economic conditions that influence market demand and the world production and distribution of goods and services.

Additionally, ever-evolving political alliances, the changing of governments and political leaders, and the development of new laws, regulations and policies have greatly affected the global market. In recent years, some former enemies have become allies, while some former friends are now embroiled in turbulent relationships. Russia, for decades seen as America's chief adversary, is now a partner in exciting new ventures in space exploration and industrial development. Former enemies in the Middle East and Asia are now economic partners in many joint endeavors with U.S. business and industry. As these changes unfold, new opportunities emerge for American firms to expand markets, establish new production sites and enhance their global presence.

The Role of Mergers, Acquisitions, and International Partnerships

Chrysler and Daimler-Benz, Pfizer, Pharmacia and Upjohn, Boeing and McDonnell Douglas, Dow Chemical and Union Carbide, General Motors and Saab are but a few examples of the thousands of mergers, acquisitions and industrial partnerships that have become commonplace in the global economy. Almost on a daily basis the media inform us of yet another joint venture, buyout, or corporate spin-off with international implications. No industry has been immune from these activities, as computer, electronic, pharmaceutical, chemical, communication, automotive and banking firms have joined forces in an attempt to combine technologies, increase revenues and manage costs.

The net effect of these activities has accelerated global opportunities for engineers, regardless of their initial interest in international careers. If engineers working for a U.S.-based corporation suddenly learn that their employer has been acquired by an international firm, by default those individuals become "global engineers." Following such an acquisition, management decisions may now originate from an overseas site. Product development, design and manufacturing may now be performed from a different cultural perspective. Expectations and goals may change significantly. Travel, communications and interactions with others in the firm now may take on a new perspective.

NAFTA

One of the most dramatic, controversial and significant influences on the globalization of the engineering profession has been the 1994 North American Free Trade Agreement (NAFTA) between the United States, Canada and Mexico. NAFTA was designed to lower tariffs and increase international competitiveness. Based on original projections it was forecast that NAFTA would eventually create the world's largest market, comprised of 370 million people and $6.5 trillion of production—from "the Yukon to the Yucatan." At present, NAFTA has created the second largest free trade market in the world, after the European Economic Area, which has a population of 375 million and a gross domestic product of $7 trillion. Large and small corporations around the world are feeling the impacts of NAFTA.

NAFTA's primary objective is to liberalize trade regulations as a way to stimulate economic growth in this region, and to give NAFTA countries equal access to each other's markets. Virtually all tariff and non-tariff barriers to trade and investment between NAFTA partners are eliminated by 2009. However, as an important side feature, each NAFTA member is able to continue to establish its own external tariffs for trade with third countries.

Other important outcomes for this agreement include an increased market access for trade, greater mobility for professional and business travelers, legal protection for copyrights and patents, and a mechanism for tariffs to reemerge if an import surge hurts a domestic industry. Almost all economic sectors are covered by NAFTA, though special rules apply to particularly sensitive areas such as agriculture, automotive industry, financial services, and textiles and apparel. Side agreements cover cooperation on labor issues and environmental protection.

Since its inception, NAFTA has been very controversial in all participating countries. Many in the U.S. felt that it would result in the loss of critical manufacturing jobs to cheaper labor markets (Mexico), the relocation of new production facilities, and a rise in unemployment and inflation. In actuality, more U.S. jobs have been lost to locales with cheaper labor than Mexico, notably Central America and Asia.

According to the 2004 NAFTA Free Trade Commission joint statement produced by trade representatives of the three participating countries, NAFTA, now in its tenth year, has achieved many positive results. Based on the data in this report, the success of NAFTA is illustrated by the following:

- Since 1994, trade between the U.S., Canada, and Mexico has more than doubled. From less than $297 billion in 1993, trilateral trade has surpassed $623 billion.
- Investment in the three economies has increased significantly, with total foreign direct investment in the NAFTA countries has increased by over U.S. $1.7 trillion.
- As a result of this growth in trade and investment, millions of new jobs have been created in all three countries, lower costs and more choice for consumers.

The report summarizes "The evidence is clear—NAFTA has been a great success for all three parties. It is an outstanding demonstration of the rewards that flow to outward looking, confident countries that implement policies of trade liberalization as a way to increase wealth, improve competitiveness and expand benefits to consumers, workers and businesses."

It should be noted, however, that many displaced U.S. workers would strongly disagree with this assessment. In several areas, notably the automotive and manufacturing sectors, union workers are fighting to preserve jobs, worker security, and plant projects as more activities are being moved to non-U.S. locales.

The opportunities, challenges and experiences available to engineers as a result of NAFTA are exciting. Engineers now find that their assignments may take them to Mexico or Canada, perhaps even for extended periods of time. Interactions between "work colleagues" may actually take place from remote distances via e-mails or teleconferences. Today's engineer may be planning production implementation for a plant opening in Mexico or Canada, with the engineering done in the U.S. Or a Canadian firm may be producing goods in a U.S. or Mexican plant. The world of engineering has truly changed in North America and for the North American engineer.

The European Union

Another factor which has played a significant role in the globalization of the engineering profession has been the development of the European Union. The European Union (EU), previously known as the European Community, is an institutional framework for the construction of a united Europe. It was created after World War II in order to unite the nations of Europe economically, with the intent that such an alliance would make another European war among them unthinkable. In May, 2004, member countries grew from 15 to 25, and in 2007 grew to 27 with the addition of Romania and Belgium. They share the common economic institutions and policies that have brought an unprecedented era of peace and prosperity to Western and Eastern Europe. The first steps toward European integration began when six founding member countries first pooled their coal and steel industries. They then set about creating a single market in which goods, services, people and capital would move as freely as within one country. The single market came into being in January 1993. The original treaties gave the EU authority in a number of areas including foreign trade, agriculture, competition and transport. Over the years, in response to economic developments, formal authority was extended to new areas such as research and technology, energy, the environment, development, education and training.

Of great relevance to world trade and economic competition, the EU cleared the way for the introduction of a single currency, which was implemented January 2002, and the creation of a European Central Bank. As a result, the United States and Europe are now more economically interdependent than ever. According to the delegation of the European Commission to the United States Report, the European Union and the United States are involved in a thriving economic relationship. Their trade and investment relationship is the largest in the world. As each other's main trading partners, they are also in position as the largest trade and investment partners for nearly all other countries, therefore shaping the picture of the global economy as a whole.

In 2007, the EU generated an estimated gross domestic product (GDP) of 16.83 trillion (US dollars), which is 31% of the world's total. It is the largest exporter of goods, and the second largest importer. The investment trends are even more impressive as both the EU and U.S. are each other's most important source of foreign direct investment with a total two-way investment amount of $1.8 trillion. Direct employment due to these investments, along with indirect employment such as joint ventures and other trade related job creation, indicates that 7 million U.S. workers are in jobs connected to EU companies.

As one of the world's largest trading powers, and as a leading economic partner for most countries, the EU is a major player on the world scene. Its scope for action extends increasingly beyond trade and economic questions. More than 130 countries maintain diplomatic relations with the EU, and the EU has over 100 delegations around the world.

The expansion of EU countries from 15 to 25 includes many in Central and Eastern Europe. Since the independence of these Central and Eastern European countries in 1989, the EU has drafted trade and cooperation accords with most of them and has been at the forefront of the international effort to assist them in the process of economic and political reform. The EU is also strengthening its links with the Mediterranean countries, which will receive some $6 billion in EU assistance over the next few years. The EU also maintains special trade and aid relationships with many developing countries. Under a special agreement, virtually all products originating from 70 African, Caribbean and Pacific countries enjoy tariff-free access to the EU market.

It is obvious that the European Union is, and will continue to be, a formidable ally as well as competitor to U.S. business interests. Many of the excellent international opportunities available to U.S. engineers have emerged and will continue to increase as a result of the partnerships, the competition and the presence of the European Union.

5.3 INTERNATIONAL OPPORTUNITIES FOR ENGINEERS

Automobile Industry

From the time the assembly-line process was developed by Henry Ford in the early 1900s until the late 1970s, the automobile industry was pretty much dominated by the "Big Three" in Detroit (General Motors, Ford and Chrysler). Products from Germany, Japan and Korea were present in the United States, but their quality was considered suspect. In the mid-1970s, Japanese auto makers began to implement quality improvement processes which resulted in products that started to become more popular in the world marketplace. The U.S. was recovering from a major escalation in gasoline prices, and high-mileage, low-cost cars were suddenly in demand. The foreign car makers manufactured products that fit this niche quite nicely. From that point in time, sales of U.S.-made vehicles dropped and those of foreign competitors gained overall market share. General Motors, which once commanded over half of the market, is now fighting to maintain a 20% share, under the auspices of federal government subsidies.

Engineering students interested in a career in the automobile industry now have many exciting choices as a result of the changes in this dynamic industry. The traditional Big Three have expanded their territories by opening new plants and operating facilities in many regions of the world. Each of the major U.S. car makers has established joint ventures, or bought or merged, with former competitors with the intent of increasing their presence and sales in new markets. It is not uncommon for their engineers to be involved in the planning and design of new products or facilities that are being developed for international customers. For these individuals, interactions with foreign colleagues is often a common occurrence.

Just as the U.S. auto makers have positioned themselves strategically in important world markets, so has the foreign competition. Corporations such as Honda, Toyota, Nissan, Mazda, BMW, Mercedes Benz and Volkswagen have all developed design, testing and manufacturing facilities in the United States and other key locales. This allows them to be in closer contact with their foreign customer base, community and workforce. While many of their employees have been transplanted from their home country, each company has implemented hiring processes to employ more American workers in the U.S., including engineers. It is now common for these firms to recruit full-time employees as well as co-op and internship candidates at many U.S. colleges.

Concurrent with this movement to locate facilities in foreign countries has been the increase in the development and expansion of firms which specialize in supplying products and parts used by the auto makers. When some foreign car makers began to establish their U.S.-based facilities, many of their foreign-based suppliers did the same. Now it is common to see many international suppliers located in close proximity to international automobile corporations. Firms that specialize in such things as electronics, engine technology, interior and exterior car products and manufacturing equipment are expanding their American presence. Many that emerged solely as suppliers to foreign auto makers have now found a strong customer base among the traditional Detroit Big Three as well. The auto supply business has become a competitive, lucrative market as all of the auto producers seek to keep costs down and maximize efficiency and profits.

As a result of this massive globalization of the automobile industry, many engineering graduates are finding rewarding opportunities with U.S.-based foreign automotive divisions and the rapidly growing base of suppliers, in addition to the Detroit Big Three. This industry has clearly emerged as one that provides many great opportunities for today's globally-minded engineer.

Manufacturing

The philosophy associated with producing the world's goods and services has changed dramatically in recent years. As multi-national corporations struggle to minimize costs while maximizing efficiency, there has been a significant increase in the location of manufacturing facilities and jobs in foreign lands. Many firms have found it more efficient and less costly to distribute products from a facility that is in close proximity to their customer base. Others have found it more cost-effective to produce their products in places where labor is cheaper, quality higher, and government policies more favorable for doing business. Therefore, it is no longer unusual in the realm of global production for a firm to have dozens of production facilities in many different countries.

Typically, much of today's research, design and development is still being done in central locations, though actual production may be occurring in many different parts of the world. Engineers may find that their work will involve communication and input from colleagues around the globe. In addition to international telephone and e-mail communication, production problems, quality control issues and meaningful design changes may necessitate frequent international travel for some engineers.

The typical foreign-based manufacturing facility is usually staffed by foreign national employees. Almost all of the production staff and skilled and unskilled labor are hired from the particular region. The corporate staff is usually involved in the initial start-up of the facility, the hiring and training of the employees, and the monitoring of the manufacturing process. Often, it is most efficient for a corporation to use local engineering talent, so many firms will recruit international students attending U.S. colleges and universities to fill these roles. The overall objective is to provide the technical expertise to the local labor force so they will be in a position to run the operation. The number of actual U.S. employees permanently assigned to a particular facility at any one time will probably be small. Their role will be to manage the operation, deal with immediate issues that arise, and serve as the liaison between the home office and the production facility.

A similar situation exists for foreign firms that have established manufacturing facilities in the United States. Most of the on-site corporate staff will be from the home country, but the objective is to try to hire and train as many U.S. employees as possible. Foreign firms will invest time and money to train these individuals to adapt to their particular style of doing business. For many employees at these facilities, it may be necessary to gain not only a technical knowledge of the engineering process, but also a proper cultural perspective. Certain methods or procedures may actually arise from a particular cultural background, belief system or age-old methodology. This can be quite an adjustment for U.S. workers hired by these firms. Often, newly-trained technical employees may be recruited from American colleges and universities to take advantage of their solid technical education and a broader understanding of cultural issues.

Many of the ideas and concepts which have revolutionized the manufacturing industry have been a result of the globalization of world production processes. Engineers and managers are adopting some of the best manufacturing technologies which have been developed in many different countries. Today's world of manufacturing is often a blend of the best practices and procedures gained from the integration of systems and concepts from engineers around the world. Manufacturing is truly a global enterprise in the life of today's engineer.

Construction Industry

The construction of roads, bridges, dams, airports, power plants, tunnels, buildings, factories, even whole cities, has created a worldwide boom for engineers involved in construction. While many U.S. firms have a long history of international construction activities, the many economic, political, and social changes around the globe have created a broad range of new opportunities for them and their engineers. Over the last 10 years, U.S. construction firms have recorded a significant increase in new foreign contracts. It is not uncommon for large construction firms to have corporate and field offices in several regions of the world. In fact, Bechtel, one of the largest global construction firms, lists on its website that it has 40 offices and 40,000 employees around the world.

The rapid growth of worldwide construction activity has been enhanced by the expansion of global economies, a loosening of trade restrictions due to NAFTA, increased competition from the European Union, and a desire by many countries to improve productivity and living conditions after years of decay and neglect of their major facilities. Many of these countries look to U.S. engineering firms for assistance with these projects, since they have demonstrated the necessary technical and managerial expertise for large-scale construction jobs. One of the most impressive projects was the construction of the city of Jubail in Saudi Arabia by Bechtel in the 1970s. Recognized by the *Guinness Book of World Records* as the largest construction project in history, this project involved construction of 19 industrial plants, almost 300 miles of roads, over 200,000 telephone lines, an airport, nine hospitals, a zoo and an aquarium.

However, as the world's economic power base is changing, these U.S. construction firms find that they are now in direct competition for new building projects with firms from such countries as Germany and Japan. In fact, many building projects are actually joint ventures between U.S. firms and those of other places. A prime example was the construction of the "Chunnel," the tunnel constructed to link Great Britain and France, that was actually developed by a team of construction companies including several from Europe and the United States.

Many republics of the Former Soviet Union and many Eastern European countries are relying on global engineering construction expertise. In addition, the rebuilding of Iraq will involve a worldwide construction effort. The end of the Cold War and the fall of the Berlin Wall have revealed massive needs for infrastructure improvement and industrial revitalization. The updating of transportation systems, restoration of buildings, renovation of power plant facilities (including nuclear-powered ones), and the conversion of industrial facilities from defense-oriented products to peacetime goods and services will require huge construction assistance. However, political instability coupled with uncertain economies may impede the actual rebuilding process.

As the world population grows, the need for the construction of new facilities to satisfy the demands of global customers should remain strong. Engineers should have ample opportunities to compete for many exciting global projects that will require a unique understanding of foreign cultures and world regions, and the ability to work as a team with others from across the globe—in addition to strong technical skills. However, competition will be strong from many new and existing construction firms around the world. Those who are prepared to compete in this environment will likely be successful.

Pharmaceutical Industry

Over time, the American pharmaceutical industry has been one of the most profitable and fastest growing segments of the U.S. economy. The nation's pharmaceutical companies are generally recognized as global leaders in the discovery and development of new medicines and products. In addition, the U.S. is the world's largest single market, accounting for an estimated two-thirds of world pharmaceutical sales. For engineering students, the pharmaceutical industry can provide some interesting international opportunities in research, development, testing and manufacturing of equipment, in construction of production facilities, and in technical sales and management around the globe.

However, this field currently faces many challenges. The growing concerns and intensifying pressure from governments and the private sector to control costs have forced many manufacturers to restrict operations. In reaction to escalating drug prices, which grew at nearly three times the overall rate of inflation during the past ten years, government and private managed care providers have instituted new policies to hold down drug costs.

For those considering a career in this industry, there are many favorable conditions that should create interesting international opportunities. However, increasing competition from firms around the globe has provided new challenges as well. While some companies have shown only modest growth, the highly successful firms have been those whose research and development efforts have generated lucrative, innovative treatment. Long-term growth for the industry is supported by a number of favorable conditions: the recession-resistant nature of the business, the aging population, and ambitious research and development efforts aimed at generating new and improved drugs (especially for the treatment of chronic, long-term conditions).

Responding to an increasingly competitive global pharmaceutical market and a more restrictive pricing environment, drug companies have tried to solidify their operations through advanced product development as well as a variety of mergers, acquisitions and alliances with international firms. Acquisitions are often viewed as the most efficient means of obtaining desirable products, opening up new market areas, and acquiring manufacturing facilities in various regions of the world. Acquiring existing products and facilities with established consumer bases is usually less expensive than developing a product from scratch. Additionally, competition in many foreign regions is not as regulated as in the United States. In the U.S., the Food and Drug Administration regulates the pharmaceutical industry. To gain commercial approval for new drugs, modified dosages and new delivery forms, a manufacturer must show proof that the medical substance is safe and effective for human consumption. Many of the competing international firms are not so tightly regulated.

In spite of the aforementioned concerns, the pharmaceutical industry has clearly become a global industry with an abundance of exciting opportunities for engineers in a variety of countries.

Food Industry

One of the basic maxims of human existence is that we all have to eat to survive. Probably few engineering-related industries are as essential as the food industry. From the earliest times, man has had to design and develop methods to plant, grow, harvest, and process food for consumption. As machinery and technology evolved, the process of food production and distribution grew more efficient. Today's food industry is involved not only with providing better products but also with developing new products which have greater shelf life.

This industry provides many global opportunities for engineers since many of the largest food corporations are involved in worldwide product distribution. Pizza Hut in Russia, Coca-Cola in China, American breakfast cereals in Africa, and M&Ms in Europe provide some popular examples of the expanding global food industry.

General Mills, one of the largest consumer food companies in the United States, has been active in developing new global products and entering joint ventures which have earned it worldwide revenues totaling $13.7 billion in 2008. Its employees work in plants and offices throughout the U.S., as well as Canada, Mexico, Central and South America, Europe and Asia. International exports represented a significant portion of General Mills' sales from its very beginning back in 1928. A more aggressive approach to the international marketplace began in the 1990s with the creation of four new strategic alliances: Cereal Partners Worldwide, Snack Ventures Europe, International Dessert Partners and Tong Want—a joint venture in China with Want Want Holdings Ltd. These joint ventures are building a strong foundation for General Mills' future growth worldwide. In recent years, General Mills has acquired companies such as Pillsbury, Haagen Dazs, and Old El Paso Mexican Foods, which continue to enhance its global presence. Sales in their international businesses grew 21% in 2008, to reach $2.6 billion.

Engineers from virtually every field of study are in demand in this growing and dynamic global industry. Technical expertise is needed for all stages of worldwide food production and distribution. Since many of the raw materials are grown throughout the world, engineering knowledge is needed to ensure that the food being grown meets the highest quality specifications. Scientists and engineers research and develop new products, while others are involved in the design of production and processing systems. Engineers also are used in the layout of production plants and facilities, and are often involved in the start-up, training and management of the local work force. Ensuring consistently high quality throughout the processing system is another essential function for engineers. Distribution in a timely and efficient manner is critical to the success of a corporation, and engineers can help ensure that the products are reaching customers safely and quickly.

Engineers entering this industry will find many interesting and challenging opportunities. Food has truly become a global commodity and the U.S. is an important contributor to the feeding of the world. Students with interest in virtually any area of this industry should be able to find ample opportunity to apply their engineering talents throughout the world.

Petroleum Industry

While many of the industries discussed in this chapter are relatively new to the global marketplace, the oil industry has long been involved on a worldwide scale. Since the vast majority of the world's oil and gas reserves are located outside the United States, the world's largest consumer, the petroleum industry has been a global operation out of necessity for many decades. However, the increasing demand and price structures of recent years have made an engineering career in this field quite challenging and demanding.

Since oil supplies in certain areas have become depleted, there is a strong demand for individuals interested in exploration of new sites. Oil companies are in need of engineers for the design and construction of new facilities, production, processing and storage. Their expertise is used primarily to make a facility operational, but the customary objective is to train and develop a local work force to run the plant once it is fully functional. Once the facility is fully operational, the engineers involved at the site function primarily as managers and troubleshooters. While new designs and operational concepts continue to be developed, they generally are being handled at U.S.-based R&D centers, with the technology adapted to the facility as needed.

With the strong emphasis on international operations, engineers involved in this industry must have an appreciation for cultural differences and must be able to work effectively with people of different backgrounds. As worldwide demand and increasing price fluctuations become more acute, it will become increasingly more important that individuals in this field possess strong managerial skills to help their companies succeed in a very competitive market. While petroleum production will continue to be critical, the exploration and development of new sites will become ever more essential.

Chemical Processing Industry

Throughout the world, thousands of products have been created by the chemical processing industry. Industries across the globe rely on a vast array of chemicals, fibers, films, finishes, petroleum, plastics, pharmaceuticals, biotechnology and composite material products developed and produced by this industry. While these firms often are not involved in the actual production of the end product, they are responsible for the key ingredients that are used by a wide range of other industries. A review of the literature from some of the leading chemical process manufacturers shows their products being used worldwide in such industries as aerospace, agriculture, automotive, chemical, computer, construction, electronics, energy, environment, food, packaging, pharmaceutical, printing and publishing, pulp and paper, textile, and transportation fields, among others.

Since these materials are used by such a diverse group of industries, it is important for the manufacturing facilities to be located in close proximity to the customers incorporating them into their products. Therefore, it is typical for these chemical processing corporations to have facilities in many regions of the world. For example, according to one of its reports, Dow Chemical Company provides chemicals, plastics, energy, agricultural products, consumer goods and environmental services to customers in 160 countries around the world. The company operates 114 manufacturing sites in 33 countries and employs more than 40,000 people who are dedicated to applying chemistry to benefit customers, employees, shareholders and society.

Dupont is another worldwide chemical company with 55,000 employees operating in more than 70 countries. The corporation is involved in a wide range of product markets including agriculture, nutrition, electronics, communications, safety and protection, home and construction, transportation and apparel.

DuPont expanded global operations significantly in the 1990s, with new plants in Spain, Singapore, Korea, Taiwan, and China, and a major technical service center opening in Japan. In 1994, a Conoco joint venture began producing oil from the Ardalin Field in the Russian Arctic—the first major oil field brought into production by a Russian/Western partnership since dissolution of the Soviet Union.

BASF is the world's largest chemical processing company with 95,000 employees on five continents. Core businesses are in chemicals, plastics, agricultural products and oil and gas. Even though Europe is their home market they have significantly increased their presence with NAFTA countries, Asia and South America.

As indicated in these examples, engineers who work for such companies must have a global perspective. Many of the pilot scale projects they work on could possibly be implemented as full production processes in several global locales. For the engineer, this may require active involvement with people from different cultures, using different languages and different methods of operation. Being successful in their endeavors will require an understanding of current global issues in addition to the technical skills needed to implement their product.

Computer and Electronics Industry

Probably no engineering-based industry has undergone more rapid, dynamic changes than the computer and electronics industry. New products are being developed, manufactured and distributed so rapidly that in many cases last year's product is quickly out of date. This is an industry where U.S. businesses must share world leadership in some product areas. Among the wide range of specialties included in this field are communications, computers, solid state materials, consumer and industrial electronics, robotics, power and energy, and biomedical applications.

Besides U.S. corporations, the world leaders in many of these technologies are companies in the Far East, notably Japan, Taiwan, and South Korea. Companies in the U.S. have been engaged in a competitive battle to maintain their position in product markets they once dominated. Engineers in this industry are often employed by a firm whose home base is outside the U.S. Those working for a foreign company at a U.S. site may face an adjustment in their method of operation. In addition to learning new applications of their technology, engineers in the global marketplace may need to learn much about foreign business culture. For many, this can be an exciting new experience.

While the U.S. no longer dominates many sectors of the electronics and computer industry, one area where the U.S. remains dominant is in the field of computer software development. U.S. computer specialists have the technical expertise that is in demand by firms throughout the world. Many software companies are now specializing in the development of products to serve clients in a multitude of foreign languages to help them in their business operations. It is not unusual for U.S. computer specialists to work on products exclusively geared for the foreign market.

Regardless of the home base of the computer or electronics manufacturer, much of the actual production of the products is now being done in Third World and developing countries. The intent is to keep costs down and profit margins up by making use of foreign labor and production. This requires engineering involvement not only in product development, but often in facility planning, plant start-up, and hiring and training of a local work force as well. For those interested in an evolving field with a variety of global applications, a career in the dynamic, challenging computer and electronics industry may be ideal.

Telecommunications Industry

Engineers have revolutionized the telecommunications industry, and as a result, have dramatically changed the world of engineering. The world around the U.S. has grown smaller, and the interaction of people worldwide has become faster and easier due to the rapid developments in this industry. Engineering developments in cellular and wireless telephone communication, voice and data transmission, video communication, satellite technology, and electronic mail have all had tremendous global impact. Engineers from around the globe now interact daily on routine matters such as program development, product design, plant facilities, operations, distribution and quality control issues. The developments in this industry have made it possible to bring fast, efficient, reliable and affordable communication and information to new markets of the world that had previously been considered unreachable.

Some common international engineering employment opportunities in this field include designing, manufacturing and distributing products such as integrated semiconductors, networks and networked computers, wireless telephone products and networks, and advanced electronic and satellite communications. Other telecommunication products include two-way voice and radio products for global applications, on-site to wide-area communication systems, messaging products (pagers and paging systems), handwriting-recognition products, image communications products, and Internet software products for worldwide distribution.

Alcatel-Lucent Technologies is an example of one major organization involved in the telecommunications field. They design, build, and service optical and wireless networks, broadband access and voice enhancement equipment to assist the world's communication service providers in building future networks combining voice, data, and video. Much of the world's telephone calls and internet sessions flow through their equipment. They employ about 31,000 employees who work with service providers on five continents.

Verizon Communications was formed by the merger of Bell Atlantic and GTE and is now one of the world's largest providers of wireline and wireless communication services for 80 million customers. Verizon had over 225,000 employees and over $94 billion in revenues in 2007. Their global operations are present in Europe, Asia, the Pacific Rim, and the Americas. Primary growth areas are projected in wireless and data transmission.

With the fast pace of telecommunications development and the increasing worldwide demand, there should be ample opportunities for global engineering in traditional markets as well as in emerging markets which have traditionally presented significant communications challenges.

Environmental Industry

One field where the technical expertise of U.S. engineers remains dominant is in the environmental arena. Given the excellent environmental training programs in the U.S., coupled with strict domestic environmental regulations, the American environmental industry is far ahead of most every other region of the world. There are many new and exciting global challenges waiting for the engineer with interest in this field.

Since many countries of the world lag far behind the U.S. in environmental policy, serious problems are surfacing at a rapid rate. Many countries are now looking to U.S. environmental firms to assist them with a wide range of projects. Many regions are in dire need of assistance with water purification systems, sanitation facilities, air pollution remediation, waste management, hazardous and toxic waste clean-up, pest control, landfill and recycling issues, and issues related to nuclear safety. In some areas of the world, the population is increasing at such a rapid rate that these problems are accelerating at a pace faster than the pace of technological development. In other areas, the funds, facilities and resources required to meet environmental challenges are not available.

Engineers who work in this field will encounter many global challenges. One industrial employer is Walsh Environmental, an environmental consulting firm providing services in the U.S. and overseas. Its engineers work on such global projects as site assessment; remediation of soil, water, and air; environmental impact assessments; underground storage tank removal; and industrial hygiene, health and safety issues. It is also involved in projects with energy and mining companies dealing with exploration, production and power generation in remote locations around the world. It has particular expertise in the rain forest environments of South America and Southeast Asia, as well as with bioremediation and the remediation of chlorinated solvents. It has project locations in such diverse areas as South America, Japan, China, Australia, Canada, Botswana, Bangladesh, Bahrain, Gabon, South Africa, Indonesia, Ireland, New Guinea, Pakistan and the Philippines.

AGRA Earth & Environmental Limited is an international network of engineering and environmental professionals now part of their parent company AMEC. Tapping the expertise of its engineers, AGRA has undertaken several large industrial construction projects throughout the world. These projects include coastal structures and power generation and distribution projects throughout North America, South America, Africa and Asia. The global experience of AGRA's engineers enables them to effectively understand the needs of clients and to develop innovative solutions to the engineering and environmental challenges.

For the environmental engineer, technical comprehension of environmental problems is not enough. This individual must also have the cultural experience and sensitivity, and motivation to work with difficult challenges in many different locales.

Consulting

Countries, governments, and firms around the globe need the expertise of American engineers. Thus there is a huge global demand for consultants with a wide range of technical backgrounds. Over time, consulting firms have evolved as a source of technical assistance for entities that do not have the time, resources or capability to address specific engineering issues from within their organization. Some of the large international consulting firms maintain a staff of engineers from diverse backgrounds which is technically and culturally prepared to assist global clients with a wide range of engineering problems. A consulting firm may be hired for a very short period of time (weeks or months), or may be involved with a project for several years, to completion.

One example of a large international consulting firm is Accenture. To its clients Accenture offers a combination of services which include strategic technical planning, business management and engineering consulting. Its mission statement emphasizes the global commitment: "Accenture is a leading global management and technology consulting firm whose mission is to help its clients change to be more successful." The firm works with thousands of client organizations worldwide. Its clients represent a wide range of industries including automotive and industrial equipment, chemicals, communications, computers, electronics, energy, financial services, food and packaged goods, media and entertainment, natural resources, pharmaceuticals and medical products, transportation and travel services, and utilities. To maintain competence in so many areas requires over 187,000 people and 200 offices in 52 countries. Its business in 2008 grew at an average annual rate of 19%. In this challenging field Accenture must have the ability to tap global resources and quickly deploy them to address customer needs. International teams are often employed to serve a rapidly growing number of global clients. Accenture's consultants often have the opportunity to work in more than one of the world's major markets during the course of their careers. Additionally, they encounter many opportunities to exchange knowledge and experiences with colleagues around the world, from diverse backgrounds, competencies and industry expertise.

For engineering students interested in a wide variety of experiences involving a broad range of global industries, working in the consulting industry can provide the unique and interesting careers they seek.

Technical Sales

Technical sales is an important function for all engineering firms, both domestic and abroad. All firms need engineers who can competently present their products or services to potential clients. These specialists articulate the technical capabilities of the product and demonstrate how it can meet the needs of a prospective customer. In addition, skilled technical sales people are trained to assist with the installation and start-up of the product, as well as troubleshooting on-site problems. In some cases, technical sales engineers work concurrently with their clients and their home-office design team to develop or modify existing products for the customer. Once the product has been implemented, the sales engineer also may be responsible for field testing and operator training. Another important responsibility is maintaining and ensuring customer satisfaction throughout the entire process, from initial purchase to installation to product implementation, and perhaps even to product phase-out.

On a global scale, technical sales takes on an even more critical role. The rapid increase in the demand for global technology translates into a critical need for trained technical people to handle international sales, installation, field engineering, and customer liaison roles with clients across the globe. These engineers must not be only technically skilled, but also must have appropriate training in language, culture, customs, and sensitivity to effectively interact with foreign customers. They must appropriately represent themselves, their company, the product, the culture, and most significantly, then must be able to demonstrate that their firm has the motivation and talent to adequately serve the global customer in a manner consistent with the cultural and technical needs of that customer.

5.4 PREPARING FOR A GLOBAL CAREER

Hopefully, the variety of exciting global opportunities available to engineers which are described in this chapter will get you thinking about your particular interests and motivation. While some engineers may drift unintentionally into an international career, most careers are the result of careful planning, experience, and preparation. There are several things that engineering students can do during their college years to better position themselves for a global engineering career. The remainder of this chapter will explore some of these factors.

Language and Cultural Proficiency

As stated earlier in this chapter, many engineering students have a difficult time seeing the value of studying a foreign language and culture. Hopefully, some of the material presented in this chapter will give you a different perspective. Even if you are studying engineering and have no specific plans for an international career, it would still be wise to anticipate that the global economy is likely, at some point, to impact your work as an engineer.

So the question remains: Are foreign language proficiency and cultural study absolutely necessary to communicate in the global workforce? The answer is, "probably not." English is commonly used as the international language among companies and project teams around the globe. Many countries now teach English as a second language to better prepare their students and workers for communication in the global workplace. It is not uncommon for U.S. engineers to deal with foreign colleagues who are fluent in several languages. However, despite its growing worldwide usage, an engineer cannot assume that English will be understood everywhere.

Therefore, it is important that engineering students take advantage of any language and cultural training available to them. Most students have had some foreign language education in high school, and this provides a good foundation for further study. Being able to speak foreign associates' language and knowing about their region's culture demonstrates a sensitivity and willingness to work together in the global workplace as an equal partner.

Unfortunately, the structure of most engineering curricula does not provide much room for the study of foreign language and culture. Most students find that they can satisfy their humanities requirement with a liberal arts course, but foreign language study generally only provides elective credit. Therefore, students who wish to add language proficiency to their education must do so by adding to their course load, or by learning the language outside their college studies.

Some schools and companies offer intensive language training designed to develop foreign language proficiency in a relatively short period of time. Other programs concentrate on teaching basic terminology, phrases and concepts that prove useful for getting around in a foreign country. Some companies also provide training programs which emphasize cultural protocol to assist those working with foreign colleagues.

While in college, there are many things you can do on your own to become more culturally sensitive. Getting to know professors and classmates from different countries, and taking part in any special cultural activity on campus, can be enlightening. Attending meetings of various international clubs in your community can also be of benefit. Try to experience anything "international" that you come across: seminars, lectures, and other events. On most campuses, you will find that you don't have to travel very far to broaden your international understanding. While knowledge of a foreign language and culture may not be required for success as a global engineer, individuals who make the effort to develop some competency in this area may find more opportunities available to them.

Study Abroad and Exchange Programs

One of the very best ways to prepare for a global engineering career is to participate in an international study program. While not all schools offer a broad selection of such opportunities, most offer some programs for their students. If the opportunities at your school are limited, you may be able to participate in such a program with a nearby college or university as a "guest" or "temporary" student. Usually you can transfer the credits you earn back to your own institution. Some schools offer exchange programs which allow a student from one institution to exchange places with a foreign student from a partner school under an established agreement.

A **foreign study** experience can provide many benefits to you. It is an opportunity to develop and expand your problem-solving skills in a different environment with new perspectives and challenges. And an international study experience can greatly facilitate the learning and application of a foreign language while enhancing your geographical and historical knowledge. You will find yourself exposed to individuals and groups who may process information differently than you. An overseas study program provides you with new professional contacts, and can give you direction with your career. It can also give you a new level of confidence with a strengthened sense of personal identity, flexibility, and creativity.

There usually are a variety of foreign study options available to students. The full-year program provides students the opportunity to study at a foreign university for the equivalent of two semesters. Typically, students are enrolled in regular courses (pre-approved as equivalent to those required in their program) that are taught and graded by faculty from the host institution. The credits are then transferred back and applied to the students' engineering program requirements. The semester-length program is structured similarly to the full-year program except that this program is limited to the equivalent of one semester. The intent of both programs is to immerse the student in a foreign setting, having them live with nationals instead of Americans, and to teach them to incorporate their knowledge of the language and culture of the host country into their daily activities. This is the best way for students to develop language and cultural skills that can be applied to their future career.

In a variation of the above model, the instruction and many of the on-site activities are under the direct supervision of faculty from the students' home institution. Typically, the U.S. institution offers an academic program, usually a semester or less in length, with regular university courses, taught by its own faculty, in rented or leased facilities that are exclusively designated for this program. The advantage of this approach is that students of varying levels of language and cultural preparation can be accommodated, and all courses are treated as though they are offered by the "home institution," thereby eliminating any potential problems with course transfer or application to degree requirements.

The **exchange program model** is founded on special agreements between two institutions in which they agree to exchange equal numbers of students for semester-length programs. Common course, credit and transfer issues are generally resolved in advance. Typically, most of these programs are designed so that students from each country pay standard tuition and fees at their home institution, with those funds then set aside to cover the costs of hosting students from the other school.

The **short-term specialty program** provides an opportunity for students to visit and see, firsthand, specialized facilities, operations, or research projects in a foreign location. Typically, these programs are for a short period of time, and the program is often incorporated as part of an existing class, almost like a special field trip. The focus of these programs is to expose students to a specialty unavailable to them through normal on-campus teaching. While this program provides little opportunity for students to experience and reflect on the local language and culture, it does provide a means of experiencing a foreign study program on a limited basis.

The **study tour model** provides an academic experience that is usually oriented more to "tour" than "study." Generally, these short-term programs are designed to help students learn about global technical areas and topics through non-traditional forms of instruction at an international location. This may include taking field trips to several international production facilities, writing journals in which the various global engineering design and development strategies the students observe are recorded, or making a series of visits to research facilities and reporting back to their classes. The intent of this program is to provide students with a short, but educational, focus on different forms of global engineering. Unfortunately, this type of opportunity is limited in its ability to incorporate the cultural and language experience into the program model.

International Work Experience: Co-ops and Internships

Another excellent way for you to prepare for a global engineering career is by obtaining an international cooperative education (co-op) or internship. This can be a very complicated task, but for students motivated enough to make the effort, it could be very beneficial. Typically, these positions emphasize the integration of responsibilities that are directly related to your particular field of study. A co-op is a paid position, monitored by your college to ensure that the experience is meeting certain established learning objectives. On the other hand, internships may be paid and may or may not be monitored by your school. The awarding of credit varies from school to school.

The benefits of obtaining an international work experience include the following:

- **Valuable work experience.** By participating in an international work experience, you have the opportunity to experience firsthand the opportunities, challenges and frustrations of working in a different cultural environment. The total experience is bound to make you more aware of, and sensitive to, the critical issues of the global workplace.
- **Guaranteed immersion in a new cultural setting.** By participating in a well-planned international program, you should enjoy a comprehensive cultural learning experience. Interacting with others in an international work setting can help you determine if the factors that initially attracted you to a certain culture and language in a particular area of the world are truly of interest to you. These experiences can help you focus your career planning.
- **Greater foreign language competency.** There is no question that a person's foreign language skills improve significantly when used regularly in the society in which it is the native language.

- **The challenges of living and learning in an unfamiliar work environment.** Many students go through a period of culture shock, which can impact their work performance. They must deal with differences in work ethics, performance expectations, and culturally-dictated ways of doing business. These can have a significant impact on a student's global career planning.
- **Developing an international network.** International work experience can provide you with a variety of new friends, professional contacts and mentors. These individuals can serve as valuable resources to you as you pursue your career.

Special Considerations

While we have discussed the positive aspects of working in an international setting, there are other issues which also should be considered.

- **Language and cultural skills.** If you cannot demonstrate language proficiency and a cultural sensitivity, many foreign organizations will be hesitant to hire you. You must be able to demonstrate that you will be an asset, not an impediment, to the daily operations of the firm.
- **It takes time and effort.** Obtaining an international work opportunity can be a time-consuming process which usually is more complicated than landing one in the U.S. A great deal of effort may be expended writing letters of inquiry; customizing your resume (perhaps in a different language); and completing application materials, visa forms, travel arrangements and other documents.
- **Confusion over job responsibilities.** It is not uncommon for students to have misconceptions about the exact nature of their work responsibilities. Many have high expectations for challenging work and a significant level of responsibility. They are disappointed when the actual assignments fail to meet their expectations.
- **International work experiences can be expensive.** There are many additional costs associated with an international work assignment that may not be readily apparent. Expenses for airfare, local and regional travel, housing, and possible administrative fees are often overlooked when planning a budget. While most co-ops and a few internships are paid positions, many others are not. Generally, any salary that is provided is scaled down to reflect local wages and the economic pay rates of the region. For most students, this will probably be much less than they expect.
- **Housing issues.** Many students find that international housing can be difficult to find, expensive, and generally not what they are accustomed to in the U.S. Students should try to work with their foreign employer to secure decent and affordable housing, if possible.

*Most students feel that the positive benefits of an international work experience far outweigh the challenges and effort. The technical experience gained, coupled with the opportunity to develop language and cultural skills, provide precisely the type of background that employers seek when selecting candidates to fill global positions.

To obtain more information about an international work experience, talk with your campus placement office, your co-op or internship coordinator, and your academic advisor. Some excellent organizations involved in providing technical students with international work assignments and general information include the International Association for the Exchange of Students for Technical Experience (IAESTE) <http://www.aipt.org/programs/iaeste/iaeste-index.htm>, the Council on International Exchange (CIEE) <http://www.ciee.org>, and the NAFSA Association for International Education <http://www.nafsa.org>.

Choosing the Right Employer or Opportunity

Most engineers fresh out of college are not likely to find a long line of employers at the campus placement office waiting to hire them just because of their interest in pursuing a global engineering career. You will probably have to thoroughly research many companies, industries, and government agencies to find the right situation for you. You must identify organizations that have a global commitment that is consistent with your background, skills, experiences and interests.

Once you have identified employers involved in the global marketplace, you should try to contact them directly through on-campus contacts or by correspondence. Since most companies do not immediately thrust new graduates into the global workforce, you must not limit your job campaign to the narrow objective of obtaining an international career. Rather, the international interests, experiences, and competencies that you have developed over time should be viewed as valuable commodities which are supplemental to your core technical skills and abilities. The goal of your initial job campaign should be to obtain a position with an organization that will be able to provide international experiences once you have been trained and are experienced in the company's culture. It is important to realize that this process takes time. It may take several years before a global opportunity emerges.

In the meantime, there are several things you can do to position yourself for a global opportunity. For example, keeping co-workers and supervisors informed of your interest of being involved in an international project could pay off in the future. It is also important to maintain your technical capabilities as well as your language and cultural skills. One never knows when an opportunity may develop. Building relationships with those who can help you achieve your objectives can be very helpful. It is important to establish a solid work reputation as one who can be trusted and is prepared for the responsibilities of a global assignment. You must continually demonstrate that you are a team player who can work well with others, that you are sensitive to cultural differences, and that you have strong communication skills. It is important to remember that sometimes the best opportunities emerge for those who demonstrate flexibility. Your goal may be to work on a project in the Far East, but if an opportunity comes up in Europe it may be advantageous to accept this opportunity as a way of gaining additional international experience that may eventually lead to an assignment in a preferred location.

The world of global engineering is here to stay. The work of engineers is reaching into all regions of the world. Companies have a need for those who are ready to accept the challenges of the global marketplace. This world economy will provide excellent opportunities for those who show the interest, motivation, capability and leadership to accept these challenges.

EXERCISES AND ACTIVITIES

5.1   Research some current articles about NAFTA at the library. Write an essay in which some of the current pros and cons of this Agreement are discussed.

5.2   Research some current articles about the evolving European Union. In an essay, discuss whether or not you feel that the EU has been a successful venture.

5.3   Find some current information about the global economy (stock markets, status of world currencies, general economic conditions, etc.). What is the state of the current world economy? Which regions of the world are doing well and which are struggling? What industries are prospering and which are having difficulty?

5.4   Visit your campus career planning or placement center and read the recruiting brochures of some employers in which you have an interest. Make a list of those who offer international opportunities in your field of interest. Write an essay describing which industry is of most interest to you in your career, and why.

5.5   Develop a list of your present international skills and abilities. Suggest how you can strengthen these skills in the next few years.

5.6   Attend a foreign study orientation session, if available, or visit your campus international center, if one is available, to find out more about some of the opportunities that are available. List some of the advantages and disadvantages of these opportunities as they relate to your personal priorities.

5.7   Attend your campus career/job fair. Talk with at least three recruiters from different firms. Write an essay comparing their views concerning the opportunities for international work experiences such as co-ops and internships.

5.8   Visit with some upper-class engineering students who have studied abroad. Write an essay that compares and contrasts their experiences. Which students do you feel gained the most from their experience, and why?

5.9   Meet with your academic advisor and develop a long-range course plan. Modify this plan to include an opportunity for international experience. Write a short essay discussing the advantages and disadvantages of including this experience in your academic program.

5.10 Browse the web, exploring the sites of employers that are of interest to you. Many will have copies of annual reports and other operations information available for review. Develop a list of those firms that are expanding global operations. What factors are cited for this expansion? Does there seem to be a pattern among organizations? Does it appear that any of these organizations are reducing their global operations? If so why?

5.11 Select a world region that is of interest to you. Using information from your foreign study office, career center, or web resources, research the study abroad and/or internship opportunities that are available through your school or another college in this area. Which programs appeal to you and why? Visit with your academic adviser, or professor, to discuss the feasibility of this opportunity in your academic program.

5.12 Foreign language study is generally not required in most engineering schools. Given the increasing expansion of global engineering, do you think the study of a foreign language should be a required part of the engineering curriculum? Why, or why not?

5.13 One of the current criteria used in accrediting engineering programs by ABET states that engineering graduates should have "the broad education necessary to understand the impact of engineering solutions in a global and societal context." What courses in your engineering program are designed to provide this exposure? If you are not sure, discuss this with your academic adviser or a professor in your program. Do you feel that the offered coursework will adequately prepare you to meet this criteria?

Chapter 6

Future Challenges

OVERVIEW

Throughout history, engineers have been problem-solvers at the forefront of change—change that has left its imprint on world history, politics, economics and global development. Now that we've entered a new millennium, it is important not only to reflect on the successful accomplishments of engineers, but also to focus on the exciting challenges and opportunities which lie ahead. Today's engineers are confronted with a world in which technology is exploding at a breakneck pace, and technical expertise is needed to solve many global issues. Some believe that the next generation of engineers will be responsible for more developments during their lifetime than all the technological advancements since the construction of the Pyramids. There are many issues and challenges that could be addressed, but we will touch on just a few in this chapter.

6.1 EXPANDING WORLD POPULATION

As can be observed in Table 6.1 and Figure 6.1, it took from the beginning of time until 1804 to reach a world population of one billion people. World population statistics show that there were approximately 1.6 billion people in 1900 and this figure reached 6 billion before the end of 1999—one century to add 4.4 billion people! According to the United Nations Population Division, this number could climb to almost 9 billion people by 2050. The concern for engineers, as well as others, is that this rapid increase in population, coupled with fast-paced industrial growth, places a tremendous strain on land, air, and water resources that are essential for human survival.

Nearly all world population growth is occurring in developing countries. The 49 least developed countries will almost triple in population over the next 50 years.

World population is growing at a rate of 1.3 percent annually, or 77 million people per year. Half the growth is in 6 countries: India, China, Pakistan, Nigeria, Bangladesh, and Indonesia. Africa is projected to double their population to 1.9 billion by 2050. This massive growth cycle coupled with extreme poverty and environmental devastation has left millions of people without adequate food and water. In contrast, U.S. population had been projected to grow from 270 million, to 298 million in 2010, 335 million in 2025, and 422 million in 2050. The population should stabilize or decline in Japan, Europe, Russia, and many other countries of the established industrial world. Recent studies indicate, however, that the overall population growth rate has slowed from these earlier projections due to increased education and use of family planning, combined negatively with increased mortality rates due to the spread of AIDS. Despite these recent projections, the overall trend of rapid population growth remains, creating numerous challenges for engineers.

The increasing global population will have significant impact on land use, air and water resources, and food production and distribution processes. In many areas of the world, water is often used faster than it can be replaced, and land suited for food production is dwindling. Forests are being destroyed for fuel and to make way for housing. More and more industrial plants are being built, which contributes to the pollution of air and water. In rapidly growing cities, the demands for housing, food, water, energy, and waste disposal are taxing available resources at an alarming rate.

The challenge for engineers will require the use of all their talents in an effort to improve these situations. As the population grows, technological solutions will be needed to maintain a proper ecological balance. Many believe that this population challenge has implications for the other challenges outlined in this chapter.

TABLE 6.1 Number of years needed to add a billion people to the world's population

- 1st billion: from beginning of time to 1804
- 2nd billion: 123 years (reached in 1927)
- 3rd billion: 33 years (reached in 1960)
- 4th billion: 14 years (reached in 1974)
- 5th billion: 13 years (reached in 1987)
- 6th billion: 12 years (reached in 1999)
- 7th billion: projected 11 years (in 2010)
- 9th billion: projected by 2050

Source: United Nations Population Division, Department of Economic and Social Affairs

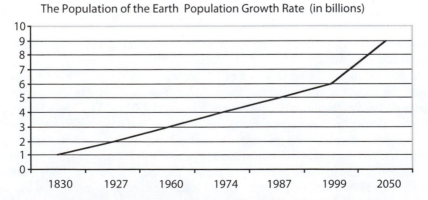

The Population of the Earth Population Growth Rate (in billions)

Figure 6.1 The Earth's population.

6.2 POLLUTION

Engineers must always be aware of the impact of their work on the environment. Sometimes small, seemingly harmless actions can have long-term, far-reaching effects. Two critical areas of pollution concern for engineers are air pollution and fresh water resources.

Air Pollution

Air pollution results when the atmosphere absorbs gases, solids or liquids. The sources of pollution can be either natural or manmade. When these contaminants enter the atmosphere, they can endanger the health of humans, animals, fish and plants. They can damage materials, create visibility problems and cause unpleasant odors.

The major cause of air pollution is burning gasoline, oil or coal in automobiles, factories, power plants and residences, without proper controls to remove the offending pollutants. Once the pollutants have entered the atmosphere, they can travel great distances and can cause problems such as acid rain in areas far from the original emissions site. Many nations of the world are burning coal and oil at record rates, which has increased atmospheric levels of carbon dioxide. Some scientists feel that this has contributed to a "greenhouse" effect in which the release of infrared radiation from the earth is inhibited, and which in theory could contribute to a possible global warming trend (which has yet to be confirmed). Some evidence seems to suggest that certain airborne pollutants are damaging the ozone layer, although this is still a scientific controversy.

While acid rain has long been recognized as a problem in the industrial countries, there is now evidence of the increasing danger of acid rain in South-East Asian nations. According to the United Nations System-Wide Earthwatch 2005 Report, the emissions of sulfur dioxide have declined significantly in Europe and North America with reduced coal use, and with the application of new emission clean-up techniques, further progress is expected. This has reduced the sulfur contribution to acid rain. However, the same level of improvement has not been achieved in nitrogen oxides and other pollutants from vehicles, where the reduction in emissions due to catalytic converters has been offset by the growing number of motor vehicles. Therefore, urban air quality in many areas has continued to deteriorate.

Some of the worst pollution problems actually appear in unexpected places, such as the Arctic, where high levels of toxins such as PCBs, DDT, mercury, and dioxin have been found. There appears to be a global process of distillation where pollutants evaporate in warmer areas, are transported by winds to the Arctic, and then condense to become concentrated in Arctic food chains.

The 2007 Energy Information Administration report "Greenhouse Gases, Climate Change, and the Economy", reports that world carbon dioxide emissions are expected to increase by 1.8% per year from 2004 levels through 2030. As indicated in Figure 6.2, total world carbon emissions are expected to reach 30 billion metric tons in 2010, 37 billion by 2020, and 43 billion by 2030. A significant portion of this increase is expected to occur in the developing countries where emerging economies, such as China and India, spur their economic development using fossil energies. Emissions from developing world countries are expected to increase greater than the world average by 2.6% per year from 2004 levels through 2030.

In the United States, several recent laws have been enacted which call for monitoring air quality and the ozone layer. The Clean Air Act, enforced by the Environmental Protection Agency, is designed to provide legally required monitoring of air quality levels. Additionally, many member countries of the United Nations have joined forces to monitor and phase out many ozone-destroying chemicals and materials. These are some beginning steps which must continue on a worldwide basis for many years. Engineers have the technical expertise to develop and implement the necessary controls and devices for the protection of delicate air resources, but must have the authority and financial support to successfully meet these challenges.

Water Pollution & Freshwater Resources

Water is a most precious resource, vital for the existence of life. Unfortunately, the pollution of lakes, streams, ponds, wells, and reservoirs has become a serious challenge for engineers. Water pollution is the contamination of a water source by any agent that negatively affects the quality of the water. The major sources of water pollution include industrial wastewater, chemicals, detergents, pesticides and weed killers, municipal sewage, acid rain, residue from metals, bacteria and viruses, and even excessive, artificially enriched plant growth.

When these wastes infiltrate a water supply, they upset the delicate balance in the ecosystem and can cause human and environmental health problems. Water quality legislation enacted in recent decades has improved conditions significantly. Modern water treatment plants and purification systems have been useful in improving the overall situation, but more comprehensive systems still are needed to remove the most elusive pollutants resulting from such matter as herbicides and pesticides.

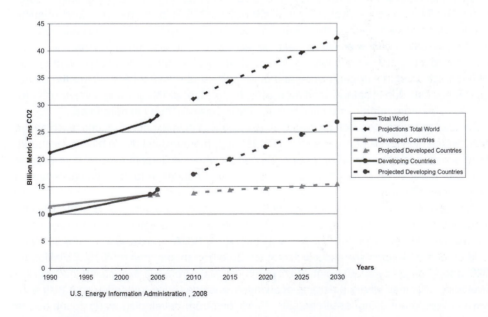

U.S. Energy Information Administration , 2008

Figure 6.2 World Carbon Dioxide Emissions

Freshwater is clearly becoming a major constraint on global development. According to the United Nations System-Wide Earthwatch 2005 Report, 40% of the world's population already faces chronic water shortages. One recent estimate suggests that more than half of available freshwater resources are already being used to meet human needs, and this could rise to 70% in 30 years. Water supplies could therefore run out in the next century if per capita consumption and excessive use in agriculture are not controlled. Another projection shows that between 1 and 2.4 billion people will live in water-scarce countries by 2050. The conflicts over sharing water in international river basins are increasing.

There are growing concerns about major regional water scarcities. Countries in North Africa and the Middle East already have 45 million people without adequate drinking water. Per capita water availability has shrunk by more than half in 30 years, and could be halved again in the next 30 years, requiring an investment of $50 billion. Three-fifths of Chinese cities are short of water and 80 million Chinese do not have adequate drinking water.

Water quality is gradually improving in most parts of Europe as a result of a massive investment program. However, there is still a problem with ground water and diffuse sources of pollution, particularly from agriculture.

The challenges for engineers of the future will be to develop more sophisticated treatment systems that can more thoroughly treat water pollution at its initial sources, while continuing to cleanse polluted water sites and facilities in order to provide fresh water resources to growing populations.

Solid Waste

The safe and efficient disposal of solid waste materials has become one of the most serious challenges facing engineers. Solid waste is usually defined as any trash, garbage, or refuse that is made of solid or semi-solid materials that are useless, outdated, unwanted, or hazardous. Generation of solid waste has increased significantly as the population has expanded and consumers have adopted a "use it and toss it" attitude.

Estimates indicate that in 1920, the U.S. was generating about 2.75 pounds of solid waste per person per day. By 1970, this figure had increased to about 5.5 pounds per person per day. By 1985, 15 years later, estimates indicate that this number had almost doubled again to about 10 pounds, per person, per day. As indicated in Fig. 6.3, in 1998 the U.S. was gen-

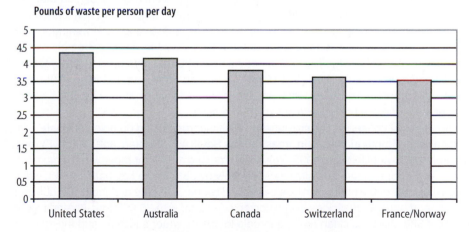

Figure 6.3 The top five trash producing countries.

erating over 1,600 pounds of trash per person per year. It is no wonder that some have called U.S. consumers the "throwaway society."

Generally, the establishment of solid waste management and disposal programs have been the responsibility of state and local governments. They typically develop policies and procedures concerning how waste will be managed, using such options as landfills, incineration, recycling and composting methods. Engineers are, and will continue to be, actively involved at all levels of governmental policy implementation as well as in roles as consultants and providers of the services needed by state and local agencies.

Until the mid-1980's, approximately 90% of the disposal of solid waste involved sanitary landfills—specially constructed facilities used to bury the nation's refuse. In a typical landfill, deep holes are dug and lined with special protective materials. The refuse is spread in thin, compacted layers and then covered by a layer of clean earth. Pollution of the surface water and groundwater is minimized by lining and sloping the fill, by compacting and planting the top cover layer, and by diverting drainage. Landfills are a fairly cost-effective method of solid waste disposal assuming an adequate supply of available land is in close proximity to the sources of the waste. Unfortunately, this has become a critical issue in recent years as land appropriate for use as landfills is disappearing.

Since 1984, 72% of U.S. landfills have closed because they became filled, or due to new restrictive disposal legislation. Many landfills have been augmented by new methodologies including "waste to energy" plants. In this process, refuse (which has about 50% of the energy content of coal) is burned in special incinerators and special thermal processes are used to recover the energy from the solid waste; the volume of the waste is reduced by 90%. This method now accounts for almost 20% of the solid waste management programs used across the country. However, in recent years, the concerns over increasing costs and environmental problems have slowed the use of these processes.

In the last decade, recycling and composting have become the fastest growing method of solid waste management, and now make up about 27% of waste management programs. An increasing number of local governments have implemented recycling systems for solid waste materials and have established composting programs for yard waste.

According to the United Nations System-Wide Earthwatch 2005 Report, there are several new areas of concern emerging as challenges related to solid waste issues. As military tensions change and disarmament agreements have been implemented, there has been a growing recognition of the enormous problem with the disposal of obsolete weapons, particularly nerve gas and chemical and nuclear weapons, which were never designed with safe disposal in mind. The combination of explosives and highly dangerous chemicals, often deteriorating and becoming increasingly unstable, makes dismantling such weapons, neutralizing their contents, and even transporting them to disposal facilities, extremely expensive and environmentally risky. The U.S. alone has over 30,000 tons of chemical weapons whose disposal could cost at least $12 billion. More than 50 ocean and inland lake sites across the U.S. contain explosive items, and harbors and beaches at the site of old battles throughout the world are riddled with unexploded bombs.

Another issue of concern is the cross-national transportation and dumping of waste. This problem has significantly increased in areas such as South Asia, Eastern and Central Europe, and the Former Soviet Union. In 1992, Poland intercepted 1,332 improper waste shipments from Western Europe alone, and similar cases grew by 35% in the first half of 1993, which illustrates the scale of the problem in less developed countries. There are about 100,000 tons of obsolete and unused pesticides in developing countries, with 20,000 tons in Africa alone that will cost $80 million to clean up. The threats to health, water supplies, and the environment from these and other dumped toxic chemicals are serious. Worldwide legislation was initiated in 1995 to ban waste exports among some participating countries, but it will take time, resources, and strong governmental commitment to achieve effective implementation.

Based on the U.N. Report, as the world economy grows so does its production of wastes. For example, U.S. production of hazardous and toxic waste rose from 9 million tons in 1970 to 238 million tons in 1995. Europe produces more than 2.5 billion tons of solid waste a year, and every day the inhabitants of New York throw away approximately 26,000 tons.

So, what possible solutions exist for the challenges of solid waste disposal? In recent years recycling has become a preferred choice of waste disposal. The British government set a target of recycling 25% of all household waste within the next few years. Likewise, the proportion of household rubbish recycled in Germany increased from 12% to 30% between 1992 and 1995. Around 75% of the average European car is already recycled, largely because the metal can be sold as scrap. However, electrical scrap accounts for merely 2% of waste produced in the European Union, and car scrap even less. While U.S. efforts in recycling have improved in recent years, as illustrated in Fig. 6.4, the U.S. lags behind many countries in recycling systems for wastepaper and cardboard.

Each method of waste disposal has drawbacks. Reusing glass bottles can require more energy than their initial manufacture as they have to be sterilized. Incineration is a source of greenhouse gases and toxic chemicals like dioxins and lead. Landfill sites are a possible source of toxic chemicals and produce large quantities of methane gas. They must be managed so that pollutants do not seep into groundwater and should therefore be kept dry, but this slows down the rate of decomposition.

Tires constitute another problem, as their resilience and indestructible nature become distinct disadvantages when it comes to disposal. Tires are virtually non-degradable and spread noxious fumes when burned. Western Europe, the US, and Japan produce around 580 million tires a year. Possible short-term solutions might include banning whole and shredded tires in any landfill. However, long-term acceptable disposal solutions continue to present significant challenges to engineers.

Further research needs to be carried out on the effects of various waste management options to determine the extent to which they positively benefit the environment. There are complex tradeoffs between costs, energy consumption, transportation, pollution, greenhouse gas production, and toxic by-products, which require the expertise of engineers.

The challenge for engineers will be to continue to develop processes for the recycling of parts and components, and solving all the intricate problems related to hazardous materials.

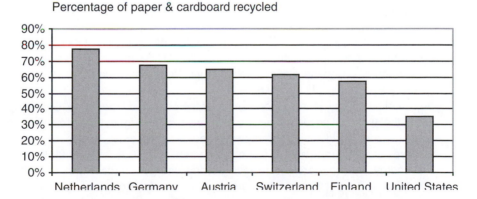

Percentage of paper & cardboard recycled

Figure 6.4 The six best countries for paper and cardboard recycling.

These "green engineers" will need to incorporate the types of methodologies that have been successful in the recycling of such things as aluminum and paper into future designs for products such as automobiles, computers, factory equipment and machinery.

6.3 ENERGY

The 2008 Annual Energy Outlook report prepared by the United States Energy Information Administration, predicts that world energy demand will continue to grow steadily, and liquid fuels will account for a substantial portion of world energy demand by 2025. (For these purposes, liquid fuels are defined as petroleum and petroleum derived fuels such as ethanol and biodiesel, coal to liquids and gas to liquids. Petroleum coke, which is a solid, is included as are natural gas liquids, crude oil consumed as fuel, and liquid hydrogen). There will be a shift in energy demand by world regions as many developing and Third World countries increasingly demand significantly greater energy resources.

Other specific highlights of the report include:

- World energy consumption is projected to increase by 50 percent from 2005 levels to 2030. Projections show that world energy consumption could increase to 695 quadrillion British thermal units (Btu) by 2030 (Fig. 6.5)
- Although high prices for oil and natural gas, which are expected to continue throughout the period, are likely to slow the demand for energy in the long term, world energy consumption is projected to increase steadily as a result of strong economic growth and expanding populations in the world's developing countries.
- China and India will be major contributors to world energy consumption in the future. Over the recent past decades, their energy consumption as a share of total world energy use has increased significantly. In 1980, China and India together accounted for less than 8 percent of the world's total energy consumption; in 2005 their share had grown to 18 percent. Even stronger growth is projected over the next 25 years, with their combined energy use more than doubling and their share increasing to one-quarter of world energy consumption in 2030. In contrast, the U.S. share of total world energy consumption is projected to contract from 22 percent in 2005 to about 17 percent in 2030.
- Energy consumption in other regions also is expected to grow strongly from 2005 to 2030, with increases of around 60 percent projected for the Middle East, Africa, and Central and South America. A smaller increase, about 36 percent, is expected for developing countries in Europe and Eurasia (including Russia and the other former Soviet Republics). (Fig. 6.6)
- As indicated in Fig. 6.7, the US Energy Information Administration reports that the U.S. relies heavily on foreign oil imports. In 2008, 62% of U.S. consumption was from international sources. Canada, Saudi Arabia, Mexico, Venezuela, and Nigeria were the top 5 countries providing oil to the U,S. in 2008.
- The use of all energy sources increases over the projected time frame. Given expectations that world oil prices will remain relatively high throughout the projection, liquid fuels are the world's slowest growing source of energy; liquids consumption increases at an average annual rate of 1.2 percent from 2005 levels to 2030. Renewable energy and coal are the fastest growing energy sources, with consumption increasing by 2.1 percent and 2.0 percent, respectively. Projected high prices for oil and natural gas, as well as rising concern about the environmental impacts of fossil fuel use, improve prospects for renewable energy sources. Coal's costs are comparatively low relative to the costs of liquids and natural gas, and abundant resources in

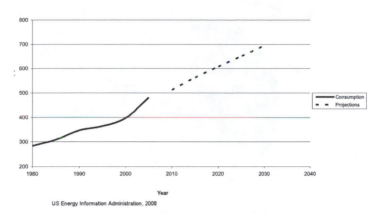

Figure 6.5 World energy consumption.

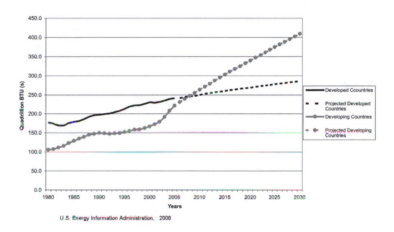

Figure 6.6 World energy consumption by region.

large energy-consuming countries (including China, India, and the United States) make coal an economical fuel choice. (Fig 6.8)

- Although liquid fuels and other petroleum are expected to remain important sources of energy throughout the projections, the liquids share of marketed world energy consumption declines from 37 percent in 2005 to 33 percent in 2030 as high world oil prices could force many consumers to switch from liquid fuels and other petroleum when feasible.
- Natural gas remains an important fuel for electricity generation worldwide, because it is more efficient and less carbon intensive than other fossil fuels. In the 2005 – 2030 projections, total natural gas consumption increases by 1.7 percent per year on average, from 104 trillion cubic feet to 158 trillion cubic feet, while its share of world electricity generation increases from 20 percent in 2005 to 25 percent in 2030.
- Coal consumption is projected to increase by 2.0 percent per year from 2005 to 2030 and to account for 29 percent of total world energy consumption in 2030. In the absence of policies or legislation that would limit the growth of coal use, the United States, China, and India are expected to turn to coal in place of more expensive fuels. Together, the three nations account for 90 percent of the projected increase from 2005 to 2030. (Fig. 6.9).

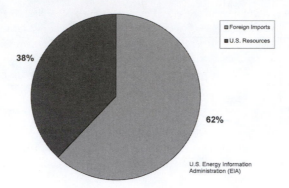

Figure 6.7 Total U.S. oil consumption: 20.8 million barrels per day

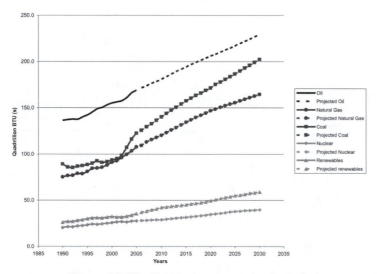

Figure 6.8 World energy consumption by fuel type

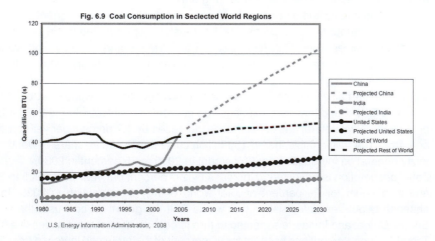

Figure 6.9 Coal consumption in selected world regions

- There is still considerable uncertainty about the future of nuclear power, however, and a number of issues could slow the development of new nuclear power plants. Plant safety, radioactive waste disposal, and the proliferation of nuclear weapons, which continue to raise public concerns in many countries, may hinder plans for new installations, and high capital and maintenance costs may keep some countries from expanding their nuclear power programs.

Despite rising levels of consumption, estimates of available world energy resources have become more optimistic in recent years. According to the January, 2003 Journal of Petroleum Technology, "if energy consumption were to remain constant at current levels, proved reserves would supply world petroleum needs for 40 years, natural gas needs for 60 years and coal needs for well over 200 years." There are other reports that are equally, if not more, optimistic about the future of the world's energy sources.

However, one theme is common to all reports: these resources appear to be finite. Even though current estimates show there to be adequate energy resources for the next generation, engineers and scientists must recognize that the long-term challenge of providing energy to a global community remains a high priority. New supplies of traditional energy resources must be discovered, while at the same time we must develop and expand options for alternative energy sources (such as wind, biobased fuels, solar, photovoltaics, geothermal, and low-head hydro).

6.4 TRANSPORTATION

Americans, as well as a growing worldwide population, continue to rely heavily on highway systems as their primary means of transportation. The ever increasing use of the automobile has created significant problems such as crowded highways, inadequate parking, and overall congestion in major urban areas of the world. Transportation problems in larger cities have increased as the number of cars and trucks on the roads has surpassed the capacities of streets and bridges built to accommodate the traffic flow of earlier decades.

According to a 2007 study by the Texas Transportation Institute, it has been estimated that traffic delay and congestion cause motorists in 437 U.S. urban areas studied to spend the equivalent of more than one work week per year in traffic jams. Researchers tried to estimate the annual economic impact of traffic congestion and determined that the value of motorists' time coupled with wasted fuel resulted in combined "congestion costs" of $78 billion in the areas studied. 28% of the energy used in the U.S. goes to transporting people and goods from place to another. In essence, trips take longer, congestion affects more of the day, weekend travel, rural areas, personal trips and freight shipments. Trip travel times have become increasingly unreliable.

Worldwide, transportation problems are also becoming serious issues. A nation's transportation system is generally an excellent indicator of its level of economic development. In many countries personal transportation still means walking or bicycling, and domestic animals are still used to move freight. According to the 2002 World Energy Projection System Report, over the next two decades, rapid growth of transportation infrastructure is expected in the developing world. Developing Asia and Central and South America are predicted to account for 52% of the increase in the world's motor vehicle population between 1996 and 2020 (Fig. 6.10).

According to the 2002 U.S. Energy Information Administration projections, energy use for transportation among developing countries is projected to grow at an average annual rate of 4.0%, nearly triple the rate of growth in the industrialized countries. Growth in the transportation sector in the industrialized countries—where modern transportation systems have been in place for many decades—is expected to average only 1.4% per year. Even in the most economically advanced countries, however, transportation energy consumption per

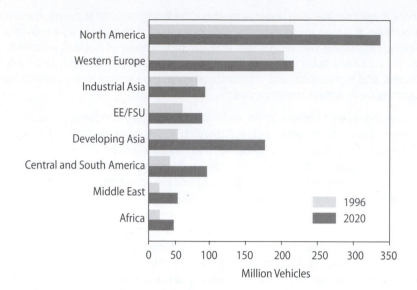

Figure 6.10 Road vehicle populations by region.

(Sources: American Automobile Manufacturers Association, World Motor Vehicle Data, Detroit, MI, 1997.
2020: EIA, World Energy Projection System, 2002.)

capita continues to increase over the projection period, as rising per capita incomes are accompanied by purchases of larger personal vehicles and by increased travel for business and vacations.

The challenge facing engineers is to continue the process of building new highways, bridges, and parking facilities that can adequately meet world demand. However, new construction should be viewed as only one solution to a very complicated problem. Engineers will need to provide their expertise to develop new ways to better control traffic flow, improve the use of "intelligent" traffic communication systems, change usage patterns, develop efficient methods for quicker accident identification and removal, encourage ride sharing and flex time programs which reduce peak-hour travel, and design and implement imaginative, appealing mass transportation systems.

6.5 INFRASTRUCTURE

Infrastructure is a term referring to entities designed to support human activities, such as roads, bridges, transportation services (including railways, airports and waterways), public service facilities, and buildings. The current state of the world's infrastructure is causing great concern as many aging facilities and systems rapidly deteriorate. This, too, presents a significant challenge for engineers.

Many of these problems have become more visible in the U.S. in recent years. Roads are deteriorating faster than they can be repaired due to a lack of available funding. Bridges and highway overpasses are slowly falling apart and pose safety hazards to the traffic loads and volume. Miles of railway tracks and beds are in dire need of maintenance or replacement. Municipal water and wastewater systems are not able to handle adequately the increasing public demands, and new environmental safeguards are needed to meet evolving legislative guidelines.

In 2001, the American Society of Civil Engineers first issued a "report card" for a variety of American infrastructure categories. The report gave an overall grade of D+ with total

TABLE 6.2 United States Infrastructure Report Card

	2005 Grade	2009 Grade
Roads	D	D-
Bridges	C	C
Transit	D+	D
Aviation	D+	D
Schools	D	D
Drinking Water	D-	D-
Wastewater	D-	D-
Dams	D	D
Levees	N/A	D-
Solid Waste	C+	C+
Hazardous Waste	D	D
Navigable Inland Waterways	D-	D-
Energy	D	D+

United States Infrastructure G.P.A. = D

Total Investment Needs = $2.2 Trillion

Source: American Society of Civil Engineers, 2009

investment needs of $1.3 trillion over five years to remediate the problems. In 2005, the report was re-evaluated; the grade dropped to a D, and total dollar investment needs climbed to $1.6 trillion. In 2009, another report was issued and the overall grade remained a D, but the total dollar needs climbed to $2.2 trillion (Table 6.2).

Because of the increased visibility of many of these problems in the U.S., local, state, and federal government agencies are working diligently to secure adequate funding for repairs. Once secured, it will then be a challenge for engineers to effectively design, develop, and implement the necessary programs and systems that will address these critical issues. It is encouraging that in some areas of the country, efforts to deal with this massive problem are underway. However, adequate funding must increase and the ingenuity, creativity, and expertise of the engineering community must be adequately utilized to solve these problems.

On a global scale, infrastructure will continue to be a critical factor in economic competitiveness. Developing nations need to, and are, investing in highways, railroads, and utilities to allow their economies to grow. The linkage between economic activity and infrastructure continues to grow stronger and more critical as economic activity becomes increasingly more complicated and global in scope.

Historically, as engineering infrastructure improvement has enhanced the transportation of people, goods, commodities, water, waste, energy, and information, this increased mobility has translated directly into better choices for people across the globe: choices of where to work, what to buy, how to communicate, and in some ways how to live. In an era of accelerating change, the infrastructure challenge for the global engineer will continue to be to expand those range of choices, and in so doing, improve the quality of people's lives.

6.6 AEROSPACE

The U.S. Space and Defense industries which have provided so many career opportunities for engineers since the 1960's have been in period of dramatic change and transition over the past several years. The Challenger disaster in 1986, coupled with the loss of the Columbia space shuttle in 2003, have caused re-evaluation of the future of the space shuttle missions. The break-up of the Former Soviet Union in the early 1990's, the war in Iraq, and the continuing tensions in many parts of the world has necessitated a major refocusing of defense spending priorities. All of these issues have created an uneven job market for engineers interested in careers in space and defense during the last 15 years.

Worldwide terrorist activity and localized conflicts in Iraq, the Middle East, Eastern Europe, and parts of the Former Soviet Union have demonstrated that the U.S. must continue to maintain a prepared and technically current military. This has resulted in increased spending and business for many sectors of the defense industry and homeland security. Until the Columbia disaster, the space industry had seen a dramatic increase in business. Much of this increase was attributable to the telecommunications industry and its increased need for the deployment of a great number of satellites. According to the United Nations System-Wide Earthwatch 2000 Report, international telecommunication companies hope to put up three times as many satellites as were launched in the last 40 years, with a predicted total of over 1,000 in the coming years. New technologies involving microscopic rocket thrusters will allow for a substantial increase in small satellites designed for telecommunication, military, and astronautical applications.

However, the commercial aviation business is struggling with the combined effects of 9/11, the war in Iraq and higher fuel costs. This includes airlines, airports, air cargo, and passenger traffic. All have experienced economic problems.

In preparation for 2011 Congressional budget hearings, NASA was asked to prepare commentary concerning their perspective of the coming years for their organization. The following provides some of the details found at www.nasa.gov.

"Although NASA's philosophy and approach to exploration will change, our fundamental goal remains the same: to send human explorers into the solar system to stay. This new era in human exploration will leverage American ingenuity and propel the nation on a new journey of innovation and discovery. Groundbreaking new technologies will enable exploration of new worlds and increase our understanding of the Earth, our solar system and the universe beyond. Further collaborations on the International Space Station will increase NASA's return on investment and provide an optimal test bed for space technology research and development. In order to place NASA's focus on this forward-looking space enterprise, the President's Budget Request proposes canceling the Constellation Program."

"NASA doesn't intend to embark on this new journey alone. Commercial and international partnerships will benefit from a collective spirit of discovery and adventure, and will reduce the cost of space exploration by employing new business practices and leveraging common goals. NASA also invites citizen stakeholders to participate and share in the excitement of space exploration through upcoming initiatives designed to educate as well as glean new, creative ideas from standard and unconventional contributors."

In order to achieve these goals, NASA has proposed the creation of six exploration study teams formed to develop strategies for the proposed new programs. These include the following with the overall objective listed for each (per www.nasa.gov):

- "Flagship Technology Demonstrations—To formulate plans for a series of in-space, large-scale demonstrations that validate capabilities key to sustainably explore deep space."
- "Commercial Crew and Cargo—To formulate plans to expedite and improve robustness of current Commercial Orbital Transportation Services solutions. The study team will also draft a human ratings requirements document and develop a plan that supports the development of commercial crew transportation providers to whom NASA could competitively award a crew transportation services contract."
- "Enabling Technology Development and Demonstration—To formulate plans for conducting small-scale development and testing of key, long-range exploration technologies."

- "Exploration Precursor Robotic Missions—To formulate plans for a series of robotic precursor missions to scout targets for future human exploration. Potential destinations include the moon, Mars and its moons, Lagrange points and nearby asteroids."
- "Human Research—To enhance current human research work and create more robust exploration-enabling projects with an increased focus in the areas of biomedical technologies, space radiation, behavioral health research, International Space Station utilization, and education. The team will continue to address human health and performance risks endorsed by National Research Council and Institute of Medicine."
- "Heavy Lift and Propulsion Technologies—To formulate plans for investigating a broad scope of research and development activities to support next-generation liquid chemical propulsion technologies, including foundational propulsion research and demonstrations of first-stage and in-space engines."

Of course, these proposals are subject to funding approvals, and priorities established by future government administrations.

The net effect of all these increased opportunities in the space and defense industries could provide many new challenging positions for engineers. The employment in aerospace-related industries was about 1.3 million workers in the late 1980's. This figure dropped to 800,000 by the mid-1990's, but has rebounded to over 1 million. The U.S. aerospace industry also affects the jobs of 15 million other workers in the American economy, including those in large and small non-governmental firms.

There are many challenges ahead for engineers in all aspects of the aerospace and defense industries. Since the nature of these industries cuts across virtually all areas of engineering, there should be opportunities for engineering students from a variety of majors and backgrounds. It appears that once the uncertainty in these industries ends, that the funding and demand for work in these areas will increase. There should be more stabilized opportunities and career challenges for those interested in these dynamic fields.

6.7 COMPETITIVENESS AND PRODUCTIVITY

The health of a nation's economy in the global marketplace can often be measured by its relative productivity and competitiveness. Global competitiveness is the ability of a nation to successfully compete in the world marketplace while improving the real income of its citizens. Measures of competitiveness include savings rates (which ultimately provide the resources for facilities and equipment which are needed to improve efficiency); financial investments in areas such as research, development, and education; and the productivity of its industries, which generates economic growth and increased wages.

During recent decades, there has been increasing concern over the declining U.S. position in global competitiveness, especially in certain vital technological areas. There is specific concern regarding those firms which once dominated world markets and are gradually losing significant market share to foreign competitors.

Advances by foreign countries in basic science and research now often surpass those of the U.S., which has implications for jobs, industry, national security, and the strength of the nation's intellectual and cultural life.

One area of intense global competition involves patents. High-quality patents are seen as strong indicators of a nation's future prosperity because they signify the emergence of important new technologies that will be under the patent holder's control for many years.

While U.S. researchers still earn a significant number of new patents, the percentage has been falling as international scholars, particularly in Asia, have become more active. The U.S. corporate share of patents has fallen from over 60% in 1980, to 49% in 2008.

Another area of concern is in published scientific papers. According to the National Science Foundation, works by Americans peaked in 1992 and have fallen 10% since. A 2007 study by the "Science Watch," which studies yearly scientific publishing activity, indicates that the European Union surpassed the United States in the mid-1990s as the world's largest producer of scientific literature.

The number of new U.S. doctorate degrees in science and engineering has been increasing the past several years, influenced by large numbers of foreign students. However, increasing numbers of doctoral students from countries such as China, India, and Taiwan are electing to return to their home countries, causing a reverse "brain drain." These declines are critical because new scientific knowledge is a significant factor in the American economy in a wide range of industries, new products and technologies.

The World Economic Forum in its annual Global Competitiveness Report for 2009-2010 ranks the top ten countries based on a variety of factors. After many years at the top of the world rankings, the United States has fallen to second place behind Switzerland. The report indicates the U.S. continues to be strengthened by many structural factors that can make the economy extremely productive and in a position to survive business cycle shifts and economic turbulence. However, a number of escalating issues hindered the US ranking in 2009.

Specifically, the report outlines several positive areas of strength and some issues of concern:
Positives/Strengths:

— Highly sophisticated and innovative companies that operate in efficient market areas
— An excellent higher education system with strong collaborations with the research and development sectors of business and industry
— Good labor markets with the ease and affordability of hiring with significant wage flexibility
— The world's largest overall domestic economy provides many opportunities for capitalizing on the points listed above

Issues of concern:

— Government interactions with the business community can be turbulent with the perception that government spends its resources wastefully
— An overall concern with the functioning of private institutions and issues related to recent turmoil and scandals in the financial sector
— Weakening assessments of the U.S. financial market capabilities
— The U.S. has built up large economic imbalances in recent years with continuing fiscal deficits and increasing levels of public indebtedness, which is magnified by significant stimulus spending

In partial response, and as part of its 2010 legislative agenda, the U.S. Congress is busy debating the final points of the America COMPETES Reauthorization Act of 2010. The intent is to continue funding programs that invest in modernizing manufacturing, spur American innovation through basic Research and Development, high-risk/high-reward clean energy research, and strengthen math and science education to prepare students for the good jobs of the 21st century.

The overall goals of the Act will focus on:

- Keeping the United States on a path to double funding for basic scientific research, crucial to some of the most innovative breakthroughs, over the next several years
- Creating jobs with innovative technology loan guarantees for small and mid-sized manufacturers and Regional Innovation Clusters to expand scientific and economic collaboration
- Promoting high-risk/high-reward research to pioneer cutting-edge discoveries through the Advanced Research Projects Agency for Energy (APRA-E)
- Creating the next generation of entrepreneurs by improving science, math, technology, and engineering education at all levels

The goals are to create prosperity through science and innovation, reassert the necessary economic and technological leadership over the next several years, and give future generations greater opportunities. Boosting U.S. competitiveness is essential to the economy, as almost half of the growth in the Gross Domestic Product (GDP) since World War II is related to the development and adoption of new technology.

EXERCISES AND ACTIVITIES

6.1   In reference to the challenge of "population," there was a statement made that said "there are many who believe that this particular "future challenge" has implications which touch most of the other challenges outlined in this chapter." Do you agree with this statement? Why or why not? Use examples to support your comments.

6.2   Further explore and examine some of the causes of air pollution discussed in this chapter. Write an essay on possible new engineering developments that might address these problem areas.

6.3   Experts claim that the water supply system in the U.S. is the safest and most drinkable in the world. Write a research report on factors that have adversely influenced water quality in other parts of the world.

6.4   Research some of the options to current methods of waste treatment and waste storage. What might be some other more desirable solutions to the waste management problem?

6.5   Explore some of the long-term projections for world energy usage over the next 25 years. Develop a list of factors that could increase the demand for energy resources. Develop another list that presents those variables which could reduce energy consumption during this period.

6.6   Propose some alternative options to the construction of new highways, bridges, and parking facilities that would help to reduce traffic congestion problems in urban areas.

6.7   Suggest some alternative solutions that could provide the necessary funding resources to improve the nation's infrastructure.

6.8   What issues do you feel are of most importance if the U.S. is to improve its competitive position in the global economy? Provide specific examples to support your comments.

6.9   Research one aspect of the U.S. space or defense programs. Write an essay that outlines your assessment of where this part of the industry will be in the next 20 years and explain why.

6.10   Select one of the "challenge" topics from this chapter. Further research the problem and prepare a paper that discusses additional negative impacts such as unemployment, economic growth, health, and global stability issues related to your topic area.

6.11   Review recent magazine and newspaper articles concerning nuclear power. Prepare a report that explores the advantages and disadvantages of developing additional nuclear power facilities.

6.12   Put together a group of 4-5 students in class to brainstorm and discuss the issues presented in this chapter. Consider what new majors, programs, or course might evolve during the next 20 years in an attempt to met these future challenges. Present your results to the class.

6.13   Some legislators have recommended further budget reductions for space and defense programs. Review recent magazine and newspaper articles on this topic. Do you agree or disagree with these recommendations? Explain your position.

6.14   Develop a list of other "challenges" that could be added to this chapter. Select one and write an essay that outlines the challenge in greater detail. Use the web and recent newspaper and magazine articles.

Chapter 7

Succeeding in the Classroom

7.1 INTRODUCTION

As an engineering student, an important goal is to succeed in the classes needed to attain an engineering degree. How this is done depends to a great deal on the individual student and the style, temperament and strengths of that student. This chapter presents strategies proven to be effective in helping students succeed. As with any tips on success, you need to look at the techniques being outlined and decide if they can be applied to you and your own individual style. If you are unsure about some of the suggestions, give them a try. You may find yourself in the same situation as one freshman engineering student after being given a book on study skills.

The student enters a professor's office very excited and asks for a short conference with the professor:

Student:	Professor! It worked!
Professor:	That's great. What worked?
Student:	The book!
Professor:	That's great. What book?
Student:	That study skills book.
Professor:	So you read it?
Student:	Yes, I read it like you suggested. It sure sounded like a lot of unworkable stuff at first. But then I decided, what the heck. I'll try it for a week. Couldn't hurt, I figured. Well, it really did work! It even worked for physics lab. I got an A on the last lab. I still can't believe it.
Professor:	That's great. I am glad it helped.
Student:	Could I get another copy of that book? I have some friends who want to read it now . . .

Just like that student, we are often reluctant to try new things. We are comfortable in what we do. You may have been very successful in high school with your study methods and wonder why you need to develop different study methods and techniques. The answer is that your environment has changed. College is very different from high school. The courses are structured differently and require significantly more preparation outside of class. The expectations of the instructors are different and the grading is different. Modifying your habits to become more effective is a key to succeeding in engineering studies. Learning to adapt to meet new challenges is a skill that will serve you well not only in college, but for the rest of your life.

The methods for excelling in the classroom outlined in this chapter have proven effective for numerous students. We suggest that you consider all of these ideas and adapt the ones that fit best with your own personality. If you are unsure whether they will work, try some out for a trial period and see. If those don't work, move on to others until you find something that fits with your own style.

The three components to succeeding in your academics are your ability, your attitude, and your effort. There is no book that can address the first item. While there is unquestionably a level of ability that is needed to succeed in engineering, there are numerous students who had the ability but failed at engineering because of poor attitudes or poor or inefficient effort. There are also many students who lacked natural ability but because of their positive attitude and extraordinary effort are now excellent engineers. This chapter focuses on techniques to improve these two components for success that are totally under your control.

7.2 ATTITUDE

> *"Ability is what you're capable of doing. Motivation determines what you do. Attitude determines how well you do it."*
>
> —Lou Holtz, South Carolina football coach

Your attitude is the first thing that you have control over when beginning any challenge, including studying engineering. You can expect success or expect failure. You can look for the positives or dwell on the negatives.

Approaching your classes, professors, and teaching assistants with a positive attitude is the first key to succeeding in your engineering studies. Each class should be looked at as an opportunity to succeed. If there are difficulties along the way, learn to deal with those and move ahead. Many students will decide that a certain class is too difficult or that the professor is a poor teacher and will expect to do poorly. Many fulfill their own prophecy and actually fail. Other students look at the same situation and overcome the hurdles and excel.

It is a distinct possibility that you may have professors you don't like or who you think teach poorly. You may even have an experience like the following example.

Example 7.1

A physics professor of mine discovered the overhead projector during a semester and moved his lecture from the blackboard to overheads. The problem was that he was very tall and stood so that he blocked the overhead. It shined on his coat and tie but we couldn't see what he was writing. When we told him in class that we couldn't see, he moved out of the way and looked at the overhead. He then adjusted the focus and asked if we could see now. After we responded yes, he moved back to write more and blocked the overhead again. When another student tried to explain that he was blocking his view of the overhead, the professor explained that no matter where he stood, he would block someone's view. He suggested that the student move to the other side of the room. A third student from the other side of the room chimed in that he couldn't see from that side either. Unfortunately he was in the back of the lecture hall, so the professor thought that was the reason. He moved out of the way of the projector and asked one of the students in the front row if she could see. She said, "Yes, but . . ." The professor cut her off before she could finish and asked the guy in the back of the room if he had glasses. When he said no, the professor recommended he get his eyes checked because the woman in front said she could see. At this point we all gave up taking notes for the rest of the lecture.

Later, I complained to an advisor who was the director of the university honors program. Rather than a strategy for communicating with the professor, I received a story about a professor he had had. It seems that he had an English professor who was about 108 years old (or *seemed* that old). He couldn't stand up without leaning on something, so he used the blackboard. His shoulder made contact with the blackboard, which wasn't a problem in itself except that he wrote at shoulder level. The result is that he erased what he wrote as he moved along the board. So if you weren't in a seat where you could see the three or four words he wrote before he erased them with his shoulder, you couldn't take notes.

The point he was making was that there are challenges in learning as well as later in life. What we have to do is to find ways to overcome obstacles, not to dwell on them. If your attitude is that it is a professor's fault that you won't learn, you won't. If you keep a positive attitude you can turn challenges into opportunities. He shared that his class became much closer and studied more together because they had to piece the lecture together from their collective notes. They learned a lot, and the positive outcome was the closeness the class developed. He challenged us to go and talk to the professor outside of class, during his office hours, and try to explain our difficulty with the lecture. Instead of just complaining, take action with a positive attitude to solve our problem. It was a real lesson for life.

I am still not sure what lesson I was supposed to learn from that story. It struck me, however, that even with professors like that, the advisor earned his Ph.D., became a professor and eventually ran the university's honors program. He had developed strategies to cope and kept a good attitude. Attitude is totally under your control.

7.3 GOALS

"Do you feel the world is treating you well?
If your attitude toward the world is excellent, you will receive excellent results. If you feel so-so about the world, your response from that world will be average. Feel badly about your world and you will seem to have only negative results from life."

—Dr. John Maxwell, founder of INJOY, Inc.

As freshmen, there were a number of us who were in an honors program. One in the group set a goal to be a Rhodes Scholar from the first day of his freshman year. The rest of us were too busy adjusting to college and taking it all in. In our senior year, when we were interviewing for jobs and deciding whether to go to graduate school, the friend who had set the goal was being toured around the state as the newest Rhodes Scholar from our university. The difference over our college careers was astounding, simply because he had established clear, elevated goals. The rest of us did okay, but he excelled, in large part because he had defined a standard and kept striving for it.

The lesson that I took away from that friendship in college was that setting clearly defined goals early is a key to success in school, in an engineering career or in life as a whole.

In setting goals, there are two essential elements—"height" and time. You might ask, "How high do I set my goals?" This will vary from individual to individual. Goals should be set high, but attainable. A common management practice is for a team to decide on a reasonable goal to accomplish. Then the bar is raised and a stretch goal is defined that encourages the team to produce more. This stretch goal is designed to push the group farther than they would have gone with only the original goal. Sometimes this stretch goal looks reasonable and sometimes it looks totally unreachable. What often happens, though, is that the stretch goal is met, realistic or not. As an individual, setting stretch goals helps us grow. When setting goals for yourself, look at what you think you can do and set base goals. Then establish stretch goals a little higher than the base goals. You may just find yourself meeting those higher goals.

The second key element in goal setting is time. When will the goal be accomplished? It is important to have long-term goals and short-term goals. Long-term goals help guide where you are headed. For example, what do you want to do with your engineering degree? An appropriate long-term goal as a freshman could involve determining the kind of entry-level job you desire. This will help answer the question "Why am I doing this?" when your classes get tough.

Short-term or intermediate goals should be stepping-stones to the achievement of your long-term goals. An advantage of an academic setting is that there are natural breaks, semesters, or quarters, for the establishment of goals within specific time frames. Goals can be set for the academic year, and for each semester or quarter. These provide intermediate goals.

Short-term goals need to be set and rewarded. Rewarding short-term goals is essential to keeping motivation high in the quest for long-term goals. Many students lose sight of their long-term goals because they didn't set short-term, "rewardable" goals. Short-term goals may be an "A" on an upcoming quiz or test, or it may be completing a homework set before the weekend.

Example 7.2

Long-term goal	Job as a design engineer for a major manufacturer of micro-processors
Intermediate goals	3.7 GPA after freshman year 3.5 GPA at graduation Hold an office in the local chapter of IEEE Internships after sophomore and junior years
Short-term goals	'B' on first calculus test 'A' in calculus 3.7 GPA after first semester Join IEEE and attend meetings Complete chemistry lab by Friday

In the list of short-term goals above, there are opportunities for rewards. Perhaps there's a compact disk you want. Set a goal of a certain grade on that calculus test. If you meet the goal, you can go buy the CD.

Clearly define your goals by writing them down. It is much more powerful to write down your goals than to simply think about them. Some people post them where they can see them daily. Others put them in a place where they can retrieve and review them regularly. Either method is effective. For short-term goals, establish weekly goals. Select a day, probably Sunday or Monday, and establish the goals for the week. This is a great time to establish the rewards for your weekly goals. Plan something fun to do on the weekend if you meet those goals.

At the end of each semester, examine your progress toward your long-term goals. This is an advantage to being in school. Every semester, there is an opportunity to assess your progress and start fresh in new classes. Take advantage of this. Examine your long-term goals and make any changes if necessary.

Determining Goals

It can be intimidating or difficult for students early in their careers to decide on appropriate goals. Fortunately, there are numerous resources on campus to help. Academic advisors can provide criteria for grade-point goals. Things to consider include:

1. What is the minimum GPA to stay enrolled in school?
2. What grades do you need to continue in engineering or to enter engineering?
3. What grades are needed to be eligible for co-op or intern positions?
4. What GPA is required to make the dean's list or honor list?
5. What GPA is needed to be eligible for the engineering honorary organizations?
6. What grades are needed for admittance to graduate school?
7. What grades are needed for scholarships?

There are also people on campus who work with the placement of graduates. They are excellent resources to get information on what employers are looking for. What credentials do you need to be able to get the kind of job you desire? If you know this as a freshman, you can work toward it. You can also go directly to the employers and ask these same questions of them. If companies come to your campus, talk to them about what credentials are needed to get a job. Job fairs are great times to do this. Approaching company representatives about establishing goals is also a great way to show initiative and distinguish yourself from your classmates.

Goals are important for outside the classroom too. It is important to have goals in all areas of your life. Later in this chapter we will discuss such goals.

7.4 KEYS TO EFFECTIVENESS

Once goals have been set regarding grades, the next step is to achieve those grades. Effort and effectiveness may be the most important components of your success as a student. There are numerous cases of well-equipped students who end up failing. There are also students who come to the university poorly prepared or start off poorly but eventually succeed. The difference is in their effort and their effectiveness. The following suggestions contain strategies to improve personal effectiveness in your studies.

 A general rule is that it takes a minimum of two hours of study outside of class for each hour of class lecture.

Take Time to Study

One of the strongest correlations with a student's performance is the time he or she spends studying.

According to the reminder above, if you are taking 15 hours, that means that you need to spend 30 effective hours a week studying outside of class. In a typical high school, the time requirement is much less than this. Developing the study habits and discipline to spend the needed time studying is a prime factor in separating successful and unsuccessful students.

Go to Class

There is also a high correlation with class attendance and performance. This may seem obvious, but you will run into numerous students who will swear that it doesn't matter if you go to class. Many instructors in college do not take attendance, so no one but you knows if you were there, which is different from high school. The classroom is where important information will be presented or discussed. If you skip class, you may miss this information. Also, by skipping class, you miss an opportunity to be exposed to the material you will need to know, and you will have to make up that time on your own. Missing class only delays the expenditure of time you will need to master the material. Most often, it may take much more than an hour of independent study to make up an hour's worth of missed class time. It is possible for you to never miss a class in your undergraduate years, except for illness or injury, serious family problems, or trips required by groups to which you belong. Never missing a class could be one of your goals.

Also, you may want to schedule classes at 8 a.m. and 4 p.m. This will get you up and going and keep you there! You will seldom, if ever, study at 8 a.m. and at 4 p.m. Do your studying in the middle of the day. This will help you to use your time effectively.

 "After class, reread the assigned sections in your textbook. Concentrate on the sections highlighted by the lecture. By doing so, you will not waste time trying to understand parts of the book that are not critical to the course."

Make Class Effective

The first component of making class effective is sitting where you can get involved. If the classroom is large, can you see and hear in the back? If not, move up front. Do you need to feel involved in the class? Sit in the front row. Do you fall asleep in class? Identify why. Ask yourself, "Am I sleeping enough at night?" If the answer is no, get more sleep.

The second step for making class time effective is to prepare for class. Learning is a process of reinforcing ideas and concepts. As such, use the class time to reinforce the course material by reviewing material for the lecture beforehand. Most classes will have a textbook and assigned reading for each lecture. Take time before the lecture to skim over the relevant material. Skimming means reading it through but not taking time to understand it in depth. This will make the class time more interesting and more understandable, and it will make note taking easier.

By following the above advice, you will have had three exposures to the material, which helps you remember and understand the material better.

Keep Up with the Class

Class is most effective if you keep up. An excellent short-term goal is to master the lecture material of each course before the next lecture. A course is structured in such a way that you master one concept and then move on to the next during the next lecture. Often in science and engineering classes the concepts build on each other. The problem with falling behind is that you will be trying to master concepts that depend on previous lectures, and you are going to hear lectures that you will not understand.

There is a very practical reason that a three-credit class meets three times per week. During a 15-week semester, there are only 45 hours of class meeting time. That amount of class time could be held in a week of consecutive nine-hour days. Arranged that way, a whole semester could be taken in a month. Then it would only take a year to get a bachelor's degree. So why don't we do it that way?

The reason is that the current educational model is designed to introduce a concept and give you, the learner, time to digest it. It is an educationally sound model. In engineering study, digesting means reviewing notes, reading the text, and working several problems related to the topic. If you are still confused about an aspect of the previous lecture, the next class provides a great opportunity to bring it up with the professor. Get your questions answered before going on to the next topic.

A note of caution should be given here. Friends of yours may be studying subjects where they can study from test to test rather than from class to class. They may insist that you don't need to keep up so diligently because they don't. Some majors cover material that is more conceptually based. That is, once you understand the concept, you have it. Engineering, math, and science courses are not this way. To be mastered they require extensive study and preparation over an extended period of time. In some courses there are only a few basic concepts, yet the entire course is dedicated to the application of these few concepts. An example is Statics, a sophomore-level course, in which there are only two fundamental equations germane to the course: the sum of forces equals zero, and the sum of moments equals zero. The entire course involves applying these two equations to various situations, and using the information in engineering applications.

Take Effective Notes

A main component in keeping up with your classes is taking effective notes for each lecture. Effective notes capture the key points of the lecture in a way that allows you to understand them when you are reviewing for the final exam three months later. Suggested note-taking strategies follow:

1. Skim the assigned reading prior to class to help identify key points.
2. Take enough notes to capture key points but don't write so much that you fail to adequately listen to the presentation in class.
3. Review your notes after class to annotate them, filling in gaps so they will still make sense later in the semester.
4. Review your notes with other students to ensure that you captured the key points.
5. Review your notes early in the semester with your professor to be certain you are capturing the key points.

In class, your job is to record enough information to allow you to annotate your notes later. You don't have to write everything down. Getting together with classmates after class is a terrific way to annotate your notes. That way you have different perspectives on what the main points were. If you really want to make sure you captured the key points, go and ask your professor. Doing so early in the semester will not only set you up for successful note taking throughout the semester, it will also allow you to get to know your professor.

Work Lots of Problems

Because the applications of the concepts are the core of most of your classes, the more problems you can work, the better prepared you will be. In math, science, or engineering courses, doing the assigned homework problems is a minimum. Search for additional problems to work. These may be from the text or from old exams. Your professor should be able to steer you to appropriate problems to supplement the homework.

Caution Using Solution Manuals and Files

For some classes, there are homework files or solution manuals available. While these tools may help get the homework done quicker, they very often adversely affect exam performance. The problem is that most professors don't allow the solutions to be used during the test. Using them becomes a crutch and can impair your ability to really learn the material and excel on exams. If they are used at all, use them only after exhausting all other possibilities of working out the answer yourself. If it takes a lot of work to figure out a homework problem, that concept will be learned well and you will have a higher probability of demonstrating your knowledge on the next test. Taking the easy way out on homework has a very consistent way of showing up on test results. Minimal learning results from copying down a solution.

Group Studying

A better model for studying than using solutions is to study in a group. Numerous studies have shown that more learning takes place in groups than when students study by themselves. Retention is higher if a subject is discussed, rather than just listened to or read. If you find yourself doing most of the explaining of the ideas to your study partners, take heart! The most effective way to learn a subject is to teach it. Anyone who has taught can confirm this anecdotally; they really learned the subject the first time they taught it.

> *"The most important single ingredient to the formula of success is knowing how to get along with people."*
>
> —Theodore Roosevelt

If you aren't totally convinced of the academic benefits of group studying, consider it a part of your engineering education. Engineers today work in groups more than ever before. Being able to work with others effectively is essential to being an effective engineer. If you spend your college years studying in groups, working in groups will be second nature when you enter the workforce.

Studying with others will also make it more bearable, and even fun, to study. There are numerous stories about students preparing for exams and spending the whole day on a subject. It is difficult to stick to one subject for an entire day by yourself. With study partners you will find yourself sticking to it and maybe even enjoying it.

Studying with others makes it easier to maintain your study commitments. Something we all struggle with is discipline. It is much easier to keep a commitment to study a certain subject if there are others who are depending on you. They will notice if you are not there and studying.

"I will pay more for the ability to deal with people than any other ability under the sun."

—John D. Rockefeller

Group studying is more efficient if properly used. Chances are that not everyone in the group will be stuck on the same problems. So right away, you will have someone who can explain the solution. The problem areas common to everyone before the group convenes can be tackled with the collective knowledge and perspectives of the group. Solutions are found more quickly if they can be discussed. More efficient study time will make more time for other areas of your life!

Choosing a group or partner may be hard at first. Try different study partners and different size groups. Ray Landis, Dean of engineering at California State University, Los Angeles and a leading expert on student success, suggests studying in pairs. "That way each gets to be the teacher about half the time." Larger groups also can be effective. The trick is to have a group that can work efficiently. Too large a group will degenerate into a social gathering and not be productive in studying. Whichever group size you choose, here are some basic tips for group studying.

1. Prepare individually before getting together.
2. Set expectations for how much preparation should be done before getting together.
3. Set expectations on what will be done during each group meeting.
4. Find a good place to convene the group that will allow for good discussion without too many distractions.
5. Hold each other accountable. If a group member is not carrying his or her weight, discuss it with that person and try to get them to comply with the group's rules.
6. If a member continues to fail to carry their own weight or comply with the group's rules, remove that person from the group.

Select a Good Study Spot

In a subsequent section, differences in learning styles will be discussed. Depending on your own learning style and personality, you may need a certain kind of environment to study efficiently. Some resources attempt to describe the "perfect study environment." In reality, there is no one perfect study environment or method of studying. What is crucial is that you find what you need and find a place that meets your needs. For some, it involves total quiet, sitting at a desk. Others may prefer some noise, such as background music, and prefer to spread out on the floor. Whatever you decide is the best environment, pick one where you can be effective.

Cynthia Tobias describes the conflict she has with her husband's view of a proper study environment in her book *The Way They Learn*:

I have always favored working on the floor, both as a student and as an adult. Even if I'm dressed in a business suit, I close my office door and spread out on the floor before commencing my work. At home, my husband will often find me hunched over books and papers on the floor, lost in thought. He is concerned.

"The light is terrible in here!" he exclaims. "And you're going to ruin your back sitting on the floor like that. Here, here! We have a perfectly clean and wonderful rolltop desk." He sweeps up my papers and neatly places them on the desk, helps me into the chair, turns on the high-intensity lamp, and pats my shoulder.

"Now, isn't that better?" he asks.

I nod and wait until he is down the hall and out of sight. Then, I gather all my papers and go back down on the floor. . . . It does not occur to him that anyone in their right mind could actually work better on the floor than at a desk, or concentrate better in 10-minute spurts with music or noise in the background than in a silent 60-minute block of time.

Different people have different ways in which they can be most efficient. You need to discover yours so that you can be efficient. If you are unsure of what works best for you, test some options. Try different study environments and keep track of how much you get done. Stick with the one that works best for you.

If you and your roommate have different styles and needs, then you will need to negotiate to decide how the room will be used. Both of you will have to be considerate and compromise, because it is a shared space.

One item that is a distraction for almost everyone is television. Because it is visual, it is almost impossible to concentrate on homework with a TV on, although many students swear they are still "productive." A few may be, but most are not. Watching TV is one of the biggest time-wasters for students. It is hard enough for first-year students to adjust to college life and the rigors of pre-engineering classes. We strongly recommend **not** having a television set in your residence hall room or apartment for at least your first year. It is not even good for study breaks, which should be shorter than 30- or 60-minute programs. This may be one temporary sacrifice you have to make to be successful in your studies.

7.5 TEST-TAKING

Most courses use written, timed exams as the main method of evaluating your performance. To excel in a course, you need to know the material well and be able to apply it in a test situation. The early part of this chapter provided tips on how to improve your understanding of the course material. The second part is test-taking skills. Ask any student who has been away from school for a few years. Taking tests is a skill and needs to be practiced. Just like a basketball player who will spend hours practicing lay-ups, it is important to practice taking tests. A great way to do this is to obtain past exams. A couple of days before the test, block out an hour and sit down and "take" one of the old tests. Use only the materials allowed during the actual test. For instance, if it is a closed-book test, don't use your book. In this way, you are doing two things. The first is assessing your preparation. A good performance on the practice test indicates that you are on the right track. The second thing you are doing is practicing the mechanics of taking the test. Again, with the basketball analogy, players will scrimmage in practice to simulate game conditions. You should simulate test conditions. What do you do when you reach a problem you can't do? Most people will start to panic, at least a little bit. It is critical that you stay calm and reason your way beyond whatever is blocking you from doing the problem. This may mean skipping the problem and coming back to it, or looking at it in another way. In any case, these are the types of things you have to do on tests, and by practicing them under test conditions, you will do better on the actual tests.

Before the first test in the course, visit your professor and ask him or her how to assemble a simulated test. It may be that old exams are the way to go, or possibly certain types of problems will be suggested. Professors don't like the question "What is going to be on the test?" but are much more receptive to requests for guidance in your preparation. The professor's suggestions will also help you focus your studying, to make it more efficient.

Taking the Test

There are some general guidelines for taking any test. The first one is to come prepared to take the test. Do you have extra pencils just in case? Can you use your calculator efficiently? Breaking your only pencil or having your calculator fail during the test can be very stressful and prevent you from doing your best, even if you are given a replacement. Proper planning for the test is the first step toward succeeding.

The second step is to skim over the entire test before beginning to work the problems. This is partially to make sure that you have the complete test. It also gives you an overview of what the test is going to be like so you can plan to attack it efficiently. Take note of the weighting of the points for each problem and the relative difficulty.

The next step is to look for an easy problem or one that you know you can do. Do this problem first. It will get you into the flow of doing problems and give you confidence for the rest of the test. Also, doing the problems you know you can do will ensure that you have made time for the problems for which you should get full credit.

Always keep track of the time during a test. Tests are timed, so you need to use the time efficiently. Many students will say they "just ran out of time." While this may prevent you from finishing the test, it should not prevent you from showing what you know. Look at the number of points on the test and the time you have to take the test. This will tell you how long to spend on each problem. If there are 100 points possible on the test and 50 minutes to complete it, each point is allotted 30 seconds. So a 10-point problem should only take 5 minutes. A common mistake is wasting too much time struggling through a difficult problem and not even getting to others on the test. Pace yourself so you can get to every problem even if you don't complete them. Most instructors will give partial credit. If they do, write down how you would have finished the problem and go on to the next when the allotted time for that one is up. If there is time at the end, you can go back to the unfinished problem. But if not, the instructor can see that you understood the material and can award partial credit.

Remember that the goal of a test is to get as many points as possible. A professor can't give you any credit if you don't write anything down. Make sure you write down what you know and don't leave questions blank. Another common mistake occurs when students realize they have made a mistake and erase much of their work. Often, the mistake they made was minor and partial credit could have been given, but none is awarded because everything was erased.

Concentrate on the problems that will produce the most points. These are either problems you can do relatively quickly and get full credit for, or ones worth a large portion of the total points. For instance, if there are four problems, three worth 20 points and one worth 40 points, concentrate on the 40-point problem.

Leave time at the end, if possible, to review your answers. If you find a mistake and there is not enough time to fix it, write down that you found it and would have fixed it if you had time. If you are taking a multiple-choice test, be careful about changing your answers. Many studies find that students will change right answers as often as they change wrong answers. If you make a correction on a multiple-choice test, make certain you know that it was wrong. Otherwise, your initial answer may have been more likely to have been correct.

Think. This may sound basic, but it is important. What is the question asking? What does the professor want to see you do? When you get an answer, ask yourself if it makes sense. If you calculate an unrealistic answer, comment on it. Often a simple math error produces a ridiculous answer. Show that you know it is ridiculous but that you just ran out of time to find the error.

After the Test

After you get your test back, look it over. It is a great idea to correct the problems you missed. It is much easier to fill in the holes in what you missed as you go along in the semester than to wait until you are studying for the final exam. Most final exams cover the material from all the semester tests. If you can't correct the problems yourself, go and see the professor and ask for help. Remember, you *will* see the material again!

7.6 MAKING THE MOST OF YOUR PROFESSORS

One of the most under-utilized resources at a university is the faculty. Many students do not take full advantage of the faculty that they are paying to educate them. The professor is the one who decides on course material, what is important and what is not, and how you will be evaluated. Yet few students take the time to get to know their professors. Here are some reasons to get to know your professors:

1. They are professionals and have experience in the fields you are studying and, therefore, have perspectives that are valuable to you.
2. Every student will need references (for scholarships, job applications, or graduate school applications) and professors can provide them.
3. They are the ones in charge of the course and can help you focus your studying to be more effective.
4. They are the experts in the field and can answer the hard questions that you don't understand.
5. They assign your grades, and at some point in your college career, you may be on the borderline between grades. If they know you and know that you are working hard, they may be more likely to give you the higher grade.
6. They are likely to know employers and can provide job leads for full-time positions or internships.
7. They may be aware of scholarship opportunities or other sources of money for which you could apply.
8. They can provide opportunities for undergraduates to work in labs, which is great experience. You may even get paid!

It is interesting to examine the history of the university. In centuries past, a student did not go to a university to get "trained" to do a job. They went to be mentored by great scholars (the faculty). In modern universities, many students come looking primarily for the training needed to get a good job, and getting mentored by faculty is low among their priorities. Consider this: a school could send you the course materials at home and have you show up for one day to take your exams. Yet students still come to a university to study. Why? A main reason is for the interaction with the people at the university—faculty, staff, and other students.

If there are so many reasons to get to know professors, why don't most students do so? Many professors don't encourage students to come see them, or don't present a welcoming appearance. Professors may remind students of their parents, people who seem to be out of touch with college students. It may also be that many students don't really understand the advantages of getting to know their professors.

Here are some things to keep in mind when getting to know professors. The first is that they have spent their careers studying the material you are covering in your courses and, therefore, they enjoy it. So it would be a very bad idea to start by telling your calculus professor that you find math disgusting. This is insulting and suggests he has wasted his career.

The second is that professors teach by choice, especially in engineering. Industry salaries are higher than academic salaries for engineering Ph.D.s, so professors teach because they want to. They also probably think that they are good at it. If you are having a problem with an instructor's teaching style, approaching him or her with a problem solving strategy such as the following can be very constructive: "I am having trouble understanding the concepts in class. I find that if I can associate the concepts with applications or examples I understand them better. Can you help me with identifying applications?"

As a group, professors also love to talk, especially about themselves and their area of expertise. They have valuable experience that you can benefit from. Ask them why they chose the field they did. Ask them what lessons they learned as students. You will get some valuable insights and also get to know someone who can help you succeed in your class, and possibly in your career. Steven Douglass, in his book *How to Get Better Grades and Have More Fun,* suggests asking each professor, "What is the main objective in the course?"

Faculty members are very busy. They have many demands on their time (research, securing grants, writing, committee assignments, etc.) in addition to your class. These other demands are what keep the university running. So respect their time. It is okay to get to know them but don't keep popping in just to shoot the breeze. Also, when you come with questions, show them the work you have been doing on your own; you only need to get past a hurdle. A busy faculty member won't mind helping a hard-working student clear up an idea or concept. However, showing up and asking how to do the homework can get interpreted as "Do my homework for me." This would not make the positive impression you want.

7.7 LEARNING STYLES

In the last several decades scientists have made many discoveries regarding the complexities of the human brain. While the brain is an important organ in each person's body, we now know that each person's brain is unique to him or her. Even before birth our genetic code, as well as our environment after birth, shapes our developing brain. Proper nutrition, stress, drugs and alcohol are some of the factors that can affect a developing brain.

Each person is born with all the brain cells, or neurons, that they will ever have. (Estimates reach as high as 180 billion neurons.) Neurons differ from other cells in that they do not reproduce. However, they do form connections with other neurons. As they are used, neurons can grow tentacle-like protrusions called dendrites which can receive signals from other neurons. In the first five years of life, with normal stimulation, a multitude of connections are created as the brain learns to move the body, speak, analyze, create, and control. However, people can continue to create more neurological connections as they continue to use their brain and learn new material. With the exception of diseases, like Alzheimer's, or physical injury, the brain is capable of creating new connections throughout our lifetime.

This is great news! None of us is ever too old or too dumb to learn something new. Another benefit of brain research is that it has identified several different ways that people think and memorize. Memorizing refers to how people assimilate, or add, new material to existing knowledge and experience. It also includes how we accommodate, or change our previous way of organizing material. Thinking refers to how we see the world, approach problems and use the different parts of our brain. One method does not fit all when it comes to how we learn best, so time spent identifying your own strengths and weaknesses will prove very beneficial. Once you identify the aspects of your own preferred learning style, information can be manipulated so that you learn more information in less time with higher rates of recall.

Memory Languages

Place a check mark by all the statements that strongly describe your preferences.

Auditory
_____ I enjoy listening to tapes of lectures.
_____ I often need to talk through a problem aloud in order to solve it.
_____ I memorize best by repeating information aloud or explaining it to others.
_____ I use music and/or jingles to memorize.
_____ I remember best when information fits into a rhythmic pattern.
_____ I would rather listen to the recording of a book than read it.
_____ I remember names of people easily.

Visual
_____ I follow pictures or diagrams when assembling something.
_____ I am drawn to flashy, colorful, visually stimulating objects.
_____ prefer books that include pictures or illustrations with the text.
_____ I create mental pictures to help me remember.
_____ I usually remember better when I can see the person talking.
_____ I remember faces of people easily.
_____ I prefer to follow a map rather than ask for directions.

Kinesthetic
_____ I can memorize well while exercising.
_____ I usually learn best by physically participating in a task.
_____ I almost always have some part of my body in motion.
_____ I prefer to read books or hear stories that are full of action.
_____ I remember best when I can do something with the information.
_____ I solve problems best when I can act them out.

You probably checked some in each of the three areas; however, most people have a preferred modality for learning new material. See which area you checked most often and develop ways to study within that area. For example, if you are an auditory learner, buy a small tape recorder and record your lectures to listen to over and over again—perhaps while going to and from class. In lecture, sit where you can hear the professor well. Try focusing on what is being said during the class and taking notes on the material from your tape recording later. Ask the professor questions during class and office hours. Read your assignments out loud to yourself, or into your tape recorder. Keep visual distractions in your study place to a minimum so you can focus on active listening, to your own voice or to a tape. Create jingles, songs, poems, or stories to help you remember. When studying with others, have them drill you on what you're learning, or spend time explaining it to them.

If you are a visual learner, you will benefit from a different set of strategies. During class make sure you sit where you can see the professor and the board or screen clearly. Write notes during lecture with plenty of pictures and meaningful doodles. Rewrite notes later in a more organized fashion, highlighting the main ideas emphasized in class. Create diagrams, lists, charts, and pictures to help clarify material. Write out questions that you have to ask the professor, either in class or during office hours. Highlight main ideas in your books as you read, and don't be afraid to write notes in the margins. Use more than one color to emphasize different things. Use mental images to help you remember how to solve problems. When studying with others, have them give you written practice problems and tests. Use mental maps to illustrate how ideas fit together or to plan upcoming papers or projects. Use video material to reinforce concepts and events.

Example 7.3

I sat down next to a Ph.D. student for a lecture. As we got ready for class, she prepared a little differently. Instead of taking out one writing instrument, she had four—she held four different-colored pens in her left hand. She would take one at a time and use them with her right hand to write. Different ideas had different colors. As I watched, she drew diagrams between points with the different colors. Her notes looked like a work of art. She was one of the most visual learners I have ever seen. Early in her academic career she struggled until she developed methods for succeeding. And she has succeeded, all the way through her Ph.D. program in engineering.

Kinesthetic learners may have the biggest disadvantage, because most formal college learning environments lack kinesthetic activities. The exception to this would be classes with labs. Take advantage of these as often as possible, since doing something with the information you are learning is easiest for you. During lecture, make connections between what is being said and similar things you have done in the past. Talk with your professor during office hours about ways to be more hands-on with the material. Perhaps you could volunteer in your professor's lab, or use models and experiments on your own at home. Try memorizing while exercising, especially repetitive activities like jogging, running stairs, jumping rope, or swimming laps. When studying with others, try acting out the material. Shorter, more frequent study times may be more productive than a single marathon session.

Thinking Skills

The way the brain works is much more complex than simply adding new material to old. Thinking refers to how we see the world, approach problems, and use the different parts of our brain. As science has rapidly improved its ability to measure the brain's activity, our understanding of brain functions has dramatically changed. As research continues, our understanding of this wondrously complex organ is continuing to evolve.

Models have been developed to try to categorize how different individuals think. Early models focused on the specialization of the right and left hemispheres of the brain. Further research has unlocked other complexities and required the development of more complex models. A common current model was developed by Ned Herrmann and shows the brain as four quadrants. Originally, these were four physiological components of the brain. As research continued, it was determined that brain function was too complex to be precisely represented in this format. However, the model has proven effective as a representation of our preferred thinking style and the four quadrants are currently used as a metaphorical model of how we prefer to process information. Currently, the four quadrants are used as a metaphorical model.

Figure 7.1 shows the evolution of the quadrant model from the hemispherical models. Quadrants A and B represent the "left brain" and quadrants C and D represent the "right brain." Herrmann's complete model is detailed in his book *The Creative Brain*. These quadrants illustrate personal preferences and a dominant thinking style. However, most people exhibit characteristics of each quadrant to some degree. Successful people are aware of their strong

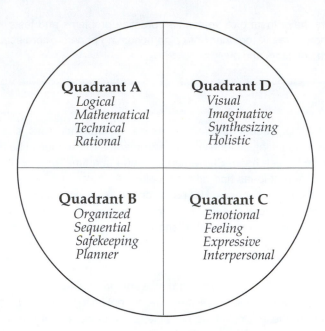

**Figure 7.1 Herrmann's "whole brain model" describing four quadrants
of the brain**

(Copyright Herrmann International.)

areas and can capitalize on them while compensating for their weaker areas, and possibly building them up.

Tables 7.1–7.4 highlight examples of the various thinking skills, preferred learning activities and practice activities for each quadrant. Examine the lists and try to place yourself in one or more quadrants.

The most successful students pick and choose a variety of strategies that work best for their learning style and specific classes. Observe what other successful students are doing and learn from their example. Remember, your one-of-a-kind brain is very capable of learning new material. Spend some time learning how you learn best and you've paved the way for successful learning.

The model can also be used to help understand professors who have a different thinking style. Most people teach like they would like to be taught. If you are in a class where your professor has a very different style than you do, use the model to problem-solve ways to get past this hurdle. It is great experience for later in life. In your professional career, you will have to deal with all different styles effectively to be successful. Being aware of the differences and being able to work with them will be a tremendous asset.

If you would like to know which quadrants apply to you, you can take an assessment profile called the Herrmann Brain Dominance Instrument. An academic advisor or testing center on campus should be able to provide you with information about the HBDI. You may also contact Herrmann International at 794 Buffalo Creek Road, Lake Lure, North Carolina 28746, or at their website at *www.hbdi.com.*

TABLE 7.1 Quadrant A

Thinking	Preferred Learning Activities	Practice Activities
Factual	Collecting data and	Develop graphs, flow charts
Analytical	information	and outlines from info
Quantitative	Listening to lectures	and data
Technical	Reading textbooks	Do a library search
Logical	Studying example	on a topic of interest
Rational	problems	Join an investment club
Critical	and solutions	Do logic puzzles or games
	Using the scientific	Play "devil's advocate"
	method to do research	in a group decision process
	Doing case studies	Break down a machine
	Dealing with things	and identify the parts
	rather than people	and their functions

TABLE 7.2 Quadrant B

Thinking	Preferred Learning Activities	Practice Activities
Organized	Following directions	Cook a new dish following a
Sequential	Doing repetitive, detailed	a complicated recipe
Controlled	homework problems	Organize a desk drawer or closet
Planned	Doing step-by-step lab work	Prepare a family tree
Conservative	Using programmed learning	Be exactly on time all day
Structured	and tutoring	Assemble a model kit
Detailed	Listening to detailed lectures	by instructions
Disciplined	Making up a detailed budget	Develop a personal budget
Persistent	Practicing new skills	Learn a new habit through
	through repetition	self-discipline
	Making time management	Set up a filing system
	schedules	for your paper work
	Writing a "how to" manual	

TABLE 7.3 Quadrant C

Thinking	Preferred Learning Activities	Practice Activities
Sensory	Sharing ideas with others	Study in a group
Kinesthetic	Learning by teaching others	Share your feelings with a friend
Emotional	Using group study opportunities	Get involved in a play or musical
Interpersonal	Enjoying sensory experiences	Volunteer in your community
Symbolic	Taking people-oriented field trips	Explore your spirituality
	Respecting other's rights and views	Become a penpal to someone
	Study with background music	from another culture

TABLE 7.4 Quadrant D

Thinking	Preferred Learning Activities	Practice Activities
Visual	Getting actively involved	Daydream
Holistic	Using visual aids in lectures	Play with clay, Legos, Skill Sticks
Innovative	Leading a brain storming session	Focus on big picture,
Metaphorical	Experimenting	not the details of the problem
Creative	Doing problems with many answers	Create a logo
Imaginative	Relying on intuition, not facts	Invent a new recipe and prepare it
Conceptual	Thinking about the future	Imagine yourself 10, 20, 50
Spatial		years in the future
Flexible		Use analogies and metaphors
Intuitive		in writing

7.8 WELL-ROUNDED EQUALS EFFECTIVE

Being an effective person goes beyond being a good student. Developing the habits that make you truly effective during your college years will set you up for success in life. Part of being effective is functioning at full capacity. To do this, you must have the various dimensions of your life in order. One analogy that is frequently used is to look at a person as a wheel. A wheel will not roll if it is flat on one side. Similarly, people cannot function optimally if there is a problem area in their life. Five key areas are graphically represented in Figure 7.2. It is important to maintain each area to become as effective as you can be.

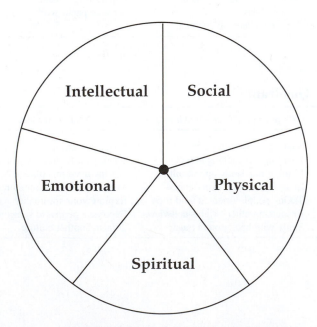

Figure 7.2   Areas of wellness for students

Intellectual

One of the main purposes of going to college is to expand the intellectual dimension of your life. Take full advantage of the opportunities at your institution. Besides engineering courses, schools will require general education courses (languages, humanities, or social science classes). These non-technical classes are a great way to broaden yourself. Many students find these classes a refreshing break from their engineering classes.

In addition to course work, colleges have a wide range of activities to enhance your intellectual development. These range from student organizations to seminars to special events. Planning some non-engineering intellectual activities will help to keep you motivated and feeling fresh for your engineering classes. Some students find that reading a novel or two during the semester also provides a refreshing, intellectual break from engineering studies.

Social

The social aspect of college is one that most students seem to master. Planning appropriate social activities will make your college experience more fun. There is a strong correlation between enjoying your college experience and being successful. Also, establishing a group of friends to socialize with will provide a support structure to help you when times get tough. College friends also provide a network that can be very beneficial later in your professional life.

If you have trouble identifying appropriate social activities, there are campus resources to help you. The first place to go is your residence hall staff. Also, most campuses have a central office for student activities. This office can provide you with a list of student organizations and activities.

Physical

All the studying and preparation is a waste of time if you are not physically able to perform when the tests or quizzes come. Also, you can waste time if you are not able to study efficiently. There are three physical areas that are essential to becoming an effective student. They are fitness, sleep, and nutrition.

People are more productive when they are physically fit, and students are no exception. Exercise is an activity which will not only help you study better but will allow you to live a longer, more productive life. College is a great time to develop a fitness habit. Fitness activities can include aerobics, jogging, biking, walking, or participating in intramural sports. On every campus there are people who can help you develop a fitness plan that is right for you.

Sleep is an area that many students abuse. Finally they are away from their parents and can go to bed any time they want. The problem is that classes still come at the same time regardless of when you fall asleep. Sleep deprivation reduces a person's productivity, which includes studying. Each person needs a different amount of sleep. Typically, most adults need six to eight hours of sleep. It is important to schedule enough time for sleep or you won't be able to be effective in the classroom. Studies have shown that student performance is actually reduced by staying up very late the night before a test to study. You cover more material the night before the test but are so tired at test time that you can't retrieve the material. Honestly evaluate how much sleep you need. Do you fall asleep in class? If so, you probably aren't getting enough rest. Another test for sleep deprivation is to not set an alarm one morning and see when you wake up. If you are on a schedule, most people will wake up at the same time without the alarm. If you sleep several hours past when the alarm would have gone off, you very possibly are sleep-deprived and need more rest.

The third physical area is diet. Again, Mom and Dad are not there to make you eat that broccoli, so you don't. Our bodies are designed to work properly with the right input. While you don't have to become a diet fanatic, eating a sensible diet will enhance your ability to succeed in your academics.

A final note on the physical dimension pertains to non-dietary substances, such as alcohol. The number one problem on most college campuses is alcohol abuse. Every year, thousands of very capable students fail due to alcohol. If you choose to indulge, make honest assessments of the impact on your studying. Don't let alcohol become the barrier to your success.

Spiritual

Few would disagree that humans are spiritual beings and function much better and more efficiently with a balanced spiritual component in their lives. Stephen Covey, in his book *The Seven Habits of Highly Effective People*, puts it this way:

> *The spiritual dimension is your core, your center, your commitment to your value system. It's a very private area of life and a supremely important one. It draws upon the sources that inspire and uplift you and tie you to the timeless truths of all humanity. And people do it very, very differently.*
>
> *I find renewal in daily prayerful meditation on the Scriptures because they represent my value system. As I read and meditate, I feel renewed, strengthened, centered and recommitted to serve.*

Search for the spiritual activities that keep you renewed. This may be through a small group study, attending campus worship services, meditating, or immersing yourself in good music for a time of personal reflection. If you aren't sure where to start, try something which is close to what you were used to before college. This will probably be the most comfortable for you. If it doesn't seem to be the right thing for you, branch out from there.

Emotional

Bill Hybels, pastor of a suburban Chicago church, explained the need to attend to this dimension when he addressed a gathering in 1996 in Detroit. He is an avid runner, eats well, and is physically sound. He has many casual friends and is socially active. He has written a number of books and articles and is sound intellectually. Yet he reached an emotional crisis that prevented him from being effective in his job and with his family.

Hybels describes emotional reserves like a gas tank. We need to put reserves into our tank so we have the ability to take it out when we need it. There will be times in your college career, and beyond, when you need those reserves. The lesson Hybels shared is the need to watch the gas gauge and not to let it run out. Students are no different.

A positive step to take to keep the emotional dimension in check is to set up a support network. This network is what you use to make deposits into your emotional tank. Friends are an integral part of the network. Do you have a person or people whom you can talk honestly with about important issues in your life? This might be a close friend, a parent, a relative, an advisor, a professor, or a counselor. Cultivate at least one friendship where you feel the freedom to share honestly.

The second aspect to maintaining emotional reserves is to schedule time that is emotionally neutral or energizing. Many students get themselves into a lot of activities that are emotionally draining and don't schedule time for themselves for recovery. Eventually this catches up with them and adversely impacts their effectiveness.

A final note on this area is that on each campus there are advisors and counselors who are available to talk to you. Don't feel that you have to have a severe psychological problem to simply talk with them. These people are trained to help you and will just listen if that is what you need. They are a great resource for maintaining the emotional dimension and thereby helping to keep your academic effectiveness high.

7.9 YOUR EFFECTIVE USE OF TIME

You may be wondering how to find time to maintain all these areas every day: studying two hours for each hour spent in class, keeping a balanced lifestyle, etc. And what happened to all that fun time you've heard about? Aren't the college years supposed to be the best years of your life?

The answer is effective time management. Time management is a skill that will help you succeed as a student and will serve you well throughout your entire career. Mastering time management will set you ahead of your peers. It is interesting that when engineers and managers are surveyed and asked why they don't take the time to manage their time, the most common response is inevitably that they don't have time. Students have the same feelings.

If we consider the academic requirements with a 15-credit load, we will have 45 hours per week of class and study time. That could be done Monday through Friday from 8 a.m. until 6 p.m., assuming you take an hour for lunch every day. Think about it. That would give you every evening and every weekend free to have fun or to do other activities.

Many students don't believe they can do this, but the hours add up. The key to making such a schedule work is to make every minute count. Treat school like a full-time job, working similar hours. This will not only help you to effectively manage your life during your academic years, but it also will provide excellent preparation for your future work life.

Let's assume that you have a schedule like the one in Table 7.5. This student has Chemistry, Math, English, Computer Science, and an Introduction to Engineering course. On Monday, the day starts at 8 a.m. with two classes back to back. Then there is an hour break before

TABLE 7.5 Sample Class Schedule

	Mon	Tues	Wed	Thurs	Fri
8:00	English		English		English
9:00	Chem 101		Chem 101		Chem 101
10:00		Chem Lab		CS Lab	
11:00	CS 101	Chem Lab	CS 101	CS Lab	
12:00		Chem Lab			
1:00				Chem Rec	
2:00	Math	Math	Math		Math
3:00		Engr 101		Engr 101	
4:00					
5:00					
6:00					
7:00					
8:00					
9:00					

Computer Science. If you are on the job, you need to be working, and we'll define working as attending class or studying. You need to find a place to study during this hour break. This is where many students go astray. They will waste an hour like this between classes. Similarly, there is a two-hour time block before Math class. This is time to grab lunch and get some more studying done.

It will take some planning to make your day effective because you will need to find suitable places to study. You can use these times to do your individual studying or meet with a study partner or group. It is much more efficient to decide in advance what you will study and when. It can be a waste of precious time to try decide which subject you will study during breaks between classes. Plan enough time for each subject so you can study from class to class and not fall behind in any way. For example, you can't do all your studying for Chemistry on Monday, Math on Tuesday, and Computers on Wednesday or you will fall behind the lectures.

Another thing that scares many students away from effective time management is that it sounds too rigid. One of the reasons for going to college is to have fun and meet people. How can you "go with the flow" if everything is scheduled? The answer is responsible flexibility. If you have your schedule planned out, you can decide when it's a good time to interrupt your "work time" to have some fun. A criterion for making this decision is recovery time. In your schedule, is there enough non-work time to convert to work time before deadlines arrive in your classes? For instance, taking a road trip the day before an exam would be irresponsible. However, right after you have midterms, there might be a window of time to take such a break and still recover.

Scheduling two hours of study per class period is usually enough to keep up during your first year. However, there are classes or times in a semester when this won't be enough. Once you fall behind, it is hard to catch up. To avoid this, a weekly make-up time is an excellent idea. This is like an overtime period at work. At an aerospace company where I worked, Saturday morning from 6 to 10 was my favorite time to schedule this overtime work. This didn't interfere with anything else in my normal schedule and I could get the work done and still have a full Saturday to enjoy. For some reason, 6 a.m. on Saturday isn't a very popular time for students. However, an evening or two or Saturday afternoon will work, too. These extra time blocks will allow you to keep up and give you the flexibility to take time off with a friend for something fun but still stay on your weekly schedule.

Scheduling Work and Other Activities

No matter how good you get at squeezing in study time between classes, there are still only 24 hours in each day. As a result, there is a limit to what anyone can do in one week. Planning a schedule complete with classes, study time (including make-up time), eating, sleeping, and other activities is essential to deciding how much is too much. If you need to work while you're in school, schedule it into your week. If you join a student organization, plan those hours into your schedule. You need to see where it all fits. A common pitfall with students is that they over-commit themselves with classes, work, and activities. Something has to give. It is easier to drop an activity than to have to repeat a class.

Ray Landis, from California State University at Los Angeles, describes a 60-hour rule to help students decide how much to work. Fifteen credit-hours should equate to 45 hours of effort. That leaves 15 hours for work, totaling 60 hours. Students may be able to work more than this, but each individual has a limit. You need to learn for yourself what your limit is.

If you need to work to make it through college financially, you may need to look at the number of credits you can handle effectively each semester. A schedule can help guide you in determining what is manageable. Before deciding to take fewer credits so you can work more hours, examine the value of the money you can make at your job compared with the starting salary you would expect by graduating a semester earlier. In some cases, it will make sense to work more while taking fewer classes. In others, it will make financial sense to take out a loan, graduate sooner, and pay the loan back with your higher salary.

Calendars

It is important to have a daily and weekly calendar. Another critical item in effective time management is a long-range schedule. For students, this is a semester calendar. This will help you organize for the crunch times, like when all five of your classes have midterms within two days, or you have three semester-projects due on the same day at the end of the semester. It is a fact of student life that tests and projects tend to be bunched together. That is because there are educationally sound times to schedule these things, and most professors follow the same model. Given this fact, you can plan your work schedule so that you spread out your work in a manageable way. This is the part you have control over.

Once, while I was in school, we complained to a professor about this on the first day of class. He offered to give the final exam in the next class period and have the semester project due the second week of the semester to free us up for our other classes. We declined his offer and decided it was best to learn the material first.

Organizers: Paper or Electronic?

Calendars and task lists can be kept in numerous forms, and you will need to determine which is the most effective for you. There has been a proliferation of electronic tools and devices that are designed to help people communicate and be organized. BlackBerries and cell phones can store tasks and calendars and be used to replace paper calendars. Laptops can be equipped with software such as Microsoft Outlook or Franklin Covey software to manage your tasks and keep appointments. Even with all the technology available, many people still find a role for paper lists or traditional calendars. Just like with the learning styles and study strategies, the method you use to organize your tasks needs to work for you. It does not have to impress your friends with how "tech-savvy" you are.

Do not feel like you have to adopt the latest technology to be effective. The media you use to organize your schedule is a personal choice and needs to work for you. It can be electronic, paper, or a combination, a little of each. The important thing is that it works for you. Our recommendation is to explore different approaches and test how they work and fit for you, your style, and needs. Stay flexible as your needs change and as new approaches and devices come on-line.

Organizing Tasks

A common method for organizing one's day is to make a "To Do" list each morning (or the evening before). This is a good way to ensure that everything needing attention gets it. Things not done on one day's list get transferred over to the next.

The one drawback to making a list is determining what to do first. You can approach the tasks in the order you write them down, but this ignores each task's priority. For instance, if you have a project due the following day for a class and it gets listed last, will it get done? A popular planning tool is the Franklin Planner produced by the Franklin Covey Co. (Material here used courtesy of Franklin Covey.) They stress the importance of prioritizing your life' activities according to your values, personal mission statement, and goals. They suggest using a two-tier approach to prioritizing activities. The first is to use A, B, and C to categorize items as follows:

A = Vital (needs attention today)
B = Important (should be taken care of soon)
C = Optional (no one will notice if it is not done today)

Keep your long-term and intermediate goals in mind when assigning priority. And be sure to write simple statements for each task on your list, beginning each item with an action verb.

Once you have categorized the items, look at all the A priorities. Select the item to do first and give it a 1. This item now becomes A1. The next becomes A2, etc. Now you can attack the items on your list starting with A1 and working through the A's. Once you have finished with the A's, you move on to B1 and work through the B's. Often, you will not get to the C priorities, but you have already decided that it is okay if they don't get done that day.

Table 7.6 shows a sample list for a typical student's day. If you are starting a time management routine, include your list. This keeps you reminded of it, and it gives you something to check off your list right off the bat!

Some lessons can be learned by looking at how this student has scheduled activities. One is that paying the phone bill is not necessarily a top priority, but it requires little time to address. A helpful approach is to look at the quickly done A items and get them out of the way first. Then move on to the more time-consuming items. Also note that a workout was planned into the schedule, right after the chemistry lab. The workout is a priority that comes after some extensive studying. This can help motivate the student to finish a difficult task and then move on to something enjoyable.

Looking at this schedule, you may think that poor Stacy is never going to get called or the letter to the parents written. It often happens that the more fun things are, the more likely

TABLE 7.6 Sample Prioritized List of Tasks

Priority	Task
A3	Finish chemistry lab (due tomorrow)
A5	Read math assignment
B3	Do math homework problems
B4	Outline English paper
A4	Workout at Intramural Building
B2	Debug computer program
A6	Annotate math notes
B1	Annotate chemistry notes
A7	Spiritual meditation
C1	Call Stacy
C2	Write parents
A2	Mail phone bill
C3	Shop for CDs
A1	Plan day

they end up a C priority. But it is okay to put relationship-building or fun activities high in priority. People need to stay balanced. Maybe today Stacy is a C, but tomorrow might have that call as an A priority.

The Franklin Covey people provide additional help for identifying priorities. They rate tasks by two criteria, urgency and importance, as shown in Table 7.7. Franklin Covey stresses that most people spend their time in Quadrant I and III. To achieve your long-term goals, Quadrant II is the most important. By spending time on activities that are in II, you increase your capacity to accomplish what matters most to you. The urgency is something that is usually imposed by external forces, whereas importance is something you assign.

You may well want to use a schedule that shows a whole week at a time and has your goals and priorities listed. An general example is shown in Table 7.8.

A weekly schedule allows you to see where each day fits. It also keeps your priorities in full view as you plan. Franklin Covey offers many versions of their Panner, including the Collegiate Edition. Contact them at 1-800-654-1776 or visit their website at www.franklin covey.com.

TABLE 7.7   Franklin Covey's 'Time Matrix' for Daily Tasks

(copyright 1999 Franklin Covey Co.)

	Urgent	**Not Urgent**
Important	I • Crises • Pressing deadlines • Deadline-driven projects, meetings, preparations	II • Preparation • Prevention • Values clarification • Planning • Relationship building • Recreation • Empowerment
Not Important	III • Interruptions, some phone calls • Some mail, some reports • Some meetings • Many pressing matters • Many popular activities	IV • Trivia, busywork • Some phone calls • Time-wasters • 'Escape' activities • Irrelevant mail • Excessive TV

Stay Effective

There are two final notes on time management that can help you stay effective. The first is to plan to tackle important activities when you are at your best. We are all born with different biological clocks. Some people wake up first thing in the morning, bright-eyed and ready to work. These people are often not very effective in the late evening. Others would be dangerous if they ever had to operate heavy equipment first thing in the morning, especially

TABLE 7.8 Sample Generic Weekly Schedule and Planner

Weekly Schedule			Sunday	Monday	Tuesday	Wednesday	Thursday	Friday	Saturday
Musts	**Goals**				**Day's Priorities**				
Academics:				Run laps		Run laps		Finish CS	Run laps
Math	Finish Prob							Finish Math	
Chem lab	Anal. data							Movie	
Comp Sci	Debug Prog								
English	Essay								
					Appointments/Commitments				
		8:00		English		English		English	
		9:00	Chapel	Chem		Chem		Chem	
Hall Dance			Synagogue						
		10:00			Chem Lab		CS Lab		
		11:00		CS		CS			
		12:00							
		1:00					Chem Rec		
		2:00		Calculus	Calculus	Calculus		Calculus	
		3:00		Meet prof		Engr 101		Engr 101	
		4:00							
		5:00							
Physical	Run and Ride	Evening							
Mental	Read a novel								
Spiritual	Devotions								
Social	Hall dance								
Emotional	Lunch w/Dave								

before that first cup of coffee. However, once the sun goes down, they can be on a tear and get a lot done. Know yourself well enough to plan most of your studying when you are at your best. It is a common tendency among students to leave studying until late in the evening. This is often because they haven't done it earlier in the day. There are always stories about how the best papers are written at 3:00 a.m. Some are, but most are not. Know yourself well enough to plan effectively.

And finally, pace yourself. You are entering a marathon, not a sprint. Plan your study time in manageable time blocks and plan built-in breaks. A general suggestion is to take a 5-10 minute break for every hour spent studying. This will help keep you fresh and effective.

7.10 ACCOUNTABILITY

The single most important concept to being successful is to establish accountability for your goals and intentions. The most effective way to do this is with an accountability partner. This is someone whom you trust and from whom you feel comfortable receiving honest feedback.

Ideally, you will meet regularly with this person and share how things are going. Regularly means about once per week. Throughout a semester you can fall behind quickly, so weekly checks are a good idea. The person you choose can be a peer, a mentor, a parent, a relative, or any one you feel comfortable with, respect, and trust.

Your accountability could also be to a group of people. Bill Hybels describes the change in his life when he entered into an accountability group with three other men. These men held each other accountable in all areas of their lives—professional, familial, and spiritual. You can do the same as a student.

The first step is to give your accountability partners the standard to which you want to be held. An easy way to do this is to share your goals with them. Let them see and hear from you where you want to be in five years, in ten years, and beyond that. Let them also see the short-term goals you have set to achieve those longer-term objectives. Then, weekly, share with them how you are doing in meeting your goals. You will find that it is much easier to stay on course with the help of at least one close friend. Success will come much easier.

7.11 OVERCOMING CHALLENGES

We all at some point in our lives encounter adversity. As a student, adversity may take the form of a bad grade on a test or for an entire course, a professor you feel is treating you unfairly, not getting into the academic program of your choice, or any of a myriad other college-related obstacles. Since everyone encounters adversity and failure at some point in their life, your personal success is not always dependent on how well you avoid failing, but rather on how you learn from it, move on, and improve. This is especially true as a student.

Almost every student will run into a course or a test where they struggle. The key is to avoid focusing on the negative and continuing to be productive. Every academic advisor has stories about good students who tripped up on a test or in one class and let that affect their performance in other classes.

Focus on "Controllables"

If something bad happens, figure out what can be done. This may mean writing down a list of actions you can take. Don't focus on things you can't change. If you receive a low grade on a test, you can't change the grade. You may, however, be able to talk to the professor about the test. You can study the material you missed and be ready for the next one.

Keep It in Perspective

Examine the real impact of the situation. What are the consequences of the situation? One class can affect your graduation GPA by about 0.05. This will not produce life-altering consequences, unless you would have graduated with a 4.0! If you are unsure about the consequences, see your professor or an academic advisor and have them help you put it in perspective.

Move to the Positive

It is very important if you hit a bump in the road to move beyond it quickly and remain effective in your studies. It is necessary to examine the problem to learn from it. Perhaps you failed a test. This is a great opportunity to examine how you are studying and take corrective action to improve your effectiveness. You can affect your present and future, but not your past.

Remember, Everyone has Setbacks

Again, keep things in perspective. If you run into difficulty, you are not alone. There are numerous examples of setbacks and even failures.

Henry Ford neglected to put a reverse gear in his first automobile. But he recovered and went on to be successful.

Abraham Lincoln, likely our greatest president, provided a tremendous example of overcoming adversity. Here is a brief list of some of his accomplishments and setbacks:

1831 – failed in business	1843 – defeated for Congress
1832 – defeated for legislature	1846 – elected to Congress
1833 – failed in business again	1848 – defeated for Congress
1834 – elected to legislature	1855 – defeated for Senate
1835 – fiancée died	1856 – defeated for Vice-President
1838 – defeated for Speaker	1858 – defeated for Senate
1840 – defeated for Elector	1860 – elected President

Lincoln never stopped achieving just because he ran into a setback. You don't have to either.

Another example comes from a grandfather who relayed a story about someone in his family failing. "We were visiting your grandmother's brother and his family. It was the time of year when college was in session, so I was surprised when my nephew, Hughey, opened the door and greeted us. He was in his first year of college at the state university. I jokingly said, "What happened to you, Hughey? Flunk out already?" I sure felt stupid when he replied that in fact he had. After that I felt obligated to try to help him out. That evening, he, his father and I talked about what he was going to do. I encouraged him to enroll in the other big state university. I had heard a lot of good things about it even though their science and engineering programs were not as prestigious. He enrolled and did very well. Not only did he graduate, he even got his master's degree before taking a job with a top company."

As a footnote to the story, Hughey continued to do well. The job he took was with a Fortune 100 company. He married and had four children and five grandchildren at last count. He stayed with the same company for his entire career and retired a senior vice president, one step below CEO. Failing in his first attempt at college didn't stop him. He learned from it, moved on and did very, very well. You can do the same with the right attitude and effort.

REFERENCES

Covey, S.R., *The 7 Habits of Highly Effective People*, Simon and Schuster, New York, 1989.

Douglass, S. and A. Janssen, *How to Get Better Grades and Have More Fun*, Success Factors, Bloomington, Indiana, 1997.

Herrmann, N., *The Creative Brain,* Brain Books, Lake Lure, North Carolina, 1988.

Landis, R., *Studying Engineering, A Road Map to a Rewarding Career*, Discovery Press, Burbank, California, 1995.

Lumsdaine, E., and M. Lumsdaine, *Creative Problem Solving, Thinking Skills for a Changing World*, 2nd ed., McGraw-Hill, New York, 1993

Maxwell, J. C., *The Winning Attitude, Your Key to Personal Success*, Thomas Nelson Publishers, Nashville, Tennessee, 1993.

Tobias, C.U., *The Way They Learn, How to Discover and Teach to Your Child's Strengths*, Focus on the Family, Colorado Springs, Colorado, 1994.

EXERCISES AND ACTIVITIES

7.1   Identify three things you can do to keep a positive attitude toward your studies.

7.2   What could you have done if you were in the Physics class as described in Section 7.2?

7.3   Actually writing goals as described is an important element in defining your goals.
   a) Write five goals to achieve by graduation.
   b) Write five goals to reach within ten years.
   c) Write ten goals to reach during your lifetime.

7.4   Write some specific goals:
   a) Set a grade goal for each class this semester.
   b) Calculate your expected semester GPA.
   c) Define a stretch goal for your semester GPA.

7.5   Find out what the following grade point averages are:
   a) Minimum GPA to stay off academic probation
   b) Minimum GPA to be admitted to Tau Beta Pi
   c) Minimum GPA to be admitted to the engineering honor society in your engineering discipline (e.g., Pi Tau Sigma for mechanical engineering, Eta Kappa Nu for electrical engineering)
   d) Minimum GPA to be on the Dean's List or Honors List
   e) Minimum GPA to graduate with honors

7.6   Identify a particular employer and an entry-level position with that employer that you would like to obtain following graduation. Find out what is required to be considered for that position.

7.7   List the barriers you see to achieving your graduation goals.
   a) Categorize them as those within your control and those beyond your control.
   b) Develop a plan to address those within your control.
   c) Develop strategies to cope with those beyond your control.

7.8   For each class you are currently taking:
   a) Map out where you sit in class.
   b) Rate your learning effectiveness (A, B, C, D, or F) for each class.
   c) For each class not rated an A, develop a plan to increase your learning effectiveness.

7.9   For one week, make a log of everything you do. Break it up into categories of attending class, studying, working, playing, eating, sleeping, physical activity, intellectual pursuits, emotional health, spiritual well-being, and other.
   a) Total your time spent on each area. Are you spending the time needed on your studies (2 hours for each hour in class)?
   b) Is this the kind of schedule that will make you successful? If not, what will you change?

7.10 Review your current class schedule and identify times and places you can study between classes.

7.11 Develop a "work" schedule. Identify the times you are "on the job," that is, studying and taking classes. Identify a make-up time for weekly catch-up.

7.12 List all your current courses. Put a "+" next to each one in which you are caught up. Put a "−" next to each one in which you are not caught up right now. ("Caught up" is defined as prepared for the next lecture with all assignments completed and all assigned reading done.)

7.13 Ask one of your professors the main objective of his or her course and report the response. Does this information help you in organizing your studying?

7.14 Annotate your notes from a lecture in one of your courses, then take those notes in to your professor to get his or her comments on your effectiveness. Report the professor's comments.

7.15 Make a list of the students you already know who are pursuing the same major as you. Then make a list of the students you know in your science and math classes. Rate each one according to their suitability as a study partner.

7.16 Identify one or more people with whom you might form a study group. Try studying together on a regular basis until the next test. Report on your experience.

7.17 List the group study rules you used in your group studying. Report on the things that went well and those that did not.

7.18 Make a list of problems you can work, in addition to the assigned homework problems, in one of your math, science, or engineering courses. Report on the method you used to identify these problems.

7.19 Visit one of your professors and report three personal things about that professor. (For example: Where did she go to school? Why did he choose the field he is in? What does she like best about her job? What one piece of advice does he have for freshmen that he didn't share in class?)

7.20 After completing Exercise 19 above, did this change your feelings about your professor? Does he or she seem more approachable now? Did you learn anything else during your visit that could help you succeed in that class?

7.21 Identify at least one opportunity for you as an undergraduate to become involved in engineering research. What advantages could you gain from this experience?

7.22 Identify the type of environment in which you study the best. Describe it and include a sketch of the location.

7.23 Test your conclusion in Exercise 22 with an experiment. Do one assignment in one environment (perhaps in your room, alone and totally quiet) and another in a different environment (perhaps in a more public place, with background noise). Report your findings.

7.24 At the halfway point in the semester, calculate your grades in each class. Compare these to the goals set in Exercise 4. Are you on track? If not, identify the changes you will make to reach these goals.

7.25 Define at least one long-term goal in each of the five wellness areas discussed in this chapter. Next, define short-term goals to achieve these long-term goals.

7.26 List at least two needs you have in each of the five wellness areas. List at least two activities you can plan to meet these needs.

7.27 Rate your roundness in the wellness wheel (on next page), using 5 for very well, 4 for well, 3 for okay, 2 for some improvement needed, and 1 for much improvement needed. Do you have a well functioning wheel? If not, identify steps to correct the problem areas. List at least two new activities for each problem area.

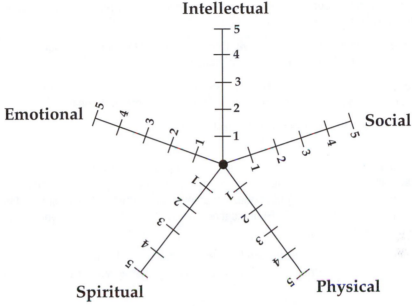

Figure for Exercise 7.27   Wellness assessment.

7.28 Write a paragraph on how each wellness area can help you succeed:
a) as an engineering student.
b) as a practicing engineer.

7.29 Make a weekly schedule with two hours of studying for every hour spent in class. Identify activities which will help develop the different areas in your life. Is this a sustainable schedule? If the answer is no, identify things which can be dropped.

7.30 Identify on your weekly schedule the times of each day when you study most efficiently. Are these the times you are actually scheduled to study? If not, explain why you've scheduled other items instead of studying at these times.

7.31 Adopt one of the weekly schedule instruments discussed in this chapter. Try using it for two weeks and report the results. Was this the right instrument for you?

7.32 Make a "To Do" list. Prioritize the list using A, B and C categories, and then number the items in each category.

7.33 Place the items of Exercise 7.32 into the four quadrants of Covey's matrix shown in Table 7.1. How many A priorities are in quadrant II? Based on this ranking of your "To Do" list, would you change any of your priorities?

7.34 List the top three things you are failing to do that would most help you succeed in your career. Briefly explain why you are currently failing to do each item.

7.35 List three people you would consider for an accountability partner.

7.36 Make a list of at least five questions you want your accountability partner to ask you at each check-up.

7.37 Meet with an accountability partner for one month. Report back on your experience. Did it help? Is this something you will continue?

7.38 Call home and tell your parents that you are developing a success strategy for college. Explain the strategy to them.

7.39 Assume that your grades will be lower than expected at the end of the semester, and write out a strategy for recovering from these low grades.

7.40 Make a list of at least three university people who can help you deal with a problem that is preventing you from succeeding. List each by name and position.

7.41 Make a list of your activities over the last 24 hours. Compare the list and the hours spent on each with your planned schedule.

7.42 List your long-term goals, including every area of your life. Fill out a schedule of what you actually did during the last seven days. Put each activity into categories corresponding to one of your long-term goals, or an extra category titled "not related." Grade yourself (using A, B, C, D, or F), indicating how well you did the past week working toward your long-term goals.

7.43 Take the inventory test discussed in Section 7.7 on memory languages. Which is your dominant mode? List three actions you can take to improve your learning based on this dominant mode.

7.44 Which of Herrmann's quadrants best describes your dominant thinking style? Which one is the weakest?

7.45 Which of Herrmann's four quadrants is best suited for engineering? Are there any dominant modes of thinking that are not ideal for engineering? Explain your answers.

7.46 List all your current professors and identify their probable Herrmann quadrant. Compare this with your own style. For the professors with styles that differ from yours, develop an action plan to deal with these differences.

7.47 Explore the availability of electronic organizers and prepare a brief report on the options for students.

7.48 Compare and contrast electronic organizers versus paper ones. What are the benefits of each and which would you recommend to your classmates?

7.49 Interview two professors and report what they use to organize and prioritize their time.

Chapter 8

Problem Solving

8.1 INTRODUCTION

Seven-year-old Dan could build all kinds of things: toy cars, ramps, and simple mechanical gadgets. Even more amazing was that he could do all this with four simple tools—scissors, hammer, screwdriver, and duct tape. One Christmas, his father asked if he'd like more tools. "No thanks, Dad" was his response. "If I can't make it with these, it's not worth making." It wasn't long, however, before Dan realized that the cool stuff he could create or the problems he could solve were dependent on the tools he had. So he began gathering more tools and learning how to use them. Today, he still solves problems and makes a lot of cool stuff—like real cars, furniture, playgrounds, and even houses. His tools fill a van, a workshop, and a garage. Having the right tools and knowing how to use them are critical components of his success.

Likewise, most people, when they are children, learn a few reliable ways to solve problems. Most fifth graders in math class know common problem-solving methods like "draw a picture," "work backwards," or the ever-popular "guess and check" (check in the back of the textbook for the correct answer, that is). Unfortunately, many people choose to stop adding new problem-solving tools to their mental toolbox. "If I don't know how to solve it, it's probably not worth solving anyway," they may think. Or, "I'll pass this problem over for now and get back to my usual tasks." This severely limits them. Engineers, by definition, need to be good at solving problems and making things. Therefore, filling your toolbox with the right tools and knowing how to use them are critical components to becoming a successful engineer.

The goal of this chapter is to increase the number of problem-solving strategies and techniques you commonly use, and to enhance your ability to apply them creatively in various problem-solving settings. Hopefully, this will be only the beginning of a lifelong process of gaining new mental tools and learning how to use them.

8.2 ANALYTIC AND CREATIVE PROBLEM SOLVING

The toolbox analogy is very appropriate for engineering students. Practicing engineers are employed to solve a very wide variety of problems. In the Engineering Majors and the Profiles chapters in this textbook we discuss how engineers can end up filling a number of different roles in daily practice. These roles can require very different problem-solving abilities. Problem solving can be broken into Analytic and Creative. Most students are more familiar with Analytic problem solving where there is one correct answer. In Creative problem solving there is no single right answer. Your analytic tools represent what's in your toolbox. Your creative skills represent how you handle your tools.

To better understand the difference between these two kinds of problem solving, let's examine the one function that is common to all engineers: design. In the Design chapter, we detail a design process with ten steps. Those ten steps are:

1. Identify the problem
2. Define the working criteria/goal(s)
3. Research and gather data
4. Brainstorm for creative ideas
5. Analyze
6. Develop models and test
7. Make the decision
8. Communicate and specify
9. Implement and commercialize
10. Prepare post-implementation review and assessment

By definition, design is open-ended and has many different solutions. The automobile is a good example. Sometime when you are riding in a car, count the number of different designs you see while on the road. All of these designs were the result of engineers solving a series of problems involved in producing an automobile. The design process as a whole is a **creative problem-solving process.**

The other method, **analytic problem solving,** is also part of the design process. Step 5, Analyze, requires analytic problem solving. When analyzing a design concept, there is only one answer to the question "Will it fail?" A civil engineer designing a bridge might employ a host of design concepts, though at some point he or she must accurately assess the loads the bridge can support.

Most engineering curricula provide many opportunities to develop analytic problem-solving abilities, especially in the first few years. This is an important skill. Improper engineering decisions can put the public at risk. For this reason, it is imperative that proper analytical skills be developed.

As an engineering student you will be learning math, science, and computer skills that will allow you to tackle very complex problems later in your career. These are critical, and are part of the problem-solving skill set for analysis. However, other tools are equally necessary.

Engineers also can find that technical solutions applied correctly can become outmoded, or prove dangerous, if they aren't applied with the right judgment. Our creative skills can help greatly in the big-picture evaluation, which can determine the long-term success of a solution. The solution of one problem can cause others if foresight is neglected, and this is not easily avoided in all cases. For instance, the guilt that the developers of dynamite and atomic fission have publicly expressed relates to a regret for lack of foresight. Early Greek engineers invented many devices and processes which they refused to disclose because they knew they would be misused. The foresight to see how technically correct solutions can malfunction or be misapplied takes creative skill to develop. What about fertilizers which boost crops but poison water supplies? Modern engineers need to learn how to apply the right solution at the right time in the right way for the right reason with the right customer.

Most people rely on two or three methods to solve problems. If these methods don't yield a successful answer, they become stuck. Truly exceptional problem solvers learn to use multiple problem-solving techniques to find the optimum solution. The following is a list of possible tools or strategies that can help solve simple problems:

1. Look for a pattern
2. Construct a table
3. Consider possibilities systematically
4. Act it out
5. Make a model
6. Make a figure, graph or drawing
7. Work backwards
8. Select appropriate notation
9. Restate the problem in your own words
10. Identify necessary, desired, and given information
11. Write an open-ended sentence
12. Identify a sub-goal
13. First solve a simpler problem
14. Change your point of view
15. Check for hidden assumptions
16. Use a resource
17. Generalize
18. Check the solution; validate it
19. Find another way to solve the problem
20. Find another solution
21. Study the solution process
22. Discuss limitations
23. Get a bigger hammer
24. Sleep on it
25. Brainstorm
26. Involve others

In both analytic and creative problem solving, there are different methods for tackling problems. Have you ever experienced being stuck on a difficult math or science problem? Developing additional tools or methods will allow you to tackle more of these problems effectively, making you a better engineer.

The rest of this chapter is a presentation of different ways of solving problems. These are some of the techniques that have been shown to be effective. Each person has a unique set of talents and will be drawn naturally to certain problem-solving tools. Others will find a different set useful. The important thing as an engineering student is to experiment and find as many useful tools as you can. This will equip you to tackle the wide range of challenges which tomorrow's engineers will face.

8.3 ANALYTIC PROBLEM SOLVING

Given the importance of proper analysis in engineering and the design process, it is important to develop a disciplined way of approaching engineering problems. Solving analytic problems has been the subject of a great deal of research, which has resulted in several models. One of the most important analytic problem-solving methods that students are exposed to is the Scientific Method. The steps in the Scientific Method are as follows:

1. Define the problem
2. Gather the facts
3. Develop a hypothesis

4. Perform a test
5. Evaluate the results

In the Scientific Method, the steps can be repeated if the desired results are not achieved. The process ends when an acceptable understanding of the phenomenon under study is achieved.

In the analysis of engineering applications, a similar process can be developed to answer problems. The advantage of developing a set method for solving analytic problems is that it provides a discipline to help young engineers when they are presented with larger and more complex problems. Just like a musician practices basic scales in order to set the foundation for complex pieces to be played later, an engineer should develop a sound fundamental way to approach problems. Fortunately, early in your engineering studies you will be taking many science and math courses that will be suited to this methodology.

The Analytic Method we will discuss has six steps:

1. Define the problem and make a problem statement
2. Diagram and describe
3. Apply theory and equations
4. Simplify the assumptions
5. Solve the necessary problems
6. Verify accuracy to required level

Following these steps will help you better understand the problem you are solving and allow you to identify areas where inaccuracies might occur.

Step 1: Problem Statement

It is important to restate the problem you are solving in your own words. In textbook problems, this helps you understand what you need to solve. In real life situations, this helps to ensure that you are solving the correct problem. Write down your summary, then double-check that your impression of the problem is the one that actually matches the original problem. Putting the problem in your own words is also an excellent way to focus on the part of the problem you need to solve. Often, engineering challenges are large and complex, and the critical task is to understand what part of the problem you need to solve.

Step 2: Description

The next step is to describe the problem and list all that is known. In addition to restating the problem, list the information given and what needs to be found. This is shown in Example 8.2, which follows this section. Typically, in textbook problems, all the information given is actually needed for the problem. In real problems, more information is typically available than is needed to do the calculations. In other cases, information may be missing. Formally writing out what you need and what is required helps you to clearly sort this out.

It is also helpful to draw a diagram or sketch of the problem you are solving to be able to understand the problem. Pictures help many people to clarify the problem and what is needed. They are also a great aid when explaining the problem to someone else. The old saying could be restated as "A picture is worth a thousand calculations."

Step 3: Theory

State explicitly the theory or equations needed to solve the problem. It is important that you write this out completely at this step. You will find that most real problems and those you are asked to solve as an undergraduate student will not require exact solutions to complete equations. Understanding the parts of the equations that ought to be neglected is vital to your success.

It is not uncommon to get into a routine of solving a simplified version of equations. These may be fine under certain conditions. An example can be seen in the flow of air over a body, like a car or airplane wing. At low speeds, the density of the air is considered a constant. At higher speeds this is not the case, and if the density were considered a constant, errors would result. Starting with full equations and then simplifying reduces the likelihood that important factors will be overlooked.

Step 4: Simplifying Assumptions

As mentioned above, engineering and scientific applications often cannot be solved precisely. Even if they are solvable, determining the solution might be cost prohibitive. For instance, an exact solution might require a high-speed computer to calculate for a year to get an answer, and this would not be an effective use of resources. Weather prediction at locations of interest is such an example.

To solve the problem presented in a timely and cost-effective manner, simplifying assumptions are required. Simplifying assumptions can make the problem easier to solve and still provide an accurate result. It is important to write the assumptions down along with how they simplify the problem. This documents them and allows the final result to be interpreted in terms of these assumptions.

While estimation and approximation are useful tools, engineers also are concerned with the accuracy and reliability of their results. Approximations are often possible if assumptions are made to simplify the problem at hand. An important concept for engineers to understand in such situations is the Conservative Assumption.

In engineering problem solving, "conservative" has a non-political meaning. A conservative assumption is one that introduces errors on the safe side. We mentioned that estimations can be used to determine the bounds of a solution. An engineer should be able to look at those bounds and determine which end of the spectrum yields the safer solution. By selecting the safer condition, you are assuring that your calculation will result in a responsible conclusion.

Consider the example of a materials engineer selecting a metal alloy for an engine. Stress level and temperature are two of the parameters the engineer must consider. At the early stages of a design, these may not be known. What he or she could do is estimate the parameters. This could be done with a simplified analysis. It could also be done using prior experience. Often, products evolve from earlier designs for which there are data. If such data exist, ask how the new design will affect the parameters you are interested in. Will the stress increase? Will it double? Perhaps a conservative assumption would be to double the stress level of a previous design.

After taking the conservative case, a material can be selected. A question the engineer should ask is: "If a more precise answer were known, could I use a cheaper or lighter material?" If the answer is No, then the simple analysis might be sufficient. If Yes, the engineer could build a case for why a more detailed analysis is justified.

Another example of conservatism can be seen in the design of a swing set (see Fig. 8.1). In most sets, there is a horizontal piece from which the swings hang, and an inverted V-shaped support on each end. If you were the design engineer, you might have to size these supports.

One of the first questions to ask is, "For what weight do we size the swing set?" One way to answer this question is to do a research project on the weight distribution of children, followed by doing research into the use of swings by children at different ages, and then analyzing those data to determine typical weights of children that would use such swings. You might even need to observe playground activity to see if several children pile on a swing or load the set in different ways—say, by climbing on it. This might take weeks to do, but you would have excellent data.

On the other hand, you could just assume a heavy load that would exceed anything the children would produce. For this case, say, 500 pounds per swing. That would assume the equivalent of two large adults on each swing. Make the calculation to determine the needed supports, and then ask if it makes sense to be more detailed.

To answer this, one has to answer the question "What problem am I solving?" Am I after the most precise answer I can get? Am I after a safe and reliable answer regardless of other concerns?

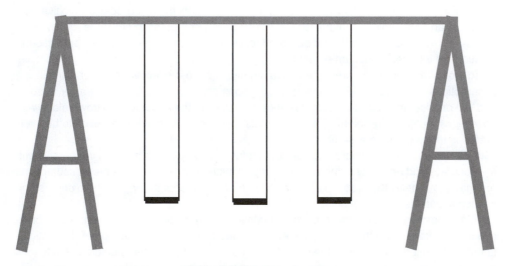

Figure 8.1 Simple swing set.

In engineering, the "other concerns" include such things as cost and availability. In the current example, the supports would most likely be made of common materials. Swing sets are typically made of wood or steel tubing. So, one way to answer the above questions is to quickly check and see if making a more detailed analysis would be justified. Check the prices of the materials needed for your conservative assumption against potential materials based on a more detailed analysis. Does the price difference justify a more detailed analysis? Another way to ask this is "Does the price difference justify spending time to do the analysis?" Remember, as an engineer, your time costs money.

The answer may be, "It depends." It would depend on how much the difference is and how many you are going to produce. It wouldn't make sense to spend a week analyzing or researching the answer if it saves you or your company only $500. If your company was producing a million sets and you could save $5 per set, it would justify that week's work.

Approximation can improve decision-making indirectly as well. By using it to quickly resolve minor aspects of a problem, you can focus on the more pressing aspects.

Engineers are faced with these types of decisions every day in design and analysis. They must accurately determine what problem is to be solved, and how. Safety of the public is of paramount concern in the engineering profession. You might be faced with a short-term safety emergency where you need to quickly prioritize solutions. Or you might be designing a consumer product where you have to take the long-term view. Swing sets are often used for generations, after all.

As an engineer, you will need to develop the ability to answer, "What problem am I really solving?" and, "How do I get the solution I need most efficiently?"

Example 8.1

A manager for an aerospace firm had a method for "breaking in" young engineers. Shortly after the new engineer came to work in his group, the manager would look for an appropriate analysis for the young engineer. When one arose, he would ask the young engineer to perform the analysis. The engineer would typically embark on a path to construct a huge computer model requiring lengthy input and output files that would take a few weeks to complete. The manager would return to the engineer's desk the following morning, asking for the results. After hearing the response that all that was really accomplished was to map out the next three weeks worth of work, the manager would erupt and explain at a very high volume that he or she had been hired as an engineer and what they were supposed to be doing was an engineering analysis, not a "science project." He would show the young engineer how doing a simplified analysis would take an afternoon to perform and would answer the question given the appropriate conservative assumptions. He would then leave the engineer to the new analysis method.

While his methods were questionable, his point is well taken. There are always time and cost concerns as well as accuracy. If conservative assumptions are made, safe and reliable design and analysis decisions can often be made.

Step 5: Problem Solution

Now the problem is set up for you actually to perform the calculations. This might be done by hand or by using the computer. It is important to learn how to perform simple hand calculations; however, computer applications make complex and repetitive calculations much easier. When using computer simulations, develop a means to document what you have done in deriving the solution. This will allow you to find errors faster, as well as to show others what you have done.

Step 6: Accuracy Verification

Engineers work on solutions that can affect the livelihood and safety of people. It is important that the solution an engineer develops is accurate.

I had a student once whom I challenged on this very topic. He seemed to take a cavalier attitude about the accuracy of his work. Getting all the problems mostly correct would get him a B or a C in the class. I tried to explain the importance of accuracy and how people's safety might depend on his work. He told me that in class I caught his mistakes, and on the job someone would check his work. This is not necessarily the case! Companies do not hire some engineers to check other engineers' work. In most systems, there are checks in place. However, engineers are responsible for verifying that their own solutions are accurate. Therefore, it is important for the student to develop skills to meet any required standard.

A fascinating aspect of engineering is that the degree of accuracy is a variable under constant consideration. A problem or project may be properly solved within a tenth of an inch, but accuracy to a hundredth might be unneeded and difficult to control, rendering that kind of accuracy incorrect. But the next step of the solution might even require accuracy to the thousandth. Be sure of standards!

There are many different ways to verify a result. Some of these ways are:

- Estimate the answer.
- Simplify the problem and solve the simpler problem. Are the answers consistent?
- Compare with similar solutions. In many cases, other problems were solved similarly to the one currently in question.
- Compare to previous work.
- Ask a more experienced engineer to review the result.
- Compare to published literature on similar problems.
- Ask yourself if it makes sense.
- Compare to your own experience.
- Repeat the calculation.
- Run a computer simulation or model.
- Redo the calculation backwards.

It may be difficult to tell if your answer is exactly correct, but you should be able to tell if it is close or within a factor of 10. This will often flag any systematic error. Being able to back up your confidence in your answer will help you make better decisions on how your results will be used.

The following example will illustrate the analytical method.

Example 8.2

Problem: A ball is projected from the top of a 35 m tower at an angle 25° from horizontal with an initial velocity $v_0 = 80$ m/s. What is the time it takes to reach the ground, and what is the horizontal distance from the tower to the point of impact?

Problem Statement: Given: Initial velocity $v_0 = 80$ m/s
                                    Initial trajectory = 25° from horizontal
                                    Ball is launched from a height of 35 m

Find: 1) Time to impact

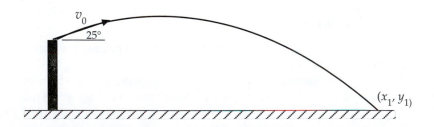

          2) Distance of impact from the tower
Equations: Found in a physics text:

$$(a)\quad x_i = x_0 + v_0 t \cos \theta + \frac{1}{2} a_x t^2$$

$$(b)\quad y_1 = y_0 + v_0 t \sin \theta + \frac{1}{2} a_y t^2$$

Assumptions: Neglect air resistance
Acceleration $= -9.81$ is constant in the vertical direction
Acceleration is zero in the horizontal direction
Ground is level and impact occurs at $y = -35$ m

Solution: Begin with equation (b):

$$-35 = 080t \sin 25° - \frac{1}{2} 9.8t^2 \text{ or } t^2 - 6.90t - 7.14 = 0$$

$$\therefore t = \frac{b \pm \sqrt{b^2 - 4ac}}{2a} = \frac{\sqrt{6.9^2 + 4 \times 1 \times 7.14}}{2} = 7.81 \text{ sec}$$

The solution must be 7.81 sec, since time must be positive.
The distance is found using equation (a):

$$x_1 = 0 + 80 \times 7.81 \times \cos 25° = 566\text{m}$$

Estimation

Estimation is a problem-solving tool that engineers need to develop. It can provide quick answers to problems and to verify complicated analyses. Young engineers are invariably amazed when they begin their careers at how the older, more experienced engineers can estimate so closely to problems before the analysis is even done.

I was amazed at one such case that I experienced first hand. I was responsible for doing a detailed temperature analysis on a gas turbine engine that was to go on a new commercial aircraft. We presented our plan to some managers and people from the Chief Engineers Office. The plan called for two engineers working full time for a month on the analysis.

One of the more senior engineers who had attended the presentation sent me some numbers the next day. He had gone back to his office and made some approximate calculations and sent me the results. His note told me to check my results against his numbers.

After the month-long analysis, which involved making a detailed computer model of the components, I checked our results with his "back of the envelope" calculations. He was very close to our result! His answers were not exactly the same as ours, but they were very close. The detailed analysis did provide a higher degree of precision. However, the senior engineer was very close.

What he had done was simply use suggestion #13 from the problem-solving methods. He solved a simpler problem. He used simple shapes with known solutions to approximate the behavior of the complex shapes of the jet engine's components.

While estimation or approximation may not yield the precision you require for an engineering analysis, it is a very useful tool. One thing that it can do is help you check your analysis. Estimation can provide bounds for potential answers. This is especially critical in today's dependence on computer solutions. Your method may be correct, but one character mistyped will throw your results off. It is important to be able to have confidence in your results by developing tools to verify accuracy.

Estimation can also be used to help decide if a detailed analysis is needed. Estimating using a best case and worst case scenario can yield an upper and lower bound to the problem. If the entire range of potential solutions is acceptable, why do the detailed analysis?

What if, for our case, the whole range of temperatures met our design constraints? It would not have justified a month of our work. In this case, the added precision was needed and did lead to a design benefit. As engineers, you will be asked to decide when a detailed analysis is required and when it is not.

Estimation can be a very powerful tool to check accuracy and make analysis decision. Senior engineers in industry often comment that current graduates lack the ability to do an approximation. This has become more important with the dependence on computer tools and solutions. Computer analyses lend themselves to typos on inputs or arithmetic errors. Being able to predict the ballpark of the expected results can head off potentially disastrous results.

Example 8.3

With all of today's sophisticated analytical tools, it may be hard to imagine that accurate measurements could be made without them. Take the size of a molecule. Today's technology makes this measurement possible. Could you have made this measurement in the 18th century? Benjamin Franklin did.

He found the size of an oil molecule. He took a known volume of oil and poured it slowly into a still pool of water. The oil spread out in a circular pattern. He measured the circle and calculated the surface area. Since the volume of the oil remained the same, he calculated the height of the oil slick and assumed that was the molecular thickness, which turned out to be true. His result was close to the correct value.

Do we have more accurate measurements now? Certainly we do. However, Benjamin Franklin's measurement was certainly better than no value. As engineers, there will be times when no data exists and you will need to make a judgment. Having an approximate result is much better than no data.

8.4 CREATIVE PROBLEM SOLVING

Many engineering problems are open-ended and complex. Such problems require creative problem solving. To maximize the creative problem-solving process, a systematic approach is recommended. Just as with analytic problem solving, developing a systematic approach to using creativity will pay dividends in better solutions.

Revisiting our toolbox analogy, the creative method provides the solid judgment needed to use your analytic tools. Individual thinking skills are your basics here. The method we'll outline now will help you to apply those skills and choose solution strategies effectively.

There are numerous ways to look at the creative problem-solving process. We will present a method which focuses on answering these five questions:

1. What is wrong?
2. What do we know?
3. What is the real problem?
4. What is the best solution?
5. How do we implement the solution?

By dividing the process into steps, you are more likely to follow a complete and careful problem-solving procedure, and more effective solutions are going to result. This also allows you to break a large, complex problem into simpler problems where your various skills can be used. Using such a strategy, the sky is the limit on the complexity of projects which an engineer can complete. It's astonishing and satisfying to see what can be done.

Divergence and Convergence

At each phase of the process, there is both a divergent and a convergent part of the process, as shown in Figure 8.2. In the divergent process, you start at one point and reach for as many ideas as possible. Quantity is important. Identifying possibilities is the goal of the divergent phase of each step. In this portion of the process (indicated by the outward point-ing arrows, $\longleftarrow \longrightarrow$) use the brainstorming and idea-generating techniques described in

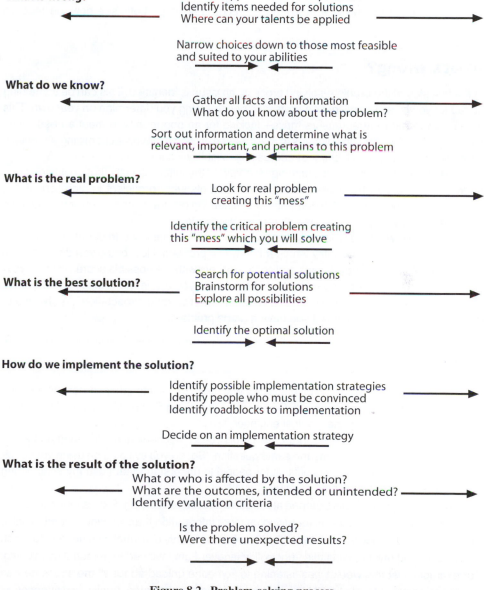

What is wrong?

Look for opportunites to fix problem
Identify items needed for solutions
Where can your talents be applied

Narrow choices down to those most feasible
and suited to your abilities

What do we know?

Gather all facts and information
What do you know about the problem?

Sort out information and determine what is
relevant, important, and pertains to this problem

What is the real problem?

Look for real problem
creating this "mess"

Identify the critical problem creating
this "mess" which you will solve

What is the best solution?

Search for potential solutions
Brainstorm for solutions
Explore all possibilities

Identify the optimal solution

How do we implement the solution?

Identify possible implementation strategies
Identify people who must be convinced
Identify roadblocks to implementation

Decide on an implementation strategy

What is the result of the solution?

What or who is affected by the solution?
What are the outcomes, intended or unintended?
Identify evaluation criteria

Is the problem solved?
Were there unexpected results?

Figure 8.2   Problem-solving process.

subsequent sections. Look for as many possibilities as you can. Often, the best solutions come from different ideas.

In the convergence phase (indicated by inward pointing arrows, $\rightarrow\leftarrow$) use analytical and evaluative tools to narrow the possibilities to the one(s) most likely to yield results. Quality is most important, and finding the best possibility to move the process to the next phase is the goal. (One common method is to use a matrix to rate ideas based on defined criteria.) If one choice fails to produce satisfactory results, go back to the idea lists for another method to tackle the problem.

Sample Problem: Let's Raise Those Low Grades!

In order to illustrate the problem-solving process, let's consider an example of a problem and work through the process. The problem we will use for illustration is a student who is getting low grades. This certainly is a problem that needs a solution. Let's see how the process works.

What's Wrong?

In the first step of the problem-solving process, an issue is identified. This can be something stated for you by a supervisor or a professor, or something you determine on your own. This is the stage where entrepreneurs thrive—looking for an opportunity to meet a need. Similarly, engineers look to find solutions to meet a need. This may involve optimizing a process, improving customer satisfaction, or addressing reliability issues.

To illustrate the process, let's take our example of the initial difficulty that beginning engineering students can have with grades. Most students were proficient in high school and didn't need to study much. The demands of engineering programs can catch some students off guard. Improving grades is the problem we will tackle.

At this stage, we need to identify that there is a problem that is worth our effort in solving. A good start would be to identify whether it truly is a problem. How could you do this? Total your scores thus far in your classes and compare them with the possible scores listed in your syllabus. Talk with your professor(s) to find out where you are in the classes and if grades will be given on a curve. Talk to an advisor about realistic grade expectations. At the end of this process, you should know if you have a grade problem.

Example 8.4

Engineers are trained problem solvers. Many engineers move into other careers where their problem-solving abilities are valued. One such case is Greg Smith, who was called into the ministry after becoming a mechanical engineer.

"As soon as people discover that I have a degree in mechanical engineering and am in the ministry, they always ask me the same question, 'So, do you still plan on using your engineering degree some day?' It is difficult for people to understand that I use my engineering training every day as a full-time minister.

"One of the greatest skills I learned at Purdue was problem solving. Professors in mechanical engineering were notorious for giving us more information than we would need to solve the problem. Our job was to sift through all the facts and only use what was needed to reach a solution. I find myself using this very skill whenever I am involved in personal counseling. For example, just this week I was listening to someone unload about all the stress he was experiencing from life. He listed his job, a second business at home, family, and finances as the primary sources of stress. After further conversation, I was able to 'solve' his problem by showing that most of his stress was related to only one of the four sources he listed. The root issue was that this person was basing much of his self-image on his success in this one area of life. I helped him regain his sense of balance in seeing his self-worth not in job performance, but in who he was as a whole person."

What Do We Know?

The second step in problem solving is the gathering of facts. All facts and information related to the problem identified in the first step are gathered. In this initial information gathering stage, do not try to evaluate whether the data are central to the problem. As will be explained in the brainstorming and idea generating sections to follow, premature evaluation can be a hindrance to generating adequate information.

With our example of the student questioning his grades, the information we generate might include:

1. Current test grades in each course
2. Current homework and quiz grades in each course
3. Percent of the semester's grade already determined in each course
4. Class average in each course
5. Professors' grading policies
6. Homework assignments behind
7. Current study times
8. Current study places
9. Effective time spent on each course
10. Time spent doing homework in each course
11. Performance of your friends
12. Your high school grades
13. Your high school study routine

Part of this process is to list the facts that you know. To be thorough, you should request assistance from someone with a different point of view. This could be a friend, an advisor, a parent, or a professor. They might come up with a critical factor you have overlooked.

What Is the Real Problem?

This stage is one that is often skipped, but it is critical to effective solutions. The difference between this stage and the first is that this step of the process answers the question *why*. Identifying the initial problem answers the question of *what*, or what is wrong. To effectively fix the problem, a problem solver needs to understand *why* the problem exists. The danger is that only symptoms of the problem will be addressed, rather than root causes.

In our example, we are problem solving low class grades. That is the *what* question. To really understand a problem, the *why* must be answered. Why are the grades low? Answering this question will identify the cause of the problem. The cause is what must be dealt with for a problem to be effectively addressed.

In our case, we must understand why the class grades are low. Let us assume that poor test scores result in the low grades. But this doesn't tell us why the test scores are low. This is a great opportunity for brainstorming potential causes. In this divergent phase of problem definition, don't worry about evaluating the potential causes. Wait until the list is generated.

Possible causes may include:

- Poor test-taking skills
- Incomplete or insufficient notes
- Poor class attendance
- Not understanding the required reading
- Not spending enough time studying
- Studying the wrong material
- Attending the wrong class
- Studying ineffectively (e.g., trying to cram the night before tests)
- Failing to understand the material (*note*: might reveal need for tutor or study group)
- Using solution manuals or friends as a crutch to do homework
- Not at physical peak at test time (i.e., up all night before the test)
- Didn't work enough problems from the book

After you have created the list of potential causes, evaluate each as to its validity. In our example, there may be more than one cause contributing to the low grades. If so, make a rank-ordered list. Rank the causes in order of their impact on class performance.

Assume the original list is reduced to:

- Poor class attendance
- Studying ineffectively
- Not spending enough time studying
- Not understanding the material

Of these things, rank them in order of their impact on grades. You may be able to do this yourself or you may need help. An effective problem solver will seek input from others when it is appropriate. In this case, a professor or academic advisor might be able to help you rank the causes and help determine which ones will provide the greatest impact. For this example, say the rank order was:

1. Not understanding the material
2. Studying ineffectively
3. Not spending enough time studying
4. Poor class attendance

We determined that while class attendance was not perfect, it was not the key factor in the poor performance. The key item was not understanding the material. This goes with ineffective studying and insufficient time.

Identifying key causes of problems is vital, so let's look at another example.

Example 8.5

"What was the confusion?" Defining the real problem is often the most critical stage of the problem-solving process and the one most often skipped. In engineering, this can be the difference between a successful design and a failure.

An example of this occurred with an engine maker. A new model engine was introduced and began selling quickly. A problem developed, however, and failures started to happen. Cracks initiating from a series of bolt holes were causing the engines to fail. A team of engineers was assembled and the part was redesigned. They determined that the cause was high stress in the area of the bolt hole. Embossments were added to strengthen the area. It was a success.

Until, that is, the redesigned engines accumulated enough time in the field to crack again. A new team of engineers looked at the problem and quickly determined that the cause was a three-dimensional coupling of stress concentrations created by the embossments, resulting in too high a stress near the hole where the crack started. The part was redesigned and introduced into the field. It was a success.

Well, it was a success, until the redesigned engines accumulated enough time in the field to crack yet again. A third team of engineers was assembled. This time, they stepped back and asked what the real cause was. Part of the team ran sophisticated stress models of the part, just like the other two teams had done. The new team members looked for other causes. What was found was that the machine creating the hole introduced microcracks into the surface of the bolt holes. The stress models predicted that further stresses would start a crack. However, if a crack were intentionally introduced, the allowable stresses were much lower and the part would not fail. The cause wasn't with cracking but with the wrong kind of cracking. The third team was a success because it had looked for the real problem. Identifying the real cause initially, which in this case involved the machining process, would have saved the company millions of dollars, and spared customers the grief of the engine failures.

What Is the Best Solution?

Once the problem has been defined, potential solutions need to be generated. This can be done by yourself or with the help of friends. In an engineering application, it is wise to confer with experienced experts about the problem's solution. This may be most productive after you have begun a list of causes. The expert can comment on your list and offer his own as well. This is a great way to get your ideas critiqued. After you gain more experience, you will find that technical experts help you narrow down the choices rather than providing more. Also, go to more than one source. This may provide more ideas as well as help with the next step.

In our example, the technical expert may be a professor or advisor who is knowledgeable about studying. Let's assume that the following list of potential solutions has been generated.

- Get a tutor
- Visit the professors during office hours
- Make outside appointments with professors
- Visit help rooms
- Form study group
- Outline books
- Get old exams
- Get old sets of notes
- Outline lecture notes
- Review notes with professors
- Do extra problems
- Get additional references
- Make a time schedule
- Drop classes
- Retake the classes in the summer
- Go to review sessions

The list now must be evaluated and the best solution decided upon. This convergent phase of the problem-solving process is best done with the input of others. It is especially helpful to get the opinions of those who have experience and/or expertise in the area you are investigating. In an engineering application, this may be a lead engineer. In our case, it could be an academic advisor, a teaching assistant, or a professor. These experts can be consulted individually or asked to participate as a group.

When refining solutions, an effective tool is to ask yourself which of them will make the biggest impact and which will require the most effort. Each solution gets ranked high or low as to impact and effort. Ideal solutions are those that have a high impact yet require a low level of effort. A visual tool which helps make this process easier is the problem-solving matrix shown in Figure 8.3. Solutions that fall in quadrant B should be avoided. These solutions require a great deal of effort for marginal payoff. Solutions that are the most effective are those in quadrant C. These are the ones that are easy to implement and will produce significant results. It may be necessary to include solutions in quadrant D, requiring high effort and yielding high impact. These should be carefully evaluated to ensure that the investment is worth the reward.

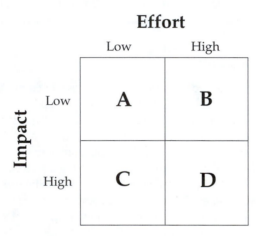

Figure 8.3 A Problem-Solving Matrix.

Implementing the Solution

Implementing the solution is a step that may seem trivial and not worth discussion. This step, however, is a critical phase of the problem-solving process. In our scenario, the solution selected may have been:

- Do extra problems

To accomplish this solution, appropriate additional problems must be selected, done, and corrected. Implementation probably requires the assistance of the instructor to help select appropriate problems. To randomly pick additional practice problems could move the effort from quadrant C to D and even B. To have the greatest chances of success, you would want to gain the assistance of your instructor. Other sources of assistance could come from a study group which poses the challenges of getting them to do extra problems with you.

As with all the other steps, a divergent phase begins the step including such activities as brainstorming. The convergent phase of the step culminates with the selection of the implementation plan.

In engineering applications, implementation can be a very critical phase of the process. Most of the solutions to problems either require additional resources including money or the cooperation of other groups that are either not directly affected by the problem or not under your control. Especially early in your career, you will not have control over the people who need to implement solutions so you will need to sell your ideas to the managers who do. An effective implementation plan will be critical to getting your solutions accomplished.

President Jimmy Carter was educated as an engineer and was one of the most respected presidents as a strong moral person. During his presidency, the country was in an energy crisis and action was needed to address the problems. The story goes that he summoned Tip O'Neill, the speaker of the House of Representatives, to the White House to explain his plan. Mr. O'Neill responded by giving the President a list of people who would need to be sold on the plan before it went public. President Carter who was new to Washington was going to skip the development of an implementation plan and enact the plan. He believed that with the country in such an energy crisis that it was obvious what was needed and cooperation was a given. Tip O'Neill was very experienced in the Washington political system and understood that they needed to spend time gaining acceptance to their plan before it could be implemented.

Engineering situations can be very similar. While most are not as political as Washington D.C., most require gaining acceptance of others to implement a solution. In all of these cases, gaining the recognition of those involved is a critical part of the problem-solving process.

Evaluating the Solution

Implementation does not necessarily end the problem-solving process. Just as the design process is a circular process with each design leading to possible new designs, problem solving can also be cyclic. Once a solution has been found and implemented, an evaluation should be performed. As with the other steps in the problem-solving process, this step begins with a divergence phase as the problem solver asks what makes a successful solution and how it will be evaluated. To effectively evaluate a solution, a criterion for success must be established. Sometimes there are objective criteria for evaluation but in many circumstances the criterion for success is more subjective.

Once the criterion has been established, the evaluation process should be defined including who will evaluate the solution. Often it is desirable to get a neutral view of someone who was not involved in the formulation of the solution process to be the evaluator.

If the solution is a success, the process may be done. Often in engineering applications, solutions are intermediate and lead to other opportunities. The software industry is a prime example. Software is written to utilize the current computer technologies to address issues. Almost as soon as it is completed, new technologies are available which open up new opportunities or require new solutions.

Sometimes a "success" does not address the true problem as was the case of the turbine disk failures described earlier. In these cases, the evaluation process may need time to deem it a true success. In other cases, solutions have unintended outcomes that require a new solution even if the initial solution was deemed a success.

Whether a solution is effective or not, there is value in taking time to reflect on the solution and its implications. This allows you as a problem solver to learn both from the process and the solution. Critically evaluating solutions and opportunities is a skill that is extremely valuable as an engineer and is discussed further in the critical thinking section of this chapter.

8.5 PERSONAL PROBLEM-SOLVING STYLES

The creative problem-solving model presented previously is one of the models that can be used to solve the kind of open-ended problems engineers face in everyday situations. You may find yourself in an organization that uses another model. There are many models for problem solving and each breaks the process up into slightly different steps. Isaken and Treffinger, for example, break the creative problem-solving process into six linear steps [Isaken and Treffinger]. These six steps are:

1. Mess finding
2. Data finding
3. Problem finding
4. Idea finding
5. Solution finding
6. Acceptance finding

Dr. Min Basadur of McMaster University developed another model, a circular one, which he calls Simplex, based on his experience as a product development engineer with the Procter and Gamble Company. This model separates the problem-solving process into eight different steps that are listed below [Basadur].

1. Problem finding
2. Fact finding
3. Problem defining
4. Idea finding
5. Evaluating and selecting
6. Action planning
7. Gaining acceptance
8. Taking action

All these problem-solving processes provide a systematic approach to problem solving. Each process has divergent phases where options need to be generated along with convergent phases when best options need to be selected.

Basadur has created and patented a unique method of helping individuals participating in the creative problem-solving process to identify which parts of the process they are more comfortable in than others. This method, called the Basadur Simplex Creative Problem Solving Profile, reflects your personal creative problem-solving style. Everyone has a different creative problem-solving style. Your particular style reflects your relative preferences for the different parts of the problem-solving process.

Basadur's Creative Problem Solving Profile method identifies four styles and each style correlates with two of the eight problem-solving steps in the eight-step circular model of creative problem-solving above. Thus the eight problem-solving steps are grouped into four quadrants or stages of the complete problem-solving process.

These four steps are shown as the quadrants in Figure 8.4. The Basadur Creative Problem Solving Process begins with quadrant one, the generation of new problems and opportunities. It cycles through quadrant two, the conceptualization of the problem or opportunity and of new, potentially useful ideas and quadrant three, the optimization of new solutions. It ends with quadrant four, the implementation of new solutions. Each quadrant requires different kinds of thinking and problem-solving skills.

Basadur describes the four different quadrants or stages as follows.

Generating

Generating involves getting the problem-solving process rolling. Generative thinking involves gathering information through direct experience, questioning, imagining possibilities, sensing new problems and opportunities, and viewing situations from different perspectives. People and organizations strong in generating skills prefer to come up with options or diverge rather than evaluate and select, or converge. They see relevance in almost everything and think of good and bad sides to almost any fact, idea, or issue. They dislike becoming too organized or delegating the complete problem, but are willing to let others take care of the details. They enjoy ambiguity and are hard to pin down. They delight in juggling many new projects simultaneously. Every solution they explore suggests several new problems to be solved. Thinking in this quadrant includes problem finding and fact finding.

Conceptualizing

Conceptualizing keeps the innovation process going. Like generating, it involves divergence. But rather than gaining understanding by direct experience, it favors gaining understanding by abstract thinking. It results in putting new ideas together, discovering insights that help define problems, and creating theoretical models to explain things. People and organizations strong in conceptualizing skills enjoy taking information scattered all over the map from the generator phase and making sense of it. Conceptualizers need to "understand": to them, a theory must be logically sound and precise. They prefer to proceed only with a clear grasp of a situation and when the problem or main idea is well defined. They dislike having to prioritize, implement, or agonize over poorly understood alternatives. They like to play with ideas and are not overly concerned with moving to action. Thinking in this quadrant includes problem defining and idea finding.

Optimizing

Optimizing moves the innovation process further. Like conceptualizing, it favors gaining understanding by abstract thinking. But rather than diverge, an individual with this thinking style prefers to converge. This results in converting abstract ideas and alternatives into practical solutions and plans. Individuals rely on mentally testing ideas rather than on trying things out. People who favor the optimizing style prefer to create optimal solutions to a few well-defined problems or issues. They prefer to focus on specific problems and sort through large amounts of information to pinpoint "what's wrong" in a given situation. They are usually confident in their ability to make a sound logical evaluation and to select the best option or solution to a problem. They often lack patience with ambiguity and dislike "dreaming" about additional ideas, points of view, or relations among problems. They believe they "know" what the problem is. Thinking in this quadrant includes idea evaluation and selection and action planning.

Implementing

Implementing completes the innovation process. Like optimizing, it favors converging. However, it favors learning by direct experience rather than by abstract thinking. This results in getting things done. Individuals rely on trying things out rather than mentally testing them. People and organizations strong in implementing prefer situations in which they must somehow make things work. They do not need complete understanding in order to proceed and adapt quickly to immediate changing circumstances. When a theory does not appear to fit the facts they will readily discard it. Others perceive them as enthusiastic about getting the job done but also as impatient or even pushy as they try to turn plans and ideas into action. They will try as many different approaches as necessary and follow up or "bird dog" as needed to ensure that the new procedure will stick. Thinking in this quadrant includes gaining acceptance and implementing.

Your Creative Problem-Solving Style

Basadur's research with thousands of engineers, managers, and others has shown that everyone has a different creative problem-solving style. Your particular style reflects your

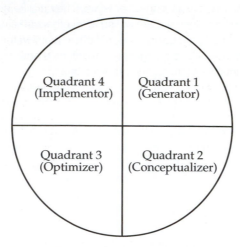

Figure 8.4 The four quadrants of Basadur's Simplex Creative Problem Solving Process (with permission from the Center for Research in Applied Creativity).

relative preferences for each of the four quadrants of the creative problem-solving process: generating/initiating, conceptualizing, optimizing, and implementing. Your behavior and thinking processes can not be pigeonholed in any single quadrant. Rather they're a combination or blend of quadrants: you prefer one quadrant in particular, but you also have secondary preferences for one or two adjacent quadrants. Your blend of styles is called the creative problem-solving style. Stated another way, your creative problem-solving style shows which particular steps of the process you gravitate toward.

Typically as an engineer, you will be working as part of a team or larger organization. Basadur's research has shown that entire organizations also have their own problem-solving process profiles. An organization's profile reflects such things as the kinds of people it hires, its culture and its values. For example, if an organization focuses almost entirely on short-term results, it may be overloaded with implementers but have no conceptualizers or generators. The organization will show strengths in processes that deliver its current products and services efficiently. But it will show weaknesses in processes of long-term planning and product development that would help it stay ahead of change. Rushing to solve problems, this organization will continually find itself reworking failed solutions without pausing to conduct adequate fact finding and problem definition. By contrast, an organization with too many generators or conceptualizers and no implementers will continually find good problems to solve and great ideas for products and processes to develop. But it will never carry them to their conclusion. You can likely think of many examples of companies showing this imbalance in innovation process profiles.

Basadur suggests that in order to succeed in creative problem solving, a team requires strengths in all four quadrants. Teams must appreciate the importance of all four quadrants and find ways to fit together their members' styles. Team members must learn to use their differing styles in complementary ways. For example, generating ideas for new products and methods must start somewhere with some individuals scanning the environment, picking up data and cues from customers, and suggesting possible opportunities for changes and improvement. Thus, the generator raises new information and possibilities—usually not fully developed, but in the form of starting points for new projects. Then the conceptualizer pulls together the facts and idea fragments from the generator phase into well-defined problems and challenges and more clearly developed ideas worth further evaluation. Good conceptualizers give sound structure to fledgling ideas and opportunities. The optimizer then takes these well-defined ideas and finds a practical best solution and well-detailed efficient plans for proceeding. Finally, implementers must carry forward the practical solutions and plans to make them fit real-life situations and conditions.

Skills in all four quadrants are equally valuable. As an engineering student, you will find yourself working on group projects. Often, these projects are designs that are open-ended and are well suited for creative problem solving. Use these opportunities to practice utilizing your problem-solving skills and preferences as well as those of your team members.

Example 8.6

What is your own preferred problem-solving style? You can find out what your preferred problem-solving style is by taking the Basadur Simplex Creative Problem Solving Profile. This profile and additional information about the Simplex model can be obtained from the Center for Research in Applied Creativity at:

Center for Research in Applied Creativity, 184 Lovers Lane, Ancaster, Ontario, Canada, L9G 1G8 or http://basadursimplex.com.

Why Are We Different?

What causes the main differences in people's approaches to problem solving? Basadur's research suggests that they usually stem from inevitable differences in how knowledge and understanding—"learning"—are gained and used. No two individuals, teams or organizations learn in the same way. Nor do they use what they learn in the same why. As shown in Figure 8.5, some individuals and organizations prefer to learn through direct, concrete experiencing (doing). They gain understanding by physical processing. Others prefer to learn

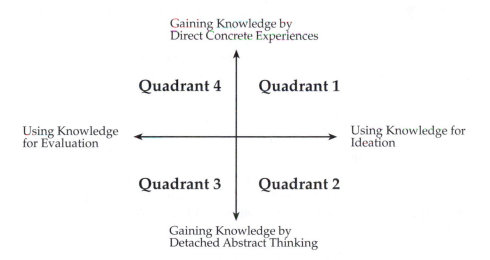

Figure 8.5 The two dimensions of the Basadur Simplex Creative Problem Solving Profile.

(The Basadur Problem Solving Model and Profile are copyrighted and are used with the permission of Dr. Min Basadur.)

through more detached abstract thinking (analyzing). They gain understanding by mental processing. All individuals and organizations gain knowledge and understanding in both ways but to differing degrees. Similarly, while some individual and organizations prefer to use their knowledge for ideation, others prefer to use their knowledge for evaluation. Again, all individuals and organizations use their knowledge in both ways but to differing degrees.

How an individual or organization combines these different ways of gaining and using learning determines their problem-solving process profile. When you understand these differences, you can shift your own orientation in order to complement the problem-solving process preferences of others. Equally important you can take various approaches to working with people. You can decide on the optimum strategy for helping someone else to learn something. And you can decide whom to turn to for help in ideation or evaluation. Understanding these differences also helps you interact with other people to help them make the best use of the creative problem solving process. For example, you can help strong optimizers discover new problems and facts or present new problems and facts to them. You can help strong implementers better define challenges or present well-defined challenges to them. You can help strong generators/initiators evaluate and select from among solutions and make plans, or present to them evaluated solutions and ready-made plans. You can help strong conceptualizer to convince others of the values of their ideas and push them to act on them, or push their ideas through to acceptance and implementation for them [Basadur].

8.6 BRAINSTORMING STRATEGIES

Alex Osborn, an advertising executive in the 1930s, devised the technique of stimulating ideas known as *brainstorming*. Since that time, countless business and engineering solutions have come about as a result of his method. Basically, brainstorming is a technique used to stimulate as many innovative solutions as possible. These then can be used in both analytic and creative problem-solving practices.

The goal of brainstorming is to stimulate your mind to trigger concepts or ideas that normal problem solving might miss. This has a physiological basis. The mind stores information in a network. Memories are accessed in the brain by remembering the address of the memory. When asked about ideas for potential solutions, some would come just by thinking about the problem. A portion of your brain will be stimulated, and ideas will be brought to the surface. However, there will be areas of your brain that will not be searched because the pathways to get there are not stimulated. In brainstorming, we get the brain to search as many regions of the brain as possible.

At times it may seem difficult to generate ideas. Isakssen and Treffinger suggest using the method of idea lists. The concept is simple. Write down the ideas you have currently. Then, use these ideas to branch out into new ideas. To help stimulate your thinking in new directions, a set of key words was developed by Alex Osborn. These words include:

Adapt – What else is like this? What other ideas does this suggest? Are there any ideas from the past that could be copied or adapted? Whom could I emulate?

Put to other uses – Are there new ways to use the object as is? Are there other uses if modified?

Modify – Could you change the meaning, color, motion, sound, odor, taste, form, or shape? What other changes could you make? Can you give it a new twist?

Magnify – What can you add to it? Can you give it a greater frequency? Make it stronger, larger, higher, longer, or thicker? Can you add extra value? Could you add another ingredient? Could you multiply or exaggerate it?

Minify – What could you subtract or eliminate? Could you make it smaller, lighter, or slower? Could you split it up? Could you condense it? Could you reduce the frequency, miniaturize it, or streamline it?

Substitute – What if I used a different ingredient, material, person, process, power source, place, approach, or tone of voice? Who else instead? What else instead? What if you put it in another place or in another time?

Rearrange – Could you interchange the pieces? Could you use an alternate layout, sequence, or pattern? Could you change the pace or schedule? What if you transpose the cause and effect?

Reverse – What if you tried opposite roles? Could you turn it backwards, upside down, or inside out? Could you reverse it or see it through a mirror? What if you transpose the positive and negative?

Combine – Could you use a blend, an assortment, an alloy, or an ensemble? Could you combine purposes, units, ideas, functions, or appeals?

Creativity expert Bob Eberle took Osborn's list and created an easy-to-remember word by adding "eliminate" to the list. Eberle's word is SCAMPER. The words used to form SCAMPER are:

Substitute?
Combine?
Adapt?
Modify? Minify? Magnify?
Put to other uses?
Eliminate?
Reverse? Rearrange?

Using the SCAMPER list may help to generate new ideas.

Individual Brainstorming

Brainstorming can be done either individually or in a group. Brainstorming individually has the advantage of privacy. There may be times when you deal with a professional or personal problem that you don't want to share with a group.

Some research even shows that individuals may generate more ideas by working by themselves. Edward de Bono attributes this to the concept that "individuals on their own can pursue many different directions. There is no need to talk and no need to listen. An individual on his or her own can pursue an idea that seems 'mad' at first and can stay with the idea until it makes sense. This is almost impossible with a group."

When brainstorming individually, select a place that is free from distraction and interruption. Begin by writing down your initial idea. Now you will want to generate as many other ideas as possible. Let your mind wander and write down any ideas that come into your head. Don't evaluate your ideas; just keep writing them down. If you start to lose momentum, refer to the SCAMPER questions and see if these lead to any other ideas.

There are different formats for recording your ideas when brainstorming. Some find generating in ordered lists works best for them. Others will write all over a sheet of paper, or in patterns. Kinesthetic learners often find that putting ideas on small pieces of paper, which allows them to move them around, helps the creative process. Auditory learners might say the ideas out loud as they are generated.

Group Brainstorming

The goal of group brainstorming is the same as with individual brainstorming—to generate as many potential solutions as possible without judging any of them. The power of the group comes when each member of the group is involved in a way that uses their creativity to good advantage.

There are a few advantages to group brainstorming. The first is simply that the additional people will look at a problem differently, and bring fresh perspectives. When brainstorming is done correctly, these different views trigger ideas that you would not have come up with on your own.

Another advantage to group brainstorming is that it gets others involved in the problem-solving process early. As you will see in the following section, a key component in problem solving is implementing your solution. If people whose cooperation you need are in the brainstorming group, they will be much more agreeable to your final solutions. After all, it was partly their idea.

To run an effective group brainstorming session, it helps to use basic guidelines. Those outlined in this section have been proven effective in engineering applications. They are also very helpful for use in student organizations and the committees that engineering students are encouraged to join. This process not only covers the generation of ideas, but also provides a quick way to evaluate and converge on a solution. In some situations, you may want to stop after the ideas are generated and evaluate them later.

The guidelines for this process include:

1. Pick a facilitator
2. Define the problem
3. Select a small group
4. Explain the process
5. Record ideas
6. Involve everyone
7. No evaluating
8. Eliminate duplicates
9. Pick three

Pick a Facilitator

The first step is to select a facilitator who will record the ideas and keep the group focused. The facilitator is also responsible for making sure the group obeys the ground rules of brainstorming.

Define the Problem

It is important that all the participants understand the problem you are looking to solve before generating solutions. Once you start generating ideas, distractions can bring the definition process to a grinding halt. Idea generation would be hampered if the group tried to define the problem they are trying to solve at the wrong time. The definition discussion should happen before the solution idea-generating phase.

Small Group

The group size should be kept manageable. Prof. Goldschmidt from Purdue University recommends groups of six to 12 members. Groups smaller than that can work, but with less benefit from the group process. Groups larger than 12 become unwieldy, and people won't feel as involved. If you need to brainstorm with a larger group, break it into smaller sub-groups and reconvene after the groups have come up with separate ideas.

Explain the Process

Providing the details of the process the group will follow is important. It gives the participants a feeling of comfort knowing what they are getting into. Once the process begins, it is counterproductive to go back and discuss ground rules. This can stymie idea generation just like a mistimed definition of the problem.

Record Ideas

Record ideas in a way that is visible to the whole group. Preferably, arrange the group in a semicircle around the person recording. A chalkboard, flip chart, whiteboard, or newsprint taped to a wall all work well. It is important to record all ideas, even if they seem silly. Often the best ideas come from a silly suggestion that triggers a good idea. By writing the suggestions in a way that is visible to all, the participants can use their sight as well as hearing to absorb ideas which may trigger additional ideas. The use of multiple senses helps to stimulate more ideas. Also, having all the ideas in front of the participants guarantees recollection and allows for new ideas, triggered by earlier ideas. Using the SCAMPER questions is a great way to keep ideas flowing.

Involve Everyone

Start with one idea from the facilitator or another volunteer and write it down. It is easier to get going once the paper or writing surface is no longer completely blank. Go around the group, allowing each person to add one idea per round. If a person doesn't have an idea, let them say "pass," and move on. It is more important to keep moving quickly than to have the participants feel like they must provide an idea every time.

By taking turns, all the participants are ensured an equal opportunity to participate. If the suggestions were taken in a free-flowing way, some participants might monopolize the discussion. The power of group brainstorming lies in taking advantage of the creative minds of every member of the group, not just the ones who could dominate the discussion. Often, the best ideas come from the people who are quiet and would get pushed out of the discussion in a free-flowing setting. It matters not when or how the ideas come, as long as they are allowed to surface.

Continue the process until everyone has "passed" at least once. Even if only one person is still generating ideas, keep everyone involved and keep asking everyone in turn. The one person who is generating ideas might stimulate an idea in someone else at any point. Give everyone the opportunity to come up with a last idea. The last idea may very well be the one that you use.

No Evaluating

This may be hard for the facilitator or some participants. Telling someone that their suggestion is dumb, ridiculous, out-in-left-field, or making any other negative judgment makes that person (and the rest of the group) less likely to speak up. The genius of brainstorming is the free generation of ideas without deciding if they are good until the correct stage. There are countless times when the wackiest ideas are brought up right before the best ideas. Wacky suggestions can trigger ideas that end up being the final solution. Write each idea down, as crazy as some may be. This makes participants feel at ease. They will feel more comfortable making way-out suggestions and will not fear being made fun of or censored.

Avoiding negative comments is one thing, but if you are facilitating, you need to watch for more subtle signals. You might be communicating indirectly to someone that his or her ideas are not good. If you are writing the ideas down, it is natural to say "good idea" or "great idea." What happens, however, if one person gets "great idea" comments and the others don't? Or what if it is just one person who doesn't get the "great idea" comments?

The participants need to feel completely at ease during brainstorming. Providing negative feedback in the form of laughter, sarcastic remarks, or even the absence of praise will work to dampen the comfort level and, therefore, the creativity. It is safest not to make any value judgments about any ideas at this point. If you are facilitating, try to respond to each idea the same way and to keep "order in your court," so to speak. If you are a participant, try not to make value statements or jokes about other ideas.

Eliminate Duplicates

After generating ideas, you may want to study the suggestions. In some cases, it is to your advantage to sort out a solution or the top several solutions. Doing this at the end of group brainstorming can be very easy, and also serves to involve the group members in the decision process.

The first step is to examine the list of ideas and eliminate duplicates. There is a fine line between eliminating duplicates and creating limiting categories. This step is designed to eliminate repeated ideas. Ask if they really are the same. If they are similar, leave both. Creating categories only reduces the possible solutions and defeats the purpose of brainstorming.

The next step is to allow the group to ask clarifying questions about suggestions. This is still not the time for evaluation—only clarification. Don't let this become a discussion of merits. Clarification may help in eliminating and identifying duplicates.

Pick Three

Once everyone understands all ideas as best they can, have them evaluate the suggestions. This can be done quickly and simply. First, ask everyone to pick their three top choices. (Each person can select more if you need more possibilities or solutions.) After each member has determined these in their own minds, let each member vote "out loud" by marking their choices. Do this with a dot or other mark, but not a 1, 2, and 3. The top ideas are selected based on the total number of marked votes they receive. The winners then can be forwarded to the problem-solving phases to see if any survive as the final solution.

Note that with this method of voting, the group's valuable time is optimized. Time is not wasted determining whether an idea deserves a "1" or a "3." Also, no time is spent on the ideas that no one liked. If all ideas are discussed throughout the process, the ideas that are not chosen by anyone or which are least preferred often occupy discussion for quite a while. It's best to eliminate these likely time sinkholes as soon as possible.

Another benefit of not ranking choices at first is that each member could think that their ideas were the fourth on everyone's list, which might frustrate them. Also, how might it affect someone if their ideas were ranked low by the other members? Such a discouraged person could be one who would have very valuable input the next time it is needed, or it could be that this person will be one whom you need as an ally to implement the group's decision. Either way, there is no need to cast any aspersions on ideas which weren't adopted.

8.7 CRITICAL THINKING

Since engineering is so dominated by systematic problem solving, it is easy to get caught up in the methodologies and look at everything as a problem needing a solution or goal. We have presented problem-solving methods that give a systematic approach for analyzing and solving problems. However, many of life's issues are more complex and have no single solution; some cannot be solved. I once assigned a project to design a program using MATLAB, to predict students' financial future and life plans, including a potential spouse. I jokingly suggested that the class use the program when dating to evaluate the financial impact of a potential spouse or life partner. ("Sorry, MATLAB says we wouldn't be compatible. . . .") The class laughed at the suggestion. Clearly, whom you marry, or if you marry, is a complex question and not one that can be solved by some stepwise methodology or computational algorithm. Life is more complex than that.

Issues and challenges you will face have a similar complexity and require a critical perspective to effectively address engineering issues with human, environmental, or ethical impacts. For example, many programs have sought to address poverty, and many programs and "solutions" involved engineering. Poverty, however, has remained a pressing social issue in this country and abroad. Solutions to these larger social issues are complex and often do not have a single or simple solution.

Many of the issues you will be exposed to in your career are problems that cannot be easily or quickly solved. It is easy to get caught up in the processes and methodologies and move to solutions and implementation plans. Sometimes this is appropriate, but other times we need to evaluate our actions and examine the larger issues of our actions. Being able to evaluate why and how you are doing something by taking a step back to gain perspective is an essential skill. At each step in your problem solving, one of the questions you can ask is "why?"

There are times when you need to evaluate your actions, to discover deeper meanings and larger issues. One example involves the design of toys for children who are physically challenged. In the EPICS (Engineering Projects in Community Service, http://epics.ecn .purdue.edu) Program, students work on engineering solutions to local community issues. One of the partnerships in the program is with a clinic that works with children with disabilities. As the students work with the children, families, and therapists, they gain an appreciation for the myriad complex issues facing these children. When the engineering students design toys, they must design them so they can be used with the given physical abilities of the children. This is fairly straightforward.

What the students have learned is that they can address many other issues, such as socialization dimensions, in their designs. Some of the toys are designed so that children with disabilities can more easiliy interact with their peers in their integrated classroom, when using the toys. The students have also learned to design so that the children with disabilities can control aspects of the play when using the toys with their peers. The engineering students have learned about the socialization of children with and without disabilities, and its impact on self-esteem. They have also come to understand the role of play in child development and the barriers children with disabilities may encounter. Armed with this knowledge, the engineering students can design toys that meet a wider spectrum of needs.

When working with issues such as children or adults with disabilities, there are not simple problems to be solved, but complex issues that need to be addressed and pondered. By working through these issues and their multiple dimensions, students can become aware of the social and emotional issues of their work. They can become better designers, and better educated about relevant issues.

As engineers, we have grave responsibilities because our work has great potential for good and for harm. While having the highest of ethical standards is essential, keeping a critical eye on the context and direction of your work is equally vital. Creating opportunities to gain perspective and evaluating your actions, experiences, and perceptions is as important as the problem-solving methodology presented in this text. At the beginning of the creative problem-solving process, as well as the design process, we discussed how crucial it was to get the question right. Similarly, when evaluating an issue, asking the right question will enable you to generate the best solution.

We use our personal background and experience to evaluate problems, and this perspective can limit our thinking. When evaluating a solution, getting an outside opinion has value. You must become aware of the perspective you bring to issues based on your background so it does not limit you. Thomas Edison wanted people around him who would not limit their thinking by jumping to conclusions. He would take every prospective engineer to lunch at his favorite restaurant, which was known for their soup, which he would recommend. Invariably, the candidate would order soup. Edison wanted to see if he or she would salt the soup before tasting it. Salting without tasting, he reasoned, was an indication of someone who jumps to conclusions. He wanted people who were open to questioning. Periodically, ask yourself if there are things that you are salting without tasting. Are there issues for which you jump to conclusions based on your background, personality, or thinking style?

Critical thinkers ask the Why? question about themselves as well as about the things around them. Periodically take a step back and ask Why? or What? about yourself. Why are you studying to become an engineer? What will you do as an engineer to find personal fulfillment? Becoming a critical thinker will make you a better problem solver and engineer, and one who is more engaged in society and more fulfilled.

REFERENCES

Basadur, M., *Simplex: A Flight to Creativity*, Creative Education Foundation, Inc., 1994.

Beakley, G. C., Leach, H. W., Hedrick, J. K., *Engineering: An Introduction to a Creative Profession*, 3rd Edition, Macmillan Publishing Co., New York, 1977.

Burghardt, M.D., *Introduction to Engineering Design and Problem Solving*, McGraw-Hill Co., 1999.

Eide, A.R., Jenison, R. D., Mashaw, L. H., Northup, L. L., *Introduction to Engineering Problem Solving,* McGraw-Hill Co., 1998.

Fogler, H. S. and LeBlanc, S. E., *Strategies for Creative Problem Solving*, Prentice Hall, Inc., Englewood Cliffs, New Jersey, 1995.

Gibney, Kate, *Awakening Creativity*, Prism, March 1998, pp. 18-23.

Isaksen, Scott G. and Treffinger, Donald J., *Creative Problem Solving: The Basic Course*, Bearly Limited, Buffalo, New York, 1985.

Lumsdaine, Edward and Lumsdaine, Monika, *Creative Problem Solving—Thinking Skills for a Changing World*, 2nd Edition, McGraw-Hill Co., 1990.

Osborn, A. F., *Applied Imagination: Principles and Procedures of Creative Problem-Solving*, Charles Scriber's Sons, New York, 1963.

Panitx, Beth, "Brain Storms," *Prism*, March 1998, pp. 24-29.

EXERCISES AND ACTIVITIES

8.1   A new school has exactly 1000 lockers and exactly 1000 students. On the first day of school, the students meet outside the building and agree on the following plan: the first student will enter the school and open all the lockers. The second student will then enter the school and close every locker with an even number (2, 4, 6, 8, etc.). The third student will then "reverse" every third locker (3, 6, 9, 12, etc.). That is if the locker is closed, he or she will open it; if it is open, he or she will close it. The fourth student will then reverse every fourth locker, and so on until all 1000 students in turn have entered the building and reversed the proper lockers. Which lockers will finally remain open?

8.2 You are stalled in a long line of snarled traffic that hasn't moved at all in twenty minutes. You're idly drumming your fingers on the steering wheel, when you accidentally started tapping the horn. Several sharp blasts escaped before you realize it. The driver of the pickup truck in front of you opens his door, gets out and starts to walk menacingly toward your car. He looks big, mean, and unhappy. Your car is a convertible and the top is down. What do you do?

8.3 Your school's team has reached the national championship. Tickets are very difficult to get but you and some friends have managed to get some. You all travel a great distance to the site of the game. The excitement builds on the day of the game until you discover that your tickets are still back in your room on your desk. What do you do? What is the problem that you need to solve?

8.4 You are a manufacturing engineer working for an aircraft company. The production lines you are responsible for are running at full capacity to keep up with the high demand. One day, you are called into a meeting where you learn that one of your suppliers of bolts has been forging testing results. The tests were never done, so no data exists to tell if the bolts in question meet your standards. You do not know how long the forging of the test results has been going on. You are only told that it has been "a while." This means that your whole production line including the planes ready to be delivered may be affected. You are asked to develop a plan to manage this crisis. What do you do?
Some background information:
  - Bolts are tested as batches or lots. A specified number are taken out of every batch and tested. The testing typically involves loading the bolts until they fail.
  - The bolts do not have serial numbers so it is impossible to identify the lot number from which they came.
  - The supplier who forged the tests supplies 30% of your inventory of all sizes of bolts.
  - Your entire inventory is stored in bins by size. Each bin has a mixture of manufacturers. (To reduce dependence on any one company, you buy bolts from three companies and mix the bolts in the bins sorted by size.)
  - Bolts have their manufacturer's symbol stamped on the head of the bolt.
  - It takes weeks to assemble or disassemble an aircraft.
  - Stopping your assembly line completely to disassemble all aircraft could put your company in financial risk.
  - Your customers are waiting on new planes and any delays could cause orders to be lost.
  - The FBI has arrested those who forged the test and their business has been closed, at least temporarily.

8.5 How much of an automobile's tire wears off in one tire rotation?

8.6 A farmer going on a trip with a squirrel, acorns, and a fox had to cross a river in a boat in which he couldn't take more that one of them with him each time he crossed. Since he often had to leave two of them together on one side of the river or the other, how could he plan the crossings so that nothing gets eaten, and they all get across the river safely?

8.7   Measure the height of your class building using two different methods. Compare your answers and indicate which is more accurate.

8.8   Pick a grassy area near your class building. Estimate how many blades of grass are in that area. Discuss your methodology.

8.9   How high do the letters on an expressway sign need to be to be readable?

8.10   A company has contacted you to design a new kind of amusement park ride. Generate ideas for the new ride. Which idea would you recommend?

8.11   Estimate the speed of a horse.

8.12   Estimate the maximum speed of a dog.

8.13   Assume that an explosion has occurred in your building. Your room is intact but sustained damage and is in immediate danger of collapsing on you and your classmates. All exits but one (as specified by your instructor) are blocked. The one remaining is nearly blocked. Quickly develop a plan to get your classmates out. Who goes first and who goes last? Why?

8.14   A child's pool is eight feet in diameter and two feet high. It is filled by a garden hose up to a level of one foot. The children complain that it is too cold. Can you heat it up to an acceptable temperature using hot water from the house? How much hot water would you need to add?

8.15   How else could you heat the water for the children in problem 8.14?

8.16   Without referring to a table or book, estimate the melting temperature of aluminum. What are the bounds of potential melting temperatures? Why? What is the actual melting temperature?

8.17   Select a problem from your calculus class and use the analytic problem-solving steps to solve it.

8.18   Select a problem from your chemistry class and use the analytic problem-solving steps to solve it.

8.19   Select a problem from your physics class and use the analytic problem-solving steps to solve it.

8.20   Write a one- or two-page paper on a historical figure that had to solve a difficult problem. Provide a description of the problem, solution, and the person.

8.21   Aliens have landed on earth and they come to your class first. They are very frustrated because they can not communicate with you. Develop a plan to break this communication barrier and show them that we are friendly people.

8.22   You are working on a project that has had many difficulties that are not your fault. However, you are the project leader and your manager has become increasingly frustrated with you and even made a comment that you are the kind of irresponsible person who locks your keys in your car. Your manager is coming into town to see first hand why the project is having problems and you are going to pick her up. You leave early to take time to relax. With the extra time, you decide to take the scenic route and stop the car next to a bubbling stream. Leaving your car, you go over to the stream to just sit and listen. The quiet of the surroundings is just what you needed to calm down before leaving for the airport. Glancing at your watch, you realize that it is time to leave to pick up your manager. Unfortunately, you discover that you indeed have locked your keys in the car. A car passes on the road about every 15 minutes. The nearest house is probably a mile from your location and the nearest town is 15 miles away. The airport is 20 miles away and the flight is due to arrive in 30 minutes. What will you do?

8.23 You are sitting at your desk in the morning reading your e-mail when your boss bursts into your office. It seems that a local farmer was not happy with your company and has dropped a truckload of potatoes in your plant's entrance. The potatoes are blocking the entrance and must be moved. You have been selected as the lucky engineer to fix this problem. What would you do to move the potatoes and how would you get rid of them? Brainstorm ideas and select the best solution. (Note: You work at a manufacturing plant of some kind that does not use potatoes in its process. The potatoes also appear to be perfectly fine, just in the way.)

8.24 You work in a rural area that is known for chickens. An epidemic has swept through the area killing thousands of chickens. The EPA has mandated that the chickens may not be put into the landfill nor can they be buried. Develop a plan to get rid of the chickens.

8.25 A fire has destroyed thousands of acres of forest. Seedlings have been planted in the area but are being destroyed by the local wildlife. Develop a way to reforest the region by preventing the wildlife from eating the seedlings.

8.26 A survey shows that none of the current high school students in your state are planning to major in engineering when they get to college. Your dean has asked your class to fix this dilemma (with no students, she would be out of a job!). Develop a plan to convince the high school students that engineering is worthwhile.

8.27 Select a problem from your math book. How many ways can you solve the problem? Demonstrate each method.

8.28 You and seven of your friends have ordered a round pizza. You need to cut it into eight pieces and are only allowed to make three straight cuts. How can you do this?

8.29 Select an innovative product and research the story of its inventors. Prepare a presentation on how the ideas were formed and the concept generated.

8.30 With which part of the Basadur problem-solving process are you most comfortable? Write a one-page paper on how your preference affects your problem solving.

8.31 Take the Basadur problem-solving profile test. Report on your scores. How will this information affect your ability to solve problems.

8.32 Organizations have problem-solving styles. Select an organization and prepare a brief report on their own problem-solving styles.

8.33 You and your classmates have discovered that your funding for college will be withdrawn after this semester. Develop ideas on how you could obtain the money for the rest of your college expenses.

8.34 Your class has been hired by a new automotive company that wants to produce a brand new car. Their first entry into the market must be truly innovative. Do a preliminary design of this car. What features would it have to set it apart from current vehicles?

8.35 An alumnus of your engineering program has donated money to build a jogging track that would encircle your campus. The track will be made of a recycled rubberized material. How much of this rubberized material will be needed?

8.36 Every Christmas season it seems that one toy becomes a "I have to have one" craze. You and some classmates have been hired to come up with next year's toy of the season. What is it?

8.37 A professional sports team has decided to locate in your town. A local businessperson is going to build a stadium for the team. Unfortunately, the site is contaminated. The project will fall through unless you can devise a way to get rid of the contaminated soil under the new arena. What do you do with the soil?

8.38 NASA has decided to send astronauts to the other planets in the solar system. Psychologists, however, have determined that the astronauts will need something to keep them occupied during the months of travel or they will develop severe mental problems that would threaten the mission. Your job is to devise ways to keep the astronauts occupied for the long journey. Remember that you are limited to what will fit in a space capsule.

8.39 Landfill space is rapidly running out. Develop a plan to eliminate your city's dependence on the local landfill. The city population is 100,000.

8.40 An appliance manufacturer has hired you to expand their market. Your job is to develop a new household appliance. What is it and how will it work?

8.41 College students often have trouble waking up for early classes. Develop a system that will guarantee that the students wake up and attend classes.

8.42 Select one of your classes. How could that class be more effective in helping you learn?

8.43 Prepare a one-page report on a significant engineering solution developed in the past five years. Evaluate its effectiveness and report on the outcomes (both intended and unintended) of this solution.

8.44 Habitat for Humanity International spearheaded an initiative to eliminate substandard housing within the Georgia county in which it is headquartered. One of the first steps was to develop a set of standards to determine if housing was substandard or not. Develop criteria for your own community for determining whether a housing unit is substandard.

8.45 If your own community were to undertake an initiative like the one presented in the previous question, what would need to be done (both from an engineering and community perspective)?

8.46 Identify one current topic related to engineering or technology that has no single simple solution and write a brief paper discussing issues related to the topic.

8.47 Brainstorm ideas for toys for physically disabled children. Narrow your choices to the top three and identify the benefits of each design. What are the important issues that you would need to consider in your designs?

8.48 In the critical thinking section in this chapter, engineers had to pass Edison's "salt test." Identify areas in your own thinking where you might tend to "salt before tasting" (where you might jump to conclusions or use your own biases to overlook potential options).

8.49 Critically evaluate why you are majoring in engineering. What do you hope to gain by studying engineering?

8.50 Write a one-page paper on your responsibilities to society as a citizen in your community and as an engineer.

8.51 Early one morning it starts to snow at a constant rate. Later, at 6:00 a.m., a snow plow sets out to clear a straight street. The plow can remove a fixed volume of snow per unit time. In other words, its speed it inversely proportional to the depth of the snow. If the plow covered twice as much distance in the first hour as the second hour, what time did it start snowing?

8.52 While three *wise* men are asleep under a tree a mischievous boy paints their foreheads red. Later they all wake up at the same time and all three start laughing. After several minutes suddenly one stops. Why did he stop?

8.53 Using just a 5-gallon bucket and a 3-gallon bucket, can you put four gallons of water in the 5-gallon bucket? (Assume that you have an unlimited supply of water and that there are no measurement markings of any kind on the buckets.)

8.54 A bartender has a three-pint glass and a five-pint glass. A customer walks in and orders four pints of beer. Without a measuring cup but with an unlimited supply of beer how does he get a single pint in either glass?

8.55 You are traveling down a path and come to a fork in the road. A sign lays fallen at the fork indicating that one path leads to a village where everyone tells the truth and the other to a village where everyone tells lies. The sign has been knocked down so you do not know which path leads to which village. Then someone from one of the villages (you don't know which one) comes down the path from which you came. You may ask him one question to determine which path goes to which village. What question do you ask?

8.56 Four mathematicians have the following conversation:

> Alice: I am insane.
> Bob: I am pure.
> Charlie: I am applied.
> Dorothy: I am sane.
> Alice: Charlie is pure.
> Bob: Dorothy is insane.
> Charlie: Bob is applied.
> Dorothy: Charlie is sane.

You are also given that:
- Pure mathematicians tell the truth about their beliefs.
- Applied mathematicians lie about their beliefs.
- Sane mathematician's beliefs are correct.
- Insane mathematicians beliefs are incorrect.
Describe the four mathematicians.

8.57 Of three men, one man always tells the truth, one always tells lies, and one answers yes or no randomly. Each man knows which man is which. You may ask three yes/no questions to determine who is who. If you ask the same question to more than one person you must count it as a question used for each person asked. What three questions should you ask?

8.58 Tom is from the census bureau and greets Mary at her door. They have the following conversation:

Tom: I need to know how old your three kids are.
Mary: The product of their ages is 36.
Tom: I still don't know their ages.
Mary: The sum of their ages is the same as my house number.
Tom: I still don't know their ages.
Mary: The younger two are twins.
Tom: Now I know their ages! Thanks!

How old are Mary's kids and what is Mary's house number?

8.59 Reverse the direction of the fish image below by moving only three sticks:

Chapter 9

Visualization and Graphics

Becoming a fully competent engineer is a long yet rewarding process that requires the acquisition of many diverse skills and a wide body of knowledge. Learning most of the concepts in engineering requires the use of visual information and visual thinking. This, in turn, requires graphical communication skills, not only as an engineering student, but also as a practicing engineer. Communicating by way of graphics requires the ability to visualize information and translate it into visual products such as sketches and drawings. This allows you to represent a three-dimensional object in a two-dimensional medium. Studying this chapter and working through the activities will help you improve your visualization ability and your understanding of common graphical communication techniques used in the engineering profession.

The purpose of this chapter is to illustrate the visualization and graphical communication skills essential for all engineering disciplines. Visual experiences such as the sketching of simple objects are used in this chapter to exercise your ability to create and edit visual information, and thereby strengthen your ability to visualize new and unfamiliar designs. You will be asked to produce sketches using standard graphical techniques to reinforce the common language engineers use for communicating technical information. Traditionally, visualization skills have been taught implicitly through engineering graphics or drafting classes; however, some disciplines no longer require drafting classes. The need for visualization and graphical communication skills has not diminished. Indeed, the prevalence of computers in the workplace has increased it. This chapter seeks to provide at a minimum a basic understanding of visualization and engineering graphics that will be useful to all engineering disciplines. It must be stressed that this chapter is only an introduction to these subjects, and is not intended to be a comprehensive reference. Many engineering disciplines rely upon more sophisticated graphical techniques and a deeper understanding of the topics described here which are more effectively covered in a course dedicated solely to engineering graphics.

The motivation for this chapter, then, is twofold. First, visualization is critical to all subjects of a graphical nature. Any medium that provides a visual input to be interpreted by the observer requires comprehension of this information. If the observer is then required to respond to this information in a graphical form, the ability to create and edit the visual input is required. The second reason is the fact that all engineering disciplines rely on graphics to communicate information—both technical and non-technical. Such skills are essential for communication with colleagues or decision makers. Often, the quality of the graphical information can be the determining factor as to whether a project wins approval. If engineering students are to become effective at graphical communication, they must also have proficient visualization ability.

9.2 THE THEORY OF PROJECTION

In order to use a sheet of paper or the computer screen which only has two dimensions to display a three-dimensional object in an accurate way, we must develop a method where we can represent the three dimensions so that someone who has the ability to read technical drawings can interpret the design and build the three-dimensional object. This is done through a method known as orthographic projection, which incorporates a series of two-dimensional views of the object. Then when the views are looked at together, the drawing will give a true representation of the three-dimensional object. To understand orthographic projection, we need to look at the theory of projection from which it is derived.

Projection theory involves four specific components: (1) an object, (2) an observer, (3) a projection plane or picture plane, and (4) visual rays. Figure 9.1 shows these four components. We start with an object. In this case our object is a tree. At some measurable (finite) distance from the tree we place the observer. Between the observer and the object we place a projection plane or picture plane. From the object to the observer we show visual rays passing through the picture plane. When these visual rays come in contact with the picture plane, a two-dimensional image of the object is shown on the picture plane. This is known as a projected image.

Now, imagine the observer and the object are stationary and unable to move, but the picture plane can move either closer to the observer or closer to the object. What happens to the image on the picture plane as the picture plane moves closer to or farther away from the observer? Of course, the image gets smaller as the picture plane gets closer to the observer and it gets larger as it moves away from the observer. More importantly, the size of the image on the picture plane is never the true size of the real object. It is always a bit smaller. This is due to the fact that the observer is a finite or measurable distance from the object and the image on the picture plane is called a central perspective. Central perspectives are very common in our world since they are called pictures or photographs and we see them every day when we read the newspaper or spend time on the internet. It would seem that using this type of image for our design documentation would not be beneficial since the size of the image depends on how close the picture plane is to the observer.

In order to get the image on the picture plane to be the same size as the real object, we must move the observer to a position where he/she is an infinite (can't be measured) distance from the object. See Figure 9.2. When this occurs, the visual rays become parallel to each other and are perpendicular to the picture plane. Now the image on the picture plane is always the same size as the object. The image on the picture plane is called an orthographic view or an orthographic projection of the object. The prefix "ortho" means perpendicular so the visual rays are perpendicular (ortho) to the picture plane and the image is a two-dimensional image that is being viewed by the observer. Orthographic projection is the fundamental principal upon which all engineering drawing is based.

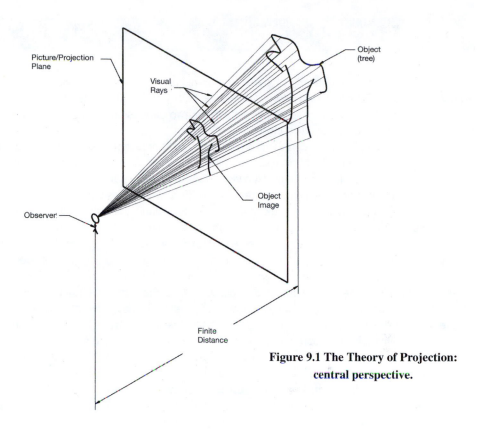

Figure 9.1 The Theory of Projection: central perspective.

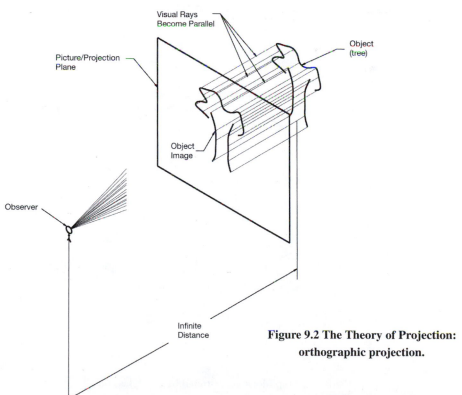

Figure 9.2 The Theory of Projection: orthographic projection.

9.3 THE GLASS BOX THEORY

The theory of projection as described above can be expanded to include viewing an object from various sides. This is known as the "glass box" theory and is the basis of how a three-dimensional object can be represented by its various two-dimensional orthographic views. Figure 9.3 shows a step block surrounded by a "glass box."

Each panel of the glass box represents a projection plane or picture plane. Note that three of the picture planes are marked Front, Top, and Side. For each view we need to imagine the observer is placed at infinity which allows the visual rays to become parallel and thus intersect each picture plane in a perpendicular manner. This yields a two-dimensional image or view of the object on each panel of the glass box. In the front picture plane, the width and height of the step block are displayed. In the top picture plane, the width and depth of the step block are displayed. And in the side picture plane, the depth and height of the step block are displayed. To see the orthographic views, we must imagine the panels of the glass box hinged between the front and top views and between the front and side views. When we open the panels of the glass box (Figure 9.4), we see the front, top, and side views of the step block. Each view shows only two dimensions of the object, but when we look at all three views, we have a complete picture of the three-dimensional shape. The front view shows the width and height of the object and the depth dimension is suppressed. The top view shows the width and depth of the object and the height is suppressed. The side view shows the depth and height of the object and the width is suppressed. The location of each view relative to the other views is extremely important in describing the object. The front view is the pivotal view. Once the front view is displayed, the top view must appear above the front view, and the side view is shown to the right. The views are never displayed out of position such as you see in Figure 9.5. This would be in violation of the "glass box" theory and thus in violation of basic orthographic projection.

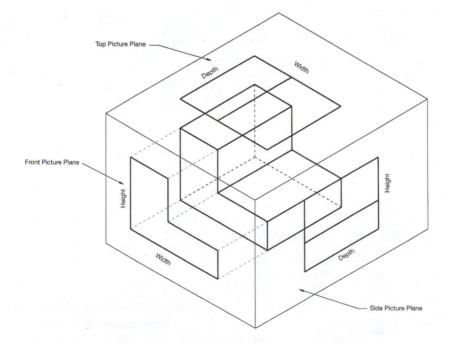

Figure 9.3 Step block surrounded by the "glass box."

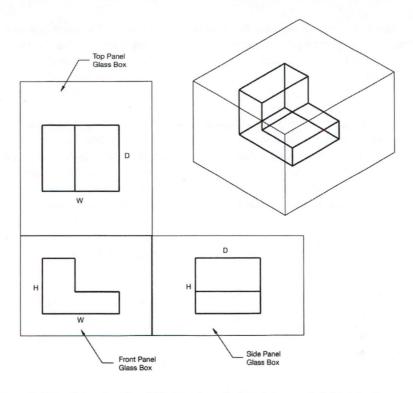

Figure 9.4 The "glass box" unfolded to show the front, top, and right side views.

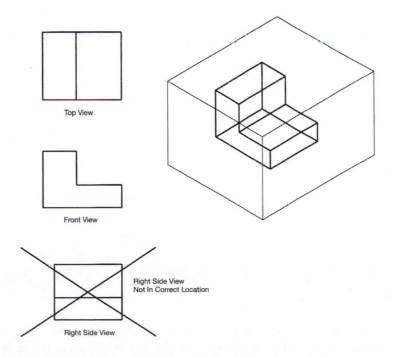

Figure 9.5 Orthographic views labeled and displayed in the wrong orientation.

Look again at Figure 9.3, the step block inside the glass box. The glass box has six sides and therefore six picture planes can be shown. We already know about the front, top, and side picture planes. The side picture plane is really the right side picture plane and captures the right side view of the object. The other picture planes of the glass box are the left side picture plane, the bottom picture plane, and the back picture plane. Figure 9.6 shows all six views of the step block in their proper orthographic positions. These views are called the principal orthographic views and there are only six principal orthographic views available for any three-dimensional object. As you can see, the left side, bottom, and back views do not yield any additional information regarding the shape of the object than what can be seen in the front, top, and right side views. If you look closely at the left, bottom, and back views, you will see some lines that appear dashed. These lines are dashed because in order to see these lines, you would have to look through the object which makes these lines "hidden," and in orthographic drawings we not only show lines that describe the parts of the object that we can see, but we also show lines that describe the parts of the object that we cannot see. These hidden lines appear as dashed lines in the orthographic views.

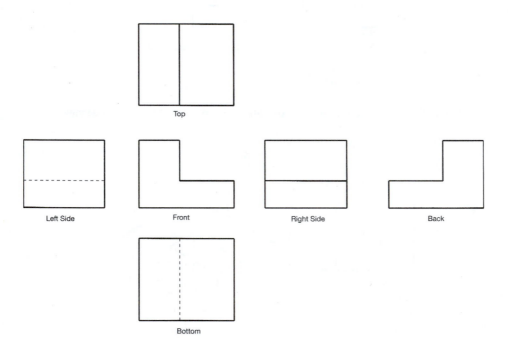

Figure 9.6 The six principal orthographic views of the step block.

9.4 FIRST AND THIRD ANGLE PROJECTIONS

Orthographic views of an object and their relative positions to each other are a function of whether the views were created as first angle projections or third angle projections. Third angle projections are most popular in the United States; however, first angle projections are used primarily in Europe. To understand the difference between first and third angle projections refer to Figure 9.7. Figure 9.7 shows a vertical plane intersecting a horizontal plane. Imagine these planes stretching to infinity in both the horizontal direction and the vertical direction. Note that in the figure, each quadrant is marked. The first quadrant is marked with I, the second with II, the third with III, and the fourth with IV. These are referred to as "angles" and in orthographic projection we either use the first (I) angle or the third (III) angle. This means we place the object within the confines of the horizontal and vertical plane in the first (I) or third (III) quadrant. When we place the object in the first quadrant, note that when applying the projection theory from above, the object is between the observer and the picture plane. When be place the object in the third quadrant, the picture plane is between the object and the observer. The major difference in the two approaches is the position of the views that are established. Note that in the first angle projection, the top view appears below the front view while in the third angle projection the top view appears above the front view. For the remainder of this chapter, we will use third angle orthographic projection since it is what is used in the United States.

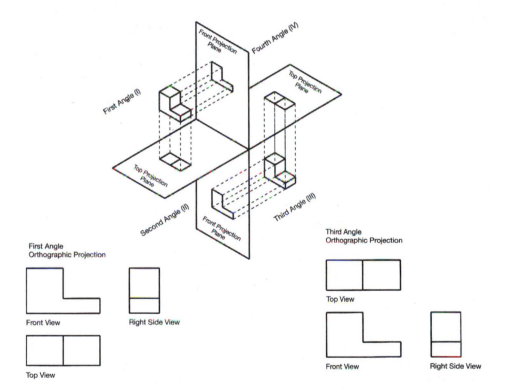

Figure 9.7 First (I) and third angle (III) orthographic projections.

9.5 THE MEANING OF LINES

Lines drawn to represent three-dimensional objects can have various meanings with regard to the object. The question becomes "What does a line represent?" When dealing with three-dimensional objects, a line can represent the edge view of a surface. Figure 9.8 shows the step block. Each line in each view represents the edge view of a specific surface. For example, surface ABCD is represented by a single line in the front view. It is also represented by a single line in the right-side view. All the lines of the step block in each of the orthographic views represent the edge of a surface.

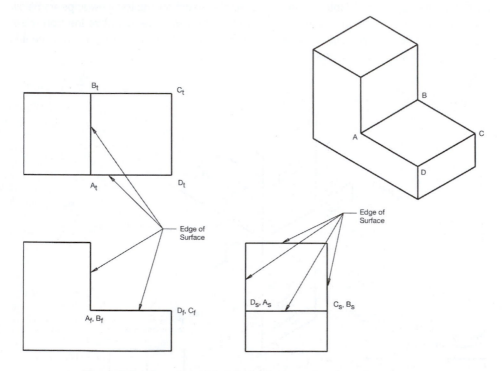

Figure 9.8 A line of an object defined as an edge view of a surface.

A line can also represent the intersection of two surfaces. This is very different from a line that represents an edge of a surface. In Figure 9.9 we see the orthographic views of a block that has an inclined surface (surface ABCD). Surface ABCD is represented in the front view by a single line that is the edge view of the surface. In the top view, line AB is not the edge view of a surface, but it is the intersection of two surfaces. In a similar manner, line CD in the right side view is the intersection of two surfaces.

Finally, a line can represent the limiting element of a curved surface. Figure 9.10 shows a cylinder. We can draw an infinite number of lines on the outside of the cylinder from the top to the bottom. These lines are parallel to the axis of the cylinder and are called elements. In the front view of the cylinder, note the lines AB and CD. Are these lines either the edge of a surface or the intersection of two surfaces? The answer to this question is obviously no. The lines represent the limiting element of a curved surface. The limiting element is the last element that one can see before the curve begins to turn back on itself. Thus a line is generated to show the limit of the cylinder.

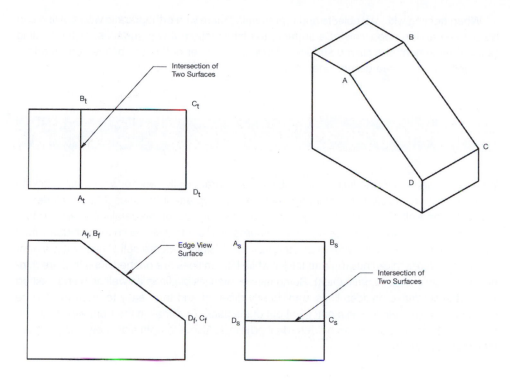

Figure 9.9 A line of an object defined as the intersection of two surfaces.

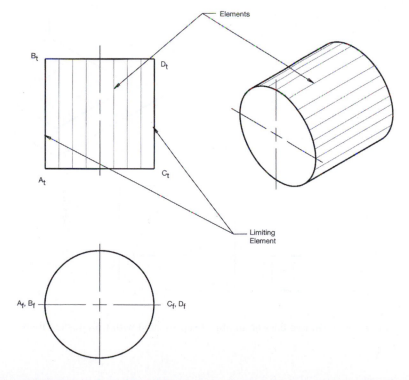

Figure 9.10 A line of an object defined as the limiting element of a curved surface.

When dealing with real objects and representing them with orthographic views, a line can be defined as the edge view of a surface, the intersection of two surfaces, or the limiting (extreme) element of a curved surface. There are no other definitions of lines on an orthographic view of a three-dimensional object.

9.6 HIDDEN LINES

When representing three-dimensional objects orthographically, we not only must show all the lines that are visible to the observer, but we must represent the lines that are hidden or not visible as well. Generally, the object itself blocks our view of the details that appear hidden, so we need a way to represent these features. In order to distinguish these hidden lines from visible lines, we display them as dashed lines. Visible lines are solid lines that are generally considered thick lines (0.7mm thick), while hidden lines are dashed lines that are thinner than visible lines (0.5mm thick). Being thinner than visible lines as well as being dashed and not solid, makes hidden lines significantly different and thus easy to interpret. Figure 9.11 shows a step block with a notch cut out of the back. Two lines in the front view that represent the notch are shown as hidden (dashed) lines. Also, the right side view has a hidden line which shows the notch.

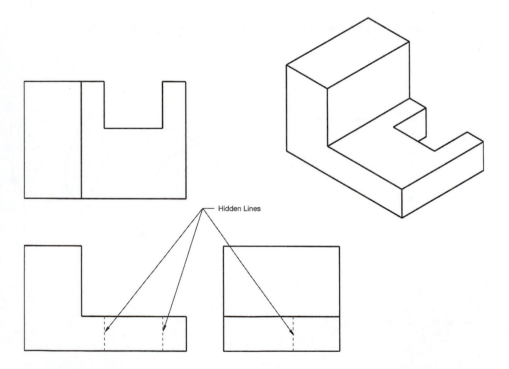

Figure 9.11 Hidden lines of an object represented with thin dashed lines.

9.7 CYLINDRICAL FEATURES AND RADII

Cylindrical features are considered either positive or negative. Positive cylinders are features that look like axles or soda cans. Negative cylinders are simply circular holes. Objects such as a hollow pipe involve both positive and negative cylinders. When dealing with cylinders, whether positive or negative, we have to include with the shape description a special treatment called a centerline. A centerline is a signal to the interpreter of the drawing that a circular feature is shown and that symmetry is involved. Whenever you show positive or negative cylinders, the centerlines should also be shown.

Figure 9.12 shows the orthographic representation of a hollow pipe. Note that there is only a front view and a right side view shown. The top view is not shown because it would be identical to the front view and would not show any additional details that are not shown in the front view or right side views. In the right side view, note the centerline. In this view where the cylinder appears as a circle, the centerline looks like a cross hair. The center point of the cylinder is represented by a small +. A small gap (about 1/16 inch) is placed on each side of the + with a longer line following that extends about 1/8 inch beyond the curve that is governed by the centerline. This gives the appearance of a cross hair.

In the longitudinal view of the cylinder (front view) we see the centerline as a long line that is broken in the middle. This line represents an axis of symmetry. People get confused and want to install a cross hair-type centerline on the longitudinal view, but this is wrong. The only thing that should appear is the long broken centerline.

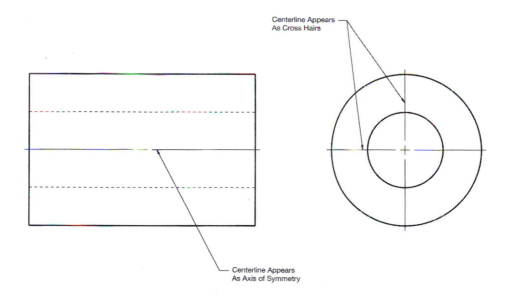

Figure 9.12 Treatment of centerlines of cylindrical objects.

Centerlines are thin lines (0.35mm) and appear thinner than both visible lines and hidden lines. Line thickness gives the appearance that these lines are light; however, this is not the case. Visible lines, hidden lines, and center lines are all the same degree of darkness, but they vary with regard to their thickness.

Radii are treated in a similar manner as cylinders. They too have centerlines associated with them. The only difference between radii and cylinders is that radii do not complete 360°, but are something less. In the view where there is a partial circle or arc, the centerline is represented as a cross hair. In the longitudinal view(s) the centerline is a long, broken line and is an axis of symmetry. See Figure 9.13.

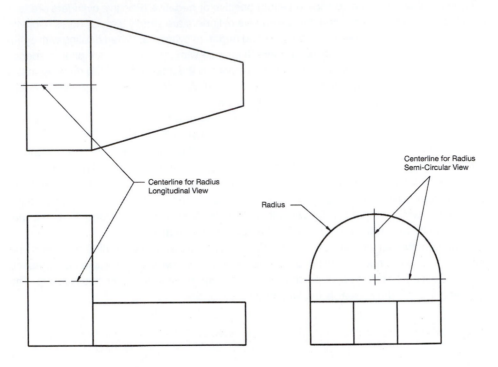

Figure 9.13 Treatment of centerlines of objects with radii.

9.8 THE ALPHABET OF LINES AND LINE PRECEDENCE

Sometimes, especially in complex shapes, various line types will get superimposed on each other in a specific view. For example, a visible line and a hidden line may occupy the exact same space in a view. This creates a small dilemma. In cases such as this, we rely on the alphabet of lines and line precedence to tell us what should be shown.

Various line types (visible, hidden, centerline) have a priority with regard to the orthographic drawing. The highest priority belongs to the visible line, the next priority belongs to the hidden line, and the lowest priority belongs to the centerline. Figure 9.14 shows the alphabet of lines from the highest priority to the lowest priority. Note that priority is given in reverse alphabetical order.

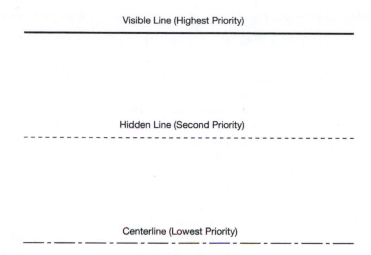

Figure 9.14 Three common standard line types used in engineering drawing and their priority.

Using this guideline, the visible line has the highest priority which means when it competes for space in a view with either a hidden line or a centerline, it has precedence and is shown before the other two line types. A hidden line has precedence over a centerline only. A centerline has no precedence. Figure 9.15 shows an example of line precedence in views. Study this figure carefully and refer to it when you have questions regarding line precedence.

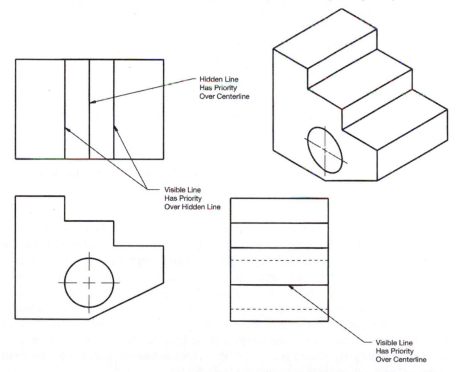

Figure 9.15 Orthographic views of an object showing line precedence.

9.9 FREEHAND SKETCHING

Freehand sketching is very important to the engineer in that it helps to make a quick graphical representation of an idea that can be easily communicated to other members of a team, a professional drafter/CAD operator, or even a client. Often sketches are an integral part of the calculations involved in the analysis of a specific problem and are used to define the object being analyzed. A product often begins with ideation sketches during brainstorming sessions; therefore, developing skills in sketching can be very beneficial to the engineer. To develop sketching skills takes practice. In this section we will explore the basics of freehand sketching so that you can practice and develop your skills in this area.

Generally, lines used in sketching include straight lines and curved lines. Straight lines are composed of horizontal, vertical, and angular lines. To draw straight lines, one should be as comfortable as possible. Often it is easier to draw horizontal lines from the right to the left if you are a right-handed person or left to right if you are a left-handed person. Whichever way is most comfortable is the best way for you. When sketching long horizontal lines, best results occur when you sketch a series of short horizontal lines from end to end. Be careful that your lines remain straight and do not arc. Arcs occur when your elbow becomes stationary and your arm than acts like a long compass. An effective way to draw straight horizontal lines is to put the pencil at the starting point and look at the finish point of the line and draw to it. Vertical lines are usually drawn from top to bottom. The technique used for drawing horizontal lines can be used for drawing vertical lines as well. Angular lines can be drawn exactly the same as horizontal and vertical lines. Simply rotate your paper to the proper angle and begin sketching your lines exactly as you would horizontal or vertical lines. See Figure 9.16.

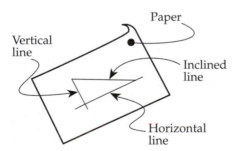

Figure 9.16 Rotate your paper to sketch inclined (angled) lines using horizontal sketching technique.

Sketching curved lines such as circles or radii can be a bit tricky. We can sketch circles most accurately by first sketching the centerline. We can then mark the radius of the circle on the centerline to use as a guide. See Figure 9.17A. Then using these marks we can construct a light construction box that will contain our circle (Figure 9.17B). It should not surprise you that this process is called "boxing in." We can draw light construction diagonals from each corner of the construction box. Mark the radius of the circle on these diagonals (Figure 9.17C). We now have defined 8 points that will be on our circle. We carefully sketch each quadrant of the circle by connecting the points that we have defined. Darken the circle so that it can be seen easily. By making your construction lines extremely light, there is no need to erase them. The finished circle is shown in Figure 9.17D.

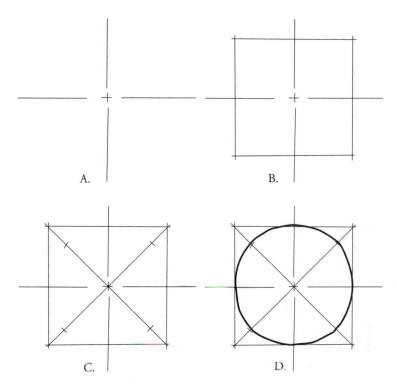

Figure 9.17 Example technique for sketching circles.

When creating a sketch, keep proportions in mind. We do not measure any distances on the sketch, but we want the width, depth, and height to remain in their proper proportions. Real world measurements will come later when working drawings, also known as production drawings, are produced. Line weights are important. For example, earlier in this chapter we discussed line thickness with regard to visible lines, hidden lines, and centerlines. When sketching, it may be rather difficult to keep the line thicknesses precise; however, your visible lines should be thick and your hidden lines and centerlines should be thin. This does not mean that the hidden lines and centerlines are light. They should be dark, but thin.

There will be additional discussion with regard to sketching as we begin pictorial drawings in the sections that follow.

9.10 PICTORIAL SKETCHING

In previous sections of this chapter we learned that orthographic drawings were composed of multiple views where each view represented two dimensions. Pictorial sketching involves creating a view of the object in which all three dimensions are shown. Pictorials are used as a method to help us with visualization which enables us to formulate in our mind's eye what the object looks like. Also, pictorials can be used to present one's case for a design recommendation among various non-technical professionals. Pictorials are relatively easy to understand and can be extremely important in selling a design.

There are generally three types of pictorial sketches that can be used. These types of sketches are known as axonometric sketches, oblique sketches, and perspective sketches.

Examples of each type of pictorial are shown in Figure 9.18. The most popular form of axonometric pictorial is the isometric, so we will limit our discussion of axonometric pictorials to isometric pictorials. The oblique pictorial is also a popular method of describing three-dimensional pictorials in engineering; therefore, we will discuss this form of pictorial as well. Since perspective is widely used in the technical illustration field as well as in the field of architecture and not engineering, we will not be discussing it.

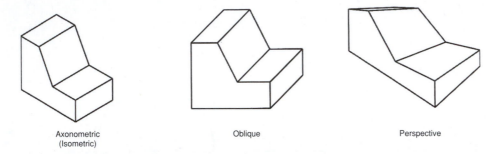

Axonometric
(Isometric)

Oblique

Perspective

Figure 9.18 Examples of pictorial sketches.

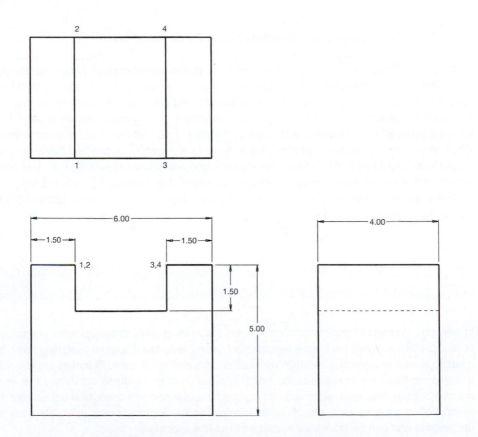

Figure 9.19 Orthographic views of the slotted block.

For consistency, we will create an isometric pictorial and an oblique pictorial using the slotted block shown in Figure 9.19. We will create each pictorial in steps so that you can follow the process and hopefully be able to create other pictorials after you learn the steps.

Isometric sketches are the quickest and easiest of all pictorials to draw and the most commonly used. The most common orientation for the isometric sketch is described when the observer would be looking down on the object (bird's eye view). In this orientation, the isometric axes are as shown in Figure 9.20A. The three axes are composed of two receding axes that measure 30° from the horizontal reference plus a vertical axis. Using these axes, we would consider the width (x measurements) along the receding axis that goes to the left, the depth (y measurements) along the receding axis that goes to the right, and the height (z measurements) on the vertical axis. We can "box in" our slotted block by plotting its height (5), width (6), and depth (4) along each appropriate axis (Figure 9.20B). Note that we have created an "isometric prism" that encompasses the entire object. This would be similar to having a block of wood (5 x 6 x 4) that we can cut in order to get the desired shape. Also note that the angles at the corner of the "isometric prism" are each 120°. Since these angles are equal, that is how the isometric pictorial got its name. The prefix "iso" means equal and the term "metric" means measure. So the word isometric means equal measure and we have equal angles of 120°. We now can develop our isometric pictorial sketch by laying out the slotted block. To do this, we will follow a step-wise procedure.

Step 1. Locate points 1,2,3, and 4 on the top surface of the slotted block. Read the orthographic drawing in Figure 9.19 in order to get the correct distances.

Step 2. Draw a line between points 1 and 2. Draw a line between points 3 and 4.

By connecting these points with lines, we have shown the top part of the slot in the block. The next steps will be followed in order to complete the slot by measuring how deep the slot is in the vertical direction.

Step 3. Sketch a vertical construction line through point 1, point 2, point 3, and point 4.

Step 4. Measure 1.5 units down each construction line to determine the locations of points 5, 6, 7, and 8. Point 5 is directly below point 1, point 6 is directly below point 2, etc.

Step 5. As in step 2, draw a line to connect points 5 and 6.

Step 6. Complete the slot by connecting 1-5, 5-6, 6-2, and 6-8.

Step 7. Erase lines 1-3, 2-4, and part of 6-8.

Note that you do not show line 7-8. This line would be hidden because it is blocked from view by the upper surface of the block. With pictorials, we are interested in what we can see. The hidden line in this case is not that important in the shape description of the slotted block, so it is omitted from the sketch. Hidden lines are omitted from pictorials unless they are absolutely necessary to define the exact shape of the object. As a result, line 6-8 has been trimmed so that the hidden portion is not shown.

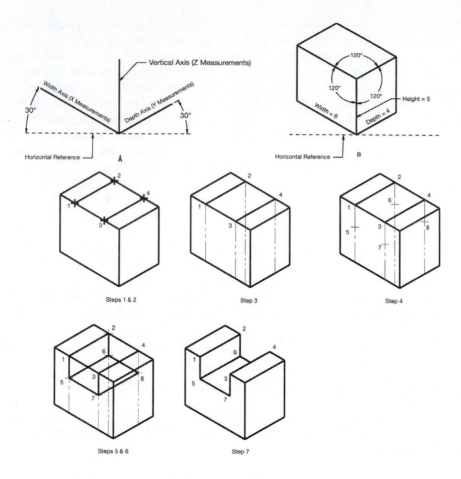

Figure 9.20 Step-by-step method of creating an isometric sketch of slotted block.

Oblique pictorials are very different from isometric pictorials. Isometric pictorials have two receding axes 30° from the horizontal reference line. Oblique pictorials have only one receding axis that shows depth. To draw an oblique pictorial, we can simply show the width and height as it appears in the front view of the orthographic drawing. Then the receding axis can be sketched in at any angle desired; however, 45° is the most common. Again, we will show the oblique pictorial construction in a step-wise manner (see Figure 9.21).

Step 1. Create the front view of the slotted block. Draw construction lines that recede at 45°. See Step 1 in Figure 9.21.

Step 2. Along each receding depth line, lay off the depth of the slotted block. See Step 2 in Figure 9.21.

Step 3. Connect the appropriate points to show the completed slotted block. See Step 3 in Figure 9.21.

Once again, hidden lines are not shown unless they are absolutely necessary to describe the shape of the object. In this case, hidden lines are not necessary, so they are omitted.

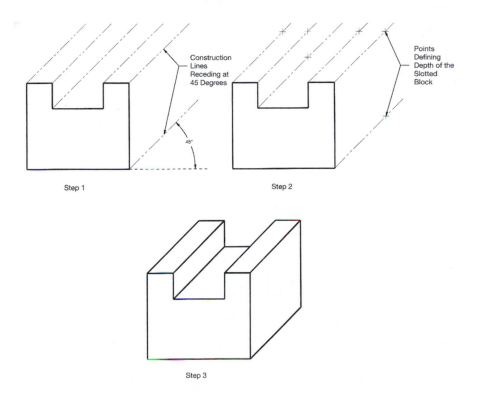

Step 1

Construction Lines Receding at 45 Degrees

45°

Step 2

Points Defining Depth of the Slotted Block

Step 3

Figure 9.21 Step-by-step method of creating an oblique sketch of slotted block.

Suppose we have an object that is a little more complex than an ordinary slotted block. Suppose our object is like what is shown as an orthographic drawing in Figure 9.22. In this object, we not only have a slot, but we also have an inclined surface. How do we make a pictorial sketch of this object? This is not difficult as long as you can plot points using three-dimensional coordinates and then connect the appropriate points. As before, we start by developing the isometric prism that encompasses our object. This is shown in Figure 9.22. Essentially, we follow the same steps as we did before to create the slotted block. The major difference in this case is we have to account for the inclined surface. To do this we must locate point A on the bottom part of the slot (Figure 9.22). This is accomplished by plotting the three-dimensional coordinates of point A. If we assume point O is the origin of our three-dimensional space, then point A is located 1.5 units down the isometric axis from O, and point B is 3.5 units up a vertical construction line and 2 units to the right along the depth construction line. From point A we draw a line to point B and we have installed our inclined surface. We erase the lines that are not needed to define our shape and we have completed the isometric pictorial of our object.

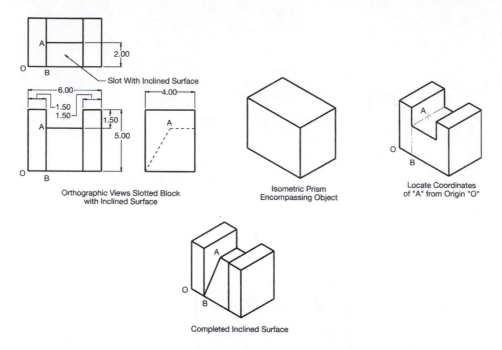

Orthographic Views Slotted Block
with Inclined Surface

Isometric Prism
Encompassing Object

Locate Coordinates
of "A" from Origin "O"

Completed Inclined Surface

Figure 9.22 Creating a sketch of slotted block with an inclined surface.

Three-dimensional pictorials can be constructed rather easily if you plot the three-dimensional coordinates and then connect the points. No matter how complex the object may be, if you can locate each point and connect the points to define the surfaces, you will be successful in creating a pictorial of the object.

Cylindrical features shown in isometric have a level of complexity that needs to be addressed. Essentially, we are talking about positive cylinders and negative cylinders or holes. In orthographic views, cylinders appear as circles and are relatively easy to draw using the "boxing in" technique. In isometric pictorials, circles will always appear as ellipses (Figure 9.23). The orientation of the ellipse depends on which surface (front, top, or side) the cylinder appears.

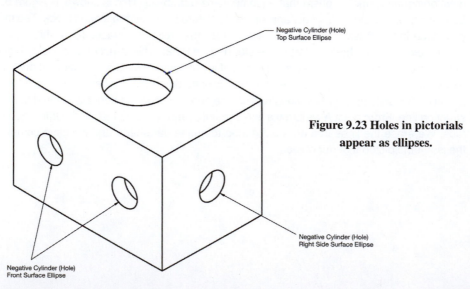

Negative Cylinder (Hole)
Top Surface Ellipse

Figure 9.23 Holes in pictorials appear as ellipses.

Negative Cylinder (Hole)
Right Side Surface Ellipse

Negative Cylinder (Hole)
Front Surface Ellipse

To construct an ellipse on the top isometric surface, follow these steps:

Step 1. Sketch in the center lines on the top surface plane (Figure 9.24).

Step 2. Plot points on the center lines that represent the radius of the circle (Figure 9.24).

Step 3. "Box in" the circle by drawing an isometric box through the points created in Step 2 (Figure 9.24).

Step 4. Sketch the long radii of the ellipse by completing arcs connecting points A and B and points C and D (Figure 9.24).

Step 5. Sketch the short radii of the ellipse by competing arcs connecting points A and C and points B and D (Figure 9.24).

Step 6. Finish the top surface ellipse.

By following these 6 steps, you will complete the ellipse. To draw an ellipse in the right face, apply the same steps to Figure 9.25. To draw an ellipse in the front face, apply the same steps to Figure 9.26.

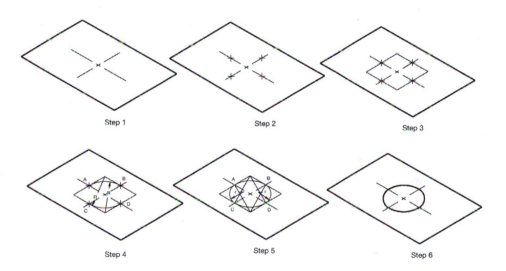

Figure 9.24 Step-by-step method of creating an ellipse on a horizontal surface.

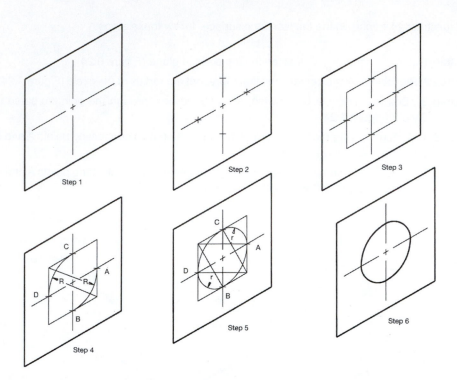

Figure 9.25 Step-by-step method of creating an ellipse on a profile (right-side) surface.

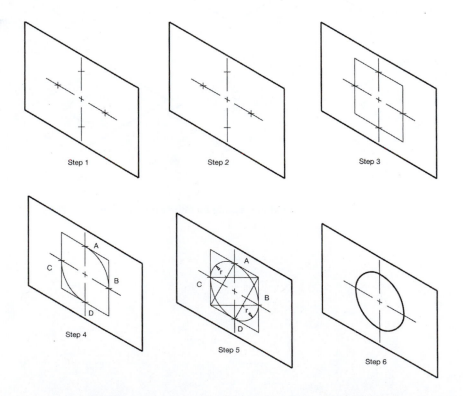

Figure 9.26 Step-by-step method of creating an ellipse on a frontal surface.

9.11 VISUALIZATION

When you see a multi-view two-dimensional orthographic drawing with your eyes and then try to read this information in your mind's eye as the three-dimensional object, it can get confusing. This requires skill that takes practice to master. Sometimes you can look at the orthographic drawing and see the three-dimensional object right away. Other times you have to think about it and perhaps employ some pictorial sketching technique as mentioned in the last section in order to visualize the object in its three-dimensional form. What we want to present in this section are methods that can be used to help you to develop your visualization skills. There is no substitute for practice, and by practicing, you can achieve the goal of seeing the object in your mind as a three-dimensional object without having to sketch it.

To be able to read and interpret shapes from orthographic drawings and to properly visualize them, it is important to understand the meaning of surfaces as they appear in the orthographic views. Recall earlier in this chapter, we discussed the meaning of lines and discussed how a line is generated in orthographic views from a three-dimensional object. Surfaces can be defined in a similar manner. There are three types of plane surfaces that can be used to create a three-dimensional object. These include principal surfaces, inclined surfaces, and oblique surfaces. A principal surface is a surface that is parallel to one of the principal planes (picture planes) shown on the glass box. Therefore we can have a frontal surface which is parallel to the frontal plane, we can have a horizontal surface which is parallel to the horizontal plane (top view), and we can have a profile surface that is parallel to the profile or right side plane. Figure 9.27 shows an object that is composed of all principal surfaces. Surface A is a frontal surface because it is parallel to the frontal plane and appears true size in the frontal view. Note that the right-side view of surface A is an edge view and the horizontal view of surface A is also and edge view. This is always true of principal surfaces. One view will appear in true size, while the other two views will be edge views. The same holds true for surface B. Surface B is true size in the horizontal view and appears as an edge in the front view and the right-side view.

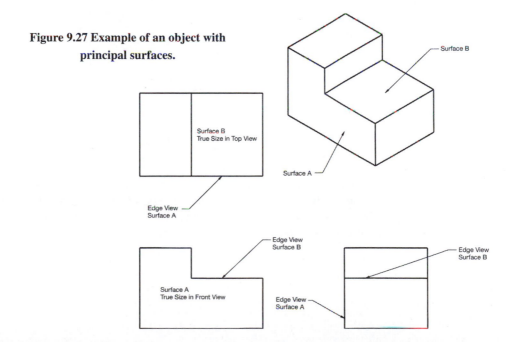

Figure 9.27 Example of an object with principal surfaces.

Surface B
True Size in Top View

Surface B

Edge View
Surface A

Surface A

Surface A
True Size in Front View

Edge View
Surface B

Edge View
Surface A

Edge View
Surface B

An inclined surface does not appear true size in any of the principal views (front, top, side), but appears foreshortened in two of the views and is an edge view in the other. Figure 9.28 shows surface A as an inclined surface. Note that surface A in the horizontal view and the right side view appears as a rectangle. This rectangle is foreshortened in both views—in other words it is not true size. Lines 1-3 and 2-4 are true length in the horizontal view, but lines 1-2 and 3-4 appear foreshortened, therefore the view of the plane is not true size. Note that the front view of surface A appears as an edge. This is always true of inclined surfaces. Two views will show a foreshortened view of the surface and the remaining view will appear as an edge.

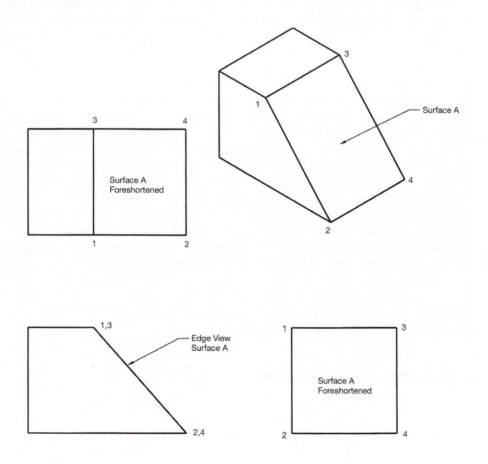

Figure 9.28 Example of an object with an inclined surface.

An oblique surface will never appear as an edge in any of the principal views. An oblique surface will always maintain its characteristic shape and show up as foreshortened in all three principal views. Figure 9.29 shows an oblique surface A in the orthographic views. Note that surface A is a foreshortened triangle in all three views.

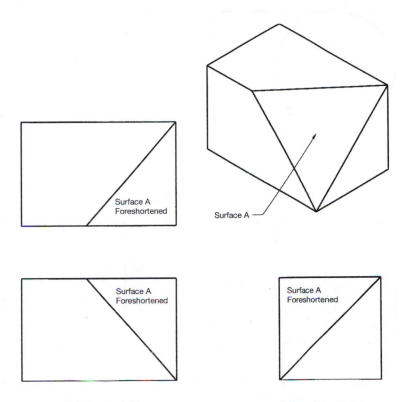

Figure 9.29 Example of an object with an oblique surface.

Regarding visualizing surfaces, there is one hard-and-fast rule that applies to any plane surface. A surface will either appear as an edge or it will maintain its characteristic shape in any view in which it appears. This does not mean that the surface will appear in true size, but it only has to maintain its characteristic shape. If it is a rectangle, it will either be an edge or it will be a rectangle. Note that in the examples of Figures 9.27 and 9.28 that the surface is either an edge or it appears as a rectangle. In Figure 9.29, the surface appears as a triangle (its characteristic shape) in all three views. This rule is very useful when visualizing objects.

Another type of surface you need to be familiar with while practicing visualization is the curved surface. Earlier in the chapter, we defined what was meant by a limiting element. Recognizing limiting elements will enable you to visualize curved surfaces. Figure 9.30 shows an object composed of plane surfaces and a cylinder. Note the limiting elements and how they appear in defining the cylindrical feature. Also note the plane surface intersects the cylinder tangentially and creates what is referred to as a run out.

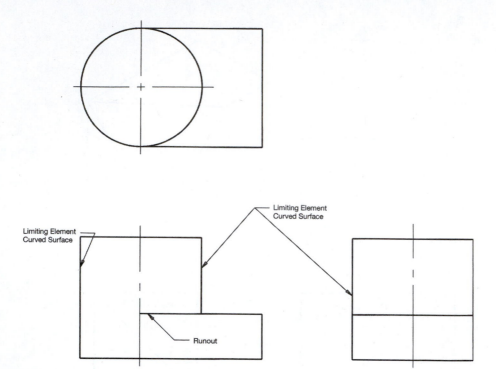

Figure 9.30 An object composed of a cylinder and a prism showing limiting elements of a curved surface and a run out where the prism meets the cylinder.

9.12 SCALES AND MEASURING

Most technical drawings are drawn to some scale. This means that there is a ratio between the length of the lines on the drawing representing the object and the length of the same line on the real world object. For example a 1:1 (one to one) scale means that one linear unit of measure on the drawing is equal to one linear unit on the object. Use of scale primarily depends on the size of the drawing media. If your object is relatively small and you are showing the orthographic views on an 8.5" x 11" (A size) page, then you probably could use a full scale or 1:1. If your object is large, and you are going to draw the orthographic views on an A size sheet, then you will have to scale it down some. The larger the object being shown on the drawing, the smaller the drawing scale has to be on the sheet. Common drawing scales used in engineering are full scale (1:1), half scale (1:2), quarter scale (1:4), and tenth scale (1:10). The scale denotes the size of the drawing. For example, a half scale drawing means the drawing appears at half the size of the real world object. Sometimes it is advantageous to make drawings larger than they are in the real world. For example, when dealing with extremely small objects, you might draw them at double scale (2:1) or larger in order to see more detail.

Drawing scale refers to the ratio that the drawing has to the real world object; however, there is an instrument called a scale as well. Sometimes this instrument is called a ruler. It is a calibrated drawing instrument used to measure linear distance according to the specified drawing scale. A full divided scale divides linear measurements along its entire length. For example, a ruler/scale that breaks an inch into 1/16 increments is a full divided scale. Although these types of rulers/scales are common to most people, engineers use other types of scales in practice because of the need to scale an object either up or down to ensure that it fits the page on which it is being drawn. Two common examples of engineering scales are the civil engineer's scale and the metric scale. The civil engineer's scale, or simply engineer's scale, is derived from the surveyor's chain that is composed of 100 links, and divides an inch into multiples of ten units. The 10 scale (Figure 9.31) divides an inch into 10 divisions, and thus each division equals 0.10 inch. The 10 scale is an example of a full scale because one inch on the scale is also on inch on the real world object.

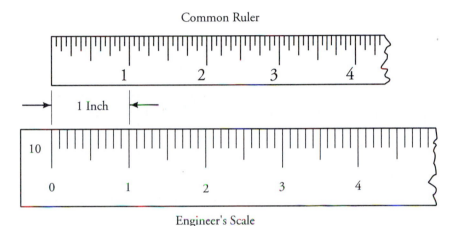

Figure 9.31 **Comparison of a common ruler and an engineer's scale.**

To measure a distance using the 10 scale, start by placing the number 0 at one of the end points of the line you wish to measure. Read each of the major graduations (1, 2, 3 ....) across the scale as full inches and add the number of minor graduations as tenths of an inch. For example, the line in Figure 9.32 measures 3.4 inches. One significant thing regarding the 10 scale of the engineer's scale is that you can also measure in multiples of ten. The scale can be multiplied by 10, 100, or 1000. When doing this, then the other graduations are multiplied by the same value. If your drawing were drawn to a scale of 1:100, then the numbers along the scale would read 100, 200, 300, etc. Likewise the minor graduations would read as 10, 20, 30, ... 90. The line shown in Figure 9.32 would measure 340 inches using a drawing scale of 1:100.

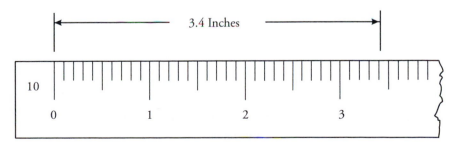

Figure 9.32 **Measuring the distance between two points using the 10 scale on an engineer's scale.**

The engineer's scale also has scales other than the 10 scale. The 20 scale has one inch divided into 20 divisions. We would use this scale to do a drawing in half size, so that every inch on the drawing represents 2 inches on the real world object. We use this scale in a similar manner as we used the 10 scale. Start by placing the number 0 at one of the endpoints of the distance you wish to measure. Read each of the major graduations (1, 2, 3, ...) across the scale in full inches and read the number of minor graduations as 0.2 of an inch. For example, in Figure 9.33, the line measures 5.4 inches. As with the 10 scale, the 20 scale is versatile and can be used to measure other multiples of twenty. The line shown in Figure 9.33 measures 5400 inches on a drawing that is produced using a scale of 1:2000. Practice measuring the lines shown in Figure 9.34 using an engineer's scale at the specified drawing scales.

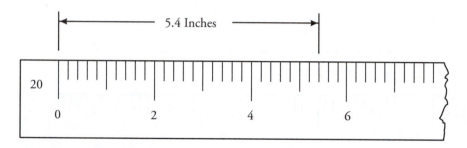

Figure 9.33 Measuring the distance between two points using the 20 scale on an engineer's scale.

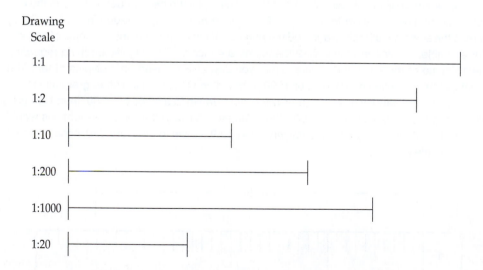

Figure 9.34 Measure the distance of each line in inches using the given drawing scale.

The engineer's scale makes measurements in decimal inches. ANSI (American National Standards Institute) has designated that mechanical parts shall be documented using decimal inches. This means that the standard for drawings of mechanical parts is decimal inches and fractional inches are no longer the standard. This does not mean that you should not know about fractional inches, for there are a number of design drawings that have been developed using fractional inches. You should realize, howere, that the modern standard is decimal inches.

In the SI (Système International) measurement system, millimeters are the base unit of measure. It is important to know that ANSI has designated that mechanical parts that use metric measurements must be done in millimeters. There is good reason for this. If a dimension on a part happens to be 50 millimeters and it is written as 50 (millimeters or inches are not written with the dimension on a drawing), then it cannot be interpreted as 50 inches. On the other hand, if the metric unit of 5 centimeters were used instead of 50 millimeters, then the number 5 could easily be misinterpreted as 5 inches. So to avoid this confusion, ANSI has designated that mechanical parts defined in SI units shall be drawn using millimeters. For drawings that deal with large areas such as topographic maps, it is permissible to use meters or kilometers.

Since a number of drawings throughout the world are drawn using metric measure, you should have some practice in measuring using a metric scale that has millimeters as its smallest unit. The 1:1 metric scale is the starting point for learning to measure using a metric scale. The 1:1 scale is divided into 1 mm graduations and is calibrated at 10 mm intervals as shown in Figure 9.35. This is an example of a full scale measurement using the metric scale.

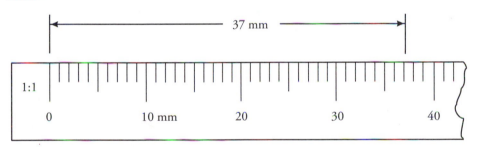

Figure 9.35 Measuring the distance between two points using the 1:1 scale on a metric scale.

Measuring distances on the metric scale is very similar to the method used to measure distances using the Engineer's scale. Start by lining up the 0 at the left end of the 1:1 scale to one endpoint of the distance you want to measure. Read the numbers of the major graduations along the scale (10, 20, 30, ...) as millimeters and add the number of minor graduations directly to the first number. The line in Figure 9.35 measures 37 mm. Like the engineer's scale, the 1:1 scale can be used to measure other multiples of 10 such as 1:10, 1:100, 1:1000, etc. Simply multiply each of the graduations by 10, 100, 1000, etc., as you measure them. This means that on a drawing scaled to 1:100, you would read 1000, 2000, 3000, etc., along the scale. For example, the line in Figure 9.36 measures 220 mm on a drawing scale of 1:100.

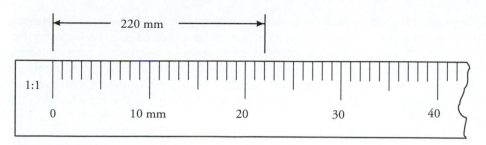

Figure 9.36 Each measurement scale can be used to measure in multiples of 10.

The 1:2 metric scale is divided into 2-mm graduations and is calibrated at 20-mm intervals as shown in Figure 9.37. This scale is a half scale, similar to the 20 scale on the engineer's scale. Using the 1:2 scale, 1 mm on the drawing represents 2 mm on the real world object. Measurements can be made using this scale just as with the 1:1 scale. Start by placing the 0 on the left side of the scale at one endpoint of the distance you wish to measure. Count the number of major graduations along the scale using the numbers provided (20, 40, 60, ...). Add the minor graduations (remember to count these by 2 since we are using the 1:2 scale) to the first number. A 76-mm line is shown in Figure 9.37. You can also use the 1:2 scale to measure other drawing scales such as 1:20, 1:200, 1:2000, etc. With some practice, learning to read a metric scale is very easy. Figure 9.38 has several lines at different metric scales on which you can practice.

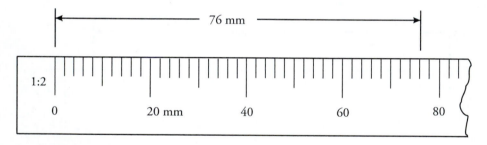

Figure 9.37 Measuring the distance between two points using the 1:2 scale on a metric scale.

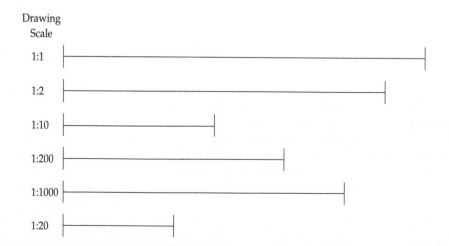

Figure 9.38 Measure the distance of each line in mm using the given drawing scale.

9.13 COORDINATE SYSTEMS AND THREE-DIMENSIONAL SPACE

It is important to have an understanding of coordinate systems as they relate to both two-dimensional space as well as three-dimensional space. Handling coordinate systems not only enhances your ability to visualize, but it is essential when dealing with three-dimensional CAD (Computer-Aided Design, also referred to here as computer graphics) systems. Recall in the first part of this chapter, we introduced the three dimensions of width, depth, and height. The link between these dimensions and computer graphics is in the x, y, z coordinates. Regarding computer graphics, the x coordinate = width, the y coordinate = depth, and the z coordinate = height. We need to explore and practice using both two-dimensional and three-dimensional coordinate systems.

In Figure 9.39, there is a horizontal line and a vertical line that meet each other at 90° (perpendicular). If we label the horizontal line as x and the vertical line as y, then we have established these two lines as the x-axis and the y-axis of our coordinate system. This coordinate system is called the Cartesian coordinate system or rectangular coordinate system. Where the two axes cross is called the origin. Note that there are four quadrants in the system. The system allows us to plot positive and negative values for both x and y. For example, a value of x = 5 and a value of y = 3 would be plotted in the upper right quadrant. When both values are positive, the resulting location of the plotted point would be in the upper right quadrant. If the values were (x = -5, y = 3), the point would be plotted in the upper left quadrant since the value of x is negative and is plotted to the left of the origin and the y value is positive and is plotted above the origin. In which quadrant would you find the point if both the x value and the y value were negative? If you said the lower left quadrant, then you understand the concepts of plotting two-dimensional coordinates. Finally if the x value was positive and the y value was negative, the point would be plotted in the lower right quadrant. This technique of plotting points is commonly used throughout mathematics, science, and engineering. By plotting points and connecting them with lines, we can generate graphical displays.

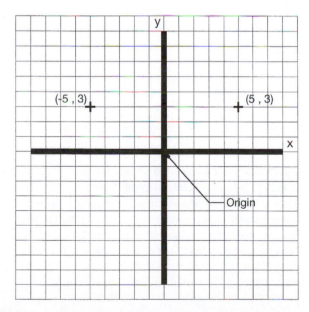

Figure 9.39 Locating points using two-dimensional rectangular coordinates.

Two-dimensional rectangular coordinates work great for two-dimensional problem solving; however, we do live in a three-dimensional world where most of the problems solved by engineers require a way to represent them in three-dimensional space. If you can plot points in two-dimensional space, you can plot them in three-dimensional space. Three-dimensional Cartesian coordinates can be defined in mathematics by attaching a third axis (z) to the x- and y-axes. Figure 9.40 shows how this axis relates to the x-axis and y-axis. In this manner, we have created an oblique (it resembles an oblique pictorial drawing) three-dimensional rectangular coordinate system. A unique characteristic of this coordinate system is the relative orientation of the axes. Take your right hand and point your fingers in the same direction of the positive x-axis. Curl you fingers toward the positive y-axis and notice the direction in which your thumb points. It should point in the direction of the z-axis. Your thumb is parallel to the positive z-axis that comes out of the page because it is perpendicular to the xy-plane at the origin. This is called a "right-handed" coordinate system and it behaves according to the right hand-rule which says as the fingers of your right hand curl from x to y, the thumb of your right hand will point in the direction of z.

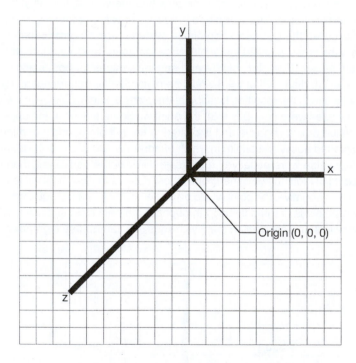

Figure 9.40 Axes for a three-dimensional rectangular coordinate system.

The coordinate system described above and shown in Figure 9.40 has been established in mathematics as the way to generate points in three-dimensional space. However, in computer graphics, it is done a little differently. In computer graphics, we want the vertical direction to indicate the z-axis instead of the y-axis, so the axis system is set up differently. Figure 9.41 shows the three-dimensional space that is defined for a CAD (computer graphics) system. It still follows the right-hand rule (check it), but instead of the xy-plane defining a wall with the z-axis coming outward, the xy-plane defines the floor with the z-axis going vertical. The xz- and yz-planes are analogous to vertical walls meeting at 90°.

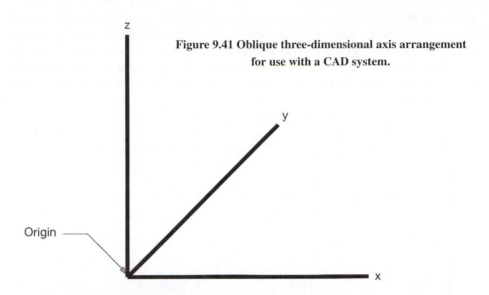

**Figure 9.41 Oblique three-dimensional axis arrangement
for use with a CAD system.**

There is good reason why CAD systems use this type of rectangular coordinate system. Industry uses lots of different computer numerically controlled machines (CNC machines) to manufacture its products. Long ago, it was established that the vertical direction with regard to computer manufacturing would be standardized as the z direction. Therefore, as CAD began to develop, it was natural to build in the same coordinate system since often times the geometry designed in a CAD system can be directly loaded into a CNC machine for manufacturing. Therefore, you will have to get accustomed to seeing this type of coordinate system when defining points in three-dimensional space for graphical solutions of problems.

What we have shown in Figure 9.41 is an oblique three-dimensional coordinate system. Sometimes it is advantageous to use an isometric three-dimensional coordinate system. This system is shown in Figure 9.42. In this system you plot points in the three-dimensional space just as in the oblique system, but you may have a better perspective to the view the object.

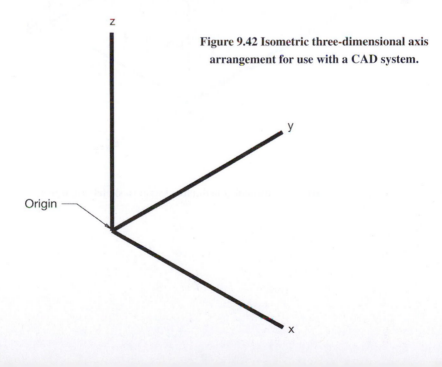

**Figure 9.42 Isometric three-dimensional axis
arrangement for use with a CAD system.**

Sketching an object in three-dimensional space is relatively easy. Figure 9.43 shows a three-dimensional rectangular coordinate system one would expect to see using a CAD representation. Table 9.1 is a list of vertices and edges. We plot each of the points V1 through V8. We then check the edge list to be sure that we connect the points correctly in order to complete the three-dimensional shape. Follow the development of this object by checking the points and the edges.

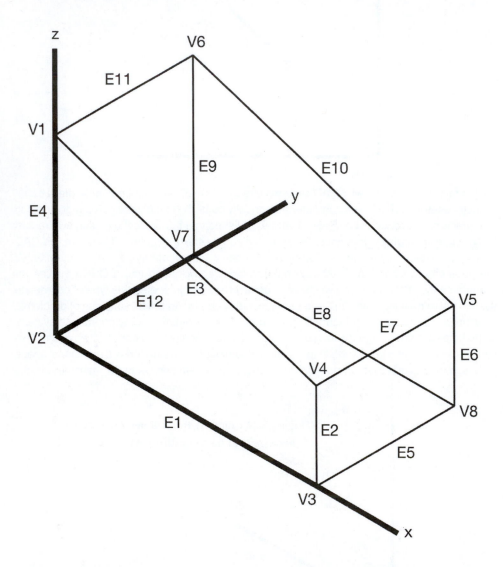

Figure 9.43 Using a CAD three-dimensional coordinate system to sketch an object.

TABLE 9.1 List of vertices and edges to be sketched

Vertex List		Edge List	
Vertex	*x, y, z Location*	*Edge*	*Vertices*
V1	(0,8,10)	E1	(V2,V3)
V2	(0,8,0)	E2	(V3,V4)
V3	(15,8,0)	E3	(V4,V1)
V4	(15,8,5)	E4	(V1,V2)
V5	(15,0,5)	E5	(V3,V8)
V6	(0,0,10)	E6	(V5,V8)
V7	(0,0,0)	E7	(V4,V5)
V8	(15,0,0)	E8	(V7,V8)
		E9	(V6,V7)
		E10	(V5,V6)
		E11	(V1,V6)
		E12	(V2,V7)

EXERCISES AND ACTIVITIES

9.1 Interview a practicing engineer or an engineering professor to find five examples of how they use visualization and five examples of how they use graphical communication in their professional lives.

9.2 Keep a sketching journal to practice your sketching technique. This idea is similar to writers who keep written journals to document their ideas in words. It is a good habit for an engineer to document ideas using sketches in order to maintain a record of when and how the idea was developed—this is especially critical in the patent application process. Suggestions for your journal include the following: use a variety of projection techniques (orthographic, oblique, isometric); use either a bound notebook or loose sheets of rectangular grid paper contained in a binder; sketch familiar objects that are in your room, apartment, or house for practice.

9.3 Practice sketching straight lines between the points given in the space below, or copy them onto a separate sheet of rectangular grid paper.

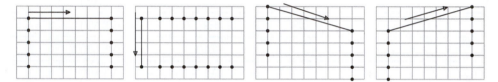

9.4 Practice sketching circles of variable radius in the space given below or on a separate sheet of rectangular grid paper.

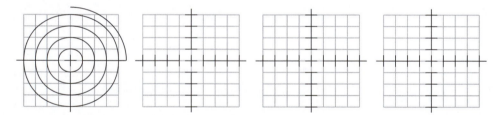

9.5 Practice sketching arcs of variable radius in the space given below or on a separate sheet of rectangular grid paper.

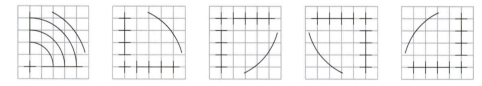

9.6 Practice sketching ellipses of variable radius in the space given below or on a separate sheet of rectangular grid paper.

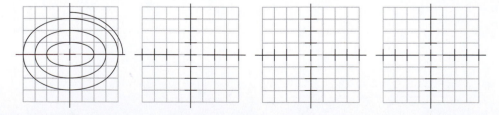

9.7 Plot the points given in the table below using either an oblique or isometric 3-D coordinate system on grid paper (rectangular or isometric). Connect the vertices according to the edge list provided in the table, and darken the visible edges to complete the drawing.

A.

Vertex	(x, y, z) Coordinates	Vertex	(x, y, z) Coordinates	Edge	Vertices	Edge	Vertices
V1	(0, 0, 0)	V7	(8, 4, 0)	E1	(V1, V2)	E9	(V3, V6)
V2	(0, 0, 2)	V8	(4, 4, 0)	E2	(V2, V3)	E10	(V4, V7)
V3	(8, 0, 2)	V9	(0, 8, 2)	E3	(V3, V4)	E11	(V9, V10)
V4	(8, 0, 0)	V10	(0, 8, 0)	E4	(V4, V1)	E12	(V10, V11)
V5	(4, 4, 2)	V11	(4, 8, 0)	E5	(V5, V6)	E13	(V11, V12)
V6	(8, 4, 2)	V12	(4, 8, 2)	E6	(V6, V7)	E14	(V12, V9)
				E7	(V7, V8)	E15	(V5, V12)
				E8	(V8, V5)	E16	(V8, V11)

B.

Vertex	(x, y, z) Coordinates	Vertex	(x, y, z) Coordinates	Edge	Vertices	Edge	Vertices
V1	(0, 0, 0)	V6	(10, 7, 2)	E1	(V1, V2)	E9	(V6, V7)
V2	(10, 0, 0)	V7	(10, 3, 2)	E2	(V2, V3)	E10	(V7, V8)
V3	(10, 7, 0)	V8	(0, 3, 2)	E3	(V3, V4)	E11	(V8, V9)
V4	(0, 7, 0)	V9	(0, 3, 5)	E4	(V4, V1)	E12	(V9, V7)
V5	(0, 7, 2)	V10	(0, 0, 5)	E5	(V4, V5)	E13	(V10, V2)
				E6	(V6, V3)	E14	(V10, V1)
				E7	(V5, V6)	E15	(V9, V10)
				E8	(V8, V5)		

9.8 Redraw the objects in problem 7 above by applying the following geometric transforms to the given vertices and edges:

    a. Translate the object by adding (–2, 1, –4) to each vertex.
    b. Rotate the object 180° about the *y*-axis.

9.9 Using the scales given below, draw lines of the specified length on a separate sheet of paper.

Drawing Scale	Line Length	Drawing Scale	Line Length
1″=10′	8.0′	1:1	15 cm
1″=10′	6.3′	1:100	5 m
1″=2.0″	8.4″	1:10	85 mm
1″=20″	54″	1:2	50 mm
1″=.01″	.035″	1:20	750 mm
1″=300′	750′	1:200	14 m
1″=1000′	3250′	1:2000	.500 km
1″=50′	245′	1:50	4350 mm

9.10 Sketch oblique drawings of the objects shown in the orthographic drawings below. Alternate between cavalier and cabinet layouts of varying angles as assigned by your instructor.

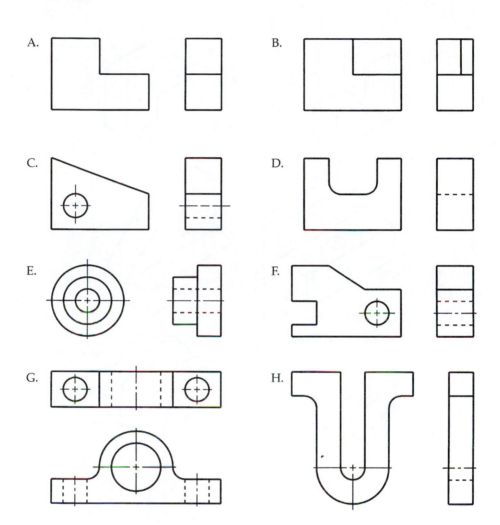

9.11   Sketch orthographic drawings of the objects shown in the isometric drawings below.
Use an appropriate drawing scale as specified by your instructor.

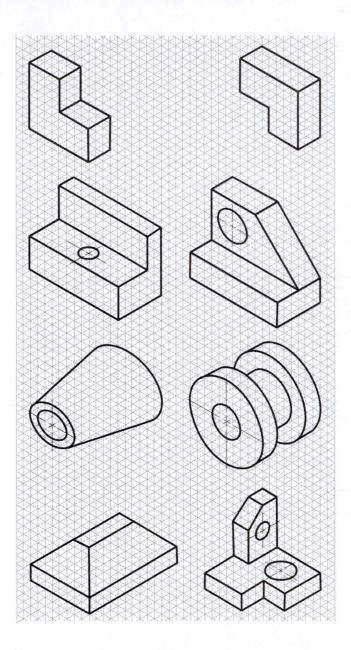

9.12 Practice sketching isometric cylinders on the isometric grid given below or on a separate sheet of isometric grid paper.

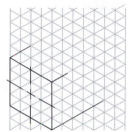  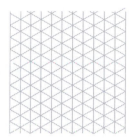

9.13 Construct isometric drawings of the objects shown in the orthographic drawings in problem 10 above. Use either rectangular or isometric grid paper, and a drawing scale specified by your instructor.

9.14 List the three things that a line can represent on a drawing, and sketch an example of each.

9.15 Compare and contrast the three types of projection described in this chapter (how are they the same and how do they differ).

Chapter 10

Computer Tools for Engineers

Assume that you are a food process engineer, and a dairy company has hired you for a project:

"Design a process to manufacture high-quality powdered milk for a profit."

This is no small undertaking! Table 10.1 shows a list of the steps that you must complete.

TABLE 10.1 Steps Required to Complete the Powdered-milk Process Design

Step	Computer Tool
Research the dairy industry	Internet
Design the process	Mathematical Software
	Design Software
	Programming Languages
Select equipment	Internet
Perform economic analyses	Spreadsheet
	Mathematical Software
Analyze experimental data	Spreadsheet
	Mathematical Software
Document the design process	Word Processor
Present findings to management	Presentation Software

With this huge project ahead of you, you realize that you need help. So you turn to the computer. Computer tools can facilitate nearly every task you need to perform. These tools include operating systems, programming languages, and computer applications. For example, word processing applications enable computer users to write papers. To be able to use the computer to solve an engineering problem, you need strong computer skills and a working familiarity with the types of computer tools and applications that are available. Table 10.1 indicates the types of computer tools you would use to complete this project.

The following sections of this chapter will highlight some of the computer tools that an engineer might use to solve problems. Many of the tools will be discussed in the context of solving the dairy design problem.

10.2 THE INTERNET

Introduction

The first step required to solve the dairy design problem is to fill gaps in your knowledge base. That is, you need to do research so that you have a better understanding of the design problem. Some things you should learn are:

- How is powdered milk manufactured?
- What are the consumer safety and health issues involved?
- Who manufactures equipment for dairy processing?
- What other dairy companies manufacture powdered milk?
- Who does dairy processing research?

Research, under the best conditions, can take a significant amount of time. So your aim is to avoid excessive trips to the library and the use of bound indexes. It is preferable to access as much information as possible via your computer. The Internet provides such capabilities.

History of the Internet

The Internet is a worldwide network of computers that allows electronic communication with each other by sending data packets back and forth. The Internet has been under development since the 1960s. At that time, the Department of Defense was interested in testing out a concept that would connect government computers. It was hoped that such a network would improve military and defense capabilities (Glass and Ables, 1999). Before the end of 1969, machines at the University of Southern California, Stanford Research Institute, the University of California at Santa Barbara, and the University of Utah were connected.

The linking of those four original sites generated much interest in the Internet. However, extensive customization was required to get those original sites linked, because each site used its own operating system. It was apparent that the addition of more sites would require a set of standard protocols that would allow data sharing between different computers. And so the 1970s saw the birth of such protocols.

Protocols are used to move data from one location to another across the network, and verify that the data transfer was successful. Data are sent along the Internet in *packets*. Each packet may take a different route to get to its final destination. The *Internet Protocol* (IP) guarantees that if a packet arrives at its destination, the packet is identical to the one originally sent. A common companion protocol to the IP is the *Transmission Control Protocol* (TCP). A sending TCP and a receiving TCP ensure that all the packets that make up a data transfer arrive at the receiving location.

Internet applications were also developed in the 1970s and are still used today. The Telnet program connects one computer to another through the network. *Telnet* allows a user on a local computer to remotely log into and use a second computer at another location. The *ftp* (file transfer protocol) program allows users to transfer files on the network. Both applications were originally non-graphical; they require that text commands be entered at a prompt.

Internet IP addresses were established in the 1970s. All organizations that become part of the Internet are assigned a unique 32-bit address by the Network Information Center. IP addresses are a series of four eight-bit numbers. For instance, IP addresses at Purdue University take the form 128.46.XXX.YYY, where XXX and YYY are numbers between zero and 255.

In the 1980s, corporations began connecting to the Internet. Figure 10.1 shows the growing number of sites connected to the Internet. With so many sites joining, a new naming system was needed. In came the *Domain Naming Service*. This hierarchical naming system allowed easier naming of individual hosts (servers) at a particular site. Each organization connected to the Internet belongs to one of the top-level domains (Table 10.2). Purdue University falls under the U.S. educational category; therefore, it has the domain name purdue.edu. Once an organization has both an IP address and a domain name, it can assign addresses to individual hosts. This author's host name is pasture.ecn.purdue.edu. "Pasture" is a large computer (server) on the Purdue University Engineering Computer Network (ecn), a network of engineering department servers. With the explosion of websites, the number of allowable domain names has increased with new top-level names and extensions.

As *Internet Service Providers* (ISP), such as America On-Line and CompuServe, came into being in the 1990s, the Internet became more mainstream. The development of the browser named Mosaic made accessing information on the Internet easier for the general public. A *browser* is software that allows you to access and navigate the Internet. Mosaic displays information (data) using both text and graphics. Each page of information is written using the *HyperText Markup Language* (HTML). The most useful aspect of HTML is the hyperlink. A *hyperlink* connects two pages of information. The beauty of the hyperlink is that it allows an Internet user to move through information in a non-linear fashion. That is, the user can "surf" the *World Wide Web* (WWW). The Web, not to be confused with the Internet, refers to the interconnection of HTML pages. Two of the most common browsers are Microsoft Internet Explorer and Mozilla Firefox. A number of other browsers have been introduced, such as Google's Chrome and Safari, that have much smaller market shares. Updates to these browsers have allowed for the construction of more complicated web pages, such as multiple-frame pages. Sound and video can now be played through the use

Year	Hosts
1981	213
1985	1,961
1990	313,000
1995	6,642,000
2000	93,047,000
2005	353,284,000
2010	768,913,036

Figure 10.1 Number of Internet hosts by year.

TABLE 10.2 Example Top-Level Domain Names

Name	*Category*
edu	U.S. Educational
gov	U.S. Government
com	U.S. Commercial
mil	U.S. Military
org	Non-profit Organization
net	Network
XX	Two-Letter Country Code
(e.g., fr, uk)	(e.g., France, United Kingdom)

of helper applications. Greater user interaction with the Web has been made possible through HTML coded references to Java applets (defined later).

Searching the Web

Let's return to the powdered-milk problem. You'd like to use the Internet to answer some of your research questions. First focus on searching for equipment vendors. Assume that you already know about one company that supplies food processing equipment, called APV. Check to see what kinds of equipment they have available for dairy processing. Using your Web browser, you access APV's website by providing the URL (*Universal Resource Locator*) for APV's home page. In other words, you tell the Web browser what Web address to use to access APV's information. Assume that the URL is "http://www.apv.com." This URL consists of the two primary components of a Web address: the protocol (means for transferring information) and the Internet address. The protocol being used here is http (Hypertext Transport Protocol) which allows access to HTML. The Internet address consists of the host name (apv) and the domain (com).

APV's URL takes you to APV's introductory (home) page. Once there, you can browse their website for information that pertains to your specific research problem. At one point, you find yourself on a page that describes a one-stage spray dryer designed specifically for temperature-sensitive materials such as milk. A spray dryer is a piece of equipment that converts liquid into a powder, or as in your application, milk to powdered milk. The URL for this page is http://www.apv.com/ anhydro/spray/onestage.htm. This Web address is longer than that for APV's home page. The address contains a path name (/anhydro/spray/) and a file name (onestage.htm). The path name tells you the location within APV's directory structure of the file that you are accessing with your Web browser, just as you store files in directories on your hard drive.

APV is just one equipment vendor. You'd like to investigate others, but you do not know any other Web addresses. To further your research, you use a Web search engine. A *search engine* matches user-specified keywords with entries in the search engine's database. The engine then provides a list of links to the search results. There are a number of search engines such as Yahoo! (www.yahoo.com) and Google (www.google.com). A keyword search performed using one search engine will yield different results than the same keyword search on another search. The reason for the different results lies in the type of information that a particular search engine tends to have in its database.

You decide to use the search engine Google to do a keyword search for "spray dryer." When your search results come up, you immediately notice that your query was not specific enough. You get back links to sites on laundry dryers! There are so many sites on the Internet that well-constructed queries are a must. Each search engine provides links to searching tips. In your case, you find that a keyword search for "spray dryer and food" improves your search results.

Home Pages

Home pages are Web pages developed by users that contain information about topics of interest to the user, and often include links to the user's favorite websites. The language of all Web pages is HTML. An *HTML document* is a text file containing information the user wants posted to the Web, and instructions on how that information should look when the Web page is accessed by a browser. HTML code can be written from scratch using a text editor, or a composer can be used to generate the code. Many word processors and desktop publishers now have the capability of saving files in HTML format so that the document is displayed by the Web browser exactly as it was in the word-processing program. The HTML code consists of instructional tags as seen in Figure 10.2. This code generates the Web page shown in Figure 10.3.

Your employer may ask you to design a Web page discussing the process you developed through the completion of the powdered-milk process design project. A personal reason for creating your own home page is to make available your resume and samples of your work to all Web users. This sort of visibility may result in employment opportunities such as internships and post graduation positions.

More information on creating home pages can be found on the Web or in any book on HTML or the Internet. You may find "HTML: The Complete Reference" by T. A. Powell (1998) to be helpful when writing HTML code.

Libraries & Databases on the Internet

Not all information is available through the Internet, but an ever-increasing amount is. New search features are coming on-line each year and databases are continually emerging and growing. The Internet can be used to locate many sources for research through search features such as Google Scholar (scholar.google.com). Information that is not available through the web can be located using the Internet, such as through a website of a university's library. Most have sophisticated search tools that can be used to access articles in electronic form or can locate hard copies.

Contacting People through the Internet via E-mail

The Internet can also be used to send electronic mail (e-mail) to individuals, groups, or mailing lists. E-mail is currently exchanged via the Internet using the *Simple Mail Transport Pro-*

```
<HTML>
<HEAD>
<TITLE>FOODS BLOCK LIBRARY – BASIC ABSTRACT</TITLE>
</HEAD>

<BODY BGCOLOR="#FFFFFF">
<H1>A Computer-Aided Food Process Design Tool for
Industry, Research, & Education</H1>
<H2>Heidi A. Diefes, Ph.D.</H2>
<H2><I>Presented at the Food Science Graduate Seminar<BR>
Cornell University, Fall 1998</I></H2>
<HR ALIGN=LEFT NOSHADE SIZE=5 WIDTH=50%>
<P>
There is a great need in the food industry to link food science research with food
engineering and process technologies in such a way as to facilitate an increase in
product development efficiency and product quality. Chemical engineers achieve
this through industry standard computer-aided flowsheeting and design pack-
ages. In contrast, existing programs for food processing applications are limited
in their ability to handle the wide variety of processes common to the food indus-
try. This research entails the development of a generalized flowsheeting and
design program for steady-state food processes which utilizes the design strate-
gies employed by food engineers.
</P>
</BODY>
</HTML>
```

Figure 10.2 HTML code for Web page shown in Figure 10.3.

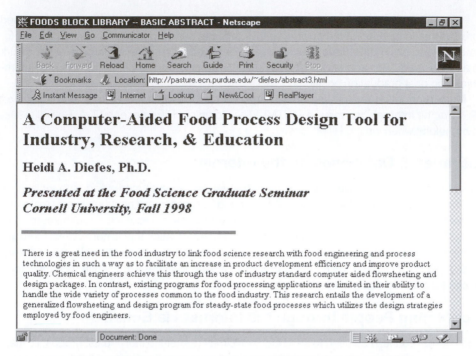

Figure 10.3 Web page generated using HTML code in Figure 10.2.

tocol (SMTP) and *Multipurpose Internet Mail Extension* (MIME). To send e-mail, two things are required: an e-mail package and e-mail addresses. There are many e-mail packages from which to choose.

Regardless of the brand, e-mail tools have certain common functions including an ability to send mail, receive mail, reply to mail, forward mail, save mail to a file, and delete mail.

To communicate via the Internet, an *e-mail address* is required. All e-mail users are assigned a unique e-mail address that takes the form username@domain.name. However, it can be difficult to locate a person's e-mail address. One method of locating an e-mail address is to use one of the many people-finding websites. For example, the Netscape People Finder can be used to find both home and e-mail addresses. In addition, most colleges and universities have methods for searching for staff and students on the school's website.

Once you have selected an e-mail address and e-mail software, and after you have signed up with an ISP, you can send and receive e-mail. As part of your dairy design problem, you may keep in touch with the dairy company contacts and other engineering consultants via e-mail. A typical e-mail message might read like the example that follows.

> **To:** dairyconsult@purdue.edu
> **cc:** dairycontacts@purdue.edu
> **Attachments:** spdry.bmp
> **Subject:** Spray dryer design model
> **Message:**
> I have been working with the spray dryer vendor on a model of
> the system we intend to use. Attached is a current picture of
> the model. Please advise.
> H. A. Diefes

The above e-mail message contains both the typical required and optional components. An e-mail is sent:

To: username@domain.name or an alias (explained in the next paragraph). A copy (**cc:**) may also be sent to a username@domain.name or an alias.

Attachments: are files to be appended to the main e-mail message.

Subject: is a brief description of the message contents.

Message: is the actual text of the message.

An *alias* is a nickname for an individual or group of individuals, typically selected by the "owner." An e-mail to an alias gets directed to the actual address. Address books, another common feature of e-mail packages, allow the user to assign and store addresses and aliases. For example, in your address book you may group all the e-mail addresses for your dairy company contacts under the alias "dairy contacts." If you put "dairy contacts" in the **To** or **cc** line of your mail tool, the e-mail message will be sent to everyone on the "dairy contacts" list.

Words of Warning about the Internet

With millions of people having access to the Internet, one must be very discerning about the information one accesses. Common sense and intuition are needed to determine the validity of material found on the Web. There are clues that you can use to ascertain the reliability of the website. One is the type of the URL domain name. Pages that have the extensions such as .edu, indicating an educational institution, or .gov, indicating some kind of governmental agency, should have more credibility than a random .com or .net site. Be aware that just because an extension such as .edu is in the URL does not mean that the institution sponsors that webpage. A "~" symbol is used to designate that a particular user is in control of the page rather than the institution. The example in Figure 10.3 showed a Web page that has the URL pasture.ecn.purdue.edu\~diefes\abstract3.html. The "~diefes" means that this is a page under the control of the user "diefes" and not the host institution.

Other clues to the trustworthiness of a website include the information on the webpage. Is there an author cited on the page? Is that author credible? What is the creation date of the site? Are there references on the site? The Internet has made information readily available but it has also made it easy to post information that is not credible, which places a responsibility on you as the researcher to verify information as credible.

In addition, care must be taken not to plagiarize materials found on the Web. As with hard copy materials, always cite your sources.

Further, computer viruses can be spread by executable files (.exe) that you have copied. The virus is unleashed when you run the program. Text files, e-mail, and other non-executable files cannot spread such viruses. Guard against viruses attached to executable files by downloading executable files only from sites that you trust. Anti-virus programs such as those from Norton or McAfee can be installed to help detect and protect against viruses.

Social media, such as Facebook and MySpace, and other websites such as YouTube, have rapidly expanded how information is transferred by text, pictures and video. Many students use these tools to share information and update others on what they are doing or have done. We recommend caution when using the Internet to share information. Once released, you lose control over that content. Once something is on the Internet, it may be there for a very long time. Employers, scholarship committees, and graduate schools can and will do Internet searches on candidates, and there are many stories of students who posted something they thought was funny to their friends but ended up being viewed as evidence of their professionalism, or lack thereof. The Internet is a powerful tool that has fundamentally changed the way we live and communicate, but like any tool it needs to be used with caution.

10.3 WORD-PROCESSING PROGRAMS

Throughout the powdered-milk process design project, you will need to do various types of writing. The largest document will be the technical report for the design. You also may have to complete a non-technical report for management. In addition, you may write memorandums and official letters, and you will need to document experimental results. The computer tool that will enable you to write all these papers is a *word processor,* such as Microsoft Word.

The most basic function of a word processor is to create and edit text. When word processors first became available in the early 1980s, that was about the extent of their ability. The user could enter, delete, copy, and move text. Text style was limited to bold, italic, and underline, and text formats were few in number.

"Word processing," as understood today, is really a misnomer, since the capabilities of word processors have become indistinguishable from what was once considered desktop publishing. Word processors such as Microsoft Word, Corel WordPerfect, and Adobe FrameMaker have features that are particularly useful to engineers. Equation, drawing, and table editors are features which make the mechanics of technical writing relatively easy. Most programs include spelling and grammar checkers and an on-line thesaurus to help eliminate common writing errors.

Figure 10.4 shows the graphical user interface (GUI) for Microsoft Word. All word processors display essentially the same components. The title bar indicates the application name and the name of the file that is currently open. The menu bar provides access to approximately 10 categories of commands such as file, edit, and help commands. Some word processors have a toolbar that gives direct access to commands that are frequently used such as "save," "bold," and "italic." A typical word processor GUI also has scroll bars and a status bar. The scroll bars allow the user to move quickly through the document. The status bar provides information on current location and mode of operation. The workspace is where the document is actually typed. The workspace in Figure 10.4 shows a snippet of the technical information related to the spray dryer of our example.

This textbook you are reading was written using word processors. Throughout these pages you can find examples of tables, equations, and figures created using a word processor. A wide variety of typefaces, paragraph styles, and formats are also in evidence. You also will see images and pictures that have been imported into this document. A user may import materials that are created using other computer applications. For instance, the photos of practicing engineers in the Profiles chapter were image files sent to the author for inclusion in his chapter.

Documenting is an ongoing process. You will need to return time and again to your word processor to record all subsequent steps of the dryer design process. Periodically, you will use the word processor to write memos and official letters and to prepare experimental reports, technical documents, and non-technical papers.

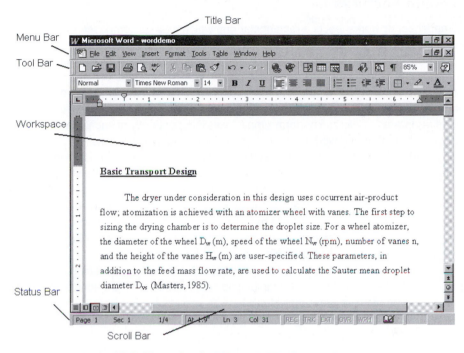

Figure 10.4 Example of a Microsoft Word paper in progress.

10.4 SPREADSHEETS

Introduction

Much data collection is performed during the research and design phases of an engineering project. Data come from a variety of sources including experiments, design calculations, product specifications, and product statistics. While designing the powdered-milk process, you have gathered cost information from your equipment vendors and collected experimental data on powder solubility. You may like to use the cost information in an economic analysis of your process (recall that your process is supposed to make a profit). You may also attempt to analyze your laboratory data. Due to the tabular nature of these data sets, this sort of information is best recorded with a spreadsheet.

Spreadsheet packages have their origins in finance. The idea for the first spreadsheet application, called VisiCalc, came in 1978 from a Harvard Business School student who was frustrated by the repetitive nature of solving financial-planning problems (Etter, 1995). VisiCalc was originally written for Apple II computers. IBM came out with its own version of VisiCalc in 1981. In 1983, a second spreadsheet package was released by Lotus Development Corporation, called Lotus 1-2-3. Now there are a number of other spreadsheet packages from which to choose, including Microsoft Excel, Corel Quattro Pro for PC, and Xess for UNIX. One thing to note about the evolution of the spreadsheet is the increase in functionality. VisiCalc centered on accounting math: adding, subtracting, multiplying, dividing, and percentages. Spreadsheets today have many, many functions.

Common Features

A spreadsheet or worksheet is a rectangular grid which may be composed of thousands of columns and rows. Typically, columns are labeled alphabetically (A, B, C, . . . , AA, AB, . . .) while rows are labeled numerically (1, 2, 3, . . .). At the intersection of a given column and row is a cell. Each cell has a column-row address. For example, the cell at the intersection of the first column and first row has the address A1. Each cell can behave like a word processor or calculator in that it can contain formattable text, numbers, formulas, and macros (programs). Cells can perform simple tasks independently of each other, or they can reference each other and work as a group to perform more difficult tasks.

Figure 10.5 shows the GUI for Microsoft Excel. Notice that many of the interface features are similar to those of a word processor, such as the title bar, menu bar, toolbar, scroll bars, and status bar. Many menu selections are similar as well, particularly the "File" and "Edit" menus. The standard features of all spreadsheets are the edit line and workspace. The edit line includes the name box and formula bar. The name box contains the active cell's address or user-specified name. Within the worksheet in Figure 10.5 is the beginning of the equipment costing information and economic analysis of our example.

Today's spreadsheets have many features in common including formatting features, editing features, built-in functions, data manipulation capabilities, and graphing features. Let's look at each one of these areas individually.

As with word processing, the user has control over the format of the entire worksheet. Text written in a cell can be formatted. The user has control over the text's font, style, color, and justification. The user can also manipulate the cell width and height as well as border style and background color.

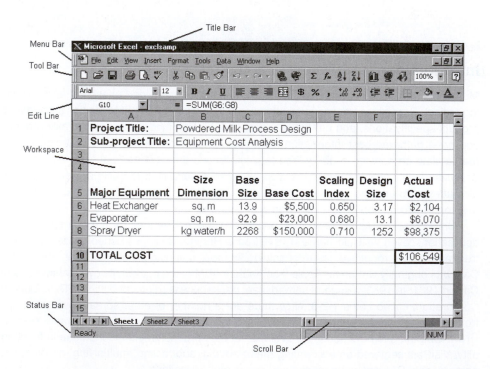

Figure 10.5 Example of a Microsoft Excel spreadsheet.

Editing features are also similar to a word processor. Individual cell contents can be cleared or edited. The contents of one or more cells can be cleared, copied, or moved. Excel even has a spell checker.

Spreadsheets have categories of built-in functions. At a minimum, spreadsheets have arithmetic functions, trig functions, and logic functions. Arithmetic functions include a function to sum the values in a set (range) of cells (SUM) and a function to take an average of the values in a range of cells (AVERAGE). Trig functions include SIN, COS, and TAN, as well as inverse and hyperbolic trig functions. Logic (Boolean) functions are ones that return either a true (1) or false (0) based on prescribed conditions. These functions include IF, AND, OR, and NOT. You might use a logic function to help you mark all equipment with a cost greater than $100,000 for further design consideration. The formula to check the heat exchanger cost would be IF(G6>100000, "re-design", "okay"). This formula would be placed in cell H6. If the value in cell G6 is greater than 100,000, the word *re-design* is printed in cell H6. Otherwise, the word *okay* is placed in cell H6.

Spreadsheet developers continually add built-in, special purpose functions. For instance, financial functions are a part of most spreadsheets. An example of a financial function is one that computes the future value of an investment. In addition, most spreadsheets have a selection of built-in statistical functions.

When a user wants to perform repeatedly a task for which there is not a built-in function, the user can create a new function called a macro. A *macro* is essentially a series of statements that perform a desired task. Each macro is assigned a unique name that can be called whenever the user has need of it.

Spreadsheets also provide a sort function to arrange tabular data. For instance, vendor names can be sorted alphabetically, while prices can be sorted in order of increasing or decreasing numeric value.

Graphing is another common spreadsheet feature. Typically, the user can select from a number of graph or chart styles. Suppose that you are interested in studying ways to reduce the energy it takes to operate the spray dryer. Figure 10.6 is a Microsoft Excel-generated *x–y* plot of such an analysis. Here you are investigating the impact that recycling hot air back into the dryer has on steam demand.

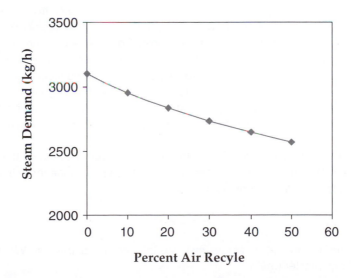

Figure 10.6 An x-y plot generated using Microsoft Excel's Chart Wizard.

Advanced Spreadsheet Features

Database construction, in a limited fashion, is possible in some spreadsheet applications such as Excel. A database is a collection of data that is stored in a structured manner that makes retrieval and manipulation of information easy. A database consists of records, each having a specified number of data fields. Typically, the user constructs a data entry form to facilitate entry of records. Once a database has been constructed, the user can sort and extract information. You could use the database feature to catalog all expenses during implementation of your powdered-milk process. This would require a record for each piece of equipment purchased. Data fields would include size, quantity, list price, discounts, and actual cost of each item.

Real-time data collection is another advanced feature of spreadsheets. This feature allows data and commands to be the received from, and sent to, other applications or computers. During the design phase of the powdered-milk process, you might use this feature to log data during pilot plant experiments. Later, you might use this feature to create a spreadsheet that organizes and displays data coming from on-line sensors in your powdered-milk process. This would enable line workers to detect processing problems quickly.

10.5 MATHEMATICS SOFTWARE

Mathematical modeling is a large component of the design process. For the powdered-milk process, you may need to develop a mathematical model of parts of the process. Problems of this nature are best solved using mathematical or computational software.

MATLAB

MATLAB is a scientific and technical computing environment. It was developed in the 1970s by the University of New Mexico and Stanford University to be used in courses on matrix theory, linear algebra, and numerical analysis (MathWorks, 1995). While matrix math is the basis for MATLAB ("MATrix LABoratory"), the functionality of MATLAB extends much farther. MATLAB is an excellent tool for technical problem solving. It contains built-in functions and a command window, used to perform computational operations. (Taken from "Introduction to Technical Problem Solving with MATLAB" by Great Lakes Press, www.glpbooks.com.)

MATLAB is best explained through an example. Consider again the powdered-milk design. As with any food process, you must be concerned about consumer safety. The best way to ensure that the dairy company's customers do not get food poisoning from this product is to pasteurize the milk before drying. The first step in pasteurization is to heat milk to a specified temperature. Milk is typically heated in a plate heat exchanger where a heating medium (e.g., steam) is passed on one side of a metal plate, and milk is passed on the other side.

The required total surface area of the plates depends on how fast the heat will be transferred from the heating medium to the milk, and how much resistance there is to that heat transfer. Mathematically, this is represented by

$$A = Q \times R$$

where A is the surface area of the plates (m^2), Q is the rate of heat transfer (W), and R is the resistance to heat transfer (m^2/W).

From your previous design calculations, assume that you know that the rate of heat transfer is 1.5×10^6 W. The resistance to heat transfer range can be any value from 4×10^{-6} to 7×10^{-6} m²/W. You may wish to study how the required surface area changes over the range of the resistance.

A typical MATLAB session would resemble that in Figure 10.7. The plot generated by these commands is shown in Figure 10.8. At this point, don't concern yourself with the details, but just get a feel for the use of MATLAB.

```
>       Q = 1.5e6;
>       R = 4e-6:1e-7:7e-6;
>       A = Q.*R;
>       plot(R,A,'s-');
>       xlabel('Area m^2');
>       ylabel('Resistance m^2/W');
```

Figure 10.7 MATLAB code for determining heat transfer surface area.

Note that MATLAB has a text-based interface. On each line, one task is completed before going on to the next task. For instance, on the first line, the value of Q is set. On the second line, the variable R is set equal to a range of values from 4×10^{-6} to 7×10^{-6} in steps of 1×10^{-7}. On the third line, the arithmetic operation $Q \times R$ is performed. On command lines 4 through 6, MATLAB's built-in plotting commands are used to generate a plot of the results.

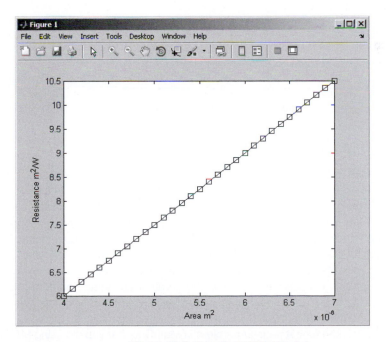

Figure 10.8 Sample MATLAB plot.

The commands shown in Figure 10.7 could be saved as a file with a ".m" extension (MAT-LAB file) and run again and again, or edited at a future date. MATLAB has a built-in editor/debugger interface that resembles a very simple word processor (Figure 10.9). The debugger helps the user by detecting syntax errors that would prevent the proper execution of the code. The editor shown in Figure 10.9 is displaying the code for the heat transfer area calculation.

MATLAB maintains, and periodically adds, toolboxes, which are groups of specialized built-in functions. For example, one toolbox contains functions that enable image and signal processing. There is also a Symbolic Math Toolbox that enables symbolic computations based on Maple (see next paragraph). SIMULINK is another toolbox, a graphical environment for simulation and dynamic modeling. The user can assemble a personal toolbox using MATLAB script and function files that the user creates.

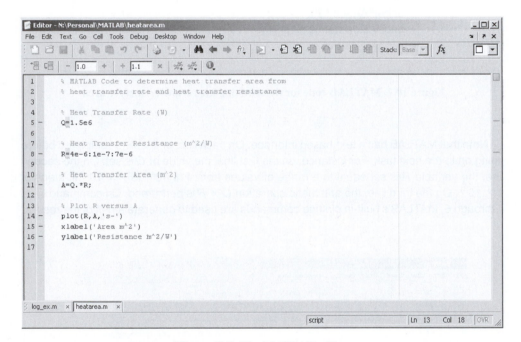

Figure 10.9 The MATLAB editor.

MathCad and Maple

MathCad and Maple are graphical user interface-based mathematics packages. Each is 100% WYSIWYG—computer jargon for "What You See Is What You Get." That is, what the user sees on the screen is what will be printed. There is no language to learn since mathematical equations are symbolically typed into the workspace. The workspace is unrestrictive—text, mathematics, and graphs can be placed throughout a document. These packages are similar to a spreadsheet in that updates are made immediately after an edit occurs. MathCad and Maple feature built-in math functions, built-in operators, symbolic capabilities, and graphing capabilities. Perhaps the greatest advantage of MathCad is its ability to handle units. In MATLAB, the user is charged with keeping track of the units associated with variables. In MathCad, units are carried as part of the assigned value.

You could work the same heat transfer problem in MathCad that you did using MATLAB. Figure 10.10 shows the GUI for MathCad and the text used to solve the problem.

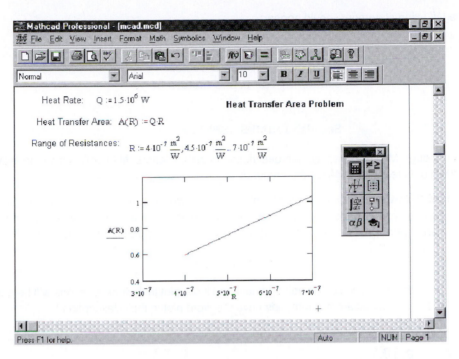

Figure 10.10 Example of a MathCad window and problem-solving session.

Example 10.1

The key to being able to use MATLAB to solve engineering problems effectively is to know how to get help with MATLAB's many commands, features, and toolboxes. There are three easy ways to access help.

Method 1. Access the hypertext documentation for MATLAB. At the MATLAB prompt type:

>> helpdesk

Method 2. Use the help command. The help command can be used in three ways. Typing help at the MATLAB prompt will generate a list of all available help topics. Try it out.

>> help

Near the top of this list, is a topic called matlab\specfun. To see the commands included in a given topic the general format is to type >> help topic. To see the commands included in specfun, type:

>> help specfun

This will generate a list of commands under that topic heading. To get help on a specific command, type >> help command. Let's see how the MATLAB command factor is used:

>> help factor

The following statement should appear:

> FACTOR Prime factors.
> FACTOR(N) returns a vector
> containing the prime factors of

N.

> See also PRIMES, ISPRIME.

Even though MATLAB displays the description in capital letters. MATLAB is case sensitive and will only recognize MATLAB commands in lower case.

Method 3. MATLAB has a means of looking for key words in the command descriptions. This enables you to discover command names when you only know what type of operation you wish to perform. For instance, I may wish to learn how to plot. So I would type:

> >> lookfor plot

A list of all the available commands that have the word plot in their descriptions will be generated. How many different commands have the word plot in their description?

Example 10.2

MATLAB is designed to work with matrices. Therefore, by default, matrix math is performed. Given the following two matrices:

$$
A = \begin{bmatrix} 1 & 0 & 4 \\ 7 & 9 & 3 \\ 3 & 6 & 9 \end{bmatrix} \qquad B = \begin{bmatrix} 9 & 0 & 1 \\ 3 & 4 & 5 \\ 5 & 9 & 7 \end{bmatrix}
$$

Problem 1. Find the product of A and B (Section 10.5):

```
% Assign each matrix to a variable name
A_matrix = [1 0 4; 7 9 3; 3 6 9]
B_matrix = [9 0 1; 3 4 5; 5 9 7]
% Perform matrix multiplication to find C
C_matrix = A_matrix *B_matrix
```

The result in MATLAB will be:

```
C_matrix =
      29      36      29
     105      63      73
      90     105      96
```

Problem 2. Find the element-by-element product of **A** and **B** and set it equal to **D** (e.g., a11 * b11 = d11):

$$\text{\% Perform element-by-element}$$
$$\text{multiplication to find D}$$
$$\text{D\_matrix = A\_matrix. *B\_matrix}$$

Note that the notation for element-by-element multiplication is .*. This same notation is used to do any element-by-element operations (e.g., .*, ./, .\, .^). The resulting matrix **D** is:

```
D_matrix =
    9   0      4
   21  36      5
   15  54     63
```

Example 10.3

MATLAB can be used to plot data points or functions. We will use both methods to plot a circle of unit radius.

Method 1: Plotting Data Points

First generate a vector of 100 equally spaced points between 0 and 2* π using linspace. Note that the semicolon is used after each command line. This suppresses the echo feature on MATLAB so the contents of the vectors are not displayed on the screen:

```
>> angle=linspace(0,2*pi,100);
```

Define vector of *x* and *y* to plot later:

```
>> x=sin(angle);
>> y=cos(angle);
```

Use the plot function to generate the plot:

```
>> plot(x,y);
```

Use the axis command to set the scales equal on the *x*- and *y*-axes. Otherwise you might produce an oval.

```
>> axis('equal');
```

Label the *x*-axis and *y*-axis and title, respectively.

```
>> xlabel('x-axis');
>> ylabel('y-axis');
>> title('Circle of Radius One');
```

Method 2: Plotting With a Function

First define the function. The name for this function is func_circle. You may choose any name, but be sure to avoid names that MATLAB already uses. Also note that the .^ is used for the exponent. Remember that MATLAB assumes everything is a matrix. The .^ function indicates that each element in the data should be raised to an exponent individually rather that using a matrix operation:

>> func_circle=('sqrt(1–x.^2)')

Plot the function using the command *fplot* instead of *plot* over the range –1 to 1:

>> fplot(func_circle,[–1,1])

This function is only positive and gives a semi-circle. To get the bottom half of the circle, define a second function that is negative:

>> func_neg=('-sqrt(1-x.^2)')

Activate the hold feature to plot the second function on the same plot:

>> hold on

Plot the negative function over the same range:

>> fplot(func_neg,[–1,1])

Make the axes equal in length to get a circle:

>> axis square

Label the x-axis and y-axis and title, respectively:

>> xlabel('x-axis');
>> ylabel('y-axis');
>> title('Circle of Radius One');

Example 10.4

MATLAB can be used to perform functional analysis. For examples, find the two local minimums of the function $y = x\cos(x + 0.3)$ between $x = -4$ and $x = 4$.

Method: First, you need to create a function assignment. Then it is best to plot the function over the specified range because MATLAB's command for finding minimums (fmin) requires that you provide a range over which to look for the minimums:

```
% Assign the function to a variable name as a string func __y = ëx.* cos
(x + 0.3)°
```

```
% Plot the function over the entire range
```

```
fplot(func_y,[-4 4])

% Find each local minimum over the specified sub-ranges
x_min1 = fmin(func_y,-2,0)
x_min2 = fmin(func_y,2,4)



% Find the y values at the minimum locations. Note that MATLAB
can only evaluate functions in terms
% of x.
x = [x_min1 x_min2]
y_mins = eval(func_y)
```

The MATLAB solution will yield:

```
x_min1 =
-1.0575
x_min2 =
3.1491
y_mins =
        -0.7683  -3.0014
```

Example 10.5

MATLAB can also be used to find the roots or zeroes of a function. As with locating the minimums in the previous example, it is best to plot the function first.

Define the function of interest. In this case, we will find the zero of $\sin(x)$ near 3:

```
>> func_zero=('sin(x)')
```

Plot the function over the interval −5 to 5:

```
>> fplot(func_zero,[-5,5])
```

Use the *fzero* command to find the exact location of the zero. An initial guess must be provided (in this case it is 3.0):

```
>> fzero(func_zero,3)
```

MATLAB returns the answer:

```
ans =
3.1416
```

Note that *fzero* finds the location where the function crosses the *x*-axis. If you have a function that has a minimum or maximum at zero but does not cross the axis, it will not find it.

Example 10.6

MATLAB can be used to write scripts to be used later. On a PC or Mac, select "New M-File" from the File menu. A new edit window should appear. On a Unix workstation, open an editor. In either case, type the following commands:

```
disp('This program calculates the square root of a number')
x=input('Please enter a number: ');
y=sqrt(x);
disp('The square root of your number is')
disp(y)
```

Save this file under the name "calc_sqrt". Note that on a PC or Mac, a ".m" extension will be added to your file name. Using Unix, you will need to add a ".m" to the end of your file name so MATLAB will recognize it as a script file. In MATLAB, type:

```
>> calc_sqrt
```

MATLAB should respond with the first two lines below. Assuming that "3" is entered and the Return key is pressed, MATLAB will display the last two lines:

```
This program calculates the square root of a number
Please enter a number: 3
The square root of your number is
   1.7321
```

If MATLAB did not recognize "calc_sqrt", check to see that it has a ".m" extension attached to the end of the file name. Alternatively, the file may not be in a directory or folder in which MATLAB is searching.

10.6 PRESENTATION SOFTWARE

At regular intervals during the design process, presentations must be made to research groups and to management. Presentation programs allow a user to communicate ideas and facts through visuals. Presentation programs provide layout and template designs that help the user create slides.

PowerPoint

Microsoft PowerPoint is a popular electronic presentation package. Figure 10.11 shows a slide view window of PowerPoint. As you have seen in other Microsoft programs, the GUI for this package has a title bar, a menu bar, a toolbar, and a status bar. The unique feature of this tool is the drawing toolbar.

PowerPoint has five different viewing windows. The "slide view" window displays one slide at a time. Typically, this is the window through which slide editing is done. The "outline view" window lists the text for each slide in a presentation. The "slide sorter view" window graphically displays all the slides. As the name implies, slide sorting is done through this window. The "notes page view" allows the user to attach text with each slide. Since bullets are used on most slides, this feature comes in very handy. It allows the user to come back to a presentation file at a future date and still remember what the presentation covered. The slides can be viewed sequentially using the "slide show view." This is the view used during a presentation.

Presentation packages have a number of features that help the user put together very professional presentations. PowerPoint has a gallery of templates. Templates specify color scheme and arrangement of elements (e.g., graphics and text) on the slide. Figure 10.11 shows a "slide view" with Clip Art. PowerPoint also has animation capabilities. With the proper computer hardware, sound and video clips also can be incorporated into a presentation.

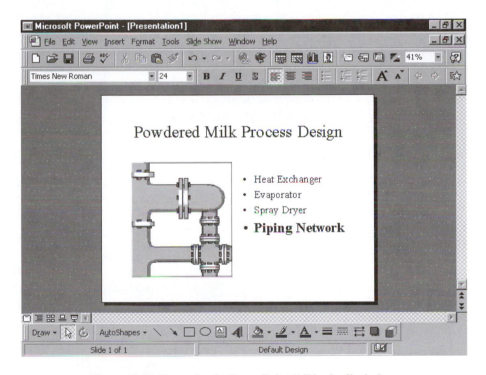

Figure 10.11 Example of a PowerPoint "slide view" window.

10.7 OPERATING SYSTEMS

An *operating system* is a group of programs that manage the operation of a computer. It is the interface between the computer hardware, the software, and the user. Printing, memory management, and file access are examples of processes that operating systems control. The most recognized operating systems are UNIX, originally from AT&T Bell Laboratories; MS-DOS, from Microsoft Corporation; Microsoft's Windows family (Windows 7, Vista, XP, etc.); and the Apple OS for Macintosh computers.

10.8 PROGRAMMING LANGUAGES

Introduction

Programming languages enable users to communicate with the computer. A computer operates using machine language, which is binary code—a series of 0's and 1's (e.g., 01110101 01010101 01101001). Machine language was used primarily before 1953. Assembly language was developed to overcome the difficulties of writing binary code. It is a very cryptic symbolic code which corresponds one-to-one with machine language. FORTRAN, a higher-level language which more closely resembles English, emerged in 1957. Other higher-level languages quickly followed. However, regardless of the language the programmer uses, the computer only understands binary code. A programming language is an artificial language with its own grammar and syntax that allows the user to "talk" to the computer.

A program is written to tell the computer how to complete a specific task. A program consists of lines of code (instructions) written in a specific programming language. The program is then compiled or converted into machine language. If the compiler detects no syntax or flow control problems in the code, the user may then execute the code. Upon execution of the code, the computer performs the specified task. If the code is well written, the computer completes the task correctly. If you are unsatisfied with the computer's execution of the task, or the computer is unsatisfied with the user's code, you must debug (fix) the program and try again.

There are a number of programming languages in use. The following sections will highlight the most popular languages in the context of their classification.

Procedural Languages

Procedural languages use a fetch-decode-execute paradigm (Impagliazzo and Nagin, 1995). Consider the following two lines of programming code:

```
value = 5
new_value = log(value)
```

In the second line of code, the computer "fetches" the variable named **value** from memory, the **log** function is learned or "decoded", and then the function log(value) is "executed." FORTRAN, BASIC, Pascal, and C are all procedural languages.

The development of FORTRAN (Formula Translating System) was led by IBM employee John Backus. It quickly became popular with engineers, scientists, and mathematicians due to its execution efficiency. Three versions are in use today: FORTRAN 77, FORTRAN 90, and FORTRAN 95. Each new version has added functionality. The American National Standards Institute (ANSI), which facilitates the establishment of standards for programming languages, recognizes FORTRAN 77 and FORTRAN 90. Thousands of FORTRAN programs and routines written by experts are gathered in standard libraries such as IMSL (International Mathematics and Statistics Library) and NAG (Numerical Algorithm Group).

BASIC (Beginner's All-Purpose Symbolic Instruction Code) was created in 1964 at Dartmouth College by John Kemeny and Thomas Kurtz. Its purpose was to encourage more students to use computers by offering a user-friendly, easy-to-learn computing language.

Pascal is a general-purpose language developed for educational use in 1971. The developer was Prof. Nicklaus Wirth (Deitel and Deitel, 1998). Pascal, named for 17th-century mathematician and philosopher Blaise Pascal, was designed for teaching structured programming. This language is not widely used in commercial, industrial, or government applications because it lacks features that would make it useful in those sectors.

Dennis Ritchie of AT&T Bell Laboratories developed the C programming language in 1972. It was an outgrowth of the BCPL (Basic Combined Programming Language) and B programming languages. Martin Richards developed BCPL in 1967 as a language for writing operating system software and compilers (Deitel and Deitel, 1998). Ken Thomas of AT&T Bell Laboratories modeled the B programming language after BCPL. B was used to create the early versions of UNIX. ANSI C programming language is now a common industrial programming language and is the developmental language used in creating operating systems and software such as word processors, database management tools, and graphics packages.

Object-Oriented Languages

Object-oriented languages treat data, and routines that act on data, as single objects or data structures. Referring back to the programming code example, **value** and **log** are treated as one single object that, when called, automatically takes the **log** of **value**. Objects are essentially reusable software components. Objects allow developers to take a modular approach to software development. This means software can be built quickly, correctly, and economically (Deitel and Deitel, 1998).

C++ (C plus plus) is an object-oriented programming language developed by B. Stroustup at AT&T Bell Laboratories in 1985. It is an evolution of the C programming language. C++ added additional operators and functions that support the creation and use of data abstractions as well as object-oriented design and programming (Etter, 1997).

Java's principle designer was James Gosling of Sun Microsystems. The creators of Java were assigned the task of developing a language that would enable the creation of very small, fast, reliable, and transportable programs. The need for such a language arose from a need for software for consumer electronics such as TVs, VCRs, telephones, and pagers (Bell & Par, 1999). Such a language also makes it possible to write programs for the Internet and the Web. Two types of programs can be written in Java—applications and applets. Examples of application programs are word processors and spreadsheets (Deitel and Deitel, 1998). An *applet* is a program that is typically stored on a remote computer that users can connect to through the Web. When a browser sees that a Java applet is referred to in the HTML code for a Webpage, the applet is loaded, interpreted, and executed.

10.9 ADVANCED ENGINEERING PACKAGES

The computer tools introduced so far are useful to all engineers. At one time or another, all engineers need to write papers, work with tabular data, and develop mathematical models. An engineer also needs to be able to solve particular classes of problems. For instance, a food process engineer may need to analyze and optimize processes. A mechanical engineer may need to perform finite element analysis. There are many software tools designed to handle specific types of engineering problems. Described here are just a few key packages that an engineer might encounter.

Aspen Plus

Aspen Plus is software for the study of processes with continuous flow material and energy. These types of processes are encountered in the chemical, petroleum, food, pulp and paper, pharmaceutical, and biotechnology industries. Chemical engineers, food process engineers, and bioprocess engineers are typical users of Aspen Plus. This tool enables modeling of a process system through the development of a conceptual flowsheet (process diagram). The usual sequence of steps to design and analyze a process in Aspen Plus is:

- Define the problem
- Select a units system (SI or English)
- Specify the stream components (nitrogen, methane, steam, etc.)
- Specify the physical property models (Aspen Plus has an extensive property models library that is user-extendible)
- Select unit operation blocks (heat exchanger, distillation column, etc.)
- Define the composition and flow rates of feed streams
- Specify operating conditions (e.g., temperatures and pressures)
- Impose design specifications
- Perform a case study or sensitivity analysis on the entire process (i.e., see how varying the operating conditions affects the design)

The output of Aspen Plus is a report that describes the entire process flowsheet. Overall material and energy balances, steam consumption, and unit operation design specifications are just a sampling of what might be included in the final output report.

GIS Products

Civil engineers, agricultural engineers, and environmental and natural resource engineers work with geographic-referenced data. The Environmental Systems Research Institute, Inc. (ESRI) is the leader in software tool development for *geographical information systems* (GIS). The purpose of these tools is to link geographic information with spatial data. For example, GIS products can be used to study water and waste water management of a particular landmass (watershed). Specifically, ARC/INFO or ArcView (GIS products) could be used to link data sets concerning drainage and weather with a geographic location as specified by latitude and longitude. GIS products also enable visualization of the data through mapping capabilities.

Virtual Instrumentation

Engineers in all disciplines of engineering have occasions that require them to acquire and analyze data. Modern software tools make it possible to transform a computer into almost any type of instrumentation required to perform these tasks. One such software tool is LabVIEW, created by National Instruments Corporation. This software allows a user to create a virtual instrument on their personal computer. LabVIEW is a graphical programming development environment based on the C programming language for data acquisition and control, data analysis, and data presentation. The engineer creates programs using a graphical interface. LabVIEW then assembles the computer instructions (program) that accomplishes the tasks. This gives the user the flexibility of a powerful programming language without the associated difficulty and complexity.

The software can be used to control data acquisition boards installed in a computer. These boards are used to collect the data the engineer would need. Other types of boards are used to send signals from the computer to other systems. An example of this would be in a control systems application. LabVIEW allows the engineer to quickly configure the data acquisition and control boards to perform the desired tasks.

The tools also can be used to make the computer function as a piece of instrumentation. The signals from the data acquisition boards can be displayed and analyzed like an oscilloscope or a filter. A sample window of a LabVIEW window performing this function is shown in Figure 10.12. In this example, data were collected and processed through a filter to yield a sine wave.

Software tools such as LabVIEW allow a single computer to take on the function of dozens of pieces of instrumentation with relative ease. This reduces cost and improves the speed of data acquisition and analysis.

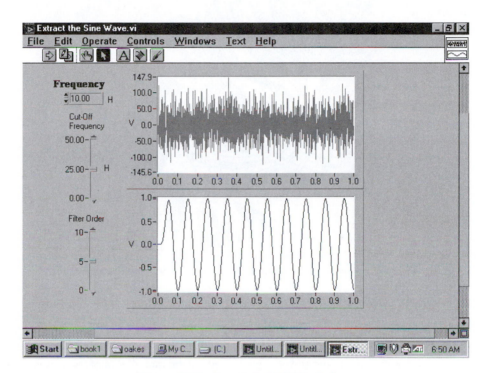

Figure 10.12   Sample LabVIEW window.

Finite Element Tools

Engineers who work with mechanical or structural components include aeronautical, agricultural, biomedical, civil, and mechanical engineers. All must verify the integrity of the systems they design and maintain. Testing these systems on a computer is the most cost-effective and efficient method. Modern finite element tools allow engineers to do just that. One such commonly used tool is ANSYS, created by ANSYS, Inc.

Finite element tools take the geometry of a system and break it into small elements or pieces. Each element is given properties and allowed to interact with the other elements in the model. When the elements are made small enough, very accurate simulations of the actual systems can be made.

The first step in using such a tool is to separate the physical component into a mesh. A sample mesh of centrifugal compressor vanes is shown in Figure 10.13. This mesh defines the physical boundaries of the elements. ANSYS contains aids to allow the user to create the mesh from a CAD drawing of the system. Other tools are also available to create the mesh to be used by ANSYS.

Figure 10.13 A sample mesh of a centrifugal compressor and diffuser vanes.

After the physical mesh is generated, the properties of the elements must be defined. The properties of the elements represent the material that the engineer is modeling. The elements can be solids or fluids and can change throughout the model. Part of the system may be one material, and another part may be different. Initial conditions such as temperatures and loads also must be defined. The engineer can use tools such as ANSYS to perform steady-state as well as transient analyses.

A final useful aspect of these tools is the post-processing capability. Tools such as ANSYS allow the engineer to examine the results graphically. The finite element model can be colored to show temperature, stress, or displacements. The engineer can then visually interpret the results, rather than searching through lists of numbers, which can highlight problems with a design.

Other Tools

We would need an entire book, or more likely a series of books, to touch on all the computer tools engineers use today. There are different sets of computer tools within each discipline and each subspecialty of each discipline. Every day more tools are produced, and those existing are improved. New versions are constantly being developed to add features to help engineers perform their jobs more efficiently and more accurately.

To a large degree, the development of new software tools is capitalizing on rapidly changing computer hardware capabilities. Microprocessors increase in speed according to Moore's Law, which states that microprocessor speed doubles every 18 months. That means that in the four and a half years from your senior year in high school to your first engineering job, processors will go through three doublings —they will be eight times faster.

This may not sound impressive, but consider what this has meant since the first microprocessor was developed in 1972. In 2007, thirty-five years after the first microprocessor was invented, 23 doublings will have taken place. The processor speed has increased over a million times. An analogy of how much faster this is would be to compare a snail inching along the ground with a commercial jet flying overhead. The 1972 computer would be the snail and the computer of 2007 would be the jet.

Considering that you will be entering a career that will last over 30 years, the computers you will be using as a senior engineer could be a million times faster than the machines you use now. Will microprocessors continue to advance at this pace? We don't know. We do know, however, that the computers of tomorrow will be much faster than the snails you are using now.

What this means in terms of computer tools is that they will continue to change. Experiment with as many tools as you can. Just as mechanics learn how to use an assortment of tools, so should engineers learn to use an assortment of computer tools. They also should always be ready to learn how to use new tools, because there always will be new tools to learn.

REFERENCES

Andersen, P.K., et al., *Essential C: an Introduction for Scientists & Engineers*, Saunders College Publishing, Fort Worth, 1995.

Bell, D. and M. Parr, *Java for Students*, 2nd Ed., Prentice-Hall Europe, London, 1998.

Chapman, S.J., *Introduction to FORTRAN 90/95*, WCB/McGraw-Hill, Boston, 1998.

Dean, C.L., *Teaching Materials to accompany PowerPoint 4.0 for Windows*," McGraw-Hill, New York, 1995.

Deiter, H.M., and P.J. Deiter, *Java: How to Program*, 2nd Ed., Prentice-Hall, Upper Saddle River, 1998.

Etter, D.M., *Microsoft Excel for Engineers*, Addison-Wesley, Menlo Park, CA, 1995.

Etter, D.M., *Introduction to C++ for Engineers and Scientists*, Prentice-Hall, Upper Saddle River, 1997.

Glass, G. and K. Ables, *UNIX for Programmers and Users*, 2nd Ed., Prentice-Hall, Upper Saddle River, 1999.

Gottfried, B.S., *Spreadsheet Tools for Engineers: Excel 97 Version*, WCB/McGraw-Hill, Boston, 1998.

Grauer, R.T. and M. Barber, *Exploring Microsoft PowerPoint 97,* Prentice-Hall, Upper Saddle River, 1998.

Greenlaw, R. and E. Hepp, *Introduction to the Internet for Engineers*, WCB McGraw-Hill, Boston, 1999.

Hanselman, D. and B. Littlefield, *Mastering MATLAB 5: A Comprehensive Tutorial and Reference*, Prentice-Hall, Upper Saddle River, 1998.

Heal, K.M, Hansen, et. al., *Maple V Learning Guide*, Springer-Verlag, New York, 1998.

Impagliazzo, J. and P. Nagin, *Computer Science: A Breadth First Approach with C*, John Wiley & Sons, New York, 1995.

James, S.D., *Introduction to the Internet*, 2nd Ed., Prentice-Hall, Upper Saddle River, 1999.

Kincicky, D.C., *Introduction to Excel*, Prentice-Hall, Upper Saddle River, 1999.

Kincicky, D.C., *Introduction to Word*, Prentice-Hall, Upper Saddle River, 1999.

King, J., *MathCad for Engineers*, Addison-Wesley, Menlo Park, 1995.

MathWorks, Inc., *The Student Edition of MATLAB, Version 4*, Prentice-Hall, Upper Saddle River, 1995.

Nyhoff, L.R. and S.C. Leestma, *FORTRAN 90 for Engineers and Scientists*, Prentice-Hall, Upper Saddle River, 1997.

Nyhoff, L.R. and S.C. Leestma, *Introduction to FORTRAN 90 for Engineers and Scientists,* Prentice-Hall, Upper Saddle River, 1997.

Palm, William J., *MATLAB for Engineering Applications,* WCB/McGraw-Hill, Boston, 1999.

Powell, T.A., *HTML: The Complete Reference*, Osbourne/McGraw-Hill, Berkeley, 1998.

Pritchard, P.J., *Mathcad: A Tool for Engineering Problem Solving*, WCB/McGraw-Hill, Boston, 1998.

Rathswohl, E.J., et. al., *Microcomputer Applications – Selected Edition – WordPerfect 6.1,* The Benjamin/Cummings Publishing Company, Redwood City, 1996.

Sorby, S.A., *Microsoft Word for Engineers*, Addison-Wesley, Menlo Park, 1996.

Wright, P.H., *Introduction to Engineering,* 2nd Ed., John Wiley, 1989.

EXERCISES AND ACTIVITIES

10.1   Access your school's home page; surf the site and locate information about your department.

10.2   Perform a keyword search on your engineering major using two different Web search engines.

10.3   Determine if your school has on-line library services. If so, explore the service options.

10.4   Send an e-mail to a fellow student and your mom or dad.

10.5   Search for the e-mail address of an engineering professional society.

10.6   Find the website of an engineering society. Prepare a one-page report on that organization based on the information found at their website.

10.7   Select the website of the publisher of this text (www.oup.com). Prepare a critique of the website itself. What could be done to improve it? How does it compare with other websites you have accessed? If you wish, you may send your suggestions to the publisher via e-mail.

10.8   Prepare an *oral* presentation on the reliability or unreliability of information obtained from the Internet. Include techniques to give yourself confidence in the information.

10.9   Prepare a *written* report on the reliability or unreliability of information obtained from the Internet. Include techniques to give yourself confidence in the information.

10.10   Do a case study on a situation where inaccurate information from the Internet caused a problem.

10.11   Write a paper for one of your classes (e.g., Chemistry) using the following word processor features: text and paragraph styles and formats.

10.12   Create a table in a word processor.

10.13   Prepare a résumé using a word processor.

10.14   Prepare two customized résumés using a word processor for two different companies you might be interested in working for someday.

10.15   Prepare a two-page report on why you selected engineering, what discipline you wish to enter, the possible career paths you might follow, and why. Create the first page using a word processor. Create the second page using a different word processor.

10.16   Prepare a brief presentation on your successes or frustrations using the two different word processors. Which would you recommend?

10.17   Use a spreadsheet to record your expenses for one month.

10.18   Most spreadsheets have financial functions built in. Use these to answer the following question: If you now started saving $200 per month until you were 65, how much money would you have saved? Assume an annual interest rate of: a) 4%, b) 6%, and c) 8%.

10.19   Assume that you wish to accumulate the same amount of money on your 65th birthday as you had in the previous problem, but you don't start saving until your 40th birthday. How much would you need to save each month?

10.20   A car you wish to purchase costs $25,000; assume that you make a 10% down payment. Prepare a graph of monthly payments versus interest rate for a 48-month loan. Use a) 6%, b) 8%, c) 10%, and d) 12%.

10.21   Repeat the previous problem, but include plots for 36- and 60-month loans.

10.22   Prepare a brief report on the software tools used in an engineering discipline.

10.23   Interview a practicing engineer and prepare an *oral* presentation on the computer tools he or she uses to solve problems.

10.24   Interview a practicing engineer and prepare a *written* report on the computer tools he or she uses to solve problems on the job.

10.25   Interview a senior or graduate student in engineering. Prepare an *oral* presentation on the computer tools he or she uses to solve problems.

10.26   Interview a senior or graduate student in engineering. Prepare a *written* report on the computer tools he or she uses to solve problems.

10.27   Select an engineering problem and prepare a brief report on which computer tools could be used in solving the problem. Which are the optimum tools, and why? What other tools could be used?

10.28   Prepare a brief report on the kinds of situations for which engineers find the need to write original programs to solve problems in an engineering discipline. Include the languages most commonly used.

10.29   Prepare a presentation using PowerPoint that discusses an engineering major.

10.30   Prepare a presentation that speculates on what computer tools you will have available to you in five, 10, and 20 years.

10.31   Use MATLAB Help to learn about the command "linspace."

10.32   What MATLAB command will put grid lines on a plot?

10.33   Perform the following operations using the matrices in Example 10.2.
a. A/B and A./B (right division)
b. A\B and A.\B (left division)
c. A^3 and A.^3
d. A^B and A.^B
Can all of these operations be performed? If not, why not?

Chapter 11

Teamwork

11.1 INTRODUCTION

More than ever, engineering schools are requiring their students to work in teams. These teams take the form of collaborative study groups, laboratory groups, design groups as part of individual classes, and design groups participating in extracurricular competitions such as the solar car, human powered vehicle, SAE formula car, concrete canoe, bridge design, Rube Goldberg projects, and many others. This team emphasis in engineering schools mirrors a management philosophy of the use of teams that has swept through the corporate world in the past few decades. Engineering employers, in fact, are the principal drivers for the use of teams in engineering schools. They seek engineering graduates who have team skills as well as technical skills. The purpose of this chapter is to impart a team vision, to show why teams and collaboration are so important to organizational and technical success, and to give practical advice for how to organize and function as a team. If you study this chapter, work through the exercises, and view some of the references, you should have most of the tools you need to be a successful team member and leader while in engineering school. Apply the principles in this chapter to your team experiences while in college and learn from and document each team experience. By so doing, you will have a portfolio of team experiences that will be a valuable part of your résumé. Above all, you will have the satisfaction that comes from being a part of great teams. Teams are fun.

11.2 ENGINEERS OFTEN WORK IN TEAMS

A full-page advertisement in *USA Today*, October 14, 1991, states clearly the current highly competitive climate in industry and what Chrysler (now DaimlerChrysler), for example, is doing about it. The use of interdisciplinary teams is an important part of the solution:

> No more piece-by-piece, step-by-step production. Now it's teams. Teams of product and manufacturing engineers, designers, planners, financing and marketing people—together from the start . . . . It's how we built Dodge Viper . . . . from dream to showroom in three years, a record for U.S. car makers. From now on, all our cars and trucks will be higher quality, built at lower cost and delivered to the market faster. That's what competition is all about.

The greater the problem, the greater the need for the use of teams. Individuals acting alone can solve simple problems, but tough problems require teams. This section lists and then discusses the reasons many organizations have embraced the use of teams. Why is the use of teams so popular today?

- Greater understanding of the power of collaboration.
- Engineers are asked to solve extremely complex problems.
- More factors must be considered in design than ever before.
- Many corporations are international in scope, with design and manufacturing engineering operations spread across the globe.
- Because time to market is extremely important to competitive advantage, "concurrent engineering," inherently a team activity, is widely employed.
- Corporations are increasingly using project management principles.

Power of Collaboration

There is a greater understanding today than ever before of the importance and power of creative collaboration. One recent book on the subject is *Organizing Genius: The Secrets of Creative Collaboration*, by Warren Bennis and Patricia Ward Biederman. Their first chapter is titled "The End of the Great Man," in which they argue that great groups are able to accomplish more than talented individuals acting alone, stating that "None of us is as smart as all of us." They cite a survey by Korn-Ferry (the worldís largest executive search firm) and *The Economist* in which respondents were asked who will have the most influence on global organizations in the next ten years. Sixty-one percent of the respondents answered "teams of leaders," whereas 14 percent said "one leader." Individual chapters of their book focus on "great groups." A chapter on the Manhattan Project discusses the development of the atomic bomb during the Second World War by a team of great scientists. One chapter discusses Lockheed's "Skunk Works," developers of the U-2 spy plane, the SR-71 Blackbird, and the F117-A Stealth fighter-bomber used in the Persian Gulf war. Another example is the development of the graphical user interface at Xerox's PARC (Palo Alto Research Center) and its adoption into the Macintosh computer by Steve Jobs and Steve Wozniak. Steve Jobs' goal was to lead his team to create something not just great, but "insanely great," to "put a dent in the universe."

Another recent book on collaboration is *Shared Minds: The New Technologies of Collaboration*, by Michael Schrage. He defines collaboration as "the process of *shared creation*: two or more individuals with complementary skills interacting to create a shared understanding that none had previously possessed or could have come to on their own." He cites examples of collaboration in science (Watson and Crick; Heisenberg, Bohr, Fermi, Pauli, and Schrodinger; Einstein and mathematician Marcel Grossman), music (Gilbert and Sullivan; Rodgers and Hart; McCartney and Lennon), art (Monet and Renoir; Van Gogh and Gauguin; Picasso and Braque), and literature (editor Maxwell Perkins and writers F. Scott Fitzgerald and Thomas Wolfe; Ezra Pound and T. S. Eliot).

James Watson and Stanley Crick, winners of the Nobel prize for discovering the double helix structure of DNA credit their collaboration for their success: "Both of us admit we couldn't have done it without the other—we were interested in what the answer is rather than doing it ourselves." Crick wrote that "our . . . . advantage was that we had evolved . . . . fruitful methods of collaboration." Sir Isaac Newton, father of classical physics, in a letter to his contemporary (and rival) Robert Hooke, recognizing his debt to fellow scientists, wrote: "If I have seen further (than you and Descartes) it is by standing on the shoulders of Giants." Modern physics also found its genesis in collaboration. Heisenberg, Bohr, Fermi, Pauli, and

Schrodinger, all Nobel laureates for their roles in founding quantum physics, worked, played, and vacationed together as they developed together their revolutionary new ideas.

Today's Complex Problems

Engineers are asked to solve increasingly complex problems. The complexity of mechanical devices has grown rapidly over the past two hundred years. David Ullman, in his text *The Mechanical Design Process*, gives the following examples. In the early 1800s, a musket had 51 parts. The Civil War–era Springfield rifle had 140 parts. The bicycle, first developed in the late 1800s, has over two hundred parts. An automobile has tens of thousands of parts. The Boeing 747 aircraft, with over 5 million components, required over ten thousand person-years of design time. Thousands of designers worked over a three-year period on the project. Most modern design problems involve not only many individual parts, but also many subsystems—mechanical, electrical, controls, thermal, and others—each requiring specialists acting in teams.

Many Design Factors

Engineering designers must consider more factors than ever before, including initial price, life cycle costs, performance, aesthetics, overall quality, ergonomics, reliability, maintainability, manufacturability, environmental factors, safety, liability, and acceptance in world markets. Satisfying these criteria requires the collective teamwork of design and manufacturing engineers, marketing, procurement, business, and other personnel. The typical new engineering student, being interested primarily in technical things, tends to believe that engineers work isolated from other factors, always seeking the "best technical solution." This is not true. The very act of engineering involves solving difficult problems, finding technical solutions while considering numerous constraints. Engineers make things happen, they are doers, they solve problems, while interacting with many entities within the culture of complex organizations. Often acting at organizational boundaries, engineers are technical experts who, to be truly effective, must understand economics, management, marketing, and the dynamics of the business enterprise.

International Scope

Many corporations are international in scope, with design and manufacturing engineering operations spread across the globe. These operations require teams to work together who may never physically meet. Instead, they meet and share data via electronic means. For example, it is not unusual for design engineers in the United States to collaborate on a design with Japanese engineers for a product that will be manufactured in China. Manufacturing personnel from the Chinese plant also contribute to the project. As another example, software engineers in the U.S., the U.K., and India may collaborate on a software project literally around the clock and around the world. At any time during a 24-hour period, programmers in some part of the world will be working on the product.

Time to Market and Concurrent Engineering

"Concurrent engineering," inherently a team activity, is widely employed in order to achieve better designs and to bring products to market more quickly. Time to market is the total time needed to plan, prototype, procure materials, create marketing strategies, devise tooling, put into production, and bring to market a new product. In the traditional company, each of these steps is done serially, one at a time. It is a relay race, with only one person running at a time, passing the baton to the next runner. The design engineer "tosses a design over the wall" to a manufacturing engineer who may not have seen it yet. She then "tosses it over the wall" to procurement personnel, and so on. Concurrent engineering, on the other hand, is a parallel operation. Like rugby, all of the players run together, side by side, tossing the ball back and forth. Marketing, manufacturing, and procurement personnel are involved from the beginning of the design phase. The use of teams, as well as new technologies such as CAD/CAM, rapid prototyping, shared data, and advanced communications have radically changed the process of engineering. By the use of teams and CAD/CAM technologies, in the face of withering international competition, U.S auto makers have cut the time needed to bring a new car to market from approximately 5 years to less than 3 years.

Speed—that is, timely delivery of products to the marketplace—is critical for companies to be profitable. Jack Welch, General Electric's widely respected CEO (his biographical sketch is included in Chapter 3), has said "we have seen what wins in our marketplaces around the globe: speed, speed, and more speed." Xilinx Corporation has stated that their research shows that a six-month delay in getting to market reduces a product's profitability by a third over its life cycle. An October 25, 1991, *Business Week* article stated: "Reduce product development time to one third, and you will triple profits and triple growth." Figure 11.1, shown below, shows graphically the losses incurred over the lifetime of a product due to delays in bringing the product to market. Early product introduction leads to higher profits because early products bring higher profit margins, are more apt to meet customer needs, and are more likely to capture a larger market share. [This figure was created by Dr. Charles J. Nuese and included in his wisdom-laden book, *Building the Right Things Right*, on product development.] The figure is fictitious, but it is based on principles Dr. Nuese has observed during his career as an engineer in and as a consultant to the semiconductor industry.

Developing new products is a key to growth and profits. Rubbermaid Corporation, during a recent year, introduced 400 new products—more than one per calendar day! At least one third of Rubbermaid's $2 billion in annual sales come from products created in the previous 5 years. 3M Corporation CEO L. D. DiSimone, facing flat revenues and stiff competition, adopted a business focus of generating 30 percent of revenues from products less than 4 years old. The creative thought processes, the innovation needed to create a steady stream

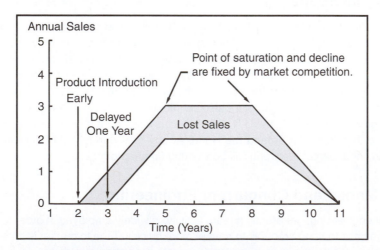

Figure 11.1   The cost of delay in bringing an equivalent new product to market. (Used by permission of Charles J. Nuese.)

of new ideas, and the expertise to manufacture and bring these to market quickly, are all functions that teams perform well.

It is important for engineering students to understand the importance of speed (without compromising quality). Delays have many negative consequences. Engineering professors who supervise teams should provide incentives to students for early completion of team projects, or at least penalties for failure to meet deadlines. Student teams should recognize the importance of speed, push themselves and their teams to early completion of projects, and document these on their resumes and/or portfolios.

Project Management is a Team Approach

Project management is widely practiced in industries and government labs. Many graduate engineers discover on their first job that engineers frequently work in project-oriented teams. Unfortunately, most engineering students never take a project management class. Project management principles were developed in the defense industry in the 1950s and 1960s as a way to productively manage Department of Defense contracts. The typical corporation is organized by vertical divisions or lines, where individuals are clustered by job functions. For example, a corporation may have separate research, manufacturing, engineering, human resources, product development, marketing, and procurement divisions. Project management is a way of organizing individuals not by function but by products or projects. Selected individuals from research, engineering, manufacturing, human resources, procurement, and other divisions are gathered together as a team for a particular purpose, for example solving a corporate problem, developing a new product, or meeting a crucial deadline. A project management approach, therefore, is inherently a cross-functional team approach and an excellent way to solve problems. Successful project management, however, is not an easy task.

The project manager's job is to complete a project, on time, within budget, with the personnel that he is given. Unfortunately, he is rarely able to hand pick his personnel, and his supervisors and/or customers typically set the project completion time and budget. In other words, project managers are virtually never given all the people, time, and money they wish for. These tough constraints are often mirrored in student design teams. Their teacher picks the team members, they don't have much money or resources to work with, and their teacher determines the time frame. These conditions are uncomfortable for students, but they prepare students for the engineering world.

Project managers use tools that any team leader would benefit from learning how to use. Project managers plan work requirements and schedules and direct the use of resources (people, money, materials, equipment). One popular project management tool that student teams should learn to use is the Gantt chart (a simple Gantt chart is pictured on page 312). A Gantt chart organizes and presents clearly information on task division and sequencing. Tasks are divided into sub-tasks, the shading conveys the sequence in which the tasks should be completed and how long a task is expected to take, and this version shows person(s) responsible for each task.

Project managers monitor projects by tracking and comparing progress to predicted outcomes. They use project management software such as Microsoft Project for planning and monitoring purposes. Project managers carefully guard the scope of the project from excessive creep. A project is successful if it is completed on time, within budget, at the desired quality, with effective utilization of resources. For more information on project management, the reader is encouraged to consult any text on Project Management and to view the publication titled "PMBOK: Project Management Body of Knowledge" online at www.pmi.org, the web site of the Project Management Institute.

Team A-9 ("Semi-Conscious Objectors"): Gantt Chart for Object Relocator: Fall 2000

Task	Leader(s)	Oct 11-17	Oct 18-25	Oct26-31	Nov1-7	Nov8-14	Nov15-21	Nov22-28	Nov 29 - Dec 5	Dec6-12
Finalize Design	Team		▓							
Procure Materials	Joe		▓							
Pick-Up Module	Jaime			▓						
Conveyor Assembly	George			▓						
Working Prototype	Team				▓	▓				
Fine Tuning Machine	Jaime/Team					▓				
Instruction Manual	George						▓			
Professor Tests Projects	Team						▓			
Report Rough Draft	Sally					▓				
Finalize CAD Drawings	Ahmad					▓				
Presentation	Ahmad/Joe							▓	▓	
Final Report Due	Team									▓
Judges test best two projects per class	Team									▓

Simply forming teams to solve problems may seem like a good idea, given the evidence and cases from the previous section, but teams are not a cure-all. Teams require nurturing. Successful team development is a process. Whether in a corporate, academic, or other setting, all teams must pass through several developmental phases before they truly become productive. These stages are titled *forming, storming, norming, performing, and adjourning*. Unfortunately, successfully completing the first four stages and becoming a productive, performing team is not guaranteed. *Many teams in practice never reach the performing stage*; they remain mired in the forming, storming, or norming stages. Every team's challenge is to grow through these stages and achieve "performance."

Stage 1: Forming

The first stage of team development is "forming." In this stage, typical dialogue may consist of the following: "Nice to meet you. Yeah, I'm not sure either why we are here. I'm afraid this might be a lot of work." In the formation stage, team members get acquainted with one another, with the leader (or choose a leader if one is not pre-appointed), with the team's purpose, and with the overall level of commitment (work load) required. Team members begin to learn of one another's personalities, abilities, and talents, but also of one another's weaknesses and idiosyncrasies. Individual team members are typically shy, reserved, self-conscious, and uncertain. A key role of the leader in this stage is to lead team ice-breaker activities, facilitate discussion, encouraging all to speak and quieting some who might tend to dominate the conversation. Another role of the leader is to help the team begin to focus on the task at hand.

Stage 2: Storming

The second stage of team development is "storming." In this stage, typical dialogue (or private thoughts) may consist of the following: "Do I have to work with this team? What did I do to deserve this? There clearly aren't any super-heroes on this team, including that dizzy leader. How are we supposed to solve this messy problem?" During the storming stage, the enormity and complexity of the task begins to sink in, sobering and discouraging the participants. "We are supposed to do what? By when?" Teams are rarely formed to solve easy problems, only very difficult and complex ones. Typically, time schedules are unrealistically short, and budgets are inadequate. Further complicating the issue, teammates have learned enough about their fellow team members to know that there are no super-heroes, no saviors they can count on to do it all. (One person doing all the work is a team failure.) Some team members may not initially "hit it off" well with the others. Cliques or factions may emerge within the team, pitted against others. Since the leader's weaknesses (all leaders have weaknesses) are by now apparent, some individuals or factions may vie for leadership of the team. Though possibly under siege, the leader's role is critical during the storming stage. The leader must help the team to focus on its collective strengths, not weaknesses, and to direct their energies toward the task. To be a successful team, it is not necessary for team members to like one another or to be friends. A professional knows how to work productively with individuals with widely differing backgrounds and personalties. Everyone must learn the art of constructive dialogue and compromise.

Stage 3: Norming

The third stage of team development is "norming." In this stage, typical dialogue (or private thoughts) may consist of the following: "You know, I think we can do it. True, there are no super-heroes, not by a long shot, but once we stopped fighting and started listening to one another, we discovered that these folks have some good ideas. Now if we can just pull these together . . ." "Norms" are shared expectations or rules of conduct. All groups have some kinds of norms, though many times unstated. Do you recall a time you joined a group or team and felt subtle influence to act, dress, look, speak, or work in a particular way? The more a team works together, they tend to converge to some common perspectives and behaviors. During the norming stage, team members begin to accept one another instead of complaining and competing. Rather than focusing on weaknesses and personality differences, they acknowledge and utilize one another's strengths. Individual team members find their place in the group and do their part. Instead of directing energies toward fighting itself, the team directs its collective energy toward the task. The key to this shift of focus is a collective decision to behave in a professional way, to agree upon and adhere to norms. Possible norms include working cooperatively as a team rather than individually, agreeing on the level of effort expected of everyone, conducting effective discussions and meetings, making effective team decisions, and learning to criticize one another's ideas without damaging the person. One suggested norm is that all team members are expected to be at all meetings, or to communicate clearly in the event that he/she cannot attend. During the norming stage, feelings of closeness, interdependence, unity, and cooperation develop among the team. The primary role of the leader during the norming phase is to facilitate the cohesion process. Some team members will lag behind the team core in embracing norms. The leader and others on the team (at this stage leadership is beginning to emerge from others on the team as well) must artfully nudge individuals along toward group accountability and focus on the task.

Stage 4: Performing

The fourth stage of team development is "performing." In this stage, typical dialogue (or private thoughts) may consist of the following: "This is a fun team. We still have a long way to go, but we have a great plan. Everyone is pulling together and working hard. No super-heroes, but we're a super team." In this stage, teams accomplish much. They have a shared, clear vision. Responsibilities are distributed. Individual team members accept and execute their specific tasks in accordance to the planned schedule. They are individually committed, and hold one another accountable. On the other hand, there is also a blurring of roles. Team members "pitch in" to help one another, doing whatever it takes for the team to be success-ful. In a performing team, so many team members have taken such significant responsibil-ity for the team's success that the spotlight is rarely on a single leader anymore. Typically, whoever initially led the team is almost indistinguishable from the rest of the team.

Stage 5: Adjourning

The fifth and final stage of team development is "adjourning." Because teams are typically assembled for a specific purpose or project, there is a definite time when the team is dis-banded when that goal is accomplished. If the team was successful, there is a definite feeling of accomplishment, even euphoria, by the team members. On the other hand, an under-performing team will typically feel anger or disappointment upon adjournment.

These stages, originally set forth by Bruce W. Tuckman in 1965, appear in some form in most books that discuss teams. Some leave off the final step, adjourning. Others switch the order and give the steps as forming, conforming (instead of "norming"), storming, and performing.

11.4 WHAT MAKES A SUCCESSFUL TEAM?

Have you ever been a member of a great team? It's a great experience, isn't it? On the other hand, how many of you have ever been part of an unsuccessful team? It was a disappoint-ing experience, wasn't it? How do you measure or evaluate a team's success or lack thereof? What specific factors make a team successful? What factors cause a team to fail?

First it is important to define what a team is and is not. A team is not the same as a group. The term "group" implies little more than several individuals in some proximity to one another. The term "team," on the other hand, implies much more. A team is two or more per-sons who work together to achieve a common purpose. The two main elements of this def-inition are "purpose" and "working together."

All teams have a purpose and a personality. A team's purpose is its task, the reason the team was formed. Its personality is its people and how the team members work together. Dif-ferent teams have their own style, approach, dynamic, and ways of communicating that are different than other teams. For example, some teams are serious, formal, and businesslike, whereas others are more informal, casual, and fun-loving. Some teams are composed of friends, others are not. Friendship, though desirable, is not a necessary requirement for a team to be successful. Commitment to a common purpose and to working together is required. Regardless of style or personality, a team must have professionalism. A team's pro-fessionalism ensures that its personality promotes productive progress toward its purpose.

Attributes of a Successful Team

The successful team should have the following attributes:

(1) A common goal or purpose. Team members are individually committed to that purpose.
(2) Leadership. Though one member may be appointed or voted as the team leader, ideally every team member should contribute to the leadership of the team.

(3) Each member makes unique contributions to the team's project. A climate exists in the team that recognizes and appropriately utilizes the individual talents/abilities of the team members.

(4) Effective team communication: Regular, effective meetings; honest, open discussion. Ability to make decisions.

(5) Creative spark. There's excitement and creative energy. Team members inspire, energize, and bring out the best in one another. A "can do" attitude. This creative spark fuels collaborative efforts and enables a team to rise above the sum of its individual members.

(6) Harmonious relationships among team members. Team members are respectful, encouraging, and positive about one another and the work. If conflicts arise, there are peacemakers on the team. The team's work is both productive and fun.

(7) Effective planning and use of resources. This involves appropriate breakdown of the tasks and effective utilization of resources (people, time, money).

Team Member Attributes

Individual team members of the successful team should have these attributes:

(1) Attendance: Attends all team meetings, arriving on time or early. Dependable, like clockwork. Faithful, reliable. Communicates in advance if cannot attend a meeting.

(2) Responsible: Accepts responsibility for tasks and completes them on time, needing no reminders nor cajoling. Has a spirit of excellence, yet is not overly perfectionistic.

(3) Abilities: Possesses abilities the team needs, and contributes these abilities fully to the team's purpose. Does not withhold self or draw back. Actively communicates at team meetings.

(4) Creative, Energetic. Acts as an energy source, not a sink. Brings energy to the team. Conveys a sense of excitement about being part of the team. Has a "can do" attitude about the team's task. Has creative energy and helps spark the creative efforts of everyone else.

(5) Personality: Contributes positively to the team environment and personality. Has positive attitudes, encourages others. Acts as a peacemaker if conflict arises. Helps the team reach consensus and make good decisions. Helps create a team environment that is both productive and fun. Brings out the best in the other team members.

11.5 TEAM LEADERSHIP

The Need for Leadership

Every successful human endeavor involving collective action requires leadership. Great teams need great leadership. Without leadership, humans tend to drift apart, act alone, and lose purpose. They may work on the same project, but their efforts, without synchronization and coordination, interfere with rather than build on one another. Without coordinated direction, people become discouraged, frustrations build, and conflicts ensue. Money is wasted, time schedules deteriorate, but the greatest loss is a loss of human potential. A failed team effort leaves a bitter taste not easily forgotten. Incomparably sweet, on the other hand, is the thrill of team success.

The Leader's Role

The leader ensures that the team remains focused on its purpose and that it develops and maintains a positive team personality. He challenges and leads the team to high performance and professionalism. With one hand he builds the team, with the other he builds the project. In support of these objectives, the leader must do the following:

1. Focus on the purpose: Help the team remain focused on its purpose.
2. Be a team builder: The leader may actively work on some project tasks himself, but his most important task is the team, not the project. He builds, equips, and coordinates the team so they can accomplish their purpose and succeed as a team.
3. Plan well and effectively utilize resources (people, time, money). He assesses and utilizes effectively the team members' abilities.
4. Run effective meetings. Ensure that the team meets together regularly and that the meetings are productive.
5. Communicate effectively. The leader effectively communicates the team vision and purpose. He also effectively praises good work and gives effective guidance for improvement of substandard performance.
6. Promote team harmony by fostering a positive environment. If team members focus on one another's strengths instead of weaknesses, conflicts are less likely to arise. Conflict is not necessarily evil, however. The effective leader must not be afraid of conflict. He should view conflicts as an opportunity to improve team performance and personality, and to refocus it on its purpose.
7. Foster high levels of performance, creativity, and professionalism. Have a high vision. Challenge team members to do the impossible, to think creatively, and to stimulate one another to high performance.

The Entire Team Leads, Too

Every team needs all the leadership it can get. Most teams have a single individual who is either voted or appointed as its leader. In the best teams, however, many members contribute leadership in ways that support and complement the appointed leader. Like the appointed leader, every member should focus on the purpose, build the team, plan, recognize the gifts of others, contribute to effective meetings, communicate well, promote a harmonious team environment, and be creative. Ideally, every member, together with the leader, should build the team with one hand and labor on the project with the other hand. Noted author John Maxwell defines leadership in this way: "Leadership is influence." Every team member, from his/her position on the team, has the opportunity to considerably influence the team's performance. How important is the leadership from the entire team? It is possible to have a truly outstanding leader, but if the team chooses to use its influence to not follow the leader, to undermine the leader, and to promote team disunity, the team will fail. Conversely, a team with a relatively weak appointed leader can be greatly successful if the team members pitch in and use their influence to build the team. Ultimately, the entire team—both the members and the appointed leader—must work together to build a great team. It takes a team to make a great team.

Leadership Styles

According to Schermerhorn, "leadership style is a recurring pattern of behaviors exhibited by a leader." It is the tendency of a leader to act or to relate to people in a particular way. There are two general categories of leadership style—task-oriented leaders and people-oriented leaders. A **task-oriented leader** is highly concerned about the team's purpose and task at hand. He likes to plan, carefully define the work, assign specific task responsibilities, set clear work standards, urge task completion, and carefully monitor results. A **people-oriented leader** is warm and supportive toward team members, develops rapport with the team, respects the feelings of followers, is sensitive to followers' needs, and shows trust in followers.

For a team, is one style of leadership better than another? A successful team needs both styles of leadership in some proportion. Earlier it was stated that a team has a purpose and a personality. The team must be task-oriented, but it also must cultivate a team personality and a positive environment in which team performance can flourish. This is further evidence for the need for individual team members to provide leadership that complements the leadership of the appointed leader. If the appointed leader tends to be more task-oriented, the team can balance this with a people-orientation. If the appointed leader is people-oriented, he will likely need individuals on the team to assist in details of task management. The best leaders welcome complementary leadership from the teams.

11.6 TEAM ORGANIZATIONAL STRUCTURES

Some engineering design teams are asked to create an organizational chart. Preparing this chart gives the team the opportunity to define roles and to give thought to the relationship between the team and the team leader. Most student teams, in part because they don't know any better, default to a traditional team structure such as is given in the first block in the figure below. This structure is easy to draw, but it may or may not accurately represent the best way the team should operate. It implies a strong leader who largely directs the actions of the group, possibly with little participation or discussion from the other team members. It suggests separation between the leader and the other team members.

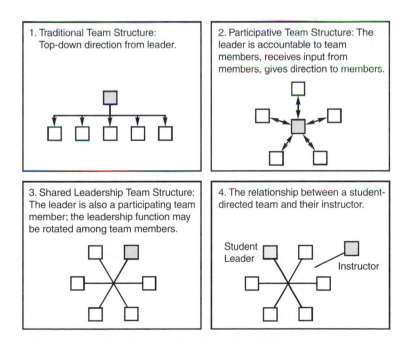

Figure 11.2 Possible Team Structures.

The second structure portrays a participative leadership/team model in which the leader is positioned in close proximity, with short, direct communication paths to all of the members. The figure implies direct accountability of the leader to all of the members and dependence of the leader on the participation of all the members.

The third structure is similar to the second except it emphasizes the leader's role as a working member of the team. It suggests a flat structure where the leader is an equal rather than a hierarchical structure where the leader is above the team. For this structure it is easy to visualize the leadership function shifting around the ring from member to member as situations arise needing the special expertise of individual members.

The fourth structure shows the relationship between a student team and its instructor. The student team may be a design group, a research team, or a collaborative study unit. The instructor, though not part of the team, will be nearby and will serve as an important resource to the team. He may be asked to advise the team on administrative issues, to act as a technical consultant, or to assist in intervention or disciplinary action for a non-performing team member. All teams—in industry, in the university, and in private organizations—ultimately operate under the authority of a leader or administrator.

There is no one "right" structure. None of these structures is inherently good or evil. Examining these example structures gives the team the opportunity to define roles and to give thought to the relationship between the team and the team leader. A team needs to choose a structure that effectively models how they want to work as a team. The ultimate success of the team then hinges upon the team and its leader(s) working together to accomplish the team's purpose. As a team works, they may find that their operational structure may shift periodically in response to different situations and challenges that they face.

11.7 EFFECTIVE DECISION MAKING

Every team makes decisions in order to accomplish its work. The ability to make high-quality decisions in a timely manner is a mark of a great team. Not surprisingly, the *process* of how decisions are made greatly affects the *quality* of these decisions. Unfortunately, most teams arrive at decisions without even knowing how they did it. In this section, we present a variety of ways in which decisions are typically made and discuss their advantages and disadvantages. Though it is recommended that a team decide in advance what decision-making process it will use, the best teams use a variety of different means of making decisions depending on the decision circumstances. The following classification of ways teams make decisions is summarized from a section in the book, *Why Teams Don't Work*, by Harvey Robbins and Michael Finley.

Consensus: A decision by consensus is a decision in which all the team members find common ground. It does not necessarily mean a unanimous vote, but it does mean that everyone had an opportunity to express their views and to hear the views of others. The process of open sharing of ideas often leads to better, more creative solutions. Unfortunately, achieving consensus can take much time and become unwieldy for larger teams.

Majority: Another way to make a decision is by majority vote. Take a vote and whatever receives the most votes wins. The advantage of this approach is that it takes less time than reaching consensus. Its disadvantage is that it provides for less creative dialogue than consensus, and there will always be a faction—the minority—who lose the vote and may become alienated.

Minority: Sometimes a small subset of the team—for example, a subcommittee—makes the decision. The advantage of this is that it may expedite the decision. Its disadvantage is that there is less overall team communication, and some team members may be prevented from making contributions to the decision.

Averaging: Averaging is compromise in its worst form; it's the way Congress and some committees arrive at decisions. Averaging is often accomplished with haggling, bargaining, cajoling, and manipulation. Usually no one is happy except the moderates. The advantage of averaging is that the extreme opinions tend to cancel one another out. The disadvantages of this are that there is often little productive discussion (as in consensus), and that the least informed can cancel the votes of the knowledgeable.

Expert: When facing a difficult decision, there is no substitute for expertise. If there is an expert on the team, he/she may be asked to make a decision. If a team lacks an expert on a particular issue, which is often the case, the best teams recognize this and seek the advice of an expert. The advantage of this is that theoretically, the decision is made with accurate, expert knowledge. The disadvantage of this is that it is possible to locate two experts who, when given the same information, disagree on the best course of action. How is a team assured that the expert gives them the best advice? Team members may be divided on which expert to consult, as well as on their assessment of the expert's credentials.

Authority Rule without Discussion: This occurs when there is a strong leader who makes decisions without discussing the details with or seeking advice from the team. This works well with small decisions, particularly of an administrative nature, with decisions that must be made quickly, and with decisions that the team is not well qualified to contribute to, anyway. There are many disadvantages to this approach, however. The greatest is that the team's trust in its leader will be undermined. They will perceive that the leader doesn't trust them and is trying to circumvent them. If one person is continually taking it upon himself to make all the decisions, or if the team is abrogating its responsibility to act together and by default, forcing the decision onto an individual, then this is a group, not a team. It is a loose aggregation of individuals, a dysfunctional team.

Authority Rule with Discussion: This can be a very effective way of making decisions. It is understood from the outset that the final decision will be made by one person—the leader or a delegated decision-maker—but the leader first seeks team input. The team meets and discusses the issue. Many, but perhaps not all, viewpoints are heard. Now, more fully informed, the appointed individual makes the decision. The advantage of this method is that the team members, being made part of the process, feel valued. They are more likely to be committed to the result. This type of decision-making process requires a leader and a team with excellent communication skills, and a leader who is willing to make decisions.

11.8 ATTITUDES TOWARDS TEAM EXPERIENCES

If you have an opportunity to become part of a team, how should you approach it? What should your attitudes be?

1. Determine to give your best to help the team grow as a team and to accomplish its purpose. Don't just tolerate the team experience.
2. Do not expect to have perfect teammates. You're not perfect; neither are they. Don't fret that your friends are not on your team. Make friends with those on your team.
3. Be careful about first impressions of your team. Some team members may seem to know all the right things to say, but you may never see them again. Others might at first look like losers, but in the long run they are the ones who always attend meetings, make steady contributions, and ensure the success of the team. Be careful about selecting a leader at the very first team meeting. Sometimes the person who seems at first to be the most qualified to lead will never make the commitment necessary to be the leader. Look for commitment in a leader.

4. Be a leader. If you are not the appointed leader, give your support to the leader and lead from your position on the team. Don't be scared off by team experiences in the past where the leader did all the work. Be a leader who encourages and grows other leaders.

5. Help the team achieve its own unique identity and personality. A team is a kind of corporation, a living entity with special chemistry and personality. No team is just like any other team. Every team is formed at a unique time and setting, with different people and a different purpose. Being part of a productive team that sparkles with energy, personality and enthusiasm is one of the rewards of teamwork. Being part of a great team is a great memory.

6. Be patient. Foster team growth, and give it time to grow. Teammates are at first aliens, unsure of one another and of the team's purpose. Watch for and help the team grow through the stages of Forming, Storming, Norming, and Performing.

7. Evaluate and grade yourself and your team's performance. Document its successes and failures. Dedicate yourself to be one who understands teams and who acts as a catalyst, one who helps the team perform at a high level. Note team experiences in your undergraduate portfolio, especially times when you took a leadership role. Prepare to communicate these experiences to prospective employers.

11.9 DOCUMENTING TEAM PERFORMANCE

Gain the Most from Team Experiences by Critical Evaluation

The primary aim of this chapter has been to give you tools and insights into teams that will significantly increase the probability of you being part of (and being an active leader in) great teams. You will likely be involved in many teams during your college years, in your professional life, and in volunteer capacities. Teams are everywhere, but few know how to make teams work. For teams, Voltaire's words are applicable: "Common sense is not so common." If you work hard to be uncommonly sensible about teams, you will be of great value to an employer. Approach every team experience as an opportunity to learn and to grow as a team member and leader. An excellent way to gain the most benefit is to critically evaluate each team experience. Some suggested dimensions for evaluating teams are as follows:

(1) Did the team accomplish its purpose? Did it get the job done?

(2) Did it do the job well? Were the results of high quality? If not, why?

(3) Did the team grow through all of the developmental stages (Forming, Storming, Norming, Performing)? Were there any detours?

(4) Reflect on the team's personality. Did the team enjoy working together? Did team members inspire one another to greater creativity and energy?

(5) Evaluate team members on a report card like the one shown on page 321.

(6) Evaluate the team leader(s). Was he/she an effective leader? Why or why not? How could he/she be effective?

(7) Honestly evaluate your contribution to the team.

Report Team Evaluations to Supervisor

In addition to learning more about teams, another reason to document team performance is to inform your instructor (when you are in college) or your supervisor (when at work) about the team's work. Virtually all teams have a supervisor to whom the team is accountable. It is important that you report data to them. Some argue that members of student teams should all get the same grade; in general, this is the best practice. But students should know from the start that if they do not participate in a significant way in the team's work, then their instructor may adjust downward the individual's team grade. In order for the instructor to make such determinations, it greatly helps if he is given (both at midterm and at the end of the semester) quantitative information on team member contributions. Sometimes instructors ask team members to grade one another, but the reasons for the grades remain unclear. The act of team members evaluating one another on specific team success criteria gives concrete evidence for team member grades. These criteria are even more valuable to the team if they are held up as a target or goal at the beginning of the team activity and if it is made clear that they will be the basis for evaluating the team. In addition to the team member report card, the following should be valuable to your instructor:

(1) Attendance records for all team meetings (usually weekly). Who attended? Was anyone habitually late or absent?

(2) Actual contributions of each member. It works well to include in the team's final report a table like the one given on page 322 noting the percentage contribution of each individual on each of the team's major projects/assignments. Sum these percentages in order to quantify a team member's total contribution. Variation is expected in the contributions of the different members because it is impossible to exactly balance effort. In the example below, four members' contributions varied from 95 to 155 percent. Also included in the table, and a sign of a heathy team overall, is the fact that each of the four active team members took a leadership role on at least one aspect of the project effort: Sally took primary lead on the final report, George led in the creation of the Instruction Manual, Ahmad led in the AutoCAD Drawings and the PowerPoint presentation, and Jaime led in the project construction. Though their contribution percentages vary, there is little evidence that their grades should be adjusted downward; they each deserve, it appears, the team grade. Joe, on the other hand, contributed nothing, and deserves a grade penalty. Chances are, the 10 percent is a gift from his teammates.

Team Member Report Card					
Criteria	**Team Member**				
	Ahmad	George	Jaime	Joe	Sally
1. Attendance: Attends all team meetings, arriving on-time or early. Dependable, faithful, reliable.					
2. Responsible: Gladly accepts work and gets it done. Spirit of excellence.					
3. Has abilities the team needs. Makes the most of abilities. Gives fully, doesn't hold back. Communicates.					
4. Creative. Energetic. Brings energy, excitement to team. "Can do" attitude. Sparks creativity in others.					
5. Personality: Positive attitudes, encourages others. Seeks consensus. Fun. Brings out best in others. Peacemaker. Water, not gasoline, on fires.					
Average Grade:					

Grading Scale: 5 - Always; 4 - Mostly; 3 - Sometimes; 2 - Rarely; 1 - Never

Team Members	Team Member Contributions on Primary Semester Projects					
	AutoCAD Drawings	Project Construction	Instruction Manual	Final Report	Power Point Presentation	Total Indiv. Contribution
Sally	0%	15%	45%	60%	15%	**135 %**
Joe	0%	5%	0%	5%	0%	**10 %**
George	10%	15%	55%	15%	10%	**105 %**
Ahmad	75%	10%	0%	10%	60%	**155 %**
Jaime	15%	55%	0%	10%	15%	**95 %**
	100%	**100%**	**100%**	**100%**	**100%**	**500 %**

REFERENCES

Bennis, Warren, and Biederman, Patricia Ward, *Organizing Genius: The Secrets of Creative Collaboration*, Addison-Wesley, 1997.

Katzenbach, Jon R., and Smith, Douglas K., *The Wisdom of Teams: Creating the High-Performance Organization*, HarperBusiness, 1993.

Nuese, Charles J., *Building the Right Things Right: A New Model for Product and Technology Development*, Quality Resources, 1995.

Robbins, Harvey A., and Finley, Michael, *Why Teams Don't Work: What Went Wrong and How to Make it Right*, Peterson's/Pacesetter Books, Princeton, New Jersey, 1995.

Schermerhorn Jr., John R., *Management,* Fifth Edition, John Wiley and Sons, 1996.

Scholtes, Peter R., Joiner, Brian L., and Streibel, Barbara J., *The TEAM Handbook,* Second Edition, Oriel Incorporated, 1996.

Schrage, Michael, *Shared Minds: The New Technologies of Collaboration*, Random House, New York, 1990.

Tuckman, Bruce W., "Developmental Sequence in Small Groups," *Psychological Bulletin*, vol. 63 (1965), pp. 384–389.

Ullman, David G., *The Mechanical Design Process,* Second Edition, McGraw-Hill, 1997.

Whetten, David A., and Cameron, Kim S., *Developing Management Skills,* Third Edition, Harper Collins, 1995.

EXERCISES AND ACTIVITIES

11.1   Some individuals consider teams as the answer to all human and organizational problems. Conversely, some individuals asked to participate on teams think that being part of a team is a pain. Comment on these propositions.

11.2   From your perspective, is it easy or hard for humans to work together as teams? Does it come naturally, like falling out of bed, or is a lot of work involved in making a great team?

11.3   Describe in your own words the growth stages of a team. Give at least two examples from your own team development experiences. Have you been a part of a team that developed to a certain point and stalled? A team that progressed quickly to a productive, performing team? A team that experienced prolonged rocky periods but ultimately made it to the performing stage? If possible, note factors that led to the success and failure of these teams.

11.4   With a team of three to five people, role play a sample of conversation(s) and team behaviors for each of the five team growth stages: Forming, Storming, Norming, Performing, Adjourning.

11.5   With a team of three to five people, role play a sample of conversation(s) and team behaviors for teams making decisions by consensus and three other decision-making methods given in the chapter.

11.6   There are two quotes (source unknown) based on the T.E.A.M acronym: "Together Excellence, Apart Mediocrity," and "Together Everyone Achieves Much." These communicate the message that by working together humans can achieve more than what they could achieve working alone. Other quotes related to this are "The whole is greater than the sum of the parts," and "None of us is as smart as all of us." Do you believe that this is true? Why, or why not? Does it depend on the situation? Are some problems more suited to teams? Give examples.

11.7   Look up the word "synergy" in a college dictionary and give its basic meaning and its root meaning from the Greek. In an introductory biology textbook look up examples of "synergism." Write and briefly describe three examples of synergism from the biological world. Discuss lessons that humans can learn from these biological examples.

11.8   There are a number of historical examples where individuals have collaborated and accomplished what likely was not possible if they acted alone. Have you ever been part of a fruitful collaboration? If so, what? Why was it fruitful? Or have you been part of a failed collaboration? If so, why?

Chapter 12

Project Management

12.1 INTRODUCTION

"Failure to plan is planning to fail."

Management courses/programs have proven this axiom true in many cases. A good plan is one of the most important attributes of successful teams and projects. Engineering projects are complex and require different tasks to be completed by different people at different times. All these tasks must come together in the end and meet specific deadlines and customer requirements. Projects that involve multiple people and steps must be organized and implemented systematically to insure success. As with the design process and problem-solving models, following systematic approaches to proper project planning will produce improved results and better solutions.

Starting a plan for your project that you have not started may seem daunting or impossible, but this discipline and skill will serve you and your classmates well. Here are eight questions that can be addressed with a plan:

1. What do you and/or your team does first?
2. What should come next?
3. How many people do you need to accomplish your project?
4. What resources do you need to accomplish your project?
5. How long will it take?
6. What can you get completed by the end of the semester or quarter?
7. When will the project be finished?
8. How will we know we are done with the project?

These questions are difficult to answer at the project's beginning, and many students will make the mistake of not using a systematic approach to project planning. They will guess what they can get done and make commitments they cannot keep. Others will assume the details will solve themselves and will conclude with too many details to work out toward the end of the semester or quarter. Stories of late nights on the final project week are common and do not have to happen. Proper planning can insure you will complete your tasks and will allow your team to distribute the workload between team members so no one is overburdened. Tasks can be spread around to leverage people's expertise and abilities and to adapt to unexpected circumstances. A good plan allows resources and materials to be in place when needed to avoid delays waiting on parts or supplies.

In a classroom setting, the ramifications of missing commitments is relatively low. Once you enter the workforce, missing deadlines can result in lost customers, revenue, and people's jobs. Taking advantage of your time in class to practice project management techniques can pay large dividends later in your career.

At the beginning of a simpler project, the discipline of developing and managing a plan may seem like a waste of time but will serve you well over the rest of your career. Good engineering practice follows systematic processes to maximize efficiency and reduce variability and failure. Spending time to plan the project may seem as if you are losing time when you could be doing something productive, but the planning you do upfront will pay off allowing your team to go more quickly and efficiently, resulting in a better product. This chapter highlights some of the tools that can be applied to planning and managing a project within an engineering classroom.

12.2 CREATING A PROJECT CHARTER

The first step in project planning is establishing a project charter. A project charter is a project summary . This may seem obvious, but you must define what the project will entail and how your team will know it has finished. If you have a customer, you should share the charter document with them to insure that the expectations are the same for your team and customer.

The elements of a charter include the following:

1. Description—Describe and summarize what you or your team will be doing. This short summary will allow all those involved or affected by the project to understand what is needed. Include the needs of the community that you are addressing.
2. Objectives—List the project objectives. Why are you doing the project? What will you or your team achieve? What issues is your team addressing that would not be addressed otherwise? What are the problems you are solving with this community project?
3. Outcomes or deliverables—What are going to be the project results? When the project is finished, what will be left behind by you and your team? Be specific. When you have achieved these outcomes, everyone should understand you have completed the project.
4. Duration—When will the project be started, and when will it meet the objectives and deliver the outcomes?
5. Community Partners—Who will benefit from the project? Who will receive the project outcomes or deliverables? Who are you serving with this project?
6. Stakeholders—Who will be affected by your project other than your customer? Who has a vital interest in the project's success? Stakeholders are others who will need to be kept informed of the project's progress, outcomes, and results.
7. Team membership and roles—If your plan involves a team, list the team members and their roles. (Some of the roles you may want to assign are outlined later in this chapter.) Listing contact information for all of the team members makes the document more valuable and useful for those who will receive the charter document.
8. Planning information—What other information do you need for the project plan? If the project involves a design, what are the steps in the design process. Do other constraints or expectations need to be summarized before planning the project?

The project charter is a useful tool. It gives the project vision and allows all team members, community partners, and stakeholders the opportunity to comment on the vision before beginning. It helps define the roles and responsibilities of your team and others involved. It forces your team to think about the project's requirements at the start and serves as a place to summarize these results. A concise project charter becomes an excellent communication tool for your team, community partners, and stakeholders.

12.3 TASK DEFINITIONS

Once you have defined the project charter, the next step is identifying the completion tasks to achieve the objectives and outcomes. In most engineering projects, many tasks will have to be completed for project completion. Your team might start this process by brainstorming what needs to be done for the project, but do not stop there. Many students will lay out their plans and stop with incomplete tasks. An example might look like this:

Plan
Design
Build
Deliver

For plans to be effective, they need to be detailed enough to manage. A detailed plan should allow you and your team to do the following:

1. Hold individual team members accountable for progress during the project. Do not wait until the project's conclusion.
2. Identify all needed resources, materials, funding, and people.
3. At any time, determine if your project is on schedule.
4. Manage your people and resources by shifting from one task to another as needed to complete the project.
5. Determine when you have finished.

Too general of a project plan defeats the planning purpose. The plan will not allow you to use it to your benefit A detailed plan will allow you and your team to identify who needs to be working on what task each week; so, each task needs to be as detailed as possible. Break larger tasks into smaller tasks whenever possible.

For effective plans, they need to be complete and this will require your team to take advantage of available resources to identify all the tasks. Chances are you are not the first ones to attempt a project like this. You will some past experience to draw upon. When I worked in a large manufacturing firm, new product plans were laid out based on what we needed to do the last time, and appropriate modifications were made for the new line. We used past experience as a guide, and your team may be able to do the same. Do you have task lists from previous semesters, quarters, or similar projects? Show your task lists to your instructors, advisors, alumni, community partners, and stakeholders for input. Gather as much data as possible to develop a comprehensive list.

Include tasks associated with other tasks. If something needs to be ordered, put in time to research for the component so you can place the order and have time for shipping. Overnight options are expensive, so initially plan on ground transportation. If you need approvals or signatures for components, incorporate those tasks. Many institutions have a process for purchasing items from outside the university (much like most companies have a purchasing department). These departments have procedures that take time, and you need account for them in the plans. If you are using outside vendors for manufacturing, assembly or testing, do you need to schedule meetings before they begin work on their tasks?

To help organize the tasks, list them in an outline format. Just like an outline from an English class, similar tasks are grouped together. This process will make later steps faster and easier to complete. In the example outline, the starting and end dates are listed along with delivery expectations.

Example

Task Definition:
Create a project schedule for your EPICS project.
Start date: September 10, 2011.
Due date: September 17, 2011.
Deliverable: List of milestones, a PERT chart, and a task timeline.

Outline View:
1. Start: Receive assignment to create a project schedule.
2. Learn about schedules
    2.1 Attend talk on proposals, including project schedules
    2.2 Review web pages on project schedules
    2.3 Read about project schedules
    2.4 Look at examples of schedules using Microsoft Project and Excel
3. Develop project plan
    3.1 Identify milestones
    3.2 Identify major components of project
    3.3 Estimate duration of each project component
4. Create project PERT chart and timeline
5. End: PERT chart and timeline submitted

12.4 MILESTONES

Along with tasks, your project plan should have milestones. A milestone is a deadline for some deliverable. Deliverables may be the completion of subcomponents for your project, or they may be when all parts have been ordered and have arrived. In a design, each phase completion of the design process could be a milestone.

Milestones are useful as checks of your plan's progress. You should not wait until the final deadlines before assessing progress. Intermediate milestones serve two purposes. First, they let your team meet short-term goals and celebrate them. It can be a big morale boost for your team to hit a milestone and get some recognition. We recommend that your team celebrates short-term successes and milestones, which will help build team dynamics.

The other benefit of intermediate milestones is you can monitor your progress to know if you are on schedule. If part of the project falls behind, you may be able to redirect people or resources from another part of the project to move that part back on schedule. When defining tasks, think of places where you could use milestones as markers or measures of your team's progress.

12.5 DEFINING TIMES

Now that you have compiled the task list, you need to determine the required task times. In full-time employment settings, tasks are measured in working days. For your project, working days are a good unit of time.

Be diligent to include the full time needed for tasks. If a component needs to be ordered, include the time it will take your institution to place the order and including shipping time. You can backorder some parts. If your school's machine shop needs to make a part, how far in advance will you need to place the order, and how long will it take to start and complete? If customers or stakeholders need to test a beta software version, how long should you leave it with them before you will get your results? It may only take an hour to perform your assessment but you may need to give them a week or two to schedule the evaluation. Student teams do not account for time lags because others will not drop what they are doing to perform the task. Give people one or two weeks and factor that into your timelines.

Student working days: Working days do not always work with students' schedules. In order to plan your tasks, you will need to develop a timescale in place of working days that works for your team. Your class probably does not meet every weekday, and you and your classmates have other commitments so this project is not a full-time job. For your class, everyone has an expectation of the number of hours each person will spend each week on the project, and you can use that as a model to develop your timescale. Some of the models used by student teams include the following:

- Work Weeks: Break tasks up into week segments (based on the number of hours devoted to the course each week). Your team should be explicit about when the week begins and when tasks are due to check milestone progress. This gives individuals the flexibility to arrange tasks around other classes and obligations during the week as long as they meet the weekly deadlines.
- Monday and Friday: Some students prefer to break the week up into two segments, the work week and weekend. This works well for teams that leave time on the weekends to do a large part of their project work. Tasks (ordering parts or communicating with outside vendors) need to occur during the week and can be assigned on Mondays with deadlines of Friday. Dedicated project work will have deadlines and milestones of Monday. This system allows your team to check on progress twice each week, but different tasks need to be planned during the week and over the weekends. Your team can pick other days of the week (e.g., Tuesday and Thursday) depending on when your team meets.
- Class periods: Some teams break up timelines by class periods. So, each time you meet, you can check on progress. Unfortunately, the time between classes is not uniform.

Your team can use any of a number of timescales, but your team members should be held accountable for their progress. They should not be expected to work full time at points of the project to achieve your objectives.

When assigning times to tasks, either by your team or outside, use prior experience or get input on your estimates. Be conservative with your time estimates. The first time you need to do something, it will take much longer than you originally thought. Simple things like ordering a part through your college's system will take some time because you must learn the system. If you need to make a circuit board, the first time will include learning how to make it. Include time for training if needed before you complete a task. In your timelines, include researching and learning new systems, etc.

We suggested you need to detail the tasks as much as possible, and the times you assigned can serve as a check for the task detail level. If tasks span several weeks, ask if they can be broken into smaller tasks. If possible, break your tasks into small enough components of not more than one or two weeks. Longer times for tasks make it easier for team members to procrastinate and your team getting behind. If you wait too long, it may be hard

to recover from something falling behind. Short timelines make holding each other accountable and project management easier.

12.6 ORGANIZING THE TASKS

With your organized task list, the next question is task order. We will systematically optimize
the project plan. The first step in this process is to determine the task relationships and
sequencing. You have grouped related tasks in your outline, but you need to relate groups.

You and your classmates have gone through this same process as you laid out your plan
of study. All your college courses have prerequisites and co-requisite classes. Some
classes can be taken at any point in your four-year degree program while others must be
taken early or be delayed. You determined what classes you needed to start with and how
many you needed to graduate. You used resources like printed sample plans as well as talking with other students, academic advisors, and faculty.

You will follow the same process in the project planning process. Take the tasks that your
team identified and prioritize what must be done. Use your outline to link tasks part of the
same task together.

12.7 PERT CHARTS

The US Navy developed Program Evaluation and Review Technique (PERT) in the 1950s
to manage the Polaris missile project. This technique is still used in many industries as the
major project tool to this day to track and order tasks.

In a PERT chart, each task is represented by a box that contains a brief description of
and duration for the task. Milestones are represented with boxes that have rounded corners.
Related tasks are connected with lines. If one task must be completed before a second task
can be started, they are connected with a line originating from the right edge of the first task
and connected the left edge of the second task, as show in Figure 12.1. In this example, the

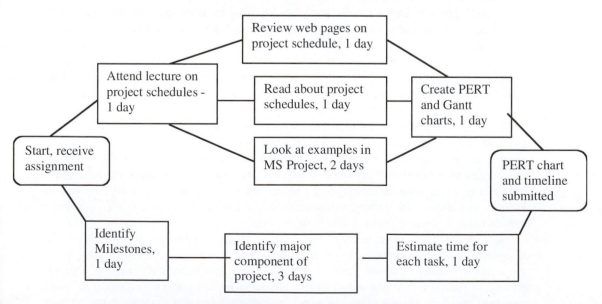

Figure 12.1 PERT Chart for Project Planning

duration is noted in terms of days, and if your team needs a component from a vendor, the task of ordering the component and shipping it have to proceed before anything else can be done. Unrelated tasks are not connected on the PERT chart.

The PERT analysis shows pathways of tasks. Some tasks have to be done in series, and others can be done in parallel. These series can be laid out as the overall project plan is laid out.

PERT charts are great "what-if" tools during the initial planning stage. Post-It notes, 3x5 cards, chalk boards or whiteboards, or project management software can be used to capture the process flow, to find ways to reduce project completion time, and to utilize the available resources and people. What needs to be done when and by whom? By mapping out the process, a team can spread its resources and allow process tracking. The planning process using this tool can help people think of parallel tasks and what can be going on at the same time.

The sum of the times along each path gives length or duration for each path. The longest path is the critical path. The critical path will set the project length unless the task times can be reduced.

12.8 CRITICAL PATHS

The critical path is a concept from industrial engineering and project management that shows the task or its series that will pace the project. The critical path is the longest string of dependent project tasks. Tasks on the critical path will hold up project completion if they are delayed. An example of a critical path is the mathematics sequence in an engineering curriculum. If a student delays a semester of calculus class early in his or her program, it typically delays graduation for one semester. Mathematics is on the critical path for graduation in an engineering program. Delaying an English or humanities class, however, could be accommodated by rearranging other classes or taking an overload one semester. The English and calculus classes differ in that other classes have calculus as a prerequisite. It must be taken before other classes and, thus, has a cascade effect on the overall study plan.

The same will be true for tasks in your project plan. Tasks that are not on the critical path may be rearranged without changing the project completion. Any delay of a critical path task will result in a project delay.

As you plan for your project, you have to know which tasks can be compressed or rearranged and which cannot. Some tasks can be accelerated by using more people while others cannot. The classic analogy is having a baby. Nine people cannot have the same baby in one month. Some tasks take time and special attention needs to be paid to tasks on the critical path, which cannot be compressed. This information will be valuable during the project if you need to rearrange tasks or team members.

12.9 GANTT CHARTS

While the PERT chart is for organizing tasks and placing priorities on timing and resources, it may be difficult to interpret for a project with many tasks, especially for people outside your team. A Gantt chart is another popular project management charting method and is easier for people to understand where your team's progress relative to your plan.

A Gantt chart is a horizontal bar chart frequently used in project management where tasks are plotted versus the time. Henry L. Gantt, an American engineer and social scientist, developed the Gantt chart in 1917, and it has become one of the most common project planning tools. In a Gantt chart, each row represents a distinct task. Time is represented on the horizontal axis. Typically, dates run across the top of the chart in increments of days, weeks, months, or a timescale that you and your classmates have determined. A bar represents the time when the task is planned for with the left end representing the expected starting time for each task and at the right end for the end of each task. Tasks can be shown to run consecutively with one starting when the previous one ends. They can be shown to run in parallel or overlap. An example of the Gantt chart, created with Excel, is shown in Fig. 12.2.

Notice that tasks and milestones are included on the Gantt chart. Significant milestones and reports are represented with different colors. As progress is made, the cells in the spreadsheet can be changed to a different color so that someone can easily see the progress at any time in the project. If a project is partially completed, that bar can be shaded in proportion of the work done on that task. If the task is half finished, half of that bar will be shaded. A vertical line can be drawn for the current date, and progress can easily be checked based on the line. Tasks that show completion to the right of the line are ahead of schedule and those to the right of the vertical line are behind schedule.

An additional column can be added to the right of the task descriptions to show who is responsible for each task. The Gantt chart is an excellent way to see who is committed to which tasks during the project to ensure that each person has responsibilities over the entire project, and no one is being asked to do too many tasks at one time.

Many teams will use a PERT chart to arrange tasks and identify the critical path. They then transfer the tasks to a Gantt chart for project management. To make a presentation of your project easier, fit your Gantt chart onto one page. If you have more tasks than can fit onto one page, break the project into subprojects with their own Gantt chart, and keep one that shows the progress of each subproject.

MagRacer 2.0 Timeline (weeks)

Project Tasks	2	3	4	5	6	7	8	9	10	11	12	13	14	15	16
Bring new team members up to speed on MagRacer	▓														
Solve FET prolem in demo track	▓	▓													
Concept of MagRacer2 cabinet	▓	▓													
Meet with IS people/ visit IS		▓													
Finalize track/coil assembly		▓	▓												
AutoCAD drawings of MagRacer2 cabinet		▓	▓												
Finalize display concept		▓													
Deliver working test track			▓												
Week 4 Demo				▓											
Milestone: Submit MR2 drawings to WP				▓											
Complete PCB layout				▓	▓										
Milestone: Submit PCB layout for fabrication					▓										
Final order of all circuit material					▓										
Construct coils					▓	▓	▓								
Construct track mounting hardware					▓	▓									
Construction of visual display						▓	▓	▓							
Week 8 Progress Report								▓							
Exected delivery of MG2 cabinet from WP (4wk)									▓						
Expected delivery of PCBs (3wk)									▓						
Spring Break									▓						
Final assembly of MagRacer2										▓					
Week 11 Design Review											▓				
Milestone: Delivery of completed MagRacer2											▓	▓			
Troubleshoot MagRacer2												▓	▓	▓	
Prep documentation for MagRacer2														▓	
Week 16 End of Semester reports due															▓

Figure 12.2 Sample Gantt Chart from an Engineering Student Team.

12.10 DETAILS, DETAILS

A common mistake of many teams is to omit too many details until the end. This will often doom your plan. The final stages of a design and assembly will take much longer than you think because you had details you did not think about when making the plan. Since you most likely have not done a project similar to this, your team could not foresee many details. These details tend to pile up at the end.

Things do not go as planned. Few plans have time built in for things to go wrong but as Murphy's Law says, Anything that can go wrong will. This is true for project planning. Your plan should have space to debug or fix things.

- Are we assuming that things will fit together the first time and nothing will have to be reworked?
- Are we assuming that we have ordered all parts with the correct parts?
- All materials will be shipped on time and nothing will be back ordered or late.
- No parts will malfunction when installed.

A suggestion to avoid a panic or failure is to push a delivery date back a week before it is due. Your team can use this week as a buffer if things go badly. Human nature is to know you have some leeway and think that the real deadline is the one to aim for. This totally defeats the purpose. Milestones should be met and the schedule adhered to up until the end so that the buffer will act as the buffer if unforeseen circumstances occur in the final stages.

12.11 PERSONNEL DISTRIBUTION

The tools that have been described can help order tasks but project management means getting the right people on the right tasks. Some models of project planning will have people identified earlier in the process based on their expertise and availability. This works when people are separated into departments with separate responsibilities and capabilities. For most student projects, we recommend assigning people after developing a draft of the plan. The people on your team will need to be assigned tasks that allow the project plan to be met. For example, one person cannot do all of the tasks in the first month. Sketching out who is the most qualified for and/or interested in each task is a great way to start. The team must look at the work distribution and make an equitable assessment. The work needs to be balanced whenever possible and must account for the team members' expertise.

When you have assigned people their tasks, add them to the Gantt charts. You can add a column next to the task description or put it on the task bar. If your timeline is detailed enough, check if each person understands what he or she is doing each week during the project. If each person does not, refine the tasks so everyone can.

12.12 MONEY AND RESOURCES

When the plan for the tasks and people is in place, you must ask what resources will be needed for the project. Almost all students will need to develop a project budget in their professional career, so practicing in a classroom setting is an excellent experience. They should learn to develop a project budget.

Setting up a budget may seem as daunting as the project plan. How can you know how much things will cost before you buy them? Though a logical question, you will be asked for budgetary estimates before you begin projects on a regular basis. In large companies, the budget request can determine if your project proceeds. If you are working for a smaller company bidding for work, your budget estimates may decide if you win a project.

Just like estimating times for tasks, past experience is a great asset to this process. Looking back at similar projects and talking with other students, alumni, faculty, corporate representatives, or community members with similar experience can provide project funding estimates. Web searches can give quick estimates on products. You may bind your cost estimates similar to how you would make analysis estimates, i.e., by estimating high (gold-plated) or cheap and low. These boundaries will give you a range with which to work when requesting project money.

Just as in estimating times, you will need to include item costs associated with your main costs. For example, shipping costs will be incurred for each ordered part. Will your team be charged tax on purchases? Will your group need to travel anywhere? Will your team be charged for small, expendable items like nails, screws, and resistors? If a vendor is making a part, include material costs and labor.

One team member should be responsible for managing the project budgets and financial aspects. A spreadsheet can be set up to monitor the project budget and track incurred expenses. Monitoring the financial project progress should be part of the teams' regular updates in monitoring task progress.

12.13 DOCUMENT AS YOU GO

Do not leave work documentation until the end. As you assemble your project plan, you should include documentation milestones of the early parts of the project. In many classes, a report is due at the end and that becomes the motivation for writing things but look at the project's needs. If you are delivering a product to a customer or the local community, you may need user manuals or maintenance procedures. These should be drafted well before the products are delivered so they can be reviewed. Documenting each step will make it easier if you or another class readdresses any of the completed tasks.

Another practical reason for documenting as you go along is work reduction. Reports are typically due at the end of a semester or quarter. Leaving all or most of the documentation until the end adds more work during project completion. There will inevitably be details that appear or were overlooked that you will have to deal with near the project's end. Leaving too many details or large tasks, like documentation, for the end will add unnecessary team burdens and will reduce your work quality.

12.14 TEAM ROLES

In project management, designate individual team members with specific responsibilities so they can be held accountable and will hold others accountable. Some of the roles that are common on student teams include the following:

- **Project Leader or Monitor**—Your team is either organized with a designated leader or you might have rotating responsibilities. In either case, someone must be designated as a project leader or monitor. This person monitors and tracks the progress of the milestones and tasks according to your plan. He or she will maintain the timelines and will be the person to shift responsibilities if tasks fall behind in part of your team. During regular meetings, the timelines should be checked and reported and, this person will be responsible for bringing this to the meetings. A diligent project leader will increase your likelihood that you meet your goals.

- **Procurement**—If your project requires much material to be ordered from outside the university, your team may designate one person to learning the purchasing system and to track the progress of team orders.
- **Financial officer**—If your project requires purchases that are tied to a budget, designate one person as a financial officer to manage your team's expenses. Going through your estimates at the beginning will give you a roadmap but you will need to adjust it. Having one person responsible for monitoring your progress will make identifying problems faster and easier. You should identify this person early in the process and he or she can be the lead person in developing the initial budgetary estimates.
- **Liaison**—If your team has a customer outside of the class, designate one person as liaison. This person is responsible for keeping everyone informed about the project plan and progress toward meeting the plan. Most outside contacts cannot track all the students and prefer one contact point. This person will share the project plan, plan progress, and any plan changes.

Project Management Software: The examples that have been shown for PERT and Gantt charts have been made with simple drawings and Excel spreadsheets. There are powerful project management tools that can allow you and your team to manage and track large and complex projects. One software products available to students is Microsoft's Project (http://www.microsoft.com/office/project/prodinfo). In Project, tasks can be entered on separate lines with start and finish dates. Projects can be grouped under headings as shown

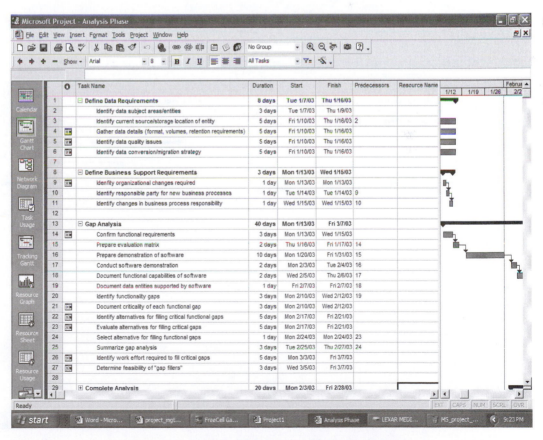

Figure 12.3 Microsoft's Project Software

in Figure 12.3. Tasks can be linked by entering the preceding task's row number in the Pre-decessor column for that task.

A Gantt chart is generated from this entered data. Clicking on the Gantt chart icon on the left will show a Gantt chart for the project. For this project, five main tasks are displayed.

Each project part can be broken into smaller subtasks to be managed. To view the sub-tasks, click on the + symbol next to the task name and an expanded view will appear as seen in Figure 12.5. This view expands the subtasks. Two of the tasks in the expanded view are shown as related, with one preceding the other. notice that as tasks are completed, they can be shown with shading to track progress.

One of the real benefits of MS Project is that you can enter the data for the tasks one time and generate several other data views. A PERT-style chart can be generated by clicking on the Network Diagram icon on the left side of the screen. This block diagram shows the same information as the PERT chart in Figure 12.6. Each task is identified and described, and it is linked with related projects. This case shows two tasks related in the subcategories.

If the entire project is shown, the relationships between subtasks can be seen in Figure 12.7.

MS Project can display a calendar, which shows the projects that should be active for given days of the month. This tool can be useful if your team has designated specific days of the week for work. You can break your tasks into components that will fit into the time blocks you have arranged to do your work outside of class. A sample view of the calendar is shown in Figure 12.8.

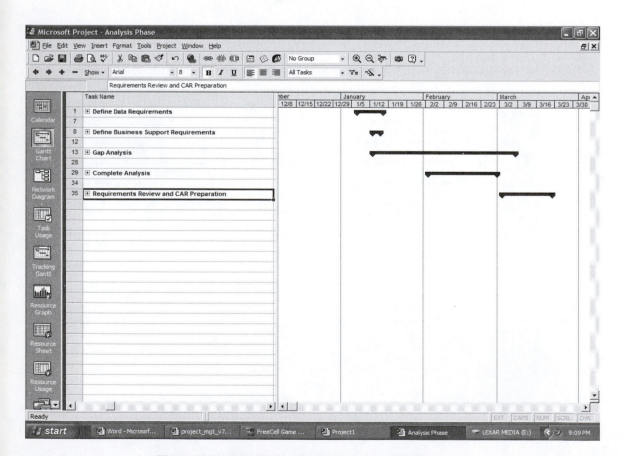

Figure 12.4 Gantt Chart with Major Project Headings

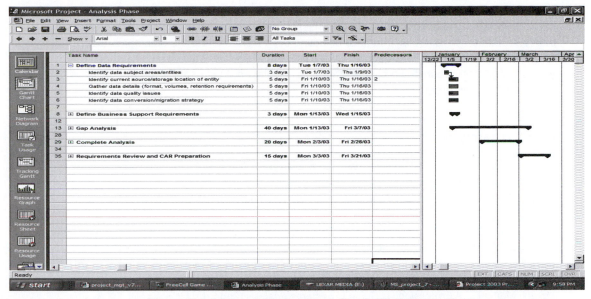

Figure 12.5 Microsoft Project Gantt Chart with Subtasks Shown

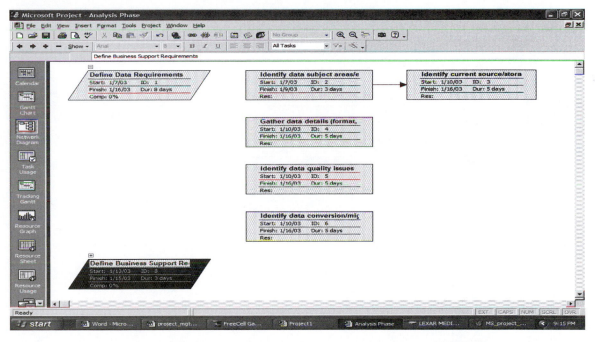

Figure 12.6 Network Diagram view of First Major Task with Subtasks

MS Project is useful for planning and managing a project and it makes it very easy to import the charts and images into a word processor for reports or presentations. Any picture in Project can be saved as an image by selecting *copy picture* under the *edit* menu. Select *to GIF image file* to save your image. In a word processor, you can now insert the image into your document for reporting. In Microsoft Word, select Insert, Picture and then from a file to find the .gif image from Project.

REFERENCES

Brooks, Frederick P. Jr., *The Mythical Man-Month,* Essays on Software Engineering, Anniversary Edition (2nd Edition), Addison Wesley, Reading, Massachusetts, 1995.

Covey, S.R., *The 7 Habits of Highly Effective People*, Simon and Schuster Inc., New York, 1989.

Fox, Terry L., and Spence, J. Wayne, "Tools of the Trade: A Survey of Project Management Tools," *Project Management*, vol. 29, no. 3, pp. 20-27.

Meredith, Jack R., and Mantel, Samuel, J., Jr., *Project Management, A Managerial Approach* (3rd Edition), John Wiley and Sons, New York, 1995.

Modell, Martin E., *A Professional's Guide to Systems Analysis* (2nd Ed.), McGraw Hill, 1996.

Naik, B., *Project Management: Scheduling and Monitoring by PERT/CPM,* Advent Books, 1984.

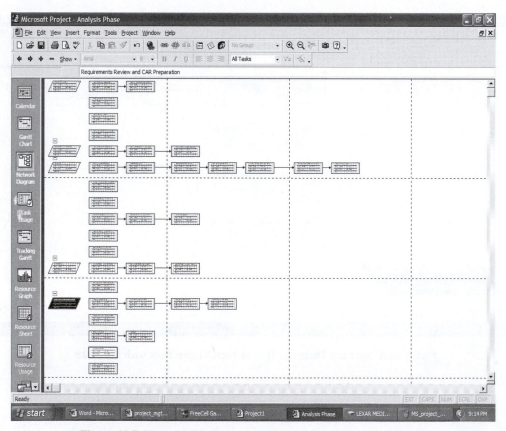

Figure 12.7 Network Diagram view of all Major Task with Subtasks

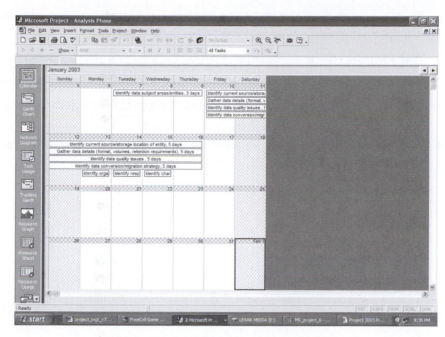

Figure 12.8 Microsoft Project Calendar View

Pyron, Tim, *Using Microsoft Project 2000,* Que, a Division of MacMillan Co., Indianapolis, IN, 2000.

Smith, Karl A., *Project Management and Teamwork*, McGraw-Hill, New York, 2000.

EXERCISES AND ACTIVITIES

12.1 Write a project charter for your project.

12.2 Deliver your project charter to one of your stakeholders, and write a paragraph on their reaction.

12.3 Develop an outline for all of the project tasks.

12.4 Select five project tasks that you anticipate doing in the first month and estimate the time required for each task. Be prepared to present your findings, including the resources you used, to the class.

12.5 Develop a PERT chart for your project and present it to the class.

12.6 Identify the critical path for your project. List the tasks on the critical path.

12.7 Write a one-page essay on ways to shorten the project's critical path.

12.8 Create a Gantt chart for your project.

12.9 Write a one-page paper that compares and contrasts the benefits and limitations of a PERT chart and a Gantt Chart.

12.10   Contact a practicing engineer and find out what methods she or he uses for project management. Write a one-page paper on her or his experience.

12.11   Prepare a short oral presentation on a project management software package.

12.12   Write a one-page paper on the benefits and uses of Microsoft's Project software.

12.13   Write a summary of details that may appear at the end of your project not included in your plan.

12.14   Write a one-page analysis of our team's progress when you are halfway through your semester or quarter.

Chapter 13

Engineering Design

13.1 WHAT IS ENGINEERING DESIGN?

Engineers create things. Engineers build things. In order to successfully perform these tasks, engineers must be involved in *design* or in a *design process*. So what is engineering design and an engineering design process? Webster's dictionary defines design as "to create, fashion, execute, or construct according to plan." It defines a natural process as "a natural phenomenon marked by gradual changes that lead toward a particular result: a series of actions or operations conducing to an end; a continuous operation or treatment, especially in manufacture."

This chapter may seem closely related to Chapter 8, "Problem Solving." Indeed, there are many similar topics and themes in Chapter 8 that are important in Design. One of the critical areas which impacts Design is the issue of "external constraints" that can influence the outcome of the process. Appropriate references back to Chapter 8 will be noted throughout this chapter.

In the work of engineers, engineering design is an important and ongoing activity. Students who graduate with an engineering degree from an accredited program have had a significant amount of design experience as part of their education. The Accreditation Board for Engineering and Technology (ABET) has traditionally defined engineering design as follows:

> *Engineering design is the process of devising a system, component, or process to meet desired needs. It is a decision-making process . . . in which the basic sciences and mathematics and engineering sciences are applied to convert resources optimally to meet a stated objective. Among the fundamental elements of the design process are the establishment of objectives and criteria, synthesis, analysis, construction, testing, and evaluation. . . . it is essential to include a variety of realistic constraints, such as economic factors, safety, reliability, aesthetics, ethics, and social impact.*

ABET then describes the basic educational components for a major engineering design experience that must be included in an engineering program:

> Each educational program must include a meaningful, major engineering design experience that builds upon the fundamental concepts of mathematics, basic sciences, the humanities and social sciences, engineering topics, and communication skills. The scope of the design experience within a program

should match the requirements of practice within that discipline. . . . all design work should not be done in isolation by individual students; team efforts are encouraged where appropriate. Design cannot be taught in one course; it is an experience that must grow with the student's development. A meaningful, major design experience means that, at some point when the student's academic development is nearly complete, there should be a design experience that both focuses the student's attention on professional practice and is drawn from past course work. Inevitably, this means a course, or a project, or a thesis that focuses upon design. "Meaningful" implies that the design experience is significant within the student's major and that it draws upon previous course work, but not necessarily upon every course taken by the student.

Many entering engineering students often confuse the "design process" with drafting or art-related work. To clarify this, ABET adds the following:

Course work devoted to developing computer drafting skills may not be used to satisfy the engineering design requirement.

Most engineering programs concentrate their engineering design coursework in the later part of the student's program, thereby allowing the student to apply much of the prerequisite background in math, science and related fields to the various engineering problems. In most engineering programs, design courses make up between 20% and 25% of the total curriculum.

13.2 THE DESIGN PROCESS

It is important to realize that there is not one uniform approach to engineering design that is followed by practicing engineers. Some firms approach engineering design as a short, simple process with only a few steps, while others use a more complex, multi-step method with several stages. No matter what process is used, it is important to realize that engineering design is always continuous. The completion of one design, or the solution of one problem, may serve to open up opportunities for subsequent designs or modifications.

A design process is used whether a product is being developed for an ongoing manufacturing process, where thousands or even millions of a certain item will be produced, or for a one-time design, such as with the construction of a bridge, dam, or highway exit ramp.

10-Stage Design

This chapter will present one design process (the 10-stage process) that is currently in use. This process will then be applied to an existing product to analyze how each of the 10 stages may have been applied in the product's realization.

The 10 stages that make up the process are as follows:

- Stage 1:   Identify the problem/product innovation
- Stage 2:   Define the working criteria/goals
- Stage 3:   Research and gather data
- Stage 4:   Brainstorm/generate creative ideas
- Stage 5:   Analyze potential solutions
- Stage 6:   Develop and test models
- Stage 7:   Make the decision
- Stage 8:   Communicate and specify

- Stage 9: Implement and commercialize
- Stage 10: Perform post-implementation review and assessment

The actual process begins with the appointment of a project manager or team leader. This individual will be responsible for oversight of the entire process to ensure that certain key elements of each stage have been satisfied before the project moves on to the next stage. This person will also be responsible for recruiting team members of varying backgrounds and expertise for each of the stages. The team will include non-engineers as well as engineers. Some of the team members will be used throughout the process, while others will be needed only for certain parts of the process.

Stage 1: Identify the Problem

Engineers are problem-solvers; and the problems they solve are often identified as the specific needs and problems of customers. For example, a new prosthesis may be required to overcome a particular handicap, or increased gas mileage standards demand higher-efficiency engines, or a new computer program is needed to monitor a modified manufacturing process, or new safety devices are required to better protect infants in automobiles. Therefore, the first stage to problem solving in engineering design is to establish the actual problem clearly, and to identify sources of information to help understand the scope and nature of the problem.

The project manager will call upon the resources of various individuals to assist with these initial stages in the process. There may be a multitude of sources outside as well as within the organization that can assist with solving the problem. Many firms have a research and development unit made up of scientists and engineers who possess the training and expertise to assist with problem evaluation. In addition, sales engineers, who maintain consistent contact with outside individuals, can provide valuable input on problem identification. If the problem is one of expanding a current product line or modifying an existing system to improve it or make it better fit customer needs, management will likely play a role in the definition of the problem. Each group would be represented on the team at this stage.

External resources may come from trade shows, conferences, technical presentations, patent listings, and publications. Ideas generated from existing or prospective clients may also be valuable. Information gathered from external research agencies, private laboratories, and government-funded foundations can be useful as well. In addition, an awareness of the competition and what products and services they are involved with can be beneficial in problem identification. At this early stage in the process, it is helpful to establish a preliminary, formal statement of the problem. For example, the need for a new automobile safety device for infants might evolve into a preliminary statement of the problem as follows: "Develop a better child restraint system that will protect children involved in automobile collisions."

It is important at this point to review the section of Chapter 8 which discusses the "real problem." Before going ahead with the design process, the team must be certain that the correct issues and background have been thoroughly explored so the most appropriate solutions can be developed.

At this point, if the project manager is satisfied that all necessary issues have been resolved, the project will likely be passed on to the next stage where project criteria and goals will be developed.

Stage 2: Define the Working Criteria and Goals

Once the problem has been identified, it is important for the team to be able to validate it throughout the design process. This requires the establishment of certain working criteria, or standards, which can be used in each of the 10 stages to measure possible solutions. The ultimate objective in this stage is to be able to establish preliminary goals which will act as the focal point of the team as it works through the process. The development of some working criteria provides a means to compare possible solutions. At this stage of the design process, everything is preliminary, so it is still possible for the team to modify the criteria if necessary.

Examples of working criteria could include answers to the following questions:

1. How much will it cost?
2. Will it be difficult to produce?
3. What will be the size, weight, strength?
4. What will it look like?
5. Will it be easy to use?
6. Will it be safe to use?
7. Are there any legal concerns?
8. Will it be reliable and durable?
9. Can it be recycled?
10. Is this what the customer truly wanted?
11. Will our customers want to purchase it?
12. Will customers want to purchase this version instead of a competitor's product?

Once some preliminary working criteria have been established, it is important to try to develop overall goals for the process. This is basically a statement of objectives which can be evaluated as the design process evolves. Using the example of new standards implemented to increase gas mileage and reduce emissions in automobiles, the goals for the design might be: "To develop an automobile engine which produces 25% less emissions while increasing gas mileage by 10%."

Having overall goals established for the project provides a means of evaluating, monitoring, and changing, if necessary, the focus of the process as it evolves through the 10 stages. For the project manager, the criteria and goals become a "checkpoint" for assessing the progress to date, and will help them determine if the project is ready to move to the next stage, or if the process needs to return to Stage 1 for re-evaluation.

Stage 3: Research and Gather Data

This stage is very important to all remaining stages of the design process. Having good, reliable background information is necessary for the team to begin exploring all relevant aspects of the problem. Consistent with the preliminary working criteria and the goals that have been established, the team members selected for this phase of the process must determine what types of information will be needed and the best sources of that information. For example, they may want to know such things as:

1. What information has been published about the problem?
2. Is there a solution to the problem that may already be available?
3. If the answer to the above is yes, who is providing it?
4. What are the advantages of their solution?
5. What are the disadvantages of their solution?
6. What is the cost?
7. Is cost a significant issue?

8.   What is the ratio of time spent compared to overall costs?
9.   Are there legal issues to consider?
10. Are there environmental concerns which must be considered?

There are many possible resources the team members can utilize to assist them in their research. A good starting point may be a simple search using the Internet. This may provide useful sources of material that can serve as the focus for additional research. Other sources of reference information may come from:

1.   libraries
2.   professional associations (technical and non-technical)
3.   trade journals and publications
4.   newspapers and magazines
5.   market assessment surveys
6.   government publications
7.   patent searches and listings (U.S. Patent Office: www.uspto.gov.)
8.   technical salespersons and their reference catalogs
9.   professional experts including engineers, professors, and other scientists
10. the competition's product (How do they construct it? Disassemble their product and study it.)

Detailed files, notes, pictures, sketches, and other supporting materials will be maintained by the team to assist them as they proceed through the remaining stages of the design process. As new supporting information is discovered by the team, the material will be added as additional reference resources.

Depending on the type of information that has been collected, at this point it may be appropriate to review the established preliminary working criteria and the overall goals. To assist the project manager throughout the process, some modifications may need to be made. For example, it may be determined that one or more of the criteria may not actually apply to the problem. Likewise, new issues may surface which may necessitate the addition of new criteria or a modification of the goals. It is important to have these issues resolved before moving on to the next stage.

Stage 4: Brainstorm/Generate Creative Ideas

The basic concept involved at this stage of the process is to creatively develop as many potential solutions to the problem as possible. The more ideas that can be generated, the better the likelihood of identifying a feasible solution to the problem. The project manager will want to gather a group of individuals from both technical and non-technical backgrounds to provide their unique perspective to the problem. This group may include engineers, scientists, technicians, shop workers, production staff, finance personnel, managers, computer specialists, and perhaps even a few clients.

A major method of generating multiple ideas to a problem is called *Creative Problem Solving*, using a technique called *brainstorming*. (Note: For a thorough discussion of creative problem solving and brainstorming strategies, please refer to Section 8.6 in Chapter 8.) With this method, a large group of individuals with varying backgrounds and training are brought together to attempt to solve a particular problem. Every idea that is spontaneously contributed from the group is recorded. The basic premise is that no idea is deemed too wild or illogical at this stage. No preliminary judgments are made about any member's idea, and no negative comments are allowed. The goal is to develop a long list of possible alternative solutions to the problem at hand. The group leader should be able to encourage participants to suggest random thoughts and ideas.

Some students may have had the opportunity to engage in brainstorming exercises. Brainstorming can be fun and highly stimulating to the creative process. For example: How many ways can you suggest to use a piece of string and a Styrofoam drinking cup? What could be created from a trash bag that contains some old magazines, tape and a ruler? Think of ways your student organization could earn extra funds for a field trip, etc.

It is conceivable that a brainstorming session would be continued on a second occasion to allow members time to consider other possible options. When the group reconvenes, members may have several new ideas or new perspectives for examining the problem.

At the conclusion of this stage, the group should have a long list of potential solutions to the problem. It is important to remember that at this stage of the process no idea has been eliminated from consideration. Each idea will be evaluated eventually, but it is important to keep all options open at this stage.

Once the project manager is satisfied that all possible solutions have been suggested, the project will likely be cleared for the next stage.

Stage 5: Analyze Potential Solutions

Note: Chapter 8.3 discussed the Analytic Method in detail. That information is directly related to this material in this stage. In the early part of the analysis stage (Phase I), it is important to try to narrow the ideas generated in the brainstorming stage to a few ideas which can be subjected to more sophisticated analysis techniques. This early narrowing could include:

- Examine the list and eliminate duplicates. As discussed earlier, it is important not to create limited categories, but only to eliminate repeated ideas. If two are similar, both should remain at this point.
- Allow the group to ask clarifying questions. This could help identify duplicate ideas.
- Ask the group to evaluate the ideas. The group members can vote for their top three ideas, and those that gain the most votes will be retained for more detailed analysis.

At this point, there will probably be a small number of ideas remaining. These can now be analyzed using more technical and perhaps time-consuming analysis techniques (Phase II).

A variety of individuals should be involved at this stage, but the engineer will be of primary importance. The analysis stage requires the engineer's time and background. It is here that one's training in mathematics, science, and general engineering principles are extensively applied to evaluate the various potential solutions. Some of the techniques in this phase can be time consuming, but a thorough and accurate analysis is important before the project moves to the next stage in the process. For example, if the problem under consideration was the development of an automobile bumper that could withstand a 20-mile-per-hour crash into a fixed object barrier, several forms of analysis could be applied, including:

Common sense: Do the results seem reasonable when evaluated in a simple form? Does the solution seem to make sense compared to the goal?

Economic analysis: Are cost factors consistent with predicted outcomes?

Analysis using basic engineering principles and laws: Do each of the proposed solutions satisfy the laws of thermodynamics? Newton's laws of motion? The basic principles of the resistance of a conductor, as in Ohm's law, etc.?

Estimation: How does the performance measure up to the predicted outcomes? If the early prediction was that some of the possible bumper solutions would perform better than others, how did they perform against the estimate? (Review Chapter 8.3 for a more thorough discussion of Estimation.)

Analysis of compatibility: Each of the possible solutions and their related mathematical and scientific principles are compared to the working criteria to determine their degree of compatibility. For example: How would each bumper solution meet the criteria of being cost-effective? What would be the size, weight, and strength of each of the proposed solutions? How easy would each one be to produce?

Computer analysis techniques: One frequently used method is *finite element analysis*. With this method, a device is programmed on a computer and then numerically analyzed in segments. These segments are then compared mathematically to other segments of the concept. In the bumper crash example, the effects of the impact could be analyzed as a head-on crash and then compared to a 45-degree angle collision or a side-impact crash. As each section is analyzed, the "worst-case scenario" can be evaluated.

Conservative Assumptions: As discussed in Chapter 8.8, this technique can be most useful in analysis. It can build safeguards into the analysis until more data is generated.

After each of the working criteria have been examined and compared to the list of possible solutions, a process eliminates those that have not performed well in the various forms of analysis. It is expected at this point that only three to five options from the original list of prospective solutions will remain. These remaining options will then be reviewed by the project manager, who will likely authorize that the project be cleared for the next stage, assuming that these remaining options meet the working criteria and overall goals. If not, the process will need to be terminated, or return to an earlier stage to correct the problem.

Stage 6: Develop and Test Models

Once each of the prospective solutions has been analyzed and the list of feasible options has been narrowed to a few possibilities, it is time to enter the phase where specific models will be developed and tested. Again, in this stage it is important to have a strong background in engineering coupled with experience and sound judgment. However, this stage will also involve team members who are computer specialists, shop workers, testing technicians, and data analysts.

There are several types of models which are commonly used by engineers and others in this stage. These include:

Mathematical models: Various conditions and properties can be mathematically related as functions and compared to one another. Often these models will be computerized to assist in visualizing the changing parameters in each of the models.

Computer models: There are various types of computer models which can be used. Typically these models allow the user to create on-screen images which can be analyzed prior to the construction of physical models. The most common computer modeling is referred to as CAD (Computer Aided Design) where models are designed and displayed as three-dimensional wire-frame drawings or as shaded and colored pictures. The computer can also be used to control equipment that can generate solid models using techniques such as "stereo-lithography," where quick-hardening liquids are shaped into models or other forms of "rapid prototyping." These on-screen models, or the prototype models they produce, can then be used in the testing process.

Scale models: Typically, these smaller models have been built to simulate the proposed design but may not include all of the particular features or functions. These models are often called prototypes or mock-ups and are useful in helping engineers visualize the actual product. Such models may be used to depict dams, highways, bridges, new parts and components, or perhaps the entire body of a prototype automobile.

Diagrams or graphs: These models provide a tool for visualizing, on a computer or on paper, the basic functions or features of a particular part or product. These diagrams or graphs could be the electrical circuit components of an operating unit of the product, or a visualization of how the components eventually may be assembled.

Once the models have been developed and created, it is time to test each of them. Performing a variety of tests on each of the models allows for comparison and evaluation against the working criteria and the overall goals that have been established. In actuality, tests are done continually throughout a project including early models, prototypes, and the testing of product quality as the product is manufactured or built. However, the results of the testing done in this stage establish the foundation for the decisions that will be made about the future of this project.

Examples of these tests include:

Durability: How long will the product run in testing before failure? If the product is a structure, what is its predicted life span?

Ease of assembly: How easily can it be constructed? How much labor will be required? What possible ergonomic concerns are there for the person operating the equipment or assembling the product?

Reliability: These tests are developed to characterize the reliability of the product over its life cycle, to simulate long-term use by the customer.

Strength: Under what forces or loads is a failure likely, and with what frequency is it likely to occur?

Environmental: Can the parts be recycled?

Quality consistency: Do tests show that product quality is consistent in the various stages? Is the design such that it can be consistently manufactured and assembled? What conditions need to be controlled during manufacture or construction to ensure quality?

Safety: Is it safe for consumer use?

Consistency of testing: This technique examines each of the testing methodologies against the various results obtained to determine the consistency among the various tests. In the bumper design example, the testing methodology might be different for a head-on impact test than for a 45-degree crash test in order to minimize testing inconsistencies.

All of this information will be evaluated by the project manager. If the manager is satisfied that these results consistently meet the working criteria and overall goal, the project will likely be cleared for the next stage.

Stage 7: Make the Decision

At this stage, it is important for the team members to establish a means to compare and evaluate the results from the testing stage to determine which, if any, of the possible solutions will be implemented. The working criteria that have been used throughout the process are the critical factors which will be used to determine the advantages and disadvantages of each of the remaining potential solutions. One of the ways to evaluate the advantages and disadvantages of each of the proposed solutions is to develop a *decision table* to help the team visualize the merits of each. Typically, a decision table lists the working criteria in one column. A second column assigns a *weighted available point total* for each of the criteria.

The team will need to determine the order of priority of each of the criteria. The third column then provides *performance scores* for each of the possible solutions. A sample Decision Table might look like the one that follows.

TABLE 13.1 A Decision Table

Working Criteria	Points Available	#1	#2	#3
Cost	20	10	15	18
Production Difficulty	15	8	12	14
Size, Weight, Strength	5	5	4	4
Appearance	10	7	6	8
Convenient to use	5	3	4	4
Safety	10	8	7	8
Legal issues	5	4	4	4
Reliability/Durability	15	7	9	11
Recyclability	5	4	3	4
Customer Appeal	10	7	8	9
Total	**100**	**63**	**72**	**84**

Based on this sample decision table, it appears that while none of the proposed solutions have scored near the "ideal" model, Solution #3 did perform better than the others. Using this information, the project manager and team leaders would make the final decision to "go" or "no-go" with this project. They may decide to pursue Solution #3, to begin a new process, or even to scrap the entire project. Assuming they decide to pursue Solution #3, the team would prepare the appropriate information for the next stage in the process.

Stage 8: Communicate and Specify

Before a part, product or structure can be manufactured, there must be complete and thorough communication, reporting, and specification for all aspects of the item. Team member engineers, skilled craft workers, computer designers, production personnel, and other key individuals associated with the proposed project must work together at this stage to develop the appropriate materials. Such materials include detailed written reports, summaries of technical presentations and memos, relevant e-mails, diagrams, drawings and sketches, computer printouts, charts, graphs, and any other relevant and documented material. This information will be critical for those who will be involved in determining final approval for the project, as well as the group involved in the final implementation of the product. They must have total knowledge of all parts, processes, materials, facilities, components, equipment, machinery, and systems that will be involved in the manufacturing or production of the product.

Communication is an important tool throughout the design process, but especially in this stage. If team members cannot adequately sell their ideas to the rest of the organization, and be able to appropriately describe the exact details and qualities of the product or process, then many good possible solutions might be ignored. At this stage of the process, it also may be important to create training materials, operating manuals, computer programs, or other relevant resources which can be used by the sales team, the legal staff, and prospective clients and customers.

At this point, if the project manager is satisfied that all necessary materials have been adequately prepared and presented, the project will likely be passed on to the next stage, the implementation stage.

Stage 9: Implement and Commercialize

The next-to-last stage of the design process is critical, as it represents the final opportunity for revision or termination of a project. At this point in the process, costs begin to escalate dramatically, so all serious issues should be resolved by this time.

In addition to the project manager and team leaders, there are a number of other individuals involved at this stage, representing a variety of backgrounds and areas of expertise. While engineers are a part of this stage, many of the activities here are performed by others. Some of those involved in this stage may include:

Management and key supervisory personnel. These individuals will make the ultimate decisions concerning the proposed project. They are concerned with the long-term goals and objectives of the organization, determining future policies and programs that support these goals, and making the economic and personnel decisions that affect the overall health of the organization.

Technical representatives. These may include skilled craft workers, technicians, drafters, computer designers, machine operators, and others involved in manufacturing and production. This group will have primary responsibility for getting the product "out the door."

Business representatives. This group may consist of:

- **human resource personnel** if new individuals must be hired
- **financial people** to handle final budget details and financial analysis questions
- **purchasing personnel** who will procure the needed materials and supplies
- **marketing and advertising staff members** who will help promote the product
- **sales people** who will be involved in the actual selling and distribution of the product

Attorneys and legal support staff. The legal representatives who will handle a variety of legal issues including patent applications, insurance, and risk protection analysis.

If all parties are in agreement that all criteria have been satisfied and the overall goal achieved, the actual production and commercialization will begin. However, there is one remaining stage where the project activities and processes will be evaluated and reviewed. (Note: Some stages of the process may be monitored differently depending on whether the project relates to a one-time product, i.e., a bridge or dam, or an ongoing manufacturing process.)

Stage 10: Perform Post-Implementation Review and Assessment

At this point, it is assumed that the project is in full production. The project manager, key supervisory personnel, and team members who had significant input with the project are gathered together for a final project review and assessment. This stage involves the termination of the project team, since the product is now considered to be a regular product offering in the firm's overall product line. The product's performance is reviewed, including the latest data on production efficiency, quality control reports, sales, revenues, costs, expenditures, and profits. An assessment report is prepared which will detail the product's strengths and weaknesses, outline what has been learned from the overall process, and suggest ways that teams can improve the quality of the process in the future. This report will be used as reference for future project managers and teams to consult.

13.3 A CASE STUDY USING THE 10-STAGE PROCESS

In this section, a simple engineering problem will be discussed, analyzed and evaluated. The intent is to provide you with a better appreciation for the 10 stages involved in getting a product to market, and the team effort required for implementation.

The following example pertains to a product which is currently being developed by several manufacturers. However, the process described in this section is speculative, intended to illustrate an example of how the 10-stage process may be used in the development of this product.

Background

The ABC Engineering Company is a small to mid-sized engineering subsidiary of the XYZ Research Corporation. XYZ is a world-renowned research and development facility in the areas of computer vision, real-time image processing and design, as well as other advanced video applications.

For many years, ABC has been a supplier of products and services for the banking industry. ABC has dominated the field in such areas as security systems, bank cards and PIN identification systems. The parent company, XYZ, an expert in computer vision systems, realizes that the future of face-to-face banking transactions is rapidly changing. The security of electronic transactions is a major problem. The challenge for XYZ and its subsidiary ABC is to develop a new electronic security system to validate banking transactions. They decide to apply this technology to the ever-growing popularity of the automated teller machine (ATM) industry.

Stage 1: Identify the Problem

A project manager is appointed by ABC to oversee the entire program intended to launch a new product and resolve a particular problem. The team from ABC first must identify the problem and attempt to generate sources of ideas to better understand the scope and nature of the problem. In these beginning stages of the design process, the team probably would include engineers, drafters and computer designers, as well as personnel from manufacturing, production, management, sales and finance. In this case, the group discusses the increasing role of automated teller machines in the banking industry. The current method of conducting transactions relies on an individual personal identification number (PIN) and a security access card. An individual approaches an ATM linked to a compatible financial institution, inserts an access card and enters a PIN number. The machine then allows the user to begin a transaction.

The team realizes that there have been increasing problems with fraudulent use of ATMs. It is possible that ABC's experience with the banking industry and the growth in its field of computer vision may provide an opportunity for them to expand their operations into this area.

Following this premise, the group begins to better understand the overall problem and tries to locate sources that will help them to narrow their focus on the perceived problem. They begin this process by working with sources from within ABC. This includes meetings with members of their research and development staff and input from sales and marketing personnel who are in constant communication with customers. In addition, managers and supervisors may have ideas regarding the expansion of current product offerings through modifications and upgrades. External sources of information include prospective banking industry customers, the Internet, libraries, trade shows, conferences and competition.

This information helps them to more narrowly define the problem. The group may develop a preliminary statement of the problem to guide them at this early stage of the process. Such a problem statement might be: "Eliminate fraudulent access to ATM accounts."

Once the project manager is satisfied that all necessary issues have been addressed at this stage, the project is passed on to the next stage.

Stage 2: Define the Working Criteria or Goals

Using the preliminary problem statement as a guideline, the team develops a list of working design criteria which will allow them to evaluate various alternative solutions. Based on the information used to identify the problem, the group establishes the following preliminary working criteria:

Cost: The machine must be affordable to banks and other buyers of ATM machines.
Reliability: The accuracy of the new solution must be near 100%.
Security: The security for the cardholder must be fraud-proof.
Technical feasibility: The solution must be easily produced with current technology.
Convenience to the customer: The solution should be easy to use.
Acceptance: The solution must not be deemed too peculiar, or the public may not use it.
Appearance: The new solution must have a pleasing appearance.
Environmental: The new solution and its parts should be recyclable.

After thorough discussions and meetings, the team decides to re-examine their preliminary problem statement. Following additional discussion and review of the working criteria, the project goal becomes: Design and implement an alternate method of identifying a cardholder at an ATM, while focusing on reducing the acceptance of fraudulent entries.

Establishing this goal for the project provides a means of evaluating, monitoring and possibly changing the focus of the process as it evolves through the 10 stages. For the project manager, these criteria and the goal become a "checkpoint," or a means for assessing the project's progress periodically, and helps them determine when the project is ready to move to the next stage.

Stage 3: Research and Gather Data

Now that the goal and working criteria have been developed, it is time to collect all information relevant to the problem. Specifically, the team needs additional information regarding the current PIN method of identification and data on ATM fraud. They will also want to investigate new developments in the field, including information available on current products already on the market. They may look at this information to determine whether there are possibilities for redesign or modification of their current product lines, or to explore the feasibility of developing a new product from scratch. They also may want to explore current public opinion and available market research.

Sources for this information may include the Internet, libraries, newspapers, magazines, journals, trade publications and patent searches. An example of a Web search on ATMs might look like that of Figure 13.1.

The team should maintain accurate notes, records and files for the information they collect. Sketches and diagrams of similar products or competitors' models may assist them when they begin to develop their solutions.

The project manager determines that all necessary conditions have been met and gives clearance for the project to move to the next stage.

Stage 4: Brainstorm/Generate Creative Ideas

Using the information that has been collected and studied, the team decides to re-evaluate, prior to the brainstorming session, the list of preliminary working criteria and to assign a weighted percentage of importance to each of the factors. The team decides to eliminate the criteria of "appearance" and "environmental" from their list. The resulting working criteria and their respective assigned weights are as follows:

Cost	=	10%
Reliability	=	25%
Security	=	10%
Feasibility	=	15%
Convenience of use	=	20%
Public acceptance	=	20%

The team then enters into a series of creative problem-solving/brainstorming sessions. At this stage the group consists of engineers, sales people, managers, supervisors, research and development staff members, production workers, computer specialists, and a few selected clients from the banking industry. Their assignment is to develop as many options as possible for uniquely identifying ATM customers. They are instructed not to make any judgments or negative comments on any of the proposed ideas. No ideas are to be considered ridiculous at this point in the process. The list they develop follows.

Creative Brainstorming

Ways to uniquely identify ATM customers:

1. Modify the current use of PIN access codes: adjust current access methods to tighten security.
2. Individualized signature verification: An ATM attendant would match the cardholder's signature with one written "on the spot" by the person standing at the machine.
3. Machine recognition signature verification: A scanner inside the ATM would match the cardholder's signature with one written "on the spot" by the person standing at the machine.
4. Voice recognition: The ATM would be equipped with a speaker phone so the person could speak their name and an internal computer could match the voice of the cardholder to a recording on file.
5. Speech pattern recognition: Similar to above, but the person would speak a complete sentence or longer phrase that could be analyzed by voice characteristics and pattern of speech.
6. Fingerprint match: The ATM would have a fingerprint pad where the customer would press on the pad and a computer would compare the print to one on file.

7. Blood match: The ATM would have some type of device to prick the customer's finger for a small blood sample which could instantly be analyzed by an internal lab testing machine for DNA compatibility.

8. Hair sample match: The individual would submit a hair sample, with root attached, into the machine. As above, it could instantly be analyzed by an internal lab testing machine for DNA compatibility.

9. Eye iris match: The ATM machine would be equipped with a camera which would take a picture of the iris structure, and perform a comparison to a profile on file.

10. Breath analyzer match: The individual would breathe onto a sensitized glass plate and the results would be processed into an internal computer that would analyze a match with that of the card holder.

11. Dental identification match: The ATM machine would be equipped with a sanitized plate that the individual would bite onto and a computer sensor would read the patterns and analyze for a match.

At this point, the project manager is satisfied that a sufficient number of ideas have been uncovered. Authorization is given for the project to move to the next stage where all these potential solutions will be analyzed.

Stage 5: Analyze Potential Solutions

The team now begins to analyze each of the 11 ideas and compare them to the working criteria. In the early part of analysis (Phase I), it is important to try to narrow the ideas to a few which can be subjected to more sophisticated analysis techniques (Phase II). It is important to remember that there are many different analysis techniques that can be used by the team. The results of this analysis are as follows:

Analysis of Alternative Ideas

1. *Modify the current use of PIN access codes:* adjust current access methods to tighten security.

   a) Similar to current solution
   b) Not reliable or secure
   c) Neutral cost

2. *Individualized signature verification:* An ATM attendant would match the cardholder's signature with one written "on the spot" by the person standing at the machine.

   a) High cost
   b) Not customer-friendly
   c) Difficult to implement

3. *Machine recognition signature verification:* A scanner inside the ATM would match the cardholder's signature with one written "on the spot" by the person standing at the machine.

   a) Technology is currently available
   b) Public would probably accept this method
   c) Higher cost to implement
   d) Long-term reliability is untested

4. *Voice recognition:* The ATM would be equipped with a speaker phone so the person could speak their name and an internal computer could match the voice of the cardholder to a recording on file.

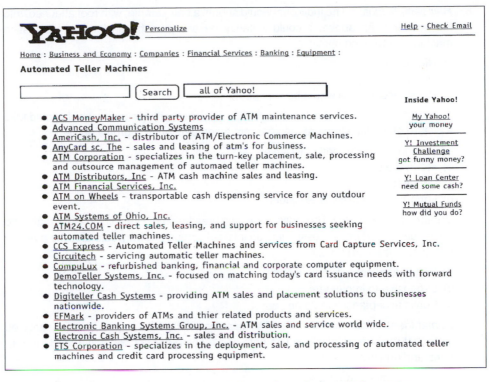

Figure 13.1 Sample Internet Web search results for ATM machines.

    a) Technology is currently available

    b) Higher cost

    c) Easy to use

    d) Long-term reliability is untested

5. *Speech pattern recognition:* Similar to above, but the person would speak a complete sentence or longer phrase that could be analyzed by voice characteristics and pattern of speech.

    a) Expensive

    b) Technology not as well developed

6. *Fingerprint match:* The ATM would have a fingerprint pad where the customer would press on the pad and a computer would compare the print to one on file.

    a) Technology is currently available

    b) Moderate cost

    c) Easy to use

    d) Long-term reliability is good

7. *Blood match:* The ATM would have some type of device to prick the customer's finger for a small blood sample which could instantly be analyzed by an internal lab testing machine for DNA compatibility.

    a) High cost

    b) Good reliability

    c) Completely unacceptable to the public

8. *Hair sample match:* The individual would submit a hair sample, with root attached, into the machine. As above, it could instantly be analyzed by an internal lab testing machine for DNA compatibility.

   a) High cost
   b) Good reliability
   c) Public acceptance would be very poor

9. *Eye iris match:* The ATM machine would be equipped with a camera which would take a picture of the iris structure, and perform a comparison to a profile on file.

   a) Technology is currently available
   b) Moderate cost
   c) Very easy to use
   d) Long-term reliability is excellent

10. *Breath analyzer match:* The individual would breathe onto a sensitized glass plate and the results would be processed into an internal computer that would analyze a match with that of the card holder.

    a) Technology not fully developed
    b) Expensive to implement
    c) Public acceptance would be poor

11. *Dental identification match:* The ATM machine would be equipped with a sanitized plate that the individual would bite onto and a computer sensor would read the patterns and analyze for a match.

    a) Technology not developed
    b) High cost
    c) Public acceptance would be extremely poor

After each of the working criteria have been examined and compared to the list of possible solutions, those that have not performed well in the various forms of analysis will be eliminated. At this point it appears that only three to five options from the original list of prospective solutions will survive. These options are now reviewed by the project manager, who determines that they satisfy the working criteria and overall goal, so the project is authorized for the next stage.

Stage 6: Develop and Test Models

At this stage, the team will take those proposed solutions that survived the analysis stage and develop models for testing and further evaluation. Based on the results from the analysis stage, the team has selected three proposals for further study: #4 (voice recognition), #6 (fingerprint match), and #9 (eye iris match).

Initially, several different diagrams and sketches of each idea concept are developed and reviewed. The team has many different styles, options and units to consider. They decide to develop different models of each, including computer-generated models and actual prototypes constructed by craft workers in the model shop. These models are shown in the photographs in Figures 13.2–13.4.

Once the various idea concepts have been created into prototypes, the actual testing begins. Based on research and information that was collected in stage #3, and the information needed for comparison to the working criteria, the team decides to concentrate on several specific tests including technical feasibility, quality performance, reliability and ease of use. The results of the tests indicate that proposed solution #9 (eye iris identification) performs very well, as do the other two. Data should be generated for all three proposals. For solution #9, the following data were generated:

Technical feasibility test: It was found that the current technology of specialty cameras and computer systems can identify a human by scanning the iris of the eye for over 400 identifying features and is able to match them to a large database.

Quality performance test: Eye scans from 512 individuals who had their iris features stored in the database system were taken. The program correctly identified all those who participated. Those whose records were not on file were correctly rejected.

Reliability test: The error rate on the eye scan reliability test was 1 in 131,578 cases. This is by far the lowest error rate among any other form of physical characteristic testing, including fingerprint and voice recognition. In addition, the iris scan provides long-term reliability, since the iris does not significantly change with age as do other body parts.

Convenience of use test: The technology is available so that a hidden camera can spot a person approaching and zoom in on the right iris. As the person draws closer, security access control can make a positive identification in a few seconds.

Cost estimation: While not a test per se, enough data have been collected to estimate the cost of this system at $5,000 per unit. While the initial estimate is high for machines which typically sell for $25,000 to $30,000, it is predicted that the cost will drop as units are mass produced.

Once all the test results have been gathered, all the information and data is evaluated by the project manager. It is determined that these results consistently meet the working criteria and overall goal, so the project is cleared for the next stage, where a decision will be made.

Stage 7: Make the Decision

A final decision table is developed and the scores from the tests and other sources of information are recorded. In this particular case, the final table could look like this:

TABLE 13.2 The Decision Table

Working Criteria	Points Available	#4	#6	#9
Cost	10	6	8	8
Reliability	25	20	22	24
Security	10	9	7	9
Technical Feasibility	15	15	14	15
Convenience of Use	20	15	16	18
Public Acceptance	20	16	17	18
Total	**100**	**81**	**84**	**96**

The results confirm that the eye scan identification system performed well. Using this information, the project manager decides to move ahead on solution #9 and the team prepares the appropriate information for the next stage in the process.

Stage 8: Communicate and Specify

Team members including engineers, skilled craft workers, computer designers, production personnel, and other key individuals associated with the proposed project work together in this stage to develop the appropriate materials. The team members prepare the necessary detailed written reports, summaries of technical presentations and memos, relevant e-mails, diagrams, drawings and sketches, computer printouts, charts, graphs, and all other relevant and documented material on the eye scan identification system. They provide complete and detailed information on all parts, processes, materials, space, facilities, components, equipment, machinery and systems that will be involved in manufacturing the product. It is also important to create training materials, operating manuals, computer programs, and other relevant resources which can be used by the sales team, the legal staff and prospective clients and customers.

The project manager is now satisfied that all necessary materials have been adequately prepared and presented. Approval is given for the project to be passed on to the implementation stage.

Stage 9: Implement and Commercialize

At this point in time, the management of ABC has given approval for the eye scan identification project to move forward into production. Questions concerning overall costs, financial and labor commitments have been resolved. The finance department has developed a project budget. The purchasing department has begun the process of obtaining bids for the needed parts and supplies. The legal staff is resolving final legal issues including patent applications and copyright materials.

A group of engineers has been selected to oversee the production start-up, which includes obtaining necessary production space, facilities, equipment, personnel, and production timetables. Once most of this is in place, a pilot production process begins.

The marketing, advertising, and public relations staffs begin to contact prospective clients and customers to promote the eye scan ATM system. The sales people involved in the actual selling and distribution of the product are trained to work with clients and are given appropriate promotional literature including training and operations material.

All the parties are in agreement that the criteria have been satisfied and the overall goal achieved. The actual production and commercialization now begins. However, there is one remaining stage where the project activities and processes will be evaluated and reviewed.

Stage 10: Perform Post-Implementation Review and Assessment

At this stage, it is assumed that the eye scan ATM system is in full production and commercialization. The project manager, key supervisory personnel and team members who had any significant input to the project gather together for a final project review and assessment. This stage involves the termination of the project team, as the product is now considered to be a regular product offering in the firm's overall product line. The eye scan ATM system's performance is reviewed, including the latest data on production efficiency, quality control reports, sales, revenues, costs, expenditures and profits. An assessment report is prepared which will detail the product's strengths and weaknesses, a report detailing what was learned during the process, and suggestions for ways that future teams can improve the quality of the process. The working criteria are examined and suggestions for future groups will be included in this report. This report is used as another reference resource for future project managers and teams to consult.

Note: The Iris Identification System is currently being piloted in several British and Western U.S. locations. The information presented in this case study is a simulated application of the 10-stage design process.

Figure 13.2 Idea Concepts for Fingerprint Identification.

Idea #1: Fingerprint Scanner with Monitor, Keypad

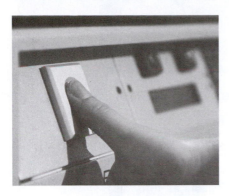

Idea #2: Fingerprint Scanner, Small

Idea #3: Fingerprint Scanner, Larger

Idea #4: Fingerprint Scanner System

Figure 13.3 Idea Concepts for Eye Iris Identification.

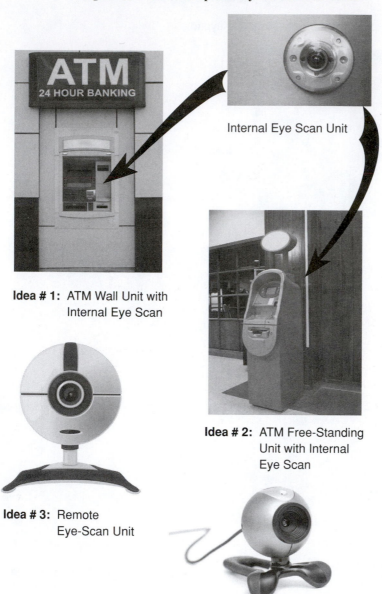

Internal Eye Scan Unit

Idea # 1: ATM Wall Unit with Internal Eye Scan

Idea # 2: ATM Free-Standing Unit with Internal Eye Scan

Idea # 3: Remote Eye-Scan Unit

Idea # 4: ATM Top-Mounted Eye Scan Unit

Figure 13.4 Idea Concepts for Voice Recognition Identification.

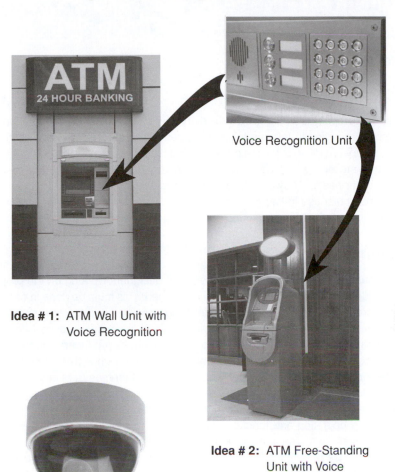

Voice Recognition Unit

Idea # 1: ATM Wall Unit with Voice Recognition

Idea # 2: ATM Free-Standing Unit with Voice Recognition

Idea # 3: Ceiling or Wall-Mounted Voice Recognition Unit

13.4 A STUDENT EXAMPLE OF THE 10-STAGE DESIGN PROCESS

by Amy Cole

Background

The ABC Engineering Company is a mid-sized subsidiary of the XYZ Research Corporation. The XYZ Research Corporation is famous for its progress in the area of athletic training research. ABC is a national supplier for the in-line skating industry. They realize that the industry is quite large, but is stabilizing. In order to keep the industry growing, they need to keep up with current consumer needs. They decide to make teenagers the focus of their studies. College students in particular use in-line skates on a regular basis as one of their primary means of transportation and, therefore, account for a large portion of the buying population.

Stage 1: Identify the Problem

A specific problem must be identified for the engineers and designers working on the project to have a clear understanding of what must be accomplished. This must be done in order to begin the design process. Those involved with the design project need to know specifically what to focus on. The team working on the early portion of the design process would probably involve engineers and drafters as well as manufacturers, producers, and marketers. The team must include members with a technical background since they are knowledgeable in the specifications and restrictions of the design. The other members of the team provide an alternate view on the project. They are responsible for answering questions pertaining to the appearance and customer satisfaction of the design. This group will be discussing, specifically, the operation and function of current in-line skate models. Currently, students use in-line skates as a means of getting to and from class, across campus, and even across town. When used in this way, students must deal with carrying their backpacks and school supplies, along with the hassle of carrying their shoes while skating. Students must also deal with the inconvenience of having their skates with them once they get to their destinations.

Realizing this, the team takes on the problem of inconvenience that comes along with the benefits of in-line skating. Considering the team's previous work with the XYZ Corporation, they should be reasonably well-suited to tackle this problem. Team members also should review old products to see if their uses could be modified to fit the project at hand. External sources could include prospective customers' needs, competition advances, the Internet, libraries, and conferences.

This should help the group to sharpen their focus on the problem and hopefully to create a preliminary problem statement. A good statement for such a problem might be: "Eliminate the inconvenience associated with in-line skates."

Stage 2: Define the Working Criteria or Goals

The team now must develop basic criteria for the design to follow. In this case, the alternative solutions should be evaluated using criteria similar to the following:

> **Cost:** The solution must be affordable to the average customer. This includes college-age and younger students.
> **Reliability:** The solution's reliability should be as close to 100% as possible.
> **Weight:** The solution should be light since it will involve mobility.

Customer Convenience: The solution should be easily operated.
Technical Feasibility: The solution needs to be produced easily and quickly.
Environmental: The components of the solution should be recyclable.
Appearance/Acceptance: The solution should appeal to the customer and make them feel comfortable while using it.

At this point, the team should discuss and re-evaluate their preliminary problem statement. A goal statement should be created in order to keep track of the progress being made and to keep the designers on the right track. A goal statement for the current project might be: "Design and implement a device to eliminate the hassles associated with in-line skating, thereby making them more convenient and increasing their usage."

Stage 3: Research and Gather Data

At this point in the design process, it is time for the team to collect information related to the problem. They need to research public opinion and market information to make sure they keep the customer in mind. They also need to research information regarding the basic operation of in-line skates and how they can use new developments as part of the solution. Researchers must decide whether to redesign an old project or start over with a completely new one.

Sources for research include libraries, the Internet, journal articles, patent searches, vendor catalogs, and trade publications. The group should produce diagrams of their preliminary ideas, along with notes and sketches of competitors' products. At this stage, it is left up to the project manager to decide when to move on to the next stage.

Stage 4: Brainstorm/Generate Creative Ideas

The team must stop now and revise their list of criteria before beginning the brainstorming session. They should review data collected and assign weighted percentages of importance to the factors they keep on the list. Assume that they decide to remove "environmental," but feel that the rest are necessary. Weights are assigned to the remaining criteria as follows:

Cost	= 15%
Weight	= 15%
Feasibility	= 15%
Convenience of use	= 25%
Appearance/acceptance	= 10%

The team is now prepared to start the brainstorming process. The group is composed of engineers, marketers, drafters, research and development members, managers, and select others who can offer a different perspective on the problem. They create as many alternatives to the current design as possible. At this stage, no negative comments or judgments should be made on any idea. The following list is produced.

Creative Brainstorming

Ways to increase the convenience of in-line skates:

1. Make the wheels slide out of the track. This makes storage of the skates easier.
2. Make the wheels flip to the sides. This creates a flat base to walk on with the wheels on the side.

3. Make the whole base of the in-line skate detachable, leaving a shoe suitable for walking.
4. Design a bag especially for carrying the skates and other accessories. Shoes can be carried in the bag during use of the in-line skates.
5. Develop a cover to go over the wheels. This allows the skates to be walked on.
6. Make the wheel assembly strap on to the skater's everyday shoes.
7. Design a handle to make the in-line skates easier to carry.

At this point, it is left again to the project manager's discretion whether there is a sufficient number of ideas. In this case, the manager is comfortable with the results and allows the team to move ahead in the design process.

Stage 5: Analyze Potential Solutions

Having already brainstormed new ideas and revised the list of criteria, the team analyzes each of the ideas generated. While evaluating, the team should keep in mind that they will need to narrow the list for future steps. The analysis proceeds as follows:

1. *Removable wheels.* Wheels slide out of the track to make storage easier.

   a) Easy to use
   b) Similar to current design
   c) Convenience level not as high

2. *Flip wheels.* Wheels flip to one side to make a base for walking.

   a) Eliminates need for shoes
   b) Moderate cost
   c) Technology is available

3. *Detachable base.* The base detaches from the skate, leaving a normal shoe.

   a) Very convenient
   b) Moderate cost
   c) Easy to use

4. *Carrying case.* A bag is designed to carry in-line skates, shoes, and accessories.

   a) Low cost
   b) More hassle/inconvenience
   c) Reliable

5. *Wheel cover.* A cover is designed to fit over the wheels of the in-line skate.

   a) Convenient
   b) Low cost
   c) Moderately reliable

6. *Strap on wheels.* The wheels of the skate strap on to the skater's shoe.

   a) Not reliable
   b) Moderate cost
   c) Not as safe

7. *In-line skate handle.* The in-line skates have a handle to make carrying them easier.

   a) Low cost
   b) Reliable
   c) Hassle involved

Figure 13.5 Option #2: Flip wheels.

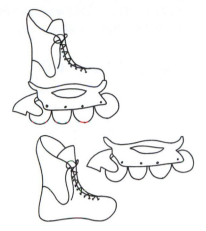

**Figure 13.6 Option #3:
Detachable base.**

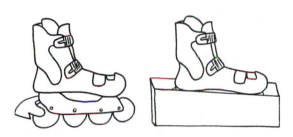

Figure 13.7 Option #5: Wheel cover.

Figure 13.8: Option #7: Handle.

The analysis is compared to the working criteria. The team eliminates solutions that do not meet the standards and retain three to four solutions to develop further. The project manager reviews the remaining solutions and gives the team permission to continue.

Stage 6: Develop and Test Models

At this point in time, the team must develop models for the remaining solutions. The models will undergo testing and more extensive evaluation. The team, after reviewing the analysis and criteria, decides to continue the process with option #2 (wheel cover), #3 (detachable base), #5 (wheel cover), and #7 (in-line skate handle).

Mathematical and computer models or diagrams should be produced for each option. The team can choose from a number of programs on the computer which offer many different options for creating the sketches. Actual working prototypes should be developed for each solution.

Once the prototypes have been developed, the team of engineers and technicians can initiate the testing process. The team performs specific tests relating to durability, reliability, and quality. The project manager reviews the data collected from the tests and gives his clearance for the next stage.

Stage 7: Make the Decision

To aid the team in the decision-making process, a final decision table is made. For this case, the table could be as follows:

Working Criteria	Points Available	#2	#3	#5	#7
Cost	20	15	16	18	18
Reliability	15	11	15	7	15
Weight	15	12	14	14	11
Convenience	25	18	25	15	18
Technical Feasibility	15	15	15	15	15
Appearance	10	9	10	5	5
Totals	**100**	**80**	**95**	**73**	**82**

The tests show that solution #3, the detachable base model, is the preferred solution. All previous information confirms this and the project manager agrees to use solution #3 to finish the design process. The team organizes their data and prepares to move on in the design process.

Stage 8: Communicate and Specify

Communication plays a critical role in engineering and really comes into play during this stage of the design process. At this point, the team consists mostly of engineers, design technicians, and production members. This is the time when all documented material is presented. Typical presentations include written reports, relevant e-mails and memos, diagrams, computer printouts, charts, graphs, and summaries of technical presentations. Everything involved in the production of in-line skates including parts, processes, materials, facilities, and machinery is described in detail in the presentations. It also is important that the team communicate how the product works through instruction manuals and videos and training materials. Again, the project manager clears the project to continue to the next phase.

Stage 9: Implement and Commercialize

The detachable-base in-line skates now have undergone all of the necessary steps to begin production and implementation. At this point, the team usually does not consist of as many engineers as in previous stages. The team usually includes members involved in the managerial, technical, business, and legal aspects of the product. The details, such as the budget and price, costs, and labor commitments, have been decided. The technical members—the craft workers, CAD operators, machinists, and technicians—handle the operations and production of the in-line skates. Select engineers must arrange production space, facilities, and start-up equipment.

The business portion of the team includes human resources personnel, finance, purchasing, marketing, advertising, sales, and technical/customer support. It is their responsibility to use the information collected and to make the in-line skates appealing to the customer. They need to figure out how to sell the product.

Everyone works together to ensure that all of the criteria are met and production runs smoothly. Production now begins, leaving only a final stage remaining in the design process.

Stage 10: Perform Post-Implementation Review and Assessment

At this point, the detachable-base in-line skates are in full production. All of the members of the team must meet one last time to review the data and the performance of the in-line skates. The production efficiency, sales, revenues, costs, expenditures, profits, and quality control reports should be reviewed. This meeting will be the last time the group members meet as a team for this project, as it is now considered a part of the firm's production line. The group produces a final report which can be referenced by future project teams.

Internet contact: www.rollerblade.com, Benneton Sports System, Rollerblade, Inc., 2000.

EXERCISES AND ACTIVITIES

13.1 Disassemble one of the devices listed below, and put it back together. Sketch a diagram of all the parts and illustrate how they fit together to make the device operate. List at least three ways you think the design could be improved. Choose one of the following devices: a) flashlight, b) lawn sprinkler, c) sink faucet, d) stapler, e) toaster, f) computer mouse.

13.2 Prepare a list of questions that should be resolved in identifying each of the following problems:
a) Develop an improved manual gearshift for a mountain bike.
b) Develop a better braking system for in-line skates.
c) Develop a recliner chair that incorporates six built-in surround sound speakers in the unit.
d) Develop a hands-free flashlight.
e) Develop a theft-proof bicycle lock.
f) Develop a baseball cap with wireless stereo headphones.
g) Develop a secure storage area on a bike for a helmet.
h) Develop a new idea to reduce overcrowded parking on campus.
i) Develop a hands-free cellphone that can be used when doing activities.
j) Develop a shockproof cellphone carrying case.

13.3 Develop a list of working criteria that could be used in deciding whether to:
a) Accept a co-op job offer from Company A or Company B.
b) Study overseas for the Fall semester or remain on campus.
c) Buy a new car or repair your old one.
d) Purchase a desktop or laptop computer.
e) Change your major or remain in engineering.
f) Purchase a new computer or upgrade your current model.
g) Live in the dormitory or lease an off-campus apartment next year.
h) Pledge a fraternity/sorority.

13.4 Identify five product, structure, or system designs you think can be improved. Pick one and write a preliminary problem statement for the engineering design process.

13.5 Using an item from your list in Exercise 13.4, develop a list of reference materials that would be used in developing possible solutions to the problem. Provide specific examples.

13.6   Get together with three other classmates and brainstorm at least 30 ways to use one of the following objects:

    a) Two-foot length of string

    b) Ping-pong ball

    c) Three plastic pop bottles

    d) Page of notebook paper and a 2-inch piece of tape

    e) Old telephone directory

    f) Trash bag

    g) Deck of playing cards

    h) Yo-yo

    i) Frisbee

    j) Metal coat hanger

    k) Empty plastic milk container

    l) 15 paper clips

    m) 2 tubes from paper towel rolls and 2 pages of newspaper

    n) Deflated balloon

    o) Bag of marshmallows

    p) Newspaper and 12 inches of masking tape

13.7   Using the information you developed in Exercise 13.4, prepare a decision table for three possible alternative solutions.

13.8   Read a current newspaper or magazine article (possibly on the Web) that describes and discusses a new product, device, or system. Prepare a three- to four-page essay which analyzes and explains, in detail, each of the stages of the design process that were probably involved in the engineering of this product or device. Apply each stage of the 10-stage design process discussed in this chapter to your product or device, and use specific examples to support your statements. Make sure you list your sources in the body of the paper or in a bibliography.

13.9   Put together a team of five students from your class. Using the 10-stage design process, have your team develop a portable, garage-like covering for a bicycle that can be stored somewhere on the bike when not in use.

13.10   Put together a team of four or five students from your class. Using the 10-stage design process, build the tallest possible tower that can support a 12-oz can of soda/pop. **Materials:** one roll of masking tape, one package of straws, and one can of soda/pop. **Constraints:** Your tower must be free standing. The base must fit on a plate. Your tower must stand for 5 seconds under load for measurement. After the design is complete, 30 minutes will be allowed for tower construction.

13.11   Put together a team of four or five students from your class. Using the 10-stage design process, build a bridge that spans three feet between supports that contact the ground. **Materials:** one roll of duct tape, ten feet of rope, a 40 inch by 75 inch piece of cardboard, and 14 cardboard slats. After the design is complete, 40 minutes will be allowed for tower construction and testing. The test: one of your team members must walk heel-to-toe across the bridge.

13.12   Put together a team of five students from your class. Using the 10-stage design process, have your team develop a carrying mechanism for Rollerblades and/or your shoes when one is not in use.

Chapter 14

Communication Skills

14.1 WHY DO WE COMMUNICATE?

What is the purpose of communication? Well, there are many purposes, actually. For starters, it might surprise you to consider that we are often the most important readers of our own writing. What we write should be valuable to us as a future resource. If you keep that in mind, written communication will be easier and more rewarding.

Of course there are a number of additional reasons for communicating with excellence: obtaining good grades, creating effective reports, convincing others of your perspective—both on a professional and personal basis. Each will require that you assume a different mindset.

Effective communication:

- transfers vital information
- provides a basis for judging your knowledge
- conveys your interest and competence
- increases the knowledge of others
- identifies to you the gaps in your information
- allows you to carry out desired changes

Engineers must possess the technical skills to complete engineering analysis, evaluation, and design. Engineers must possess strong communication and teaming skills. One of the biggest complaints of employers of entry-level engineers is lack of communication and teaming skills. Possessing these strong skill sets will enable you to excel in engineering. Most engineers in industry, research, and government complete design work in collaborative teams and routinely communicate their discoveries and progress to peers, managers, other engineers, and the community.

In this chapter, you will learn about oral and written communication skills. First, two golden rules of communication for engineers:

- Brevity is best.
- Using a shorter, simpler word is typically preferable to a longer, more difficult word.

As you consider your own communication, there are a number of other things to keep in mind:

Is the language you are using suited to your audience?

- Do you know how to communicate effectively to your audience?
- Do you need to expand your communication skills to communicate more effectively to your audience?

Is what you are saying clear to your readers?

- Will your readers draw the same conclusions you draw?
- Do you use words, phrases, or ideas that could lead to multiple interpretations?

Do you know your audience?

- Does your audience have any unique preferences?
- Do they have any hidden expectations?
- How do they feel about the things you are communicating?

Do you know how your audience perceives things?

- Do all engineers see things the same way?
- If you are addressing an audience of non-engineers, how might they interpret things differently?

Typically, you will be communicating with people who are making decisions concerning your work. These decision-makers have minimal time, and you want to have a maximal impact. While other types of communication place great emphasis on complex language and nuance, technical communication is all about getting it said and done quickly and simply.

14.2 ORAL COMMUNICATION SKILLS

In an engineering context, you will make presentations to many audiences. The purpose of this section is to share information on planning and executing an effective oral presentation.

There are two types of presentations: formal and informal. Formal presentations involve a preset time and date and usually involve a reasonable amount of preparation. They may be made to colleagues, to expert panels, at design competitions, at the conclusion of a cap-stone design experience, at conferences in which engineering information is shared, or in the workplace when you communicate your information to your design team, management, the public, or funding and decision makers. For formal presentations, you will likely use a computer program such as PowerPoint; you might hand out printed copies of your slides. Formal presentations usually require formal business attire.

Informal presentations tend to be shorter, and do not require as much preparation time or formal business attire. You do not usually need to prepare a PowerPoint presentation either. Sometimes, you will have to make impromptu, or spontaneous presentations. You might make such presentations in a business context, technical meeting, or at a meeting involving a design team.

A good presentation renders a service to your audience. A presentation is an opportunity to take the audience from where they are to where you want them to be. The best presentations will do the following:

- give valuable information the audience probably would not have had otherwise.
- be in a form that the audience can put into immediate use.
- motivate and inspire the audience to want to put your presented information into immediate use.

With proper preparation in advance, you can concentrate on how you deliver information and on audience impact during your presentation. More specific information regarding these sections is included below.

For formal presentations, you will likely use an overhead projector or a computer presentation program such as PowerPoint. Each PowerPoint image or each overhead is referred to as a slide (or an overhead) in this discussion.

If You Are Nervous

Being nervous about doing a presentation is normal. In fact, the two most common fears of people living in the United States are public speaking and flying in airplanes. The following list of techniques is intended to help you combat your nervousness.

- Breathing is important; when you get nervous, your respiration rate tends to speed up and your body can go into adrenaline overdrive (a fight or flight mechanism). Controlling your breathing is one way to stay calm.
- Take three deep breaths before you start.
- Practice square breathing, which is inhaling seven counts, holding seven counts, exhaling seven counts, and holding seven counts.
- Visualize yourself successfully presenting your talk.
- Work with your nervousness by tailoring the talk to your strengths.

Most people are most nervous during the initial stages of speaking. Use one or more of these tips, which can be used throughout the presentation.

- *Use humor:* This makes the atmosphere less formal. Start your presentation with a joke, especially one that pertains to your presentation.
- *Use props:* It gives you something to do with your hands. A show and tell will interest the audience.
- *Use your nervous habit:* Work with that habit to your audience's benefit. If you like jingling change (coins) in your pockets, pull something visual out of your pocket to start your presentation.
- *Use trivia:* Focus your audience's attention or teach the audience something interesting that pertains to your talk.

Have your presentation ready at least a day in advance. Being finished usually helps to cut down on your nervousness. Student Alicia Abadie says, "The more you procrastinate, the more nerve-wracking it can be."

Before the Presentation

Preparation is the key to successful oral presentations. To prepare for a presentation, you need to do the following:

- Identify your purpose and audience
- Organize the information
- Gather the information you need to present
- Create presentation material with maximal audience impact
- Practice your presentation after it is prepared

Identify Your Purpose and Audience

To give a presentation with maximum impact, know your audience:

- How many people will you be speaking to?
- How well do you know your audience? If you are speaking to a small audience with whom you are familiar, you can use their names during your talk. This focuses the attention of your audience.
- What is the background of your audience (age, professional background, personal background, etc.)? You must tailor your talk, using appropriate vocabulary and information the audience will understand and put into action.
- What is the audience expecting from you in terms of information, tone, level of formality, etc.?
- What does the audience know about your topic? You should know what they know and concentrate on the information that they do not know. The idea is to avoid being either repetitive or out of the knowledge base of your audience. If you are not sure what the audience knows, ask your contact person before your presentation, so you can prepare accordingly.
- How long is your allotted time to speak? Speak for less than your limit to give the audience a chance to ask questions.

You need to keep the purpose of your talk in mind when you are preparing. In what capacity are you speaking? For example, are you giving a report to your peers? Are you presenting your final design and recommendations from your project? Focus the information in your presentation around your purpose, and your audience. This will ensure that your audience is in the best position to take your information and act accordingly.

Organization

You should organize your presentation so the audience can digest the delivered information easily. The following tips will help you to organize your information.

- The key to a well-organized speech is realizing that even the most engaged audience will not pay attention all the time. Thus, you should limit your speech to three to five major points. Your talk will be successful if the audience understands and retains the information. How much your audience understands is more important than how much information you present.
- *Have a one-page summary.* For a presentation longer than five minutes, an oral one-page summary at the beginning of the talk is useful. This lets the audience know what is coming and in what order, and will help your audience to follow your presentation.
- Use the power of repetition.
- Repeat your main ideas; the audience will remember only three to five of the ideas you present.
- Repeating key points makes it easier for the audience to stay with you.
- Repeat your main message at least three times, in three different ways.

Did you know that people speak 100-125 words per minute, but the human brain can handle up to 500 words per minute? This is how you can pay attention to what is going on when someone is speaking while your mind is on something else.

- An adage from an accomplished speaker: "Tell them what you're going to tell them, tell them, then tell them what you told them."
- Provide a written supplement if appropriate. A written supplement helps people to focus on what you are saying instead of writing it down. This is a security thing; if people already have the information in front of them, they are more likely to listen and focus on your presentation. The more ways in which you present your message, the more likely the audience is going to remember it.

In terms of organizing the information itself, a sample outline of a presentation could be as follows:

- *Introduction*: Who you are and why you did this
- *Need*: Why your engineering design is needed
- *Solution*: You have the design that addresses the need, and here it is. Include major design features and why you did what you did (explain how your design specifically addressed the needs).
- *Bottom line*: What are the benefits? How much it will cost?
- *Conclusion*

Gather the Information You Need to Present

Although this point seems self-explanatory, remember these few tips when gathering information for an oral presentation:

- The information you include in your presentation should pertain to three to five main points you plan to share with the audience. This information will support your main points and will help with the organization of your presentation.
- Make use of graphics, pictures, and photographs because they communicate a wealth of information in a concise way.
- Make sure that you do not include too much or too little information in your presentation. You should have approximately one slide for every minute that you speak. Some speakers can include more slides, but you should not exceed one slide per 30 seconds. Thus, if your presentation is ten minutes long, you should not have more than 20 slides.

Create Your Presentation for Maximal Audience Impact

Your impact can be visual (what people see), auditory (what people hear), or kinematic (what people do or hands-on experience). Remember, keep it simple. Your audience will stay with you if you use accessible language, and the audience will retain information better if you provide information in more than one of the ways mentioned previously.

The following tips will assist you in maximizing the visual impact of your talk:

- Present one central idea per slide.
- Be as brief as possible; make two clear, understandable slides rather than one cluttered slide with too much information.
- Font size (the size of your type) is important. Use at least 18-point font size for overheads. You can use smaller font size for slides but not much smaller (12-point, the size of the printed words in this book, is too small for slides).

- Slides do not need to be complete and self-explanatory; the speaker can add details (handouts are helpful here).
- Orally identify or locate colors to help listeners who are color blind (10% of all males are color blind).
- Use high-contrast colors (white and blue, yellow and navy, etc.) with slides or overheads; this helps people with color blindness.
- The more you must use a pointer, the worse the slide; keep it simple.
- If you have many numbers in a table, use multiple slides with the emphasized number circled or highlighted. If you are discussing only five numbers, why show more than five on a slide?
- Do not use all capital letters. They are difficult to read. Use capital letters sparingly for emphasis of an important point.
- Use graphs or charts instead of tables. Most people can comprehend graphs and charts more easily than tables.

The following tips will assist you in maximizing the auditory impact of your presentation:

- Be an enthusiastic speaker.
- Be enthusiastic about your message and your audience; everyone enjoys listening to speakers who care about their subject.
- You do not have to be a cheerleader. Show with your voice, eyes, and body language what you are saying is important and meaningful.
- Another adage: "There are no boring topics, only boring presentations."
- Master your subject. You need to be able to concentrate on reaching the audience, which means you should make eye contact with the audience, and concentrate on "reading" your audience (are they engaged, bored, etc., and what can you do to get and keep their attention?). This takes energy. You should know your material well enough so that you appear confident.
- Never read or memorize your presentation. Reading reduces eye contact and memorization makes your talk appear canned and boring.
- Use pauses to your advantage. When you make an important point or when you have presented something complicated, give the audience time to think about and digest your point. You can build pauses into your presentation to slow down the pace if you tend to speak more quickly when doing public talks.

Using *kinematic methods* will engage your audience. The methods can be direct, where your audience does something, or indirect, where they watch as you do something hands-on. Make use of the following tips for kinematic methods:

- If you are demonstrating something, make sure that your audience can see the entire demonstration from their viewpoints. Make sure that you point out the major concepts or ideas.
- If you have the audience do something, especially if it involves multiple steps, make sure that the entire audience is with you as you step through the process. You can ask, "has everyone completed this step?" before proceeding to the next one.

Practicing Your Presentation

Practice your presentation out loud because it will give you insight into your technique. First, most presentations tend to be scheduled, and you do not want your presentation to go longer than its allotted time. Long-winded speakers will often alienate their audiences or will lose their attention. Most people tend to finish their presentations ahead of time because they speak more quickly when nervous. If you practice your presentation, you will know how long it takes to complete. If you are nervous, practice will enable you to build in several places in your presentation in which you can check yourself and slow down if necessary. Second, the more you practice, the more familiar you become with the material that you are presenting. This familiarity can be important if your information is technical. Showing the audience you are familiar and comfortable with the presented information will help them (and you) gain confidence in your subject. Most likely, you will explain the concepts better as you understand them more fully.

Now that we have talked about everything you can do before a presentation to maximize its effectiveness, think about the presentation itself.

Getting Started

The first 15 to 30 seconds of a presentation are critical; establish eye contact and rapport if you are going to turn down the lights for slides or PowerPoint. You can establish a connection with the audience in several ways:

- You can start by saying, "Good morning (or afternoon or evening), my name is <name>, and I am going to discuss <subject>."
- You can establish the level of formality. For example, you can tell the audience it is okay to ask questions at any time.
- You can establish your tone with humor (telling a joke), or asking a trivia question with a small reward (e.g., a piece of candy) for the winner.
- Interacting with the audience is an excellent way to establish good rapport. After you introduce yourself and the subject of your talk, you may involve the audience by asking a general question. Those interested can raise their hands or interact with an answer. This information could be useful to you (you can learn something about the audience for a better presentation) and can provide you with a good lead into the rest of your talk.

During your presentation, monitor your audience. Do they look engaged, confused, or bored? If they look engaged, do not change anything that you are doing. If your audience is confused, you could ask them a question, or say, "I know that this information is difficult and some of you look confused. Can I clear up anything for you?" If your audience appears bored, you might re-engage your audience with another trivia question or by asking the audience to become active.

The conclusion of your talk is important. You need to let your audience know that you have finished, e.g., "This concludes my presentation. Do you have any questions?"

Presentation Hardware

Know how the equipment (PowerPoint, overhead projector, etc.) works beforehand. For overhead projectors, if you can read the overhead when you place it on the projector, the audience can read it. You may have to focus but your overhead will not be backward or upside down.

Ensure that the proper video projection system is available and that the computer and system are compatible. You can bring your own just in case.

14.3 WRITTEN COMMUNICATION SKILLS

In your engineering travels, you will volunteer or be called upon to communicate your ideas, designs, findings, and concepts in writing. Many technical writing books contain detailed information about this. In fact, most of you will take an entire course in technical writing, during which time you will get experience writing technical papers, reports, cover letters, executive memos, engineering design notebooks, proposals, etc. This section of the chapter is intended to provide conceptual information with respect to the writing you will encounter in engineering, including reports, proposals, memorandums (memos for short), résumés, and cover letters. These next sections include brief descriptions on how to best craft these writing pieces.

Before we begin discussing the formats of written communication, recall what was said at the beginning of this chapter about the two golden rules of communication for engineers:

- Brevity is best. Simplify whenever possible.
- Using a shorter, simpler word is typically preferable to a longer, more difficult word.

And here are a couple of common grammatical issues to consider:

- Each paragraph should address a *single topic* and have a single primary focus (stated in a topic sentence) accompanied by supporting details.
- Avoid clichés, which tend only to confuse the reader of technical writing.
- Its and it's: "Its" (without an apostrophe) is used to denote possession. For example: The dog chased its tail. "It's" is a contraction that means "it is." For example, It's sunny outside. If you are not sure which to use, simply insert the words "it is" in place of "it's." If the sentence makes sense, keep "it's." If not, use "its."
- Affect and effect: "Affect" is a verb meaning to influence, e.g., Her leadership positively affected the situation. "Effect" is a noun meaning result, e.g., One effect of securing the contract was the hiring of more student workers. The tricky part is that effect can be used as a verb meaning to bring about, e.g., He effected considerable change when he became manager.

The Audience

Before you begin to write, ask yourself, "Who is going to read this, anyway?" You must consider the needs and perspectives of your audience. Their level of education or experience may dictate the language and style of your writing. They may use words or expressions that are unique to their part of the country or their part of the world.

Always remember that technical material needs to be clear and precise. You cannot simply write as you please and hope that your readers will "figure it out on their own." It's as if you are the screen writer and science requires a very clear script. You must prepare your readers, give the required information to help them understand, and repeat information as necessary to bring them to full understanding.

Technical Writing

Because your engineering career most likely will involve a great deal of technical communication, it is vital that you understand the qualities of effective technical writing. In general, technical writing:

- identifies the main premise early on
- is clear, objective, and economical

- follows a specific format (i.e., memos, lab reports, abstracts)
- takes a problem-solving approach
- involves a specialized vocabulary
- often incorporates signs, symbols, formulas, graphs, and tables
- documents completed work

Let's look at some of the unique aspects of technical writing:

Content—Technical writing is typically based on research. Therefore, it must be factual and objective.

Methodical Construction—The way technical engineering material is written is specific and consistent throughout the profession. In general, paragraphs begin with a topical sentence, which is followed by supporting statements. Technical writers like to present the most essential information first and then present supporting and clarifying information. In this way there is no mystery as to the purpose of the communication.

Clarity—Technical communications must be clear and complete.

Objectivity—Technical writing should be unemotional and unbiased. It should not convey the feelings of the writer or others.

Importance in Workplace—Technical documents provide the information necessary for the proper functioning of an engineering-related business. Such documents supply information related to design, construction, research, and operations. Improper or incomplete technical communication can lead to design flaws, hazardous developments, waste of resources, production delays, loss of competitive advantage, and other undesirable outcomes. Such failure to effectively communicate can have an adverse effect on one's career.

Formats of Written Communication

As an engineer, your written communication will take a variety of forms. These include brief notes to a superior or to a peer, hand-written reminders about a project, informal reports, memos, formal reports that may document months of work, and visual presentations that require significant preparation, yet little text. Much of your writing will need to conform to one standard format or another. In this section, a number of these formats are presented.

Formal Reports

The principal purpose of the formal report is to present information gathered from an experiment or simulation in a document that will be useful to your employer and colleagues.

Title

The title should be as brief as possible, clear, and appropriate. Seven or eight well-chosen words is typical. The title doesn't need to do any more than introduce the report.

Summary

Though the Summary (or Abstract) is presented early in a formal report, it should not be written until all other parts of the report have been completed. It should state, in simple, declarative sentences, *what* was attempted, *how* it was accomplished (including a discussion of any special techniques used), and what the *implications* are. That is, it should state the results and main conclusions.

This is both the shortest and most difficult section of the report to write. It is also the most important. In technical publications, the abstract often is the only portion of the report that many people will read. Therefore, it should be written to stand alone. It must communicate all the relevant ideas and results in one to two paragraphs (250-350 words). And be sure to discuss the details, which occurred in the past, with past-tense verbs.

Table of Contents

In the Table of Contents, list each heading found throughout your report along with its page number. However, it is customary not to list the Abstract among the contents.

Nomenclature

List and define all symbols used in the report. They should be listed alphabetically—Arabic symbols first, then Greek. Such a listing will enable your readers to find definitions of symbols used in the Analysis portion of your report with ease. Note that abbreviations should not be listed in the Nomenclature section.

Introduction

The essential task of the Introduction is to orient the reader to the substance of the experiment or project and the context in which it was executed. The Introduction should state the *motivation* for the experiment and any relevant *background* information. It should introduce the material contained in the report by noting what is presented in each of the sections that follow.

Analysis

The Analysis should begin with the basic, general (and well-known) relationships and proceed to the specific formulas to be used in the interpretation of the data. It is usually appropriate to use an explanatory sketch. All the symbols used should appear in the listing of Nomenclature, described above. Analytical results which have been previously derived and are readily available (equations from a handbook, for example) can be quoted and appropriately referenced. Their derivation need not be repeated unless the derivation is critical to the engineer's understanding of the experiment.

All relevant mathematical steps should be presented in the Analysis as well. Mathematical explanations and commentary are crucial to a good Analysis. Be careful not to force the reader to consult lab handouts or textbooks to understand the specific analysis involved in the experiment. If it is important enough to mention, it is important enough to explain.

Experimental Procedure

A schematic representation (a diagram) of the experimental equipment or simulation program, including detailed views of unusual or important components, is a valuable aid in informing the reader about the experiment. The diagram can be used to document pertinent dimensions of the apparatus, and it can be used to specify the specific experimental equipment used for the study. If the procedure used in the experiments is not a familiar one, it will be necessary to include details of the techniques used. Your information should be clear

enough that someone familiar with the general area of investigation should be able to reproduce your experiments with precision, based on the information given in this section. The
report can be hindered by an incomplete or incorrect discussion of the experiment's procedure and equipment.

Though you may be tempted to slip into use of the present tense when describing the
procedure, be careful to describe it in the past tense.

Some major points to remember are:

- When first referencing a figure, be sure to put the figure on the same page (or the next
  page). Your reader shouldn't have to hunt through your report to find the figure.
- Make figures at least one-third of a page in size. Figures that are too small are harder
  to read.
- Figures and tables should have clear and complete titles. Beneath the title, a short
  explanation of the figure's contents is normally necessary.

Results

The answers derived from the analysis of an experiment are presented in the Results section of the report. This section should contain short, declarative statements of the results
which summarize specific data presented in graphs or tables. The purpose is to tell the
reader (without discussion) your interpretation of the results. Keep in mind that the results
could be interpreted in more than one way. Thus, the importance of your stated results.

Realize that your readers may start reading your report in midstream, meaning that they
may start with the Abstract and then go directly to the Results. Never assume that the report
will be read from page 1 to the end, which is seldom the case.

You also should distinguish between "Figures," which include schematic drawings, photographs, graphs, etc., and "Tables," which are tabular presentations of data or computational results. Each type of data presentation should be numbered sequentially with its title:
for example, "Figure 1. Schematic of the Combustion Chamber."

Though you are to present your data without discussion, do not make this presentation
without explanation. A Results section containing only tables and graphs would be difficult
to understand. Remember, you want to make comprehending your report as easy as possible for your readers.

Discussion

The Discussion section starts with a brief summary statement of the results and then proceeds to a discussion of these results. The principal task in this section is to interpret the
results, to note what went "as expected," what was unexpected, and what is of technical
interest. The interpretation of the results in terms of the motivation for the experiment and its
correlation to the current company projects should be the focus of the discussion.

The discussion could involve a comparison with other, similar investigations or comparison with expected results. The strong points of the work should be brought out here along
with any limitations, because if the writer does not point out the limitations of his work, someone else surely will later. It may also be legitimate to comment on possible future investigations. Discuss the specific results of the experiment, using references to the accuracy of the
measurements. It is useful to note the estimated uncertainties and their effect on the calculated values. For example: "The pressure was 2.0 ± 1.0 kPa and the velocity was 30 ± 10
m/s." Note that the "information content" of this statement is much greater than in the statement: "The manometer liquid was fluctuating and the pressure could not be accurately

measured." It is extremely important to provide specifics and avoid vague modifiers such as greater than, about, and higher than.

Keep in mind that good paragraph construction will present a thesis statement or idea and then give supporting details for that statement. When new points need to be made, make sure you make them in new paragraphs. You should make an outline for the Discussion section (just as you should for all parts of the report) so it follows a logical progression which will support the conclusions presented in the next section, Conclusions. Carefully judge the information that you are providing to the readers to make sure it satisfies their specific expectations. For instance, they expect to find information presented in a pattern that presents what is known first and then what is new second. Sentences should begin with known information and then progress to new, related information.

Conclusions

A typical way to begin the Conclusions section is to make this statement: "The following conclusions are supported by the results of this study," and then list the conclusions in one or more simple, declarative sentences using numbers to differentiate each separate conclusion. Remember that in your Conclusions section, engineers will be looking for concise statements that clearly communicate what your results indicate—not additional explanatory material or any plans for further investigation. They want the mass of data synthesized into the briefest conclusions you can present.

References

References are to be listed at the end of the report. In general, be sure to list: author's last name, then first and middle initials (and then list co-authors with initials first), title of source being quoted (in quotes), edition of source, volume of source (if applicable), city of publication (from source's title page), publisher, date of publication, and pages referenced. Note the variety of possible formats in the following samples:

Walker, R.E., A.R. Stone, and M. Shandor. "Secondary Gas Injection in a Conical Rocket Nozzle," *AIAA Journal*, Vol. 1, No. 2, Feb. 1963, pp. 334-338.

Turner, M.J., H.C. Martin, and R.C. Leible, "Further Development and Applications of Stiffness Method," *Matrix Methods of Structural Analysis*, 1st ed., Vol. 1, New York, Macmillan, 1964, pp. 203-206.

Segre, E., ed., *Experimental Nuclear Physics*, 1st ed., Vol. 1, New York: Wiley, 1953, pp. 6-10.

Book, E. and H. Bratman, *Using Compilers to Build Compilers*. SP-176, Aug. 1960, Systems Development Corp., Santa Monica.

Soo, S.L. "Boundary Layer Motion of a Gas-Solid Suspension." *Proceedings of the Symposium on Interaction between Fluids and Particles*, Institute of Chemical Engineers, Vol. 1, 1962, pp. 50-63.

Always give inclusive page numbers for references from journal articles and a page or chapter number for books. Each reference in the text must be cited.

Communications Checklists

One way to make sure that you have successfully completed an assignment is to evaluate your work with a checklist. You can create custom checklists for each of the various communications formats you use. These checklists will guarantee that you have included everything that is required to inform the reader.

Formal Report Checklist

Title Page
1. Title of paper
2. Course/project
3. Date due
4. Section meeting time
5. Name

Abstract
6. Purpose of the lab/experiment
7. How the lab was performed
8. What was discovered, achieved, or concluded
9. Use of past tense
10. Reference to experiment, not paper
11. No personal pronouns (I, we)

Nomenclature
12. Listed in alphabetical order
13. Upper case, then lower case (A a B b c G g1 g1a)
14. Arabic and Greek listed separately (Arabic first)
15. Only symbols are listed

Table of Contents
16. All sections represented
17. Abstract and Table of Contents are not listed
18. Observations as headings; analysis, equipment, procedures, results as sub-headings
19. All columns aligned

Introduction
20. Motivation for the experiment stated
21. Sufficient information to orient reader to the substance of experiment
22. Sufficient information to excite reader
23. Listing of sections which follow

Observations
24. Mathematical model used to predict system behavior presented with ample explanation and lead-in
25. Equations numbered
26. Proper punctuation in equations
27. Equations properly spaced
28. Schematic of equipment used

29. Tables oriented correctly, labeled and referenced
30. Figures oriented correctly, labeled and referenced
31. Highlights of equipment used
32. Highlights of the procedure (not specific steps)
33. Data presented with clear indication of what it applies to
34. Clear indication of what data refers to
35. Trends in data stated (to be discussed in the Discussion section)
36. Clear indication of what reader should observe in the data

Discussion
37. Complete discussion of the results
38. Connection of data and objectives is clearly stated
39. Comparison to similar experiments presented
40. Strong points of the study given
41. Weak points of the study given
42. Statements are specific
43. Logical progression which supports Conclusions that follow

Conclusions
44. Begins with, "The following conclusions are supported by this study"
45. Conclusions are numbered
46. Conclusions are concise and highly specific
47. Vague statements are not used
48. Conclusions flow directly from Discussion

References
49. Initials for first names
50. All necessary information included
51. References numbered in text like this: [1]

A Sample Abstract

The following abstract satisfies the checklist features listed in the previous section. By making sure that the required items are included for each part of your report, you have a chance to review what you've included and what still needs to be addressed before the paper is submitted. (Note: The numbers that appear in the following abstract refer to the preceding checklist. References aren't numbered in a real abstract.)

Abstract

Experiments {10} were conducted {9} to assess the appropriateness of digital signal analysis to the design, development, and testing of Whirlwind Corporation's new light aircraft gas turbine. Such analysis could be used to monitor the transient and steady state property variation of the new power plant, predict potentially catastrophic failure, and pinpoint sources of extraneous noise generation {6}.

Several elementary sinusoidal and square wave signals were generated {9} by a commercially available Waveteck function generator. These signals were then converted {9} to the frequency domain by LabView's "Spectrum Analyzer" via the Fast Fourier Transform. Various combinations of sampling frequency and sample size were investigated {9}. When deemed appropriate, these signals were {9} also low-pass filtered {7,10}.

The Discrete Fourier Transform accurately represented {9} only components less than half the sampling frequency. Higher frequency components were reflected {9} across the Nyquist frequency or its integer multiples. This aliasing was eliminated {9} by low-pass filtering, but on occasion, important signal components were discarded {9}. Whenever the input signal contained components that were not integer multiples of the frequency resolution, the magnitude of the corresponding spectrum peaks were diminished {9}. This leakage was reduced {9} by increasing the frequency resolution by increasing the sample size {8,10}. These signal analysis techniques proved {9} their utility and applicability to the new gas turbine project.

In this abstract, the author has addressed the important elements of the assignment, which was to tell the reader why the experiment was performed, how it was done, and what the principal results were.

Notice some of the methods that the writer has used to make the text flow smoothly. Line 11 begins the paragraph with "Several elementary sinusoidal and square wave signals," which prompts the reader to wonder what is to follow. Lines 37–38 show the writer relating "These signal analysis techniques" back to the previous sentence. Be sure to help your readers by presenting a logical, flowing report.

Appendices

Lengthy calculations and side issues that are not really related to the main theme of the report should be placed in the Appendix. To decide whether something should go in the body of the text or in the Appendix, think about whether or not its inclusion in the main body of the report is required as part of the description of the investigation. If the answer is No, the item should be either left out or put in the Appendix.

14.4 OTHER TYPES OF COMMUNICATION

Executive Summaries

Some assignments may require that you submit an Executive Summary instead of a full formal report. This type of report condenses the work you have done into the briefest document possible. It assumes a managerial perspective instead of a technical one, and will contain enough information that an individual with limited technical expertise will be able to make clear decisions based on your report. The Executive Summary presents just the simple facts and describes the key elements of your work in non-technical language—concise and straight to the point. A good Executive Summary contains the following information:

- the background of the situation or problem
- cost factors
- conclusions
- recommendations

If you have constructed Abstracts before, you will understand that they are similar to the Executive Summary. Both are quite brief. They focus on what was investigated, the conclusions your work has produced, and the recommended course of action. In some instances, this Summary will be all that is read before a decision is made either to proceed or to halt further action.

Proposals

Proposals are documents that address a specific need and usually describe the need or problem at issue, define a solution, and request funding or other resources to solve the problem. Proposals can be solicited (as when an organization asks for proposals) or unsolicited (as when you send in a proposal and ask for funding or for a project to be implemented). Solicited proposals involve a Request For Proposal (RFP) with the following guidelines:

- what the proposal should cover
- what sections it should have
- when it should be submitted
- to whom it should be sent
- how it will be evaluated with regard to other proposals

For Proposals, follow the ABC format:

A: The **Abstract** gives the summary or big picture for those who will make decisions about your proposal. The abstract usually includes some kind of hook or grabber, which will interest the audience to read further. The abstract will include the following:

- the purpose of the proposal
- the reader's main need
- the main features you offer and related benefits
- an overview of proposal sections to follow

B: The **Body** provides the details about your proposal. Your discussion should answer the following questions:

- What do you want to solve and why?
- What are the technical details of your approach?

- Who will do the work, and with what?
- When will it be done, and how long will it take?
- How much will it cost?

Typical sections in the body portion of the report will be given to you in the RFP, or else you will have to develop them yourself, making sure you address the following:

- a description of the problem or project and its significance
- a proposed solution or approach
- personnel
- schedule
- cost breakdown of funds requested

Give special attention to establishing need in the Body. Why should your proposal be chosen? What makes it unique? How will your design or recommendations help the community?

C: The **Conclusion** makes your proposal's main benefit explicit and will make the next step clear. This section gives you the opportunity to control the reader's final impression. Be sure to:

- emphasize a main benefit or feature of your proposal
- restate your interest in doing the work
- indicate what should happen next

Memos

The *Memo* provides a convenient, relatively informal way to communicate the existence of a problem, propose some course of action, describe a procedure, or report the results of a test. Though informal, memos are not to be carelessly written. They must be carefully prepared, thoughtfully written, and thoroughly proofread for errors.

Memos contain introductory information that summarizes the memo's purpose. The memo's conclusion restates the main points and contains any recommendations. They begin by clearly presenting all relevant information right up front, as follows:

To: **name, job title**
            department
            organization
From: **name, job title**
            department
            organization
Subject: **(or "Re:") issue addressed in the memo**
Date: **date**

The following may be used as well, if applicable:

Dist: **(Distribution) list of others receiving the memo**
Encl: **(Enclosure) list of other items included with the memo**
Ref: **(Reference) list of related documents**

The format of the text of the memo is also simple and contains the following information:

Foreword—statement of the problem or important issues addressed
Summary (or *Abstract*)—statement of results, findings, or other pertinent information

Discussion or Details—technical information, discussion of the problem or issues, or support for the claims in the Summary

Activity

Prepare an answer to the following memo. Keep in mind the appropriate format and approach for a good memo.

Situation: Assume that the following memo has been received in your office at Engo-Tech Consultants (an engineering consulting firm specializing in technology compatible with the request in the memo). Please read it and address its concerns.

> To: Engo-Tech Consultants
> From: Michelle De LaBaron, General Manager, Euro-Flags
> Date: August 1, 2012
> Re: Increased activity for Euro-Flags
>
> To ensure the continued success of our operation at Euro-Flags we believe it is essential that we add a new technical emphasis to the variety of entertainment opportunities we currently offer.
>
> Our mission includes more than simply existing as a provider of "fun and games." We also want to stimulate our patrons intellectually, and thrill them with the wonder of current technology.
>
> With this in mind, we ask that you present us with a proposal which outlines how your firm could assist us in this endeavor. Please explain not only what your firm can do, but also how your proposed plans would appeal to our patrons.

E-mail

Electronic mail, or e-mail, has become a part of life in almost every office. Messages gather in computer mailboxes at a phenomenal rate, and the time required to respond to these e-mails continues to grow. It is, therefore, vitally important that we look closely at what we receive and what we send. E-mail must be read carefully and the responses we send must be well written. Just as with more formal reporting, if your communication includes spelling errors, poor punctuation, and sloppy construction, it will likely hamper your reader's comprehension. And it projects general incompetence—even if you are quite competent technically. Besides, e-mail provides an easily preserved record of your communication, so it's worth doing right.

E-mail's instant transmission and spontaneous feel can tempt us into serious communications errors. It's easy to get sloppy. Given the format, e-mail can be easily misunderstood—which is one reason why the variety of "smiley face" punctuation tags evolved. And once we have sent a message, it is out of our hands and we can do nothing to change the content or presentation. Here are a few simple rules to follow:

- Use proper grammar—appropriate sentence structure, subject/verb agreement, spelling, flow, etc.
- Carefully proofread, edit, and spellcheck your e-mail. Are there gaps in your train of thought, or opportunities for confusion, that may cause a reader to misunderstand your message?
- Have you thought about your response before you hit "Send"? Once it is sent, it can't be retrieved. With professional e-mail, it's often best to wait as long as possible to send your reply, even if you only have an hour. Let it rest, then re-read your reply before sending.

E-mail is a fantastic tool, but it has limitations. Never use it when a face-to-face dialog actually is needed. Do not use it when a formal document is required. E-mail does not carry the status of a formal report. And remember that e-mail is not quite as private as you think. A message may be downloaded and left on the screen of your recipient, in plain view of any-one entering his office. And in truth, many e-mail logs and data archives are more public than you may realize. It is vital, therefore, that you make every effort to use the tool wisely, and avoid letting it embarrass you or cause you undue grief.

Progress Report

At some point in every job, you will be asked to indicate how you are progressing toward some goal. Managers, co-workers, stockholders—even you yourself—will want to know what you've accomplished. The Progress Report provides the means by which you can report your status.

The reports need to be attractive and easy to understand. Commonly, the progress report will have sections much like any other report prepared for management:

- Introduction
- Project description
- Summary of what you have done

Each section is discussed briefly below:

Introduction—Here you capture the interest of the readers by informing them of what you are going to do in the report. You expose readers to the scope of the work being done, the purpose of the work, and any major changes that have been required in the project.

Project Description—Here the reader is acquainted with the time-frame of the project and the current progress. Phases that have been completed and the time it took to complete them are presented in this section. It is also important to explain the tasks that remain to be completed.

Summary—This is where you draw the whole report together. Summarize the main points of action and reiterate where you are along the road to completion.

Problem Statements

Problem solving is a critical aspect of engineering. Before you begin to work on a project you may be asked to present a Problem Statement pertaining to the project which explicitly states the problem to be investigated and outlines the course you intend to take. Correctly defining the problem is vital to the solution. If you are not on the right track from the begin-ning, you may well fail to obtain the appropriate solution. So here are some basics:

- Work with data that you already have collected—evaluate all reasonable avenues of pursuit in finding a solution as you define your problem.
- Consult your contacts who have knowledge of the problem.
- Investigate the problem first-hand.

Here's a silly anecdote which may help you remember to properly identify a problem.

A man who had a chronically sore neck went to the doctor. He told the doctor of his suffering, that he was unable to turn his head without severe pain. Even visiting the doctor caused him pain. The doctor asked him what he had done to try to solve the problem. The man listed a variety of attempted remedies—none of which had worked. The doctor looked at the man, analyzed the situation, and calmly said, "When you get up in the morning, try taking the hanger out of the shirt you're going to wear before you put it on."

Thus the problem was properly identified and solved.

Cover Letters

Cover letters will be especially important as you attempt to land your first job in engineering. Such letters are sent along with résumés and transcripts. They "cover" your other material—thus the name. You should highlight information about yourself that separates you from the rest of the applicants and demonstrates your knowledge of and interest in the company. A great cover letter can open doors, but a poor one can slam them shut.

The cover letter functions as your introduction to a prospective employer. This one-page document introduces you to the company in a more conversational manner. In many instances, they will represent the only personal impression you will get to make on a potential employer. View the letter as an opportunity to communicate what you have to offer them. If they don't like what they read, you may never get the chance to talk to them in person.

In the first paragraph, introduce yourself and describe how you became familiar with the position for which you are applying. In the next paragraph(s), detail your qualifications and what makes you qualified for the job. You should include how your personal character and interests will enable you to excel at the job and to fit in with the company. Your concluding paragraph should include a request for an interview (or for the company to follow up), your contact information, and a thank you for the company's time and consideration.

Specifics you should include in a cover letter:

- the date, your address and phone number, and the name and address of the person to whom you are writing
- 1st paragraph—the reason you are writing the letter, the source of your information about the employer, and what you would like to do for the employer (the position for which you would like to apply)
- 2nd paragraph—a brief discussion of your résumé, hitting the highlights
- 3rd paragraph—"current" information which may not yet be appropriately included in your resume (courses you are currently taking, research you may be involved in, activities you are about to engage in)
- 4th paragraph—summary paragraph in which you thank the reader for his or her consideration "in advance" (You may also include your phone number, and mention that you will call to confirm receipt of the letter.)

Remember, you want your cover letter to present the most positive impression possible. A good deal of time should be spent perfecting it. Cover letters and résumés should be printed on the same kind and color of paper. Always use a high-quality laser printer to print these items.

Sample Cover Letter

February 10, 2012

Ms. Lydia Baron
Human Resources Director
Harrigan Corporation
120 Rollaway Road
Chicago, IL 60606

Dear Ms. Baron:

I am very interested in the summer internship program offered at Harrigan Corporation. I know from my research that Harrigan's rapidly expanding automotive parts operation has an outstanding reputation. The opportunity to gain valuable experience through working in that division would be of tremendous benefit to me in my professional development.

My interest in engineering began at an early age, and I look forward to a lifetime of continual engineering education. I am completing my junior year at Michigan State University, having taken a number of courses toward my degree in mechanical engineering. As my resume shows, I have concentrated a great deal on developing my computer skills as part of my engineering background.

This semester I am taking two courses—Advanced Mechanics and Finite Element Analysis—which have been particularly valuable in preparing me for working at Harrigan Corporation. I have studied the current work schedule of the Brighton facility and understand that these courses fit well with your plans for the new turbo charger assemblies.

A copy of my resume is enclosed for your consideration. I will contact you during the week of February 20 to see if you require any additional information. If you have any questions for me in the interim, please contact me at (517) 555-1212, or at smith@pilot.msu.edu. I look forward to discussing with you the job opportunities at Harrigan Corporation.

Sincerely,

Jason D. Smith
222 Landon Hall
Michigan State University
East Lansing, MI 48824
(517) 555-1212
smith@pilot.msu.edu

Enclosure

Résumés

A résumé is one of the most important documents you will ever create. It sells you and your qualifications. There are two main types of résumés: skills résumés and experience résumés. A skills résumé is for people who have not yet completed significant work experience. The skills résumé highlights the skills and talents to benefit the potential employer, even if the applicant has little or no technical work experience. Experience résumés highlight prior work experience related to the job for which a person is applying. In general, most engineering students in the first two years of study and those who graduate without technical work experience will use a skills résumé; engineering students with co-op or internship experience in an engineering context would use an experience résumé.

There is a wide variety of appropriate formats for résumés, so ultimately you have to decide upon a format that you will be happy with. Regardless of the specific format, every résumé must present your name, address and phone number; your educational background; your previous employment (if any); and extracurricular activities. The following simple format is merely one of many. Key elements are highlighted in bold type.

Make sure that you get as many people as you can to read and critique your résumé. Listen to and evaluate all the comments you get, and then customize the résumé to fit your needs.

Résumé Checklist

Information
- Objective is clearly stated.
- Clearly convey how your education, experience, activities, and honors support the objective.
- Experience, education, and/or skills segments are effective.
- All activities, honors, and other data are appropriate for the employment and the reader.

Organization
- Name and key headings stand out.
- Information within each heading is ordered from most to least important.
- Experience segment is arranged to highlight your strengths and career objective.

Style
- Language is simple, direct, and precise.
- Noun phrases are consistently used for headings.
- Strong verb phrases or clauses are consistently used to describe experience, skills, and activities.
- Parallel structure is used effectively.
- Have no errors in grammar, punctuation, or spelling (no typos whatsoever).

Important things to remember about résumés
- A résumé should be crafted for a specific position or job. This shows the employer that you care about the position. Your résumé's objective section should reflect some specificity as a result.
- You may need to change your résumé for "scannable format," if you apply for jobs with larger engineering companies. These companies use a computer to scan each résumé submitted to them for keywords. If so, you should consult your career services office or look at the company's recruitment information to determine these keywords and use them on your résumé. The scanner will target you and your résumé as a potential employee more quickly.
- If possible, keep your résumé to one page. Many employers frown upon résumés longer than one page.
- The format of a résumé is important. Your information should be arranged to be visually pleasing and readable. Use the following strategies in terms of format:
- List dates in reverse chronological order (starting with your most recent information first)
- Use at least 11-point font size (preferably 12) and font types such as Times New Roman
- Use one-inch margins around the page
- Headings such as Education, Work History, Skills, Honors and Activities, etc., should be in bold, can be in capital letters, and should be the only information on that line of the page
- Your résumé should be printed using a high-quality laser printer on excellent bond paper (at least 40-pound ivory-colored or off-white paper)

Here is a list of recommended action verbs you may want to use categorized by topic: (See actual Sample Résumé on next page.)

Human Relations	devised	**Management**
worked with	evaluated	managed
volunteered	observed	rated
interacted	assessed	evaluated
sponsored	designed	devised
taught	created	planned
served	**Communications**	organized
counseled	drafted	coordinated
directed	designed	contracted
helped	composed	maintained
assisted	published	established
trained	presented	negotiated
guided	edited	controlled
Research and Design	revised	purchased
researched	prepared	saved
analyzed	taught	scheduled
solved	instructed	sold
discovered	addressed	verified
investigated	interpreted	produced
experimented	lectured	improved
tested	conducted	
verified	published	

Sample Résumé Format

Your Name (centered)

College Address   **Permanent Address**

College Phone Number   **Permanent Phone Number**

Objective: An entry-level position in (state field or position sought).

Education:
- your degrees
- your school
- graduation date (or anticipated date)
- grade point average and basis (e.g., 3.4/4.0)

Work Experience:
- your job titles (list most recent first)
- the companies for which you worked
- dates for which you worked for those companies
- responsibilities in those positions

Activities:
- meaningful activities in which you have participated

Sample Actual Résumé

Jason D. Smith

222 Landon Hall	37 Dilgren Avenue
Michigan State University	Port Anglican, VT 12225
East Lansing, MI 48824	(234) 555-1212
(517) 555-1212	

smith@pilot.msu.edu

Objective: An internship in the manufacturing area of a mid-sized company

Education: Bachelor of Science in Mechanical Engineering
Michigan State University, East Lansing, Michigan
Expected Date of Graduation: May 2006;
GPA: 3.6/4.0

Computer Skills: MS Office, C++, MATLAB

Employment: May 2005–August 2005; Production Engineer
Labardee and Sons, Aberdeen, Texas
- Developed new production plans for the Labardee grinding machine
- Consulted with primary Labardee clientele

June 2004–August 2004; Machinist
Rawlins Machining, Deer Park, New York
- Produced updated schematics for all standing jobs
- Instituted changes in assembly line procedures

Honors/Activities: Ames Scholarship, Dean's List (4 semesters), MSU Swim Team, racquetball

References: Available upon request

Thank-You Letters

After interviewing for a new job or co-operative position, you should send thank-you letters to the individuals who interviewed you. You shouldn't wait more than 48 hours to send these letters. They should be businesslike and concise. And remember that a strong follow-up letter may be just the thing that puts you ahead of the competition.

Here's what you should include in your thank-you letter:

1st paragraph

- Thank the interviewers for their time, and reiterate your interest in working for the company.

2nd paragraph

- Briefly restate your qualifications. This is the time to address any positive qualities you may have failed to mention during the interview.

3rd paragraph

- Close the letter with a final thank-you, and express your interest in hearing back from the interviewer. You may even want to give a specific time frame within which you will follow up with a phone call. Provide the interviewer with phone numbers where you can be reached and your e-mail address, if you have one.

Additional advice:

- Write or modify each thank-you letter separately. Don't use generic form letters. Try to personalize your summary of the interview in each letter.
- Even if you are rejected for the position, sending a thank-you letter may open up other doors or future opportunities with that company.

14.5 RELEVANT READINGS

Here are just a few of the many technical communication books that are on the market. They all contain pretty much the same information, presented in varying formats. Having such a resource on your shelf will be of great value.

Eisenberg, A., *Effective Technical Communication*, 2nd Edition. New York: McGraw-Hill. 1992.
(General overall coverage of the major topics of technical writing; includes proposals, letters, and reports.)

Fear, D., *Technical Writing*. Glenview, Illinois: Scott, Foresman and Company. 1981.
(For the writer who wants instruction in clear, concise steps.)

Hirschorn, H., *Writing for Science, Industry, and Technology*. New York: D. Van Nostrand. 1980.
(Particularly helpful appendix. An approach to writing that takes the writer from the beginning of the process to the end product.)

Houp, K. and T. Pearsall, *Reporting Technical Information*, 6th Edition. New York: Macmillan. 1988.
(The best of the bunch, both for future use and present needs. Gives lots of examples, and helps greatly in the writing process.)

Huckin, T., and L. Olsen, *English for Science and Technology*. New York: McGraw-Hill. 1991.
(Good book for the non-native English speaker about the process of report construction.)

Mathes, J. C., and D. Stevenson, *Designing Technical Reports*. New York: Macmillan. 1991.
(A process approach to the writing of a technical presentation—written from the information perspective, not from the "form of the report" perspective.)

Michaelson, H., *How to Write and Publish Engineering Papers and Reports*. Phoenix: Oryx Press. 1990.

(Primarily aims at the writing of papers for publication. Does a good job of looking at quality of writing and the concern for the reader.)

Miles, J., *Technical Writing—Principles and Practices*. Chicago: SRA Research Associates. 1982.

(A general text to help with basic problems. Includes a great deal on the process of getting started.)

Turner, M., *Technical Writing*. Reston, Virginia: Reston Publishing. 1984.

(A teaching text with a good visual format. Nice section on memo writing.)

Also:

Beer, D. *A Guide to Writing as an Engineer*. New York: John Wiley & Sons. 1997.
Eisenberg, A. *A Beginner's Guide to Technical Writing*. New York: McGraw Hill. 1998.
Finklestein, L. *Pocket Book of Technical Writing*. New York: McGraw Hill. 2000.
Pfeiffer, W. *Proposal Writing*. New Jersey: Prentice-Hall. 2000.
Pfeiffer, W. *Technical Writing*. New Jersey: Prentice-Hall. 2000.

EXERCISES AND ACTIVITIES

14.1   Prepare a five-minute presentation to be given to a group of your fellow students concerning an engineering topic of interest to you. Prepare overheads for your presentation. You may not need more than one overhead per minute of presentation.

14.2   Prepare a presentation, similar to the one above, to be given to a group of elementary students. What information did you change?

14.3   List at least five clichés that you have used or heard that could make comprehension difficult for a foreign student.

14.4   Write a memo informing your fellow students of a seminar you are presenting.

14.5   Write a memo to your supervisor requesting that your work activity be reviewed for a possible raise.

14.6   Write a formal report about a lab that you have completed, using the suggestions presented in this chapter.

14.7   Write an Executive Summary of a lab experiment that you have completed.

14.8   Create a checklist similar to the one presented in Section 14.3 which you could use in conjunction with a paper you are required to write for another course. Evaluate how well your checklist improved the quality of your paper.

14.9   Write an e-mail to your local newspaper responding to an issue (deer hunting, guns, smoking, abortion, air bags, etc.) for the "Letters to the Editor" section.

14.10   Engineering is not recognized as a profession by federal agencies since it does not have any entrance requirements. In fact, a college degree is not even required. Write an e-mail to your state representative supporting or opposing this position of the federal agencies.

14.11   a) Write your résumé.
        b) Write a cover letter to the OLG Manufacturing Company seeking employment in the engineering department.
        c) Write a thank-you letter to a Dr. Jan Harris, as though she interviewed you for an engineering position.

14.12   Locate a grammar or writing textbook in your library. Carefully read through the Table of Contents. Write down any topics with which you are unfamiliar. (For instance, your knowledge of the definition and use of verbs may be satisfactory, but perhaps you are much less familiar with subordinate conjunctions.) Read through the material in the book related to any unfamiliar or forgotten parts of speech, and write a brief report detailing new information you learned.

Chapter 15

Ethics and Engineering

15.1 INTRODUCTION

In addition to technical expertise and professionalism, engineers are also expected by society and by their profession to maintain high standards of ethical conduct in their professional lives. This chapter will cover the most important issues of ethics insofar as they are relevant to careers in the various engineering disciplines. A brief concluding section will also summarize the various legal guidelines and requirements with which engineers are expected to be familiar.

Summary

To begin, here is a summary of the basic areas and kinds of questions to be covered.

1. Since engineering ethics is a branch of ethics, we start with a clarification of the nature of ethics, and how ethical values and concerns differ from other kinds of values and concerns.
2. Next, the specific concerns of engineering ethics will be examined, and distinguished from other concerns about values which engineers may also have. Such non-ethical concerns include personal preferences, social or political values, and more technical professional values related to becoming and remaining an excellent engineer from a purely engineering or scientific point of view.
3. After the preliminary clarifications and distinctions above, we then examine the basic issues, concepts and special topics of engineering ethics. Most of these are also included in or assumed by the codes of ethics provided by various professional engineering societies.

It will be useful to examine one of these codes in more detail, namely the National Council of Examiners for Engineering and Surveying (NCEES) Model Rules of Professional Conduct. The study of this code will give engineers a good understanding of the kind of real-world requirements which professional engineering societies uphold for their members. The discussion will also be supplemented with some model answers to typical ethics questions, which provide a useful test of familiarity with and understanding of the ethical issues covered in such codes. In addition, more wide-ranging discussion and essay ethics questions are included to encourage more principled and independent thinking about ethical issues by engineers.

15.2 THE NATURE OF ETHICS

Generally, ethics is concerned with standards, rules, or guidelines for moral or socially approved conduct such as being honest or trustworthy, or acting in the best interest of a society.

Not all standards or values are ethical standards, though. For example, personal preferences and values such as individual choices of food or clothing aren't important enough to qualify as ethical values or choices. Generally speaking, ethical standards apply only to conduct which could have some significant effect on the lives of people in general.

Thus for an engineer, using a substandard grade of steel in the construction of a bridge would definitely violate ethical standards because of the potential for safety hazards for people in general. But on the other hand, using an inferior brand of ketchup on one's fries does not violate any ethical standards, since at worst it would only violate personal, non-ethical standards of what tastes good to oneself.

Ethics and ethical standards should also be distinguished from matters of legality and legal standards. Roughly speaking, the distinction is that legal standards are defined in legal documents by some properly appointed legal body, and those documents and legal experts determine what the law is and who should obey it.

In ethics, by contrast, it is instead assumed that ethical standards exist independently of any particular group of experts (being accessible to all through the exercise of their own thinking capacities), and that a codification or written-down form of the standards merely describes or summarizes what those pre-existing ethical standards are, rather than (as in the case of the law) defining their very nature.

For example, if there were a legal requirement that any load-bearing beam in a bridge must be able to bear five times the average real-world stress on the beam, then it is a simple matter of calculation to determine whether a beam meets the legal standard or not. However, an ethical standard for what would be an adequate safety factor in such a case is a very different matter, and it cannot be settled by calculation or appeal to a rule book, since an ethically concerned engineer could convincingly argue that conventional or routine safety factors are really ethically inadequate and unacceptable. Such a person cannot be answered merely by appeal to some rule book or to current legal standards for such matters.

Another vital point about ethics is that its standards are always more important than any other standards. Thus in the contrast between personal and ethical standards discussed above, if one's personal standards conflict with ethical standards, then one must suppress one's personal standards to resolve the conflict in favor of the ethical standards. For example, if one personally values finding the highest-paying job, but it turns out that the job in question involves some ethically wrong conduct, then it is one's ethical duty not to accept that job, but to take a lower-paying, more ethical job instead.

It is generally held that even legal standards must give way to ethical standards in the case of conflicts. For example, suppose that a law governing motor vehicles did not currently require a vehicle recall in the case when engineering defects were found in its construction. Nevertheless, if a good case could be made that this was an ethically unacceptable situation, then the law would have to be changed to conform to the ethical standard. So here too, other kinds of values or standards must be suppressed or adjusted in the case of a conflict with ethical standards, which are always of overriding importance.

The Subject Matter of Ethics

The subject of ethics includes both theories about the nature of ethics (this part of ethics is sometimes called *meta-ethics*), and recommendations of appropriate standards or guidelines for morally right or good behavior by people (sometimes described as *normative ethics*). *Engineering ethics* mainly deals with such normative or ethical standards as to how responsible engineers should behave.

Three main varieties of meta-ethical theories are ethical skepticism, relativism, and absolutism. Skeptics believe that ethical beliefs and standards are merely a matter of subjective personal opinion or biased feelings, so that there really are not any justifiable, reliable ethical standards. However, we can ignore this theory here, since professionals in any field, including engineering, are required as part of their professional obligations to adhere to high ethical standards. So ethical skeptics should seek some other line of work than engineering or the other professions.

Ethical relativism claims that in some way ethical standards are relative to a society or culture. In contrast, *ethical absolutism* claims that the same ethical standards apply to all societies and cultures. Applied to engineering ethics, this would be a dispute about whether or not it would be acceptable for different societies to have different engineering ethics standards for their professional engineers.

Fortunately, we can avoid this dispute here, at least at the present introductory level, in a similar manner to that in which we avoided ethical skepticism. As already noted, professionals in any field, including engineering, are required as part of their professional obligations to adhere to high ethical standards. But inevitably those high standards in practice are those approved or debated within the specific society in which an engineers beginning their own professional lives. So for present purposes we can put aside as peripheral or irrelevant the question of whether or not other societies should adopt codes of engineering ethics similar to our own.

Normative Ethics

Returning to normative ethics, recall that it involves recommendations of appropriate standards or guidelines for morally right or good behavior by people, and that engineering ethics mainly deals with such normative or ethical standards as to how responsible engineers should behave. We will be examining these standards as applied to engineering in succeeding sections. But for the present, here is a brief summary of some standard approaches to normative ethics.

There are two main approaches to normative ethics. The first views the subject primarily in terms of rules or principles for right conduct, developed on various different bases such as that of intuition, reason or social tradition. Fortunately for us, in spite of the disagreements as to the basis or source of normative ethical principles, there is very general agreement on what the resulting rules should be—rules such as "Always tell the truth," "Keep your promises," "Protect innocent life," and so on.

The other main approach to normative ethics focuses instead on ethical behavior viewed as whatever actions will best promote the general welfare or benefit of all people. This is often called a "*utilitarian*" approach to normative ethics, and it can be summed up in the slogan "The greatest good for the greatest number." Another common description for this approach is as a "*consequentialist*" ethic, in that it morally judges an action in terms of whether or not it would have better consequences than any other action one could perform.

Conflicts in Ethics

Potentially at least, it seems there could be a conflict between the rule-based versus the utilitarian or consequentialist approaches to normative ethics. For example, following the principle "Always tell the truth" may apparently not have the best consequences in a particular case, while the action which does seem to have the best consequences (possibly a case of lying) may violate a moral principle.

However, fortunately, most utilitarians agree that a deeper analysis will almost always show that what their consequentialist ethic says is morally correct is substantially the same as what conforms to the ethical standards recommended by the rule-based ethical theories. So we will find that in practice both ethics in general and engineering ethics in particular contain elements from both approaches, which generally co-exist without any fundamental difficulty.

Another example of this practical convergence in spite of initial theoretical differences is found in the issue of intentions versus actions. Rule-based theories tell us that being ethical involves following a morally correct rule or principle, whatever the consequences of so doing.

However, the following concern could be raised: Since human beings are fallible and imperfect, the best they can do is to sincerely intend to follow a morally correct rule. So whatever happens as a result of their actions in doing so, no matter how bad it is, nevertheless they will be morally blameless because their intentions were good. (Example: a well-intentioned person tries to push a passerby out of the way of a speeding vehicle, but only succeeds in pushing him or her more directly into its path.)

On the other hand, from the utilitarian or consequentialist approach, a person's intentions are morally irrelevant; only the actual consequences of their actions are relevant to judging the morality of their actions. So as before, it initially looks as if there could be a fundamental conflict between the rule-based and utilitarian, consequence-based approaches.

Fortunately, though, here too a deeper analysis can resolve the apparent clash between these approaches. Though it is true that humans are imperfect, nevertheless we also have enough intelligence and rationality so that we can legitimately be held responsible for cases in which we actually fail to follow the ethically correct rule. Though mismatches between good intentions and good actions can occur, it is the responsibility of ethical persons to eliminate or avoid all potentially serious cases where such clashes might occur.

A standard example of this is driving while drunk. Even if a person sincerely did intend to drive well and not injure anyone while he/she was drunk, nevertheless they can be held responsible for any injuries they cause in spite of their good intentions. Since their moral duty is to always follow ethical rules such as "Cause no harm to innocent persons," they have a duty to avoid getting into situations of this kind where they know they may be unable to avoid breaking such rules. Thus they should either not drink before driving, or not drive after drinking, to ensure that they do not violate the moral rule.

Hence in practice, and on a deeper analysis, both the rule-based and consequence-based theories of ethics would agree that "good intentions are not enough" in ethics, and that ethical behavior requires actually doing what is right—sincere intentions or wishful thinking have no special ethical value, nor are they legitimate excuses for ethically bad behavior.

Special Topics to Consider

Before concluding this section, a couple of special topics should be mentioned which do not fit easily into the standard classification of rule versus utilitarian approaches, but which are nevertheless generally regarded as being ethically important.

The first of these is the concept of *justice*. Intuitively, it is ethically wrong to treat people in unjust or unfair ways. People should be given what they deserve, and also all people should be treated equally. But it is difficult to reduce justice to a series of rules, and justice may even conflict sometimes with a utilitarian approach, since what is best for most may be unfair to minorities.

The second special topic is that of *human rights*. For example, the U.S. Declaration of Independence talks about the rights to life, liberty and the pursuit of happiness. As with justice, moral rights are not easily reducible to rules, and they may conflict with the greatest good for most people in some cases.

Descriptive Ethics

In addition to the term "ethics" being used to refer to meta-ethics or normative ethics, sometimes it is used more descriptively (as in "the ethics of a society") to refer to the actual ethical or moral customs or guidelines that people do follow in the given society. To some extent engineering ethics is also descriptive in this way, in that it includes the actual standards which professional engineers have adhered to and continue to adhere to as registered engineers.

Nevertheless, the standards of engineering ethics should always be viewed as having normative force as well—as part of normative ethics, not just descriptive ethics. Thus they should be viewed as standards which one is required to follow as a necessary guide to one's future conduct, not merely as standards which as a matter of fact engineers have followed up to now.

Example 15.1

A person's behavior is always ethical when one:

    A) Does what is best for oneself
    B) Has good intentions, no matter how things turn out
    C) Does what is best for everyone
    D) Does what is most profitable

Solution: Since ethics is concerned with standards (other than purely legal ones) for socially approved conduct and with promoting the general welfare, C is the only acceptable answer. The other three answers (doing what is best for oneself, acting with good intentions, and doing what is most profitable) may or may not involve socially approved conduct or promote the general welfare, depending on the individual case, and therefore they cannot guarantee ethical behavior.

Example 15.2

Which of the following ensure that behavior is ethical?

    I. Following the law
    II. Acting in the best interest of a society
    III. Following non-legal standards for socially approved conduct

    A) All of the above
    B) II and III only
    C) None of the above
    D) I only

Solution: Given that ethics is concerned with standards (other than purely legal ones) for socially approved conduct and with promoting the general welfare, the choices rate as follows:

Following the law (I) cannot ensure ethical behavior, because a given law might itself be unethical (example: earlier laws legitimizing slavery). But both II and III do ensure that behavior is ethical. Hence the correct option is B.

15.3 THE NATURE OF ENGINEERING ETHICS

Much of engineering ethics is an applied or more specific form of what could be called "general ethics," that is, ethical standards which apply to any human activity or occupation. For example, ethical duties of honesty, fair dealing with other people, obeying the relevant laws of one's country or state, and so on apply in any situation. Thus some of the ethical obligations of engineers are of this general kind also found in many other occupations and activities.

Other kinds of standards found in engineering ethics are more specific but not unique to engineers, such as the principles governing the ethical activities of any professionals in their contacts with clients or customers. These too are an important part of engineering ethics, and they show up in the large overlap in the contents of many codes of ethics derived from many different professions (engineering, law, medicine, academia, etc.).

Then finally there are some standards applying primarily to the profession of engineering, such as those dealing with the proper, ethical manner of approval of designs or plans by managers which require them to have professional engineering qualifications and expertise. These standards are what make engineering ethics distinctive relative to other kinds of applied ethics, but they are only one small part of the whole of engineering ethics, and so should not be overemphasized.

Now to the question of how ethical standards for engineers differ from other standards of value which engineers might use. This brief discussion will parallel that given above for ethics. The aim is to get clear about which of the choices an engineer might make that are relevant to engineering ethics, as opposed to other choices which are instead based on other kinds of values or standards.

As with ethics in general, purely personal choices or values are not relevant to engineering ethics. For example, an engineer who is submitting a bid to a potential client on an engineering project might personally decide to factor in a somewhat higher rate of profit on the project for him/herself than is usual in such projects.

However, as long as there is an open bidding process (where others are free to submit possibly lower bids), and as long as the personal pricing would not affect the quality of the work which would be done, then the pricing level remains a purely personal or economic decision with no ethical implications.

On the other hand, if an engineer were to decide to maximize his/her profit in another way—by submitting a low bid, but then secretly drastically compromising on the quality of materials used so as to achieve maximum personal profit—that would be a violation of engineering ethics standards, which generally require the use of high-quality materials and construction methods acceptable to all parties signing the initial project agreement. This would be a case in which a conflict between personal standards (maximum profit) and engineering ethical standards (honoring the contract and using high-quality materials and methods) must be resolved in favor of supporting the relevant engineering ethics standards in this case.

These contrasts between personal versus ethical values apply also to the related contrast of corporate or economic values versus ethical values. Thus the examples just given of legitimate versus ethically illegitimate maximizing of profits would apply just as well if the motivation were the interest of some corporation or economic group—it would be equally wrong for a corporation or engineering firm to seek profit at the risk of lower engineering quality of products or services. Thus it is the duty of an engineer to uphold engineering ethics standards even if (to take an extreme case) his/her job is at risk, such as when doing the ethically right action would conflict with the non-ethical interests of a boss or corporate structure.

The relation of engineering ethics issues to legal issues applying to engineers should also be discussed. Although (as we saw previously) legal and ethical values are basically different kinds of values which should be distinguished, nevertheless it is generally true that one has an ethical obligation to obey the prevailing laws in a given jurisdiction. Also, persons voluntarily entering the engineering profession agree to abide by codes of ethics which explicitly or implicitly require compliance with existing laws, so here too good engineering ethics standards will generally be is in close harmony with existing legal codes governing the practice of engineering.

Another area of close harmony between different kinds of values in engineering ethics is found in the case of technical or scientific values. Any scientific community dealing with pure sciences or applied sciences such as engineering will have technical values concerning the proper and correct ways to construct theories, carry out calculations, make statistically satisfactory estimates where exact results are not possible, and so on. These themselves are not ethical values (good science is not the same as good ethics). But nevertheless engineers do have an ethical obligation to use good scientific methods at all times, and so here too (as with the law) there is a close correlation between practicing good science and being an ethical engineer.

One other topic should be mentioned here. Naturally, most of engineering ethics concerns an engineer's behavior while engaged in professional engineering activities. But one should be aware that engineering ethics codes also generally include prohibitions on unethical behavior while off the job as well, if those activities would affect public perceptions of one's professional integrity or status. This would include activities such as gambling which might tend to bring the profession of engineering into disrepute, deceptive or other inappropriate forms of advertising of services, or any other activities suggesting a lack of integrity or trustworthiness in an engineer.

Example 15.3

Engineers should follow their professional code of ethics because:

A) It helps them avoid legal problems, such as getting sued
B) It provides a clear definition of what the public has a right to expect from responsible engineers
C) It raises the image of the profession and hence gets engineers more pay
D) The public will trust engineers more once they know engineers have a code of ethics

Solution: These choices are a little harder than those in previous questions, because even the wrong choices do have some connection with ethics. But as long as you follow the initial instruction (choose the best and most relevant answer), you shouldn't have any difficulty answering it.

Choice A: Avoiding legal problems is generally a good thing, but it's not the most relevant reason why the code should be followed (and in some cases, strictly following the code might make it more likely that you'd get sued, e.g., by a disgruntled contractor if you refuse to certify shoddy workmanship).

Choice C: Raising the image of the profession is a good thing, but not for the reason of getting more pay; that would be to act from self-interest rather than ethical motivations.

Choice D: Increased public trust is generally a good thing, but the mere knowledge that there is such a code means little unless engineers actually follow it, and for the right reasons.

Choice B: The best and most relevant answer. The code of ethics is designed to promote the public welfare, and hence the public has a right to expect that responsible engineers will follow each of its provisions, as clearly defined by the code.

Example 15.4

Engineers should act ethically because:

- A) If they don't, they risk getting demoted or fired
- B) The boss wants them to
- C) It feels good
- D) That's the way responsible engineers behave

Solution: As with Example 15.3, some of the wrong answers here do have some connection with ethics, so as before the task is to pick the best and most relevant answer.

Choice A: Even though ethical behavior may usually improve one's job security, it's not the most relevant reason for being ethical.

Choice B: Doing what the boss wants is a self-interested reason for acting, not an ethical reason.

Choice C: Even though, hopefully, being ethical will make engineers feel good, it's not the main reason why they should be ethical.

Choice D: The right answer. There's a very close connection between acting in a responsible manner and acting ethically.

Example 15.5

The first and foremost obligation of registered professional engineers is to:

- A) The public welfare
- B) Their employer
- C) The government
- D) The engineering profession

Solution: As before, the best and most relevant answer should be chosen.

Choice B: Engineers do have ethical obligations to their employer, but it's not their foremost or primary obligation.

Choice C: As with the employer, engineers do have some obligations to the government, but again, not the first or primary one.

Choice D: Obligations to the engineering profession are probably even more important than those to employers or the government, but still they are not the foremost obligation.

Choice A: Correct. The foremost, primary obligation of engineers is to the public welfare.

Example 15.6

Registered professional engineers should undertake services for clients only when:

- A) They really need the fees
- B) Their own bid is the lowest one

C) They are fully technically competent to carry out the services

D) Carrying out the services wouldn't involve excessive time or effort

Solution: Choice A: Personal financial need is a matter of self-interest, not of ethics.

Choice B: The competitive status of a bid is a matter of economics, not ethics.

Choice D: The amount of time involved in carrying out services is also a matter of economics and self-interest, rather than of ethics.

Choice C: Correct. Ethical engineers will not undertake to provide services for clients unless they are fully technically competent to carry them out.

15.4 THE ISSUES AND TOPICS OF ENGINEERING ETHICS

Now we will examine the basic issues, concepts, and special topics of engineering ethics. Recall that most of these are also included in or assumed by the codes of ethics provided by various professional engineering societies. Hence as noted in the Introduction, we will use the most relevant one—the NCEES Model Rules of Professional Conduct—as the basis for the discussion. Its study will provide a good understanding of the typical actual ethical requirements imposed by professional engineering societies on their members.

The Preamble

The Preamble (the Introductory section) to the Rules is important in that it describes their purpose, which is to safeguard life, health and property, to promote the public welfare, and to maintain a high standard of integrity and practice among engineers. Where it is helpful, we will slightly paraphrase the statements taken from the code so as to clarify their meaning.

This part of the code makes explicit its connection with general ethical concerns which all ethical persons should adhere to. We will examine each of the items covered in turn in this statement of the purpose of the code.

1.) First, normative ethical theories universally agree that it is ethically wrong to cause harm to people. This is why the Preamble mentions the purpose of "safeguarding" life, health, and property. Thus ethical persons in general, including engineers, should avoid doing anything which would damage or adversely affect other people. More positively, one should also take measures which will safeguard or preserve people from possible future harm.

For engineers, the more positive measures would include such things as building devices with extra "fail-safe" features included which make harmful consequences of their use as unlikely as possible. Overall, then, ethical persons take great care not to cause harm to others, and they take whatever extra steps are necessary to minimize risks of potential harm to others as well.

2.) The second purpose of the code is to " . . . promote the public welfare." Here too, standard normative ethical theories fully support this rule. It states a duty or obligation not simply to act in a harmless or safe way (as previously discussed), but in addition to take active steps so that one's professional activities will result in definite benefits and improved conditions for the general public.

For example, as an engineer planning a new highway this rule would require one not only to plan and build it in a safe manner (as previously discussed), but also to do such things as choose the shortest feasible route between its endpoints, or to choose that route which would permit the most efficient road-construction techniques, hence maximizing the utility of the highway to the general public and minimizing its cost to them as well.

3.) The third and last purpose of the code, according to the Preamble, is " . . . to maintain a high standard of integrity and practice among engineers." Thus the code also has the purpose of ensuring that engineers will continue to be honest and trustworthy, and to maintain high standards of professional conduct and scientific expertise in their work.

Note that this Preamble makes it explicit that the reason why engineers should adhere to the code isn't simply because it is ethical to do so, or because their professional organization tells them to do it. Instead, it emphasizes that there are vital practical benefits to society which can only be achieved by engineers committing themselves to rigorously follow the code at all times.

Other Issues

Other issues covered in the Preamble are as follows:

> *"Engineering registration is a privilege and not a right. This privilege demands that engineers responsibly represent themselves before the public in a truthful and objective manner."*

A privilege is a socially earned license to do something which is granted by society only under certain conditions. For example, a driving license requires the passing of a driving test and other requirements. Similarly, engineering registration must be earned by recipients (they do not automatically have a right to that status) and the granting of it requires them in return to adhere to ethical standards such as the stated one.

> *"Engineers must compete fairly with others and avoid all conflicts of interest while faithfully serving the legitimate needs and interests of their employers and clients."*

This statement sums up a range of ethical requirements which are more fully covered in the body of the code, so the various issues which it raises will be discussed along with those requirements below.

The Engineer's Obligation to Society

The first group of rules in the code address **the engineer's obligation to society**. As before, we will slightly paraphrase the statements taken from the code so as to clarify their meaning:

> 1.) While performing services, the engineer's foremost responsibility is to the public welfare.

This rule is also featured in a related form in the Preamble, as a duty to promote the public welfare. The idea here of responsibility to the public welfare includes the idea of safeguarding the public from harm, as more fully spelled out in the next rule:

> 2.) Engineers shall approve only those designs that safeguard the life, health, welfare and property of the public while conforming to accepted engineering standards.

These two rules together imply a much broader context of responsibility for engineers than those arising from any one task or project. Designs and materials that seem perfectly adequate and ethically acceptable within the bounds of a given project may nevertheless be unacceptable because of wider issues about the public interest.

For example, until recently a refrigeration engineer could have specified Freon (a chlori-nated fluorocarbon or CFC product) as the prime refrigerating agent for use in a product, and defended it as an efficient, inexpensive refrigerant with no risks to the purchaser of the appliance. However, we now know that there are significant risks to the public at large from such chemicals because of the long-term damage to the environment they cause when they leak out, perhaps many years after the useful life of the product is over. Rules 1 and 2 tell engineers that they must always keep such wider, possibly longer-term issues in mind on every project they work on.

> 3.) If an engineer's professional judgment is overruled resulting in danger to the life, health, welfare or property of the public, the engineer shall notify his/her employer or client and any authority that may be appropriate.

An important rule, which may place the engineer in a difficult position if his/her employer or client is among those contributing to the problem. But the engineer's duty in such cases is clear: ". . . any authority that may be appropriate" must be notified, even if the employer/client tries to prevent it. (Cases of this kind are popularly referred to as "whistle blowing.")

> 4.) Engineers shall be objective and truthful in professional reports, state-ments, or testimonies and provide all pertinent supporting information relat-ing to such reports, statements, or testimonies.

> 5.) Engineers shall not express a professional opinion publicly unless it is based upon knowledge of the facts and a competent evaluation of the sub-ject matter.

Rules 4 and 5 together implement the general ethical requirement that one ought to tell the truth in the specific context of an engineer's professional duties. Note that the duty as mentioned in Rule 4 is not simply to be truthful in what one says, but also to be forthcoming about all pertinent or relevant information in reports, etc.

Rule 4 also mentions being "objective," which adds the element of being unbiased and basing one's beliefs and reports only on objective, verifiable matters of fact or theory.

Rule 5 enlarges on the idea of being objective in one's reports—others should be able to rely upon one's professional opinion, and they can do this only if one knows all of the rele-vant facts and is completely competent to evaluate the matter being dealt with.

> 6.) Engineers shall not express a professional opinion on subject matters for which they are motivated or paid, unless they explicitly identify the parties on whose behalf they are expressing the opinion, and reveal the interest the parties have in the matters.

Rule 6 expresses what is sometimes called the duty of full disclosure. Even if one hon-estly seeks to be truthful and objective (as in Rules 4 and 5), still doubts might be raised about one's motivation or objectivity unless one reveals on whose behalf one is expressing an opinion, and the interests that such persons have in the case. This rule is also related to the issue of conflicts of interest (see Rules 6 and 8 in the second section, below).

> 7.) Engineers shall not associate in business ventures with nor permit their names or their firms' names to be used by any person or firm which is engaging in dishonest, fraudulent, or illegal business practice.

This might be called the "clean hands" rule (shake hands only with those whose hands are as ethically clean as your own). It isn't sufficient to be completely ethical in one's own (or one's own company's or firm's) practices; one must also ensure that others do not profit from one's own good name if their own activities are unethical in some way. This Rule 7 is clearly related to Rule 1 concerning the public welfare—one must promote this in one's external dealings just as much as in one's own activities.

> 8.) Engineers who have knowledge of a possible violation of any of the rules listed in this and the following two parts shall provide pertinent information and assist the state board in reaching a final determination of the possible violation.

Rule 8 generalizes Rule 3 (a duty of disclosure when one's professional judgment is over-ruled) to a duty of disclosure in the case of any of these rules, when one has knowledge of possible violations of them. In terms of the public welfare, it is very important that each profession regulates itself in this way, so as to minimize or eliminate future infringements of its rules. Strict adherence to this rule also will lead to wider appreciation and respect for the profession of engineering because of its willingness to "clean its own house" in this way.

The Engineer's Obligation to Employers and Clients

The second group of rules in the code address **the engineer's obligation to employers and clients.**

> 1.) Engineers shall not undertake technical assignments for which they are not qualified by education or experience.

> 2.) Engineers shall approve or seal only those plans or designs that deal with subjects in which they are competent and which have been prepared under their direct control and supervision.

Rules 1 and 2 require an engineer to be professionally competent, both in undertaking technical assignments and in approving plans or designs. Rule 2 in fact requires a double kind of knowledge, both technical competence in the matters to be approved and that one has had direct control and supervision over their preparation. Only thus can one be sure that one's approval is legitimate and warranted.

> 3.) Engineers may coordinate an entire project provided that each design component is signed or sealed by the engineer responsible for that design component.

Rule 3 in effect invokes Rule 2. As long as each component of a project is satisfactorily approved as per Rule 2, then it is permissible for an engineer to coordinate an entire project. This also underlines the importance of Rule 2, as project managers have to rely heavily on the validity of the approvals for each prior part of a project.

> 4.) Engineers shall not reveal professional information without the prior consent of the employer or client except as authorized or required by law.

This confidentiality requirement is the other side of the coin of duties to tell the truth (as in Rules 4 and 5 of the first Obligations to Society section). Just as one must not lie or misinform, so also must one restrict to whom one reveals professionally relevant information. A professional engineer must be trustworthy as well as honest, and being trusted not to reveal confidential information is an important kind of trust.

Confidentiality is a central factor in assuring employers and clients that one's professional services for them are indeed for them alone, and that they can rely upon one's discretion in not revealing to others any private information without their full consent. This rule is also related to Rules 5 through 8 below, in that any revealing of information to others would probably create conflicts of interest or other "serving more than one master" problems of those kinds.

> *5.)* Engineers shall not solicit or accept valuable considerations, financial or otherwise, directly or indirectly, from contractors, their agents, or other parties while performing work for employers or clients.

This is the first of four rules dealing with problems of "conflicts of interest." These are cases where one has some primary professional interest or group of interests—generally, to carry out some project for an employer or client—but where other factors might enter into the picture which would activate other, non-professional interests one also has which would then conflict with the professional interests.

In the case of Rule 5, the concern is that soliciting or accepting such things as gifts, hospitality, or suggestions of possible future job offers for oneself would activate non-job-related, personal interests of yours (for additional pay, career advancement, etc.) which would then be in conflict with one's primary professional interests and duties concerning a current project.

Special attention should be paid to this and the other conflict-of-interest rules, and they should be strictly observed. People are sometimes tempted to think that breaking these rules is ethically harmless, on the grounds that if one has a strong enough character, then one will not actually be professionally influenced in a detrimental way by gifts, etc., and hence that accepting such inducements cannot do any harm.

However, it is important to realize that even the *appearance* of a conflict of interest (however careful one is to avoid actually undermining one's professional interests) can create serious ethical problems. One basic concern about this is the potential loss of trust it could cause in an employer or client. Just as clients need to know that the engineer will keep their information confidential (as in Rule 4), so also they need to know that the engineer is single-mindedly working with only their interests at heart.

Any doubts raised because of the appearance of a conflict of interest could be very damaging to the client/engineer professional relationship. This is why it is necessary for the professional engineer to avoid doing anything which would create conflicts of interest or even the appearance of them.

Further rules are necessary to deal with conflicts of interest, because unfortunately in some situations the appearance or possibility of conflicts of interest may be virtually unavoidable, no matter how ethically careful everyone is. However, fortunately at the same time there is a powerful method available for minimizing any ethically bad effects of such situations. This is the method of full disclosure of potential conflicts to all interested parties, and it is addressed in the following two rules:

> *6.)* Engineers shall disclose to their employers or clients potential conflicts of interest or any other circumstances that could influence or appear to influence their professional judgment or the quality of their service.

7.) An engineer shall not accept financial or other compensation from more than one party for services rendered on one project unless the details are fully disclosed and agreed to by all parties concerned.

Both these rules address issues of full disclosure, or keeping all the relevant parties fully informed as to areas of potential conflict or potentially undue external influences. The basic idea behind full disclosure is that it can maintain trust and confidence between all parties in several important ways, as detailed below.

First, if engineer A informs other party B about a potential conflict or influence, then A has been honest with B about that matter, hence maintaining or reinforcing B's trust in A. Furthermore, if B is not further concerned about the matter once he/she knows about it, then the potential problem (namely, the apparent conflict or potentially bad influence on A's professional conduct) has been completely defused.

Suppose on the other hand that B is initially concerned about the issue even after it was honestly revealed to B by A. Even so, the problem is already lessened: at least A has honestly revealed the area of concern. Things would be much worse if B later discovered the problem for him/herself, in a case when A had not fully disclosed it—that would be very destructive of trust between A and B.

Furthermore, now that B knows about the area of concern, and knows that A is fully cooperating with him/her in disclosing the potential problem, both of them can proceed to work out mutually acceptable ways of minimizing or disposing of the problem to their joint satisfaction. Thus even if the full disclosure does lead to an initial problem which needs to be resolved, nevertheless it is one which does not break down the trust between A and B. In fact it may even reinforce trust, in that A's willingness to fully disclose a potential conflict/influence and negotiate with B about it is good evidence for B of A's professional honesty and sincerity.

8.) To avoid conflicts of interest, engineers shall not solicit or accept a professional contract from a governmental body on which a principal or officer of their firm serves as a member. An engineer who is a principal or employee of a private firm and who serves as a member of a governmental body shall not participate in decisions relating to the professional services solicited or provided by the firm to the governmental body.

Rule 8 deals with a special case of potential conflicts of interest, namely, when one of the interested parties is a governmental body. In such a case, a somewhat stricter rule is required than for the more usual cases involving only non-governmental agencies.

In non-governmental cases, it is ethically sufficient to fully disclose potential conflicts of interests to all parties, and then to negotiate with the other parties as to how to deal with the potential conflicts. For example, if one's engineering firm has an official who was also on the board of directors of a bank, it would be ethically acceptable to accept a professional contract from that bank, as long as all parties were fully informed about the official's joint appointment prior to the agreement, and also as long as they could come to agree that the joint appointment was not an impediment to their signing a contract.

Returning to the governmental case, the reason why the stricter Rule 8 is required when one of the parties is a governmental body is as follows: In the case of agreements among private persons or businesses, they themselves are the only parties having a legitimate interest in the negotiations, and hence whatever they freely decide among themselves (of course assuming that no other ethical rules or laws are being broken) is acceptable.

On the other hand, in the case when a governmental body is involved, there is another interested party which is not directly represented in negotiations, namely the electorate or citizens of the jurisdiction covered by that governmental body. The governmental body must act only in ways which fully respect the interests and concerns of the electorate. In such a case, it is impossible to ensure that full disclosure to all of the citizens of the electorate of the potential conflicts of interest would be made in a case such as that envisaged in Rule 8, and hence there is a need for the stricter rule which completely prohibits conflicts of interest of the kinds defined in Rule 8. Only then can public trust both in the engineering profession and in governmental bodies be preserved.

An Engineer's Obligations to Other Engineers

The third group of rules in the code primarily address **an engineer's obligations to other engineers**. These rules can be considered as more precise specifications of various rules already introduced, when applied to the specific context of obligations to other engineering professionals. The first rule also covers obligations to potential employers (whether they are engineers or not) when one is seeking employment.

> *1.)* Engineers shall not misrepresent or permit misrepresentation of their or any of their associates' academic or professional qualifications. They shall not misrepresent their level of responsibility nor the complexity of prior assignments. Pertinent facts relating to employers, employees, associates, joint ventures or past accomplishment shall not be misrepresented when soliciting employment or business.

This Rule 1 is an application or more specific form of Rules 4 and 5 of the first section (concerning an engineer's obligation to society), which require objectivity and truthfulness in all professional reports, statements, and opinions.

Rule 1 requires one not to misrepresent one's own qualifications, or those of associates, which is a very important kind of truthfulness and objectivity. The rule further specifies that this duty not to misrepresent applies also to issues of prior levels of responsibility and to the complexity of previous assignments as well. The rule concludes with a statement about any and all pertinent facts in one's previous history, and it requires that they too should not be misrepresented.

It might be thought that this Rule 1 is unnecessary, since it simply applies some general ethical principles already in the code to special cases and circumstances centering around issues of employment and qualifications. However, as with issues of conflicts of interest, there are some special temptations in these areas which are best addressed by explicitly spelling out what is ethically required for dealing with such situations.

A typical temptation in this area might go something like this: Even a well-meaning, otherwise generally honest engineer might be tempted to make his/her qualifications seem more impressive to a potential employer. One might rationalize that one really would be the best person to do the required job, and that therefore one is doing the employer a favor by making one's qualifications seem more impressive and thereby inducing them to hire oneself, rather than some less-qualified person whose paper qualifications might misleadingly look as good as one's own.

Clearly there could be many variants on this manner of thinking. But all would involve rationalization (the inventing of dubious reasons for what one wants to believe anyway) and wishful thinking, rather than the objectivity and true rationality and truthfulness which are required of professional engineers. So Rule 1 serves a useful function in explicitly stating requirements which some might otherwise be tempted to ignore or overlook.

> 2.) Engineers shall not directly or indirectly give, solicit, or receive any gift or commission, or other valuable consideration, in order to obtain work, and shall not make a contribution to any political body with the intent of influencing the award of a contract by a governmental body.

Rule 2 continues the prohibitions against conflicts of interest which were found in Rules 5 through 8 in the previous section (rules covering the engineer's obligations to employers and clients). Those previous rules mainly covered cases where an engineer was already employed, while Rule 2 here specifically applies to attempts to obtain future work, including the award of a contract by a governmental body.

This Rule 2 also emphasizes that it is just as wrong to attempt to unduly influence someone else (a potential employer, for instance) as it would be to allow others to unduly influence oneself. Thus the rule underlines that it is just as ethically unacceptable to try to cause conflicts of interest in others as it is to allow oneself to be enmeshed in improper conflicts of interest.

It should be noted that the second part of Rule 2, ". . . and shall not make a contribution to any political body with the intent of influencing the award of a contract by a governmental body," specifically mentions the intent of the person making the contribution. Contributions are not prohibited, only contributions with the wrong intent.

This part of the rule could be difficult to apply or enforce in practice, because it may be very hard to establish what an engineer's actual intent was in making a political contribution. Also, the freedom to make political contributions to organizations of one's own choice is generally viewed as an ethical right which should be limited as little as possible. So this part of the rule very much depends on and appeals to the ethical conscience of the individual engineer, who must judge his/her own intentions in such cases and avoid such contributions when their own intent would be self-interested in the manner prohibited by the rule.

Note also that the first part of Rule 2, "Engineers shall not directly or indirectly give, solicit, or receive any gift or commission, or other valuable consideration, in order to obtain work . . . ," also mentions the reason or intention behind giving or receiving gifts, etc., in the phrase "in order to obtain work." However, in practice it is much easier to judge when gift-giving is ethically unacceptable than when political contributions are unacceptable, since there are more behavioral and social tests for suspicious inducements to obtain work than there are for suspicious political support. So it is much easier to police and regulate infringements of this first part of Rule 2 than it is for the second part.

> 3.) Engineers shall not attempt to injure, maliciously or falsely, directly or indirectly, the professional reputations, prospects, practice or employment of other engineers, nor indiscriminately criticize the work of other engineers.

Rule 3 as it stands is somewhat unclear, so some discussion is required to bring out its ethically legitimate core.

First, how should the subordinate phrases "maliciously or falsely" and "directly or indirectly" be interpreted? On one possible interpretation, Rule 3 says outright that engineers should never attempt to injure in any way or for any reason the professional reputations of other engineers. On this interpretation those phrases just give examples of possible modes of injury which are prohibited, leaving unmentioned any other possible modes of injury which are nevertheless also assumed to be prohibited.

However, another interpretation is possible, according to which it is only certain kinds of injury which are prohibited by Rule 3, namely those spelled out by those same phrases interpreted so that the "directly or indirectly" part modifies the "maliciously or falsely" part. On this interpretation, Rule 3 prohibits only malicious or false attempts (whether carried out directly or indirectly) to injure the reputations of other engineers.

This second interpretation would ethically permit attempts to injure the reputations of other engineers, as long as the attempts were carried out in a non-malicious and honest, truthful way (and presumably with the public welfare in mind as well).

Some support for this second interpretation can be derived from the final clause of Rule 3, "nor indiscriminately criticize the work of other engineers." Clearly, in this case it is not all criticism of the work of other engineers which is being prohibited, but only indiscriminate criticism. Thus the last section of Rule 3 outlaws criticism which is over-emotional, biased, not well reasoned or factually inaccurate, and so on, but it does not prohibit well-reasoned, careful, accurate, factually-based criticisms of other engineers.

Thus in support of the second interpretation, it might be said that there too, it is not all attempts to injure reputations, which are being prohibited, but only those which are malicious, false, indiscriminate, or otherwise highly ethically questionable in the methods they employ.

However, there is still something to be said in favor of the first interpretation as well (which, it will be recalled, involves a blanket condemnation of all attempts to injure reputations). Those defending it might do so as follows: It is arguable on general ethical grounds that any attempt to injure someone's reputation must be viewed as going too far and therefore becoming unethical. Even if one is convinced that someone else's work is shoddy, dishonest, and so on, at most (it could be argued) one has a duty to point out the problems and shortcomings in their work. It is a big leap from criticizing an engineer's actions, on the one hand, to condemning the engineer in a way designed to injure his/her reputation, on the other hand.

Thus (on this line of thinking) one should criticize an engineer only out of a disinterested desire to let the truth be known by all, not with the aim of injuring someone's reputation—it is for others to judge whether the truth of what one has pointed out will diminish the reputation and so on of the engineer in question.

Fortunately, it is not necessary to definitively decide between these different interpretations of Rule 3. What is important is to become sensitive to the ethical issues involved in each interpretation. And for the purposes of conforming to Rule 3, both sides can agree that if criticism of other engineers ever becomes necessary, it should be done in a very cautious and objective manner, and with all due respect for the professional status of the person being criticized.

Example 15.7

With respect to the Moral Rules of Professional Conduct for engineers:

A) The rules are a bad thing because they encourage engineers to spy on and betray their colleagues

B) The rules are a useful legal defense in court, when engineers can demonstrate that they obeyed the rules

C) The rules enhance the image of the profession and hence its economic benefits to its members

D) The rules are important in providing a summary of what the public has a right to expect from responsible engineers

Solution: Remember here that the best/most relevant answer is to be chosen, because each answer has some truth to it, but only one has the most truth.

Choice A: It is true that the rules require those who have knowledge of violations of the rules to report such cases to the relevant State Board. And this could involve one in collecting more information on the possible infringements, and hence exposing those involved to the scrutiny of the State Board. However, the generally negative connotations of "spying" and "betraying colleagues" do not make the rules bad—the activities in question are a necessary part of responsible reporting of possible violations to the authorities, and hence are morally fully justified.

Choice B: It is true that proof in court that one has followed the rules may be a useful legal defense. However, this is only a secondary, indirect effect of the ethical value of the rules themselves, and so B does not provide the best answer.

Choice C: Again, it is true that the adoption of the rules by the engineering profession will enhance its image and economic benefits. But like B, this is only a secondary and derivative effect of the rules.

Choice D: The right answer. Since the basic function of the rules is to provide a guide for ethical conduct for engineers, the rules also provide a useful summary for the public of what they have a right to expect from responsible engineers.

Example 15.8

The Model Rules of Professional Conduct require registered engineers to conform to all but one of the following rules—which rule is not required?

A) Do not charge excessive fees
B) Do not compete unfairly with others
C) Perform services only in the areas of their competence
D) Avoid conflicts of interest

Solution: There may be one or two problems of this kind in the exam, which can be solved by memorizing the rules or by checking them directly to see which are or are not included. However, it is better to acquire a good understanding of the basic ethical concerns behind the rules, in which case the right answer here will be clear immediately.

Rule A is not required because fees are a matter of free negotiation between engineers and clients. Hence a fee which might seem excessive to some may be acceptable to others because of an interest in extra quality, or unusually quick delivery time, and so on. The other three rules are required.

Example 15.9

You are a quality control engineer, supervising the completion of a product whose specification includes using only U.S.-made parts. However, at a very late stage you notice that one of your sub-contractors has supplied you with a part having foreign-made bolts in it—but these aren't very noticeable, and would function identically to U.S.-made bolts. Your customer urgently needs delivery of the finished product—what should you do?

A) Say nothing and deliver the product with the foreign bolts included, hoping this fact won't be noticed by the customer
B) Find (or, if necessary, invent) some roughly equivalent violation of the contract or specifications for which the customer (rather than your company) is responsi-

ble—then tell them you'll ignore their violation if they ignore your company's violation

C) Tell the customer about the problem, and let them decide what they wish you to do next

D) Put all your efforts into finding legal loopholes in the original specifications, or in the way they were negotiated, to avoid your company's appearing to have violated the specifications

Solution: Choice A: This is wrong because it is dishonest—it violates the requirement of being objective and truthful in reports, etc.

Choice B: This is wrong because "two wrongs don't make a right." Negotiations with clients should always be done in an ethically acceptable manner.

Choice D: This would violate at least the spirit of the initial agreement. The ethical requirement of being objective and truthful includes an obligation not to distort the intent of any agreements with clients.

Choice C: The correct answer. Being honest with a client or customer about any production difficulties allows them to decide what is in their best interest given the new disclosures, and provides a basis for further good-faith negotiations between the parties.

Example 15.10

You are the engineer of record on a building project which is behind schedule and urgently needed by the clients. Your boss wants you to certify some roofing construction as properly completed even though you know some questionable installation techniques were used. Should you:

A) Certify it, and negotiate a raise from your boss as your price for doing so
B) Refuse to certify it
C) Tell the clients about the problem, saying that you'll certify it if they want you to
D) Certify it, but keep a close watch on the project in future in case any problems develop with it

Solution: There are some temptations and half-right elements in some of these, but they must be resisted as not being completely ethical.

Choice A: Even if one informs one's boss of the problems, and negotiates his/her consent to your certifying it, nevertheless it is always wrong to certify work which does not measure up to the highest professional standards.

Choice C: Wrong for the same reason as A. Even if one honestly reveals the problems to the clients, and gets their consent, nevertheless it is the professional duty of an engineer not to certify dubious work.

Choice D: The initial certification would be wrong, no matter how carefully one monitors future progress in the hope of minimizing any future problems.

Choice B: The right choice. Whether or not one's boss is happy with this, it is one's professional duty to refuse certification in such cases, even if as a result one is reassigned or fired.

Example 15.11

You are an engineer and a manager at an aerospace company with an important government contract supplying parts for a space shuttle. As an engineer, you know that a projected launch would face unknown risks, because the equipment for which you are responsible would be operating outside its tested range of behaviors. However, since you are also a manager you know how important it is to your company that the launch be carried out promptly. Should you:

A) Allow your judgment as a manager to override your judgment as an engineer, and so permit the launching

B) Toss a coin to decide, since one's engineering and managerial roles are equally important, so neither should take precedence over the other

C) Abstain from voting in any group decision in the matter, since as both a manager and an engineer one has a conflict of interest in this case

D) Allow your judgment as an engineer to override your judgment as a manager, and so do not permit the launching

Solution: Note that a real-life case similar to this problem occurred with the Challenger space-shuttle disaster.

Choice A: Wrong, because engineers have special professional duties and ethical commitments which go beyond those of corporate managers.

Choice B: Wrong for the same reason as A; ethically, engineering responsibilities are more important than managerial responsibilities.

Choice C: Wrong because one's duties as a professional engineer require appropriate action even if other factors may seem to point in the opposite direction or toward abstention.

Choice D: The correct choice. Whatever other duties an engineer has, his/her professional engineering responsibilities must always be given first priority.

Example 15.12

Your company buys large quantities of parts from various suppliers in a very competitive market sector. As a professional engineer you often get to make critical decisions on which supplier should be used for which parts. A new supplier is very eager to get your company's business. Not only that, but you find they are very eager to provide you personally with many benefits—free meals at high-class restaurants and free vacation weekends for (supposedly) business meetings and demonstrations, and other more confidential things such as expensive gifts that arrive through the mail, club memberships, and so on. What should you do?

A) Do not accept any of the gifts that go beyond legitimate business entertaining, even if your company would allow you to accept such gifts

B) Report all the gifts, etc., to your company, and let them decide whether or not you should accept them

C) Accept the gifts without telling your company, because you know that your professional judgment about the supplier will not be biased by the gifts

D) Tell other potential suppliers about the gifts, and ask them to provide you personally with similar benefits so you won't be biased in favor of any particular supplier

Solution: Choice B: Wrong, because even if one's company finds such gifts awkward, it is still one's duty as a professional engineer not to become involved in such conflicts of interest.

Choice C: Also wrong. It doesn't matter whether you believe you can remain unbiased, because you still have the conflict, and the possible perception by others that you might be biased by it also remains an ethical problem.

Choice D: Wrong. Remember, two wrongs don't make a right, and the same principle applies no matter how many wrong actions are involved.

Choice A: The right answer. As with the reasoning on choice B, it makes no difference whether others (having less demanding ethical standards than engineers) would find such things acceptable or not. You must not accept any gifts the acceptance of which would involve you in conflicts of interest.

15.5 ENGINEERING ETHICS AND LEGAL ISSUES

This section will provide an introduction to the various ways in which the law regulates and affects the profession of engineering. The impact of laws and regulations on various engineering ethics issues will also be discussed.

Engineers, like any citizens, are of course expected to obey the general legal rules and regulations of the societies in which they live. But here we will concentrate on those laws and legal concepts which have special relevance to engineers and the engineering professions. Every engineer needs to have a good basic grasp of these legal matters.

First is a very important area of the law which is central not only to engineering but to many professions and businesses as well. It concerns the regulation of the basic transactions between professionals and their clients or customers.

Contract Law

A *contract* is a mutual agreement between two or more parties to engage in a transaction which provides benefits to each of them. Here is a breakdown of this concept into its elements, each of which is required for a valid contract to exist. For simplicity we will concentrate on two-party contracts.

1) Mutual consent. Each of the parties to a contract must consent to the contract and agree to be bound by its terms.
2) Offer and acceptance. There must be an offer of some kind by one of the parties (e.g., to manufacture some equipment or carry out some services), and acceptance of that offer by the other party.
3) Consideration. Each party must provide something of value to the other party.

The first two conditions, mutual consent and offer/acceptance, are needed because merely discussing a possible contract in provisional terms establishes no legally binding agreement. A contract exists only when each party has established that the other consents to it, and that one is offering something which the other has decided to accept. In other words, a legally enforceable agreement requires a definite promise by each party to do something specific; vague talk about possible actions by each is not sufficient.

The issue of consideration is also important, but in more complex ways. The idea behind it is that there cannot be a legally enforceable contract unless each side stands to benefit in some way from the contract, so that after the contract is fulfilled, each has received some benefit which they did not have prior to being fulfilled.

Without such evidence of definite benefits for each party, it would often be impossible to decide whether or not each party had fulfilled their side of the agreement. For example, if an engineer were to hire a relative to do some work, and paid them for doing it, but did not specify in any way exactly what the work would be, then the courts would likely rule that no valid contract existed and so the payments to his/her relative were illegal or invalid.

Another kind of example where lack of consideration raises legal and ethical questions is as follows: If an engineering firm provided certain services to a client, but never asks for nor receives any payment for them, there would be strong suspicion that the services provided were some kind of bribe or illegal payback for other hidden services or benefits rather than part of a valid contract between the parties.

Some cases involving the issue of consideration are less clear-cut. For example, if an engineer decides to undertake a difficult and complex project, yet does so at a very low rate of pay far below standard prices for such work, could such a contract later be held to be invalid because of the very inadequate reward provided to the engineer? Generally the courts have held that it is up to the parties to a contract themselves to decide what constitutes adequate consideration or reward, so the law will not provide protection to engineers against any possibly self-destructive intentions or other errors of judgment in which they may become involved.

On the other hand, issues about specific amounts of consideration or reward are often central in settling contractual disputes between parties; see the section on breach of contracts below.

It should be noted that a contract does not have to be in writing to be valid. For example, an engineer could make a contract over the phone with a client. But a written contract is always desirable for clarity, and it provides documentation should any questions about the terms and conditions of the contract arise later. So engineers are well advised to prepare specific documentation on any agreements they make as soon as possible after the agreements are finalized.

Breach of Contract

One of the main values of having a legally enforceable system of contracts is that in the case of any disputes between the parties, the contract provides an independent check or test of the validity of the claims of each party to the dispute. For example, if a client orders some parts from an engineering firm, but later is dissatisfied with them and wants replacements or a refund of payments, then the issue can be resolved by checking in the contract to see whether the parts conform to what is specified for them in the contract. If the engineering firm has carried out its part of the contract by supplying parts as specified, then legally the client has no case against it. Each party to a contract must decide ahead of time whether they really want what they are agreeing to in a contract, and therefore they cannot blame the other party to the contract even if they are unhappy about the outcome of the agreement later.

In order for an actual breach of contract to occur, there must be some actual violation of the terms of the contract by one or both of the parties involved. For example, if specified items were not supplied, or were supplied but were of substandard quality, or if items were not supplied until long after a deadline had expired, then the other party could claim that a breach of contract had occurred and that they are entitled to compensation or termination of the contract.

Generally speaking, those suing for breach of contract will try to obtain sufficient damages or compensation to recover the value that they would have obtained under the contract had it not been violated. In other words, the other party is being required to provide an equivalent value to that which they had previously offered under the provisions of the contract.

One further distinction should be noted here: that between a material and an immaterial breach of a contract. Material breaches concern vital elements of the contract, the presence of which would justify termination of a contract and a suit for damages. But there may be less severe deviations from a contract (such as a contractor not cleaning up a site at the end of a day's work, or being a few days late in delivering a product) which would be handled differently. Usually immaterial breaches will require the offending party only to make reparations, or to accept a reduced fee for their work.

Often the distinction between a material versus immaterial breach of a contract will depend on issues about consideration (which is, you will recall, the value being received by each party to the contract). For example, suppose engineering firm A fails to supply client B with certain parts X as specified in a contract, but supplies substitute parts Y instead.

If the contract specified parts of X, then there has certainly been a breach of the contract. But whether it is a material or immaterial breach will depend on whether the substitute parts Y are similar enough in value to parts X so that client B should be willing to accept them as a substitute for parts X. Intuitively, if B is just as well off (or nearly so) with the parts Y as he would have been with parts X, then the breach of the contract would only be an immaterial one. However, if parts Y are substantially worse than parts X, there would be a material breach of the contract, requiring stronger legal action.

This material/immaterial distinction has ethical implications as well. Generally, any party to a contract who is unable to fulfill the exact provisions of the contract is under an ethical as well as a legal imperative to do everything possible to provide at least an equivalent value to the other party to a contract.

The Letter versus the Spirit of the Law

This is an important distinction. It comes about because laws or contracts typically are not completely specific in their terms, and so are open to varying degrees and types of interpretation in problematic cases.

How should such interpretations, or attempts to "read between the lines," be carried out? A standard view is that we should try to do so in terms of the intent or desired interpretations of those documents as understood by those who formulated them.

In this way ethical standards enter as an important element in understanding the "spirit" of laws and contracts. Because typically, lawmakers or contracting parties wrote their rules or agreements with general ethical standards in mind as to what would constitute the best and most ethically acceptable laws or contracts, given the purposes they were trying to achieve. Thus basic issues of engineering ethics are relevant even in cases when legal documents are being dealt with, insofar as such documents leave any room at all for disputes about their proper interpretation.

EXERCISES AND ACTIVITIES

Introductory Note

As with any form of ethics, engineering ethics demands more than an impersonal, spectator point of view from us. It is important to consider ethical problems not simply as some other person's problem—to be handled with impartial advice or citing of ethical rules—but instead as a vital problem for oneself which must be resolved in terms that you personally can live with.

To encourage this perspective, many of the cases are presented in the second person, i.e., as problems for you rather than some abstract person. In answering them it is important to maintain this perspective, and honestly explain how you would actually deal with each situation yourself.

15.1   Your company urgently needs a new manager in another division, and you have been invited to apply for the position. However, one of the job requirements is competence in the use of some sophisticated CAD (computer aided design) software. Currently you do not have that competence, but you are enrolled in a CAD course, so eventually you will have that competence (but not before the deadline for filling the new job).

The application form for the job requires you to state whether or not you have the necessary CAD software competence. Is it ethically permissible to claim that you have (in order to get the job for which otherwise you are well qualified), even though strictly speaking this is not true?

Would your answer be any different if your current boss informally (off the record) advised you to misrepresent your qualifications in this way, giving as his/her reason that you are overall the best qualified person for the new managerial position?

15.2   You are a supervisor for a complex design project, and are aware of the NCEES code rule according to which engineers may coordinate an entire project provided that each design component is signed or sealed by the engineer responsible for that design component.

You are very overworked, and so have got into the habit of, in effect, rubberstamping the design decisions of the subordinate engineers working on your project. You do no testing or checking of your own on their projects, and routinely approve their plans (after they have signed or sealed them) with only the most superficial read-through of their work. In your mind this policy is justifiable because the NCEES rule says nothing about the supervisory responsibilities of someone in your position. Besides which, you trust the engineers to do a good job, and so see no need to check up on them.

Is the supervisor's position here ethically justifiable?

15.3   Newly hired as a production engineer, you find a potential problem on the shop floor: workers are routinely ignoring some of the government-mandated safety regulations governing the presses and stamping machines.

The workers override safety features such as guards designed to make it impossible to insert a hand or arm into a machine. Or they rig up "convenience" controls so they can operate a machine while close to it, instead of using approved safe switches, etc., which requires more movement or operational steps. Their reason (or excuse) is that if the safety features were strictly followed then production would be very difficult, tiring, and inefficient. They feel that their shortcuts still provide adequately safe operation with improved efficiency and worker satisfaction.

Should you immediately insist on full compliance with all the safety regulations, or do the workers have enough of a case so that you would be tempted to ignore the safety violations? And if you were tempted to ignore the violations, how would you justify doing so to your boss?

Also, how much weight should you give to the workers' clear preference for not following the regulations: ethically, can safety standards be relaxed if those to whom they apply want them to be relaxed?

15.4   You and an engineer colleague work closely on designing and implementing procedures for the proper disposal of various waste materials in an industrial plant. He is responsible for liquid wastes, which are discharged into local rivers.

During ongoing discussions with your colleague, you notice that he is habitually allowing levels of some toxic liquid waste chemicals, which are slightly higher than levels permitted by the law for those chemicals. You tell him that you have noticed this, but he replies that, since the levels are only slightly above the legal limits, any ethical or safety issues are trivial in this case, and not worth the trouble and expense to correct them.

Do you agree with your colleague? If not, should you attempt to get him to correct the excess levels, or is this none of your business since it is he rather than you who is responsible for liquid wastes?

If he refuses to correct the problem, should you report this to your boss or higher management? And if no one in your company will do anything about the problem, should you be prepared to go over their heads and report the problem directly to government inspectors or regulators? Or should one do that only in a case where a much more serious risk to public health and safety is involved?

15.5 Your automotive company is expanding into off-road and all-terrain vehicles (ATVs). As a design engineer you are considering two alternative design concepts for a new vehicle, one having three wheels and another the four.

Engineering research has shown that three-wheeled vehicles are considerably less stable and safe overall than four-wheeled ones. Nevertheless, both designs would satisfy the existing safety standards for the sale of each class of vehicles, so either design could be chosen without any legal problems for the company.

However, there is another factor to consider. Market research has shown that a three-wheeled version would be easier to design and produce, and would sell many more vehicles at higher profit margins than would a four-wheeled version.

Is it ethically acceptable for you as an engineer to recommend that your company should produce the three-wheeled version, in spite of its greater potential safety risks? Or do engineers have an ethical obligation always to recommend the safest possible design for any new product?

15.6 Your company has for some time supplied prefabricated wall sections, which you designed, to construction companies. Suddenly one day a new idea occurs to you about how these might be fabricated more cheaply using composites of recycled waste materials.

Pilot runs for the new fabrication technique are very successful, so it is decided to entirely switch over to the new technique on all future production runs for the prefabricated sections. But there are managerial debates about how, or even whether, to inform the customers about the fabrication changes.

The supply contracts were written with specifications in functional terms, so that load-bearing capacities and longevity, etc., of the wall sections were specified, but no specific materials or fabrication techniques were identified in the contracts. Thus it would be possible to make the changeover without any violation of the ongoing contracts with customers.

On the other hand, since there is a significant cost saving in the new fabrication method, does your company have an ethical obligation to inform one's customers of this, and perhaps even to renegotiate supply at a reduced cost, so that one's customers also share in the benefits of the new technique?

More specifically, do you have any special duty, as a professional engineer and designer of the new technique, to be an advocate in one's company for the position that customers should be fully informed of the new technique and the associated cost savings?

15.7 Your company manufactures security systems. Up to now these have raised few ethical problems, since your products were confined to traditional forms of security, using armed guards, locks, reinforced alloys which are hard to cut or drill, and similar methods.

However, as a design engineer you realize that with modern technology much more comprehensive security packages could be provided to your customers. These could also include extensive video and audio surveillance equipment, along with biometric monitoring devices of employees or other personnel seeking entry to secure areas which would make use of highly personal data such as a person's fingerprints, or retinal or voice patterns.

But there is a potential problem to be considered. A literature search reveals that there are many ethical concerns about the collection and use of such personal data. For example, these high-tech forms of surveillance could easily become a form of spying, carried out without the knowledge of employees and violating their privacy. Or the data collected for security reasons could easily be sold or otherwise used outside legitimate workplace contexts by unscrupulous customers of your surveillance systems.

Your boss wants you to include as much of this advanced technology as possible in future systems, because customers like these new features and are willing to pay well for them.

However, you are concerned about the ethical issues involved in making these new technologies available. As an engineer, do you have any ethical responsibility to not include any such ethically questionable technologies in products which you design and sell, or to include them only in forms which are difficult to misuse? Or is the misuse of such technologies an ethical problem only for the customers who are buying your equipment, rather than it being your ethical responsibility as an engineer?

15.8   Can you think of any issue which should be of ethical concern to engineers but is not adequately addressed by current codes of engineering ethics?

If so, should it be added to the current rules, or is it something which cannot adequately be summed up in an ethical rule?

15.9   Can you think of any rule included in current codes of engineering ethics which is unnecessary, too restrictive, or even ethically wrong? If so, should it simply be removed from the current rules, or is it something which serves a useful function even though it is not ethically required?

If you don't think there are any superfluous rules, instead pick one which you think is one of the least ethically important, and explain why you think it is not important.

15.10   As an engineering expert on a state planning board, you have to decide which traffic safety projects (involving installation of traffic lights, road-widening, etc.) should be funded by the state. Previously these matters were decided politically: each region received roughly equal funding, which tended to maximize voter satisfaction and ensure re-election for the politicians. But now non-elected engineering experts such as yourself will decide these matters.

On what grounds should decisions such as these be made? What would be the soundest and most ethically justifiable way for you to allocate the funds?

In particular, would a utilitarian approach (the greatest good for the greatest number) be the best way to allocate funds? But there is a potential problem with this approach. Since the greatest number of people live in cities and suburbs and very few in rural areas, it is likely that scarce funds would always tend to be allocated to cities with this approach, with almost none going to sparsely-populated areas. This policy may tend to save the greatest number of lives or minimize accidents overall, but is it not very unfair to those living in or traveling through rural areas?

But on the other hand, how could you justify spending scarce resources in rural areas when you know that this would benefit far fewer people than if you spent the money in cities instead?

Chapter 16

Units and Conversions

16.1 HISTORY

"Weights and measures may be ranked among the necessaries of life to every individual of human society. They enter into the economical arrangements and daily concerns of every family. They are necessary to every occupation of human industry; to the distribution and security of every species of property; to every transaction of trade and commerce; to the labors of the husbandman; to the ingenuity of the artificer; to the studies of the philosopher; to the researches of the antiquarian, to the navigation of the mariner, and the marches of the soldier; to all the exchanges of peace, and all the operations of war. The knowledge of them, as in established use, is among the first elements of education, and is often learned by those who learn nothing else, not even to read and write. This knowledge is riveted in the memory by the habitual application of it to the employments of men throughout life."

—Secretary of State, John Quincy Adams
Report to Congress, 1821

Primitive societies used crude measures for many of their tasks. Length was measured with the forearm, the hand or the finger. Time was measured with reference to the periods of the sun and moon. The volumes of containers such as gourds or clay vessels were measured with the use of plant seeds. A container would be filled with seeds and the number of seeds would determine the volume of the container. Seeds and stones also served as a measure of weight. The "carat," used to measure the weight of a gem, was derived from the use of the carob seed.

With the evolution of society, it became more and more necessary to accurately measure various things. Commerce, land division, taxation, and scientific research made it mandatory that measurements could be reproduced with accuracy time after time and in various places. However, with limited international trade, different systems were developed for the same purposes in various parts of the world.

To ensure uniform standards for weights and measures in the United States, the Constitution gave power to Congress to establish the National Bureau of Standards, which assures uniformity throughout the country. The British system was instituted because of the dominance of Great Britain throughout the world and the position that she had in the affairs of the United States.

But the need for a single, worldwide measurement system soon became obvious. Because the British system uses 12 (12 inches in a foot) and 16 (16 ounces in a pound) as bases, it was not selected as the preferred system. The system selected, because of its simplicity was the metric system, first developed by the French Academy of Sciences in 1790. The unit of length was to be a fraction of the Earth's circumference—one ten-millionth of the distance from the North Pole to the Equator—the meter. The unit of measurement for volume was to be the cubic decimeter—the liter. The unit of mass was to be the mass of one cubic centimeter of water—the gram—at the temperature of maximum density. Larger and smaller values for each unit were to be found by multiplying or dividing by multiples of 10, thereby greatly simplifying calculations. In 1866, by an Act of Congress, the United States made it lawful to employ the weights and measures of the metric system in all contracts, dealings, or court proceedings.

In the late 19th century the Metric Convention was established, due to the need for better metric standards as required by the scientific community. This treaty was signed by 17 nations including the United States. It set up well-defined standards for length and mass, and established a method to further refine the metric system when such refinements were required. Since 1893, the metric standards have served as the fundamental weights and measures standards of the United States. The metric system is now either obligatory or permissible throughout the world. In 1971 the Secretary of Commerce recommended that the U.S. move toward predominant use of the metric system through a coordinated program. This change is occurring and has been instituted in many sectors of the U.S., but several sectors continue using the U.S. customary system, most notably those sectors associated with the construction industry.

The International Bureau of Weights and Measures (BIPM), located in Sevres, France, continues to serve as a permanent secretariat for the Metric Convention, coordinating the exchange of information about the use and refinement of the metric system. As scientific advances occur and the need for refinements arise, the General Conference of Weights and Measures—the diplomatic organization made up of members of the Convention—meets periodically to ratify changes in the system and standards.

In 1960 the General Conference ratified an extensive revision and simplification of the metric system. The French name *Le Système International d'Unités* —the International System of Units, abbreviated SI after the French name—was adopted as the modernized metric system. The SI system is a special metric system. It does not allow many of the metric units commonly used. The CGS system is a metric system, but many of the units, such as grams/cubic centimeter, are not permitted in the SI system.

Further additions and improvements have been made by the General Conference in more recent years. As an example, the second was originally defined as 1/86,400 of the mean solar day. But irregularities in the rotation of the Earth did not provide the desired accuracy. In 1967 the second was defined to be based on the transition between two energy levels of an atom. Such improvements in the definitions of the base units continue to be made by the General Conference as science dictates.

16.2 THE SI SYSTEM OF UNITS

The International System of Units—the SI system—is a modernized metric system adopted by the General Conference, a multi-national organization which includes the U.S. The SI system is built upon a foundation of seven base units, which are presented in Table 16.1. All other SI units are derived from these seven units. Multiples and sub-multiples are expressed using a decimal system. The base units for time, electric current, amount of a substance, and luminous intensity are the same in both the English and SI systems. English units, such as the yard and the pound, are officially defined in the U.S. in terms of the SI units, meter and kilogram.

The symbols used for the base and supplementary units are included in Table 16.1. In general, the first letter of a symbol is capitalized if the name of a symbol is derived from a person's name, otherwise it is lowercase. One exception is the symbol for the liter, which is recommended to be "L," so the lowercase "l" is not confused with the numeral "1." Examples are included in the following table.

Unit Name	Unit Symbol
meter	m
liter	L or l
kilogram	kg
ampere	A
pascal	Pa

Names of units and prefixes are not capitalized except at the beginning of sentences and in titles and headings in which all main words are capitalized. We would say, "Temperature is measured in kelvins," or, "Meters are longer than feet." Note that the temperature unit "kelvin" is not capitalized, but we do capitalize "Celsius." So we would write, "four degrees Celsius."

Unit symbols are printed in roman (upright) letters, and symbols for quantities or variables are printed in italic (slanted) letters. We would print $L = 10$m to mean that a length L is 10 meters in length. Or $V = 15$ L to mean that the volume V contains 15 liters. A period is never used after a symbol except at the end of a sentence.

A note regarding temperature is worth mention. In addition to the temperature T expressed in kelvins, we also can express the temperature in Celsius, often denoted t. The two temperatures are related by the equation

$$t = T - T_0$$

where $T_0 = 273.15$ K (usually 273 K) by definition. The Celsius temperature is expressed in degrees Celsius (i.e., 20°C) but we do not use the small degree symbol with K (i.e., 100 K).

Names of units are often plural for numerical values equal to zero, greater than 1, or less than −1. The singular form is used for other values. We would say 1.1 meters, 0.1 meter, or 0 meters. We could, however, say 1.1 meter or −1 degree Celsius. In symbolic form, 100 m would be read "100 meters."

When writing units after a numerical value, there are three recommendations for expressing those units.

- The product of two or more units is indicated by a dot, although it is possible to omit the dot if no confusion results. We would write N·m but never Nm.

- The division of units can be written as m/s, or m·s^{-1}, or $\dfrac{m}{s}$, for example.

- In more complicated groupings of units we could write ^{s}m / s^2 or m·s^{-2}, but not m/s/s. A more complicated example would be m · kg / (s^2 · A) or m · kg · s^{-2} · A^{-1}, but we would not write m · kg / s^2 / A.

16.3 DERIVED UNITS

Derived units are expressed algebraically in terms of base and supplementary units using the three recommendations listed at the end of Section 16.2. Several derived units have been given special names and symbols, such as the newton with symbol N. These special names and symbols may be used to express the units of a quantity in a simpler way than by using the base units. The following three tables list many such derived units. Table 16.2 lists quantities whose derived units are expressed in terms of the base and supplementary units. Table 16.3 lists the quantities whose derived units have special names. And Table 16.4 lists those quantities whose units are expressed in terms of the derived units that have special names.

There are a number of observations concerning the following three tables.

- Any measurable property is called a quantity. Symbols that represent quantities, such as *F* for force, are always printed in italic letters. The viscosity μ is italic but the SI unit prefix μ is roman.
- Although a derived unit can be expressed in several ways by using base units or special names, it is acceptable to use certain combinations to distinguish more easily between quantities with the same units. For example, we may use the newton-meter for the moment of a force and not the equivalent joule, even though 1 N·m = 1 J.
- Plane angles in degrees are usually expressed with decimal subdivisions rather than minutes and seconds.
- We never refer to mass as weight. Weight is a force measured in newtons.
- Viscosity was often measured in poises or centipoises; those terms are obsolete.
- Kinematic viscosity was often measured in stokes or centistokes; those terms are obsolete.
- The kilowatt-hour is a measure of energy (like the joule). Eventually it will be obsolete.

There are a number of units outside the SI system of units that are often used. These are listed in Table 16.5. Such units are usually avoided, however, when writing quantities of interest. Purchased electric energy is currently measured in kilowatt-hours, an example of a mixture of units that is commonly used.

Finally, there are a number of units that are being tolerated for a limited time. Eventually, the use of such units will be discouraged. These units are listed in Table 16.6. It also should be noted that the CGS system of metric units is not used in the SI system. Such units as erg, dyne, poise, gauss, and maxwell are of the CGS system and are not to be used. Other units to avoid are fermi, torr, calorie, micron, stere, and gamma, to name a few.

TABLE 16.1 The SI base and supplementary units

Unit (Symbol) Quantity	Definition	Comments
meter (m) length	1650 763.73 wavelengths in a vacuum of the orange-red line of the spectrum of krypton-86.	An interferometer is used to measure wavelength by means of light waves.
kilogram (kg) mass	A cylinder of platinum-iridium alloy kept by the International Bureau of Weights and Measures at Paris.	This is the only base unit defined by an artifact, and the only base unit whose name contains a prefix.
second (s) time	The duration of 9 192 631 770 periods of the radiation associated with a specified transition of the cesium-133 atom.	The number of periods or cycles per second is called frequency. The SI unit for frequency is the **hertz** (Hz).
ampere (A) electric current	That current which, if maintained in each of two parallel wires separated by one meter in free space, would produce a force of 2×10^{-7} N/m.	The force is due to the magnetic field.
kelvin (K) temperature	1/273.16 of the temperature of the triple point of water.	On the Celsius scale, water freezes at 0°C and boils at 100°C.
mole (mol) amount of substance	The amount of substance of a system that contains as many elementary entities as there are atoms in 0.012 kg of carbon-12.	When the mole is used, the elementary entities must be specified: atoms, molecules, ions, electrons, etc.
candela (cd) luminous intensity	The luminous intensity of 1/ 600 000 of a square meter of a blackbody at the temperature of freezing platinum.	A black body absorbs all radiation incident upon it and reflects none.
radian (rad) plane angle	The plane angle with its vertex at the center of a circle that is subtended by an arc equal in length to the radius.	This is a supplementary unit.
steradian (sr) solid angle	The solid angle with its vertex at the center of a sphere that is subtended by an area of the spherical surface equal to that of a square with sides equal in length to the radius.	This is a supplementary unit.

TABLE 16.2   Quantities whose units are expressed in terms of base and supplementary units

Quantity	SI Unit	SI Symbol
area	square meter	m^2
volume	cubic meter	m^3
speed, velocity	meter per second	m/s
acceleration	meter per second squared	m/s^2
density	kilogram per cubic meter	kg/m^3
specific volume	cubic meter per kilogram	m^3/kg
magnetic field strength	ampere per meter	A/m
concentration	mole per cubic meter	mol/m^3
luminance	candela per square meter	cd/m^2
kinematic viscosity	square meter per second	m^2/s
angular velocity	radian per second	rad/s
angular acceleration	radian per second squared	rad/s^2

TABLE 16.3 Quantities whose units have special names

Quantity	SI Name	SI Symbol	Other SI Units
frequency	hertz	Hz	cycle/s
force	newton	N	$kg \cdot m/s^2$
pressure, stress	pascal	Pa	N/m^2
energy, work	joule	J	N·m
power	watt	W	J/s
electric charge	coulomb	C	A·s
electric potential	volt	V	W/A
capacitance	farad	F	C/V
electric resistance	ohm	Ω	V/A
conductance	siemens	S	A/V
magnetic flux	weber	Wb	V·s
magnetic flux density	tesla	T	Wb/m^2
inductance	henry	H	Wb/A
luminous flux	lumen	lm	
illuminance	lux	lx	

TABLE 16.4 Quantities whose units are expressed in terms of derived units with special names

Quantity	SI Unit	SI Symbol
viscosity	pascal second	$Pa \cdot s$
moment of force, torque	newton meter	$N \cdot m$
surface tension	newton per meter	N/m
heat flux density	watt per square meter	W/m^2
entropy	joule per kelvin	J/K
specific heat	joule per kilogram kelvin	$J/(kg \cdot K)$
specific entropy	joule per kilogram kelvin	$J/(kg \cdot K)$
specific energy	joule per kilogram	J/kg
thermal conductivity	watt per meter kelvin	$W/K \cdot m$
electric field strength	volt per meter	V/m
electric charge density	coulomb per cubic meter	C/m^3
electric flux density	coulomb per square meter	C/m^2
permittivity	farad per meter	F/m
permeability	henry per meter	H/m

TABLE 16.5 Units used with the SI system

Name	Symbol	Value in SI units
minute (time)	min	$1\ min = 60\ s$
hour	h	$1\ h = 3600\ s$
day	d	$1\ d = 86\,400\ s$
degree	°	$1° = \pi/180\ rad$
minute (angle)	'	$1' = \pi/10\,800\ rad$
second	"	$1'' = \pi/648\,000\ rad$
tonne	t	$1\ t = 1000\ kg$

TABLE 16.6 Units that are being used temporarily

Name	Symbol	Value in SI Units
nautical mile		1 nautical mile = 1852 m
knot		1 knot = 0.5144 m/s
ångström	Å	$1 \text{ Å} = 10^{-10}$ m
are	a	$1 \text{ a} = 100 \text{ m}^2$
hectare	ha	$1 \text{ ha} = 10\,000 \text{ m}^2$
barn	b	$1 \text{ b} = 10^{-28} \text{ m}^2$
bar	bar	$1 \text{ bar} = 10^5$ Pa
standard atmosphere	atm	1 atm = 101 325 Pa
gal	Gal	$1 \text{ Gal} = 0.01 \text{ m} / \text{s}^2$
curie	Ci	$1 \text{ Ci} = 3.7 \times 10^{10} \text{ s}^{-1}$
röntgen	R	$1 \text{ R} = 2.58 \times 10^{-4}$ C/kg
rad	rad	1 rad = 0.01 J/kg

16.4 PREFIXES

Rather than write extremely large numbers or very small numbers, we use prefixes that have been defined for the SI system. Their symbols and pronunciations are listed in Table 16.7. Some are capitalized, while others are lowercase to avoid confusion. Note that G stands for giga while g stands for gram, K for kelvin and k for kilo, M for mega and m for milli, and N for newton and n for nano. The need for the capital letters is obvious. We do not leave a space between the prefix and the letters making up a symbol or a name. We write "milliliter" and "mL."

Hecto, deka, deci, and centi are metric prefixes that are to be avoided, except for unit multiples of area and volume and for the non-technical use of centimeter, as for body and clothing measurements.

In general, prefixes that result in numerical values between 0.1 and 1000 should be selected, except that the same prefix should be used for all items in a given context or for the entries of the same quantity in a table.

Notation in powers of 10 is often used rather than a prefix. Rather than writing 20 MJ, we often write 2×10^7 J. Either way is equally acceptable.

Avoid mixing prefixes. We would dimension an area as 40-mm wide and 1500-mm long, not 40-mm wide and 1.5-m long. However, a wire would be described as 1500 meters of 2-mm-diameter wire since the difference in size is extreme.

Never use two units for one quantity. We would say a board is 3.5-m long, not 3-m 50-cm long, or 3-m 500-mm long. A volume contains 13.58 L, not 13 L 580 mL.

TABLE 16.7 Prefixes for metric units

Multiplication Factor	Prefix	Symbol	Pronunciation	Term (USA)
$1000\ 000\ 000\ 000\ 000\ 000 = 10^{18}$	exa	E	Texas	one quintillion
$1000\ 000\ 000\ 000\ 000 = 10^{15}$	peta	P	petal	one quadrillion
$1000\ 000\ 000\ 000 = 10^{12}$	tera	T	terrace	one trillion
$1000\ 000\ 000 = 10^{9}$	giga	G	giggle	one billion
$1000\ 000 = 10^{6}$	mega	M	megaphone	one million
$1000 = 10^{3}$	kilo	k	kilowatt	one thousand
$100 = 10^{2}$	hecto	h	heck	one hundred
10	deka	da	deck	ten
$0.1 = 10^{-1}$	deci	d	decimal	one tenth
$0.01 = 10^{-2}$	centi	c	sentiment	one hundredth
$0.001 = 10^{-3}$	milli	m	military	one thousandth
$0.000\ 001 = 10^{-6}$	micro		mike	one millionth
$0.000\ 000\ 001 = 10^{-9}$	nano	n	Nancy	one billionth
$0.000\ 000\ 000\ 001 = 10^{-12}$	pico	p	peek	one trillionth
$0.000\ 000\ 000\ 000\ 001 = 10^{-15}$	femto	f	feminine	one quadrillionth
$0.000\ 000\ 000\ 000\ 000\ 001 = 10^{-18}$	atto	a	anatomy	one quintillionth

In general, when forming the multiple of a compound unit only one prefix is used, usually in the numerator. For example, we would write MN/m^2, not N/mm^2, or km/s, not m/ms.

Also, do not use double prefixes. Use 10 nm, not 10 mµm; and use 23.2 mg, not 23.2 µkg.

Finally, when using SI prefixes we do not use M to indicate thousands (as in MCF for thousands of cubic feet). Nor do we use MM to indicate millions or C to indicate hundreds.

16.5 NUMERALS

A space is always left between the numeral and the unit name or symbol, except when we write the degree symbol, as in 30°, or the symbol for temperature, as in 30°C. It would be incorrect to write 30 °C, with a space before the degree symbol. Likewise, it is incorrect to write 30m without a space before the meter symbol.

When a quantity is used in an adjectival sense, a hyphen is used between the number and the symbol or unit name (except for angle degree). An example would be: The length of the 20-mm-diameter pole is 30 m. When names are used rather than the symbols, the hyphens are still used.

The decimal point is used in the U.S. and Canada; however, in many countries a comma is used rather than a decimal point. Because of this, the comma is replaced in the SI system with a space (sometimes a small space as in Table 16.7). Decimal notation is preferred to fractions (3.75 rather than $3\frac{3}{4}$). Simple fractions, such as 1/3, are acceptable.

A zero is used before the decimal point for numbers between −1 and 1. This prevents a faint decimal point from being missed.

When grouping digits in long numbers, it is customary to separate the digits in groups of three. This is optional when there are four digits as in 3876 or 0.3219. We could write 3 876 or 0.321 9. It is not optional for numbers such as 34 627 or 0.236 98.

In calculations a number must often be rounded off. If the number 8.3745 is rounded to three digits, it would be 8.37. If it is rounded to four digits, it would become 8.374 since the digit before the "5" is an even number (other authors, or your calculator, may round this to 8.375). If it were an odd number, it would be rounded up. The number 8.3755 would be rounded to 8.376. If the original number were 8.37451, it would be rounded to 8.375 if four digits were desired, since 51 is greater than a half.

Finally, a word is in order regarding significant digits. With the advent of the calculator and the computer, each calculation may be carried out to eight or more digits. In most problems a material property (e.g., density, viscosity, conductivity) is used in the calculations. Material properties are either given in the problem statement or are found in an appropriate table. A material property, which has been measured in some laboratory, is seldom known to more than four significant digits. Most are known to only three significant digits. It is impossible to calculate an answer to more significant digits than are present in the data used to arrive at that answer. We usually assume that information given in a problem statement, such as the number 6 in a 6-mm diameter pipe, is known to four significant digits. To present an answer with six significant digits, when at most four significant digits are known in the data provided, is an error made by many students. Note: the number 4.53 has three significant digits, as does the number 1.463; the leading numeral 1 is not counted as a significant digit. Also, the numbers 0.00324, 324 000, 3.24×10^{-6}, and 3.24×10^{6} each have three significant digits. The number 3.2400 has five significant digits.

16.6 CONVERSIONS

We often encounter quantities that have units that are not appropriate for the problem at hand. The need exists to convert the units to those desired, usually to SI units. Table 16.8 presents such units. For example, to convert 40 acres to square meters we simply multiply

TABLE 16.8 Conversion Factors

To convert from	To	Multiply by
Acres	Square meters	4046.86
Acres	Square feet	43 560
Acres	Square meters	100
Atmospheres	Bars	1.0132
Atmospheres	Feet of water	33.90
Atmospheres	Inches of mercury	29.92
Atmospheres	psi	14.700
Atmospheres	Torrs	760
Bars	psi	14.504

TABLE 16.8 Conversion Factors (*continued*)

To convert from	To	Multiply by
Btu	Foot-pounds	777.6
Btu	Joules	1054.4
Btu	kWh	0.000 292 9
Btu/min	Ton of refrigerations	0.004 997
Btu/lb	kJ/kg	2.324
Btu/sec	Horsepower	1.4139
Btu/sec	kW	1.0544
Calories	Btu	0.003 968
Coulombs	Ampere-seconds	1
Cubic feet	Cubic meters	0.028 317
Cubic feet	Liters	28.317
Cubic feet	Gallons	7.4805
Cubic inches	Milliliters	16.387
Degrees	Radians	0.017 453
Dynes	Newtons	0.000 01
Electron volts	Ergs	1.60219×10^{-12}
Ergs	Joules	10^{-7}
Hectares	Acres	2.4711
Hectares	Square meters	10 000
Horsepower	Btu/hr	2545
Horsepower	Ft-lb/sec	550
Horsepower	kW	0.745 70
Inches	Centimeters	2.54
Inches of mercury	psf	70.7
Kilowatts	Horsepower	1.341 02
Kilowatt-hrs	Btu	3409.5
Knots	Feet/sec	1.6878
Knots	Meters per second	0.514 44
Maxwells	Webers	10^{-8}
Meters	Feet	3.2808
Miles	Feet	5280

TABLE 16.8 Conversion Factors (*continued*)

To convert from	To	Multiply by
Miles	Meters	1609.34
Miles/hr	kilometers/hour	1.609 34
Newtons	Pounds	0.224 81
Pounds	Kilograms	0.453 59
Pounds	Slugs	0.031 08
Pounds per cubic foot	kg/m^3	16.018
Pounds per square foot	N/m^2 (Pa)	47.88
Pounds per square inch	N/m^2 (Pa)	6895
Radians	Revolutions	0.159 15
Radians	Degrees	57.296
Revolutions	Radians	6.2832
Siemens	Mhos	1
Slugs per cubic foot	kg/m^3	515.38
Spans	Feet	0.75
Square feet	Square meters	0.092 90
Square kilometers	Acres	247.10
Square meters	Square feet	10.764
Steres	Cubic meters	1
Tons (metric)	Kilograms	1000
Tons (short)	Pounds	2000
Tons of refrigeration	Btu/hr	12 000
Tons of refrigeration	Horsepower	4.716
Tons of refrigeration	Lb of ice melted/day	2000
Torrs	mm of mercury	1
Watts	Horsepower	0.001 341
Webers	Maxwells	10^8

40 by 4046.86 and obtain 161 874 square meters if we desire six significant digits, or 161 900 if we desire three significant digits.

Also, the units for a particular quantity may not appear to be consistent with other units in a given equation. For example, the energy equation contains a work term, force times distance, as well as a kinetic-energy term, mass times velocity squared divided by two. The units on the work term are N · m and the units on the kinetic-energy term are kg · m^2 / s^2. How are these two units equal, which they must be if they are on terms in the same equation? We show that the units are equal by referring to Newton's second law, which states that force equals mass times acceleration:

$$F = ma$$

The units on each term in this equation must be equal. This leads to the very important relationship

$$N = kg \cdot m \, / \, s^2$$

This relationship between units can be used to show that the units on the work term and the kinetic-energy term are indeed the same:

$$N \cdot m = (kg \cdot m \, / \, s^2) \cdot m = kg \cdot m^2 \, / \, s^2$$

REFERENCES

"Metric Editorial Guide," 5th edition, 1993, published by the American National Metric Council, (301) 718-6508.
"The International System of Units (SI)," 1974, published by the U.S. Department of Commerce/National Bureau of Standards.
"Brief History of Measurement Systems," 1974, Special Publication 304A, published by the U.S. Department of Commerce/National Bureau of Standards.

EXERCISES AND ACTIVITIES

16.1 State both the primitive and the modern definitions of time. Which do you find the more reasonable?

16.2 List at least five metric units that are not acceptable in the SI system.

16.3 How many base and supplementary units in the SI system can one actually touch? Name such units.

16.4 Which of the following units or variables are printed incorrectly?
   a) for time
   b) T for temperature
   c) L for liter
   d) V for volt
   e) *V* for electric voltage
   f) N for force
   g) *N* for newton
   h) μ for friction coefficient

i)   μ for micro
j)   T for magnetic flux density
k)   *T* for tesla

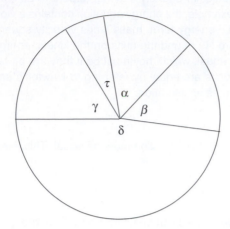

16.5   Select the angle that is approximately equal to a radian.

16.6   Select the correct statement.
a)   The temperature is 10°K.
b)   The temperature is 283°C.
c)   The temperature is 10 kelvins.
d)   The temperature is 283 Celsius.

16.7   Write each of the following in a correct form without using any exponents:
a)   12300 Nmkg^{-1} K^{-1}
b)   109.433 Cs m^{-1} K^{-1}
c)   0.000433 W s m^{-3} K^{-1}

16.8   Write each of the following in a correct form without using any negative exponents:
a)   0.004562 kNm m^{-3} K^{-1}
b)   10^{-5} MNs^{-1} m^{-3} K^{-1}
c)   0.000473001 mJs^{-1} m^{-3} kg^{-1}

16.9   Write each of the following in an acceptable form:
a)   324 kJ/kg/K/m/s/m
b)   43.098 Pa kg/K/m/s/s
c)   100 C/kg/K/m/s

16.10   Write each of the following in an acceptable form without using a decimal point:
a)   0.0000546 kW/K/m/s/m
b)   65.3207 kH/kg/m/s/s
c)   54.6367 × 10^{-6} MF/kg/m/s

16.11   Write the viscosity $\mu = 0.001$ Pa · s using newtons and not pascals. Also, write the viscosity without using either newtons on pascals.

16.12   Write the heat flux density 300 W/m^2 with the units in the following forms:
a)   one that includes joules
b)   one that includes newtons
c)   one that includes kilograms

16.13 Write the permeability 0.231 H/m using base units only.

16.14 Write the permittivity 10^{-5} F/m using base units only, and without the use of a negative exponent or a decimal point.

16.15 Write the electric charge density 3×10^{-4} C/m^3 using base units only, and without the use of a negative exponent or a decimal point.

16.16 Write each of the following using base units only, and without the use of a prefix:
  a) 5643 kN/t · d
  b) 453000 μ H/(kg · h)
  c) 654.0004 nS · m/(h · kg)

16.17 Write each of the following using base units without the use of a prefix:
  a) 4×10^{-6} GF/m
  b) 65.2 fF/m
  c) 0.00347 pPa/kg

16.18 Write each of the following using base units without the use of an exponent:
  a) 4×10^{-6} GF/m
  b) 43.64×10^6 mPa/mm^2
  c) 0.000453×10^{-4} TT/mm^3

16.19 Write each of the following in an acceptable form:
  a) 12.3 npkg
  b) 0.0035 GfMcm
  c) 6548 nTmMK

16.20 Write each of the following in an acceptable form:
  a) 23.2 N/nm^3
  b) 0.0054 kH/d · nm^3
  c) 6571 MF/ks · nm^3

16.21 The viscosity is measured as follows. Write each in acceptable form using base units only:
  a) 17/9800 Pa · s   b) 32/8741 N · s/m^2 c) 61/34 982 kN · h/mm^2

16.22 Round off 6743.865 to:
  a) six significant digits
  b) five significant digits
  c) four significant digits
  d) three significant digits
  e) two significant digits

16.23 Express each of the following to four significant digits:
  a) 1/3
  b) $89\dfrac{2}{3}$
  c) 2300.71
  d) 3/1000
  e) 1/3000
  f) 9/882

16.24   Find the force needed to accelerate a 4-kg mass at:

a) $\frac{1}{3}$ m/s²

b)   10 m/s²

c)   56 $\frac{4}{9}$ m/s²

16.25   Express each of the following conversions to four significant digits:
a)   340 Btu/s to watts
b)   3458 degrees to radians
c)   25 hectares to square meters
d)   240 horsepower to watts
e)   67 851.22 kWh to joules
f) 2×10^{-7} miles to meters
g)   40 mph to km/s

Chapter 17

Mathematics Review

INTRODUCTION

Mathematics is used as a tool to help solve the problems encountered in the analysis and design of physical systems. We will review those parts of mathematics that you may have already studied and that are used fairly often. The topics include: algebra, trigonometry, analytic geometry, linear algebra (matrices), and calculus. The review here is intended to be brief and not exhaustive. There may be some subjects included in this chapter that you have not yet encountered. If that is the case, your instructor will not require that you review that material.

17.1 ALGEBRA

It is assumed that the reader is familiar with most of the rules and laws of algebra as applied to both real and complex numbers. We will review some of the more important of these and illustrate several with examples. The three basic rules are:

Commutative Law:	$a + b = b + a$	$ab = ba$	(17.1.1)
Distributive Law:	$a(b + c) = ab + ac$		(17.1.2)
Associative Law:	$a + (b + c) = (a + b) + c$	$a(bc) = (ab)c$	(17.1.3)

Exponents

Laws of exponents are used in many manipulations. For positive x and y we use

$$x^{-a} = \frac{1}{x^a}$$

$$x^a x^b = x^{a+b}$$

$$(xy)^a = x^a y^a$$

$$x^{ab} = (x^a)^b$$

(17.1.4)

Logarithms

Logarithms are actually exponents. For example if $b^x = y$ then $x = \log_b y$; that is, the exponent x is equal to the logarithm of y to the base b. Most applications involve common logs which have a base of 10, written as $\log y$, or natural logs which have a base of e ($e = 2.7183$ . . .), written as $\ln y$. If any other base is used it will be so designated, such as $\log_5 y$.

Remember, logarithms of numbers less than one are negative, the logarithm of one is zero, and logarithms of numbers greater than one are positive. The following identities are often useful when manipulating logarithms:

$$
\begin{aligned}
\ln x^a &= a \ln x \\
\ln(xy) &= \ln x + \ln y \\
\ln(x/y) &= \ln x - \ln y \\
\ln x &= 2.303 \log x \\
\log_b b &= 1 \\
\ln 1 &= 0 \\
\ln e^a &= a \\
\log_a y &= x \text{ implies } a^x = y
\end{aligned}
\tag{17.1.5}
$$

The Quadratic Formula and the Binomial Theorem

We often encounter the quadratic equation $ax^2 + bx + c = 0$ when solving problems. The *quadratic formula* provides its solution; it is

$$
x = \frac{-b \pm \sqrt{b^2 - 4ac}}{2a}
\tag{17.1.6}
$$

If $b^2 < 4ac$, the two roots are complex numbers. Cubic and higher order equations are most often solved by trial and error.

The *binomial theorem* is used to expand an algebraic expression of the form $(a + x)^n$. It is

$$
(a + x)^n = a^n + na^{n-1}x + \frac{n(n-1)}{2!} a^{n-2}x^2 + \cdots
\tag{17.1.7}
$$

If n is a positive integer, the expansion contains $(n + 1)$ terms. If it is a negative integer or a fraction, an infinite series expansion results.

Partial Fractions

A rational fraction $P(x) / Q(x)$, where $P(x)$ and $Q(x)$ are polynomials, can be resolved into partial fractions for the following cases.

Case 1: $Q(x)$ factors into n different linear terms,

$$
Q(x) = (x - a_1)(x - a_2) \ldots (x - a_n)
$$

Then

$$
\frac{P(x)}{Q(x)} = \sum_{i=1}^{n} \frac{A_i}{x - a_i}
\tag{17.1.8}
$$

Case 2: $Q(x)$ factors into n identical terms,

$$Q(x) = (x - a)^n$$

Then

$$\frac{P(x)}{Q(x)} = \sum_{i=1}^{n} \frac{A_i}{(x - a)^i} \qquad (17.1.9)$$

Case 3: $Q(x)$ factors into n different quadratic terms,

$$Q(x) = (x^2 + a_1 x + b_1)(x^2 + a_2 x + b_2) \cdots (x^2 + a_n x + b_n)$$

Then

$$\frac{P(x)}{Q(x)} = \sum_{i=1}^{n} \frac{A_i x + B_i}{x^2 + a_i x + b_i} \qquad (17.1.10)$$

Case 4: $Q(x)$ factors into n identical quadratic terms,

$$Q(x) = (x^2 + ax + b)^n$$

Then

$$\frac{P(x)}{Q(x)} = \sum_{i=1}^{n} \frac{A_i x + B_i}{(x^2 + a x + b)^i} \qquad (17.1.11)$$

Example 17.1

The temperature at a point in a body is given by $T(t) = 100e^{-0.02t}$. At what value of t does $T = 20$?

Solution: The equation takes the form

$$20 = 100e^{-0.02t} \text{ or } 0.2 = e^{-0.02t}$$

Take the natural logarithm of both sides and obtain

$$\ln 0.2 = \ln e^{-0.02t}$$

Using a calculator, and Eq. 17.1.5, we find

$$-1.6094 = -0.02t$$

$$\therefore t = 80.47$$

Example 17.2

Solve for V if $3V^2 + 6V = 10$.

Solution: Use the quadratic formula, Eq. 17.1.6. In standard form, the equation is

$$3V^2 + 6V - 10 = 0$$

The solution is then

$$V = \frac{-6 \pm \sqrt{36 - 4 \times 3 \times (-10)}}{2 \times 3} = \frac{-6 \pm 12.49}{6}$$

$$= 1.082 \text{ or } -3.082$$

Note: Some physical situations may not permit a negative answer, so $V = 1.082$ would be the selection.

Example 17.3

Resolve $\dfrac{3x - 1}{x^2 + x - 6}$ into partial fractions.

Solution: The denominator is factored:

$$x^2 + x - 6 = (x + 3)(x - 2)$$

Using Case 1 there results

$$\frac{3x - 1}{x^2 + x - 6} = \frac{A_1}{x + 3} + \frac{A_2}{x - 2}$$

This can be written as

$$\frac{3x - 1}{x^2 + x - 6} = \frac{A_1(x - 2) + A_2(x + 3)}{(x + 3)(x - 2)}$$

$$= \frac{(A_1 + A_2)x - 2A_1 + 3A_2}{(x + 3)(x - 2)}$$

The numerators on both sides must be equal. Equating the coefficients of the various powers of x provides us with two equations:

$$A_1 + A_2 = 3$$

$$-2A_1 + 3A_2 = -1$$

These are solved to give $A_2 = 1$, A, = 2. Finally,

$$\frac{3x - 1}{x^2 + x - 6} = \frac{2}{x + 3} + \frac{1}{x - 2}$$

17.2 TRIGONOMETRY

The primary functions in trigonometry involve the ratios between the sides of a right triangle. Referring to the right triangle in Fig. 17.1, the functions are defined by

$$\sin\theta = \frac{y}{r}, \qquad \cos\theta = \frac{x}{r}, \qquad \tan\theta = \frac{y}{x} \qquad (17.2.1)$$

In addition, there are three other functions that find occasional use, namely,

$$\cot\theta = \frac{x}{y}, \qquad \sec\theta = \frac{r}{x}, \qquad \csc\theta = \frac{r}{y} \qquad (17.2.2)$$

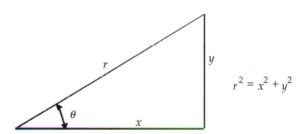

Figure 17.1 A right triangle.

The trig functions $\sin\theta$ and $\cos\theta$ are periodic functions with a period of 2π. Fig. 17.2 shows a plot of the three primary functions.

In the above relationships, the angle θ is usually given in radians for mathematical equations. It is possible, however, to express the angle in degrees; if that is done it may be necessary to relate degrees to radians. This can be done by remembering that there are 2π radians in 360°. Hence, we multiply radians by ($180 / \pi$) to obtain degrees, or multiply degrees by ($\pi / 180$) to obtain radians. A calculator may use either degrees or radians for an input angle.

Most problems involving trigonometry can be solved using a few fundamental identities. They are

$$\sin^2\theta + \cos^2\theta = 1 \qquad (17.2.3)$$

$$\sin 2\theta = 2\sin\theta\cos\theta \qquad (17.2.4)$$

$$\cos 2\theta = \cos^2\theta - \sin^2\theta \qquad (17.2.5)$$

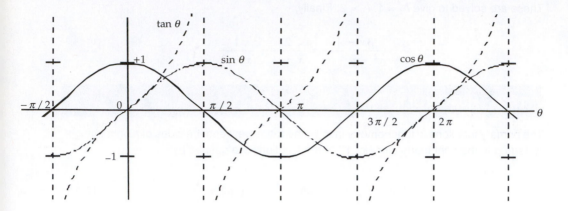

Figure 17.2 The trig functions.

$$\sin (\alpha \pm \beta) = \sin \alpha \cos \beta \pm \sin \beta \cos \alpha \qquad (17.2.6)$$

$$\cos(\alpha \pm \beta) = \cos \alpha \cos \beta \pm \sin \alpha \sin \beta \qquad (17.2.7)$$

A general triangle may be encountered such as that shown in Fig. 17.3. For this triangle we may use the following equations:

$$\text{law of sines:} \quad \frac{\sin \alpha}{a} = \frac{\sin \beta}{b} = \frac{\sin \gamma}{c} \qquad (17.2.8)$$

$$\text{law of cosines:} \quad a^2 = b^2 + c^2 - 2bc \cos \alpha \qquad (17.2.9)$$

Note that if $\gamma = 90°$, the law of cosines becomes the *Pythagorean Theorem*

$$c^2 = a^2 + b^2 \qquad (17.2.10)$$

The hyperbolic trig functions also find occasional use. They are defined by

$$\sinh x = \frac{e^x - e^{-x}}{2}, \quad \cosh x = \frac{e^x + e^{-x}}{2}, \quad \tanh x = \frac{\sinh x}{\cosh x} \qquad (17.2.11)$$

Useful identities follow:

$$\cosh^2 x - \sinh^2 x = 1 \qquad (17.2.12)$$

$$\sinh (x \pm y) = \sinh x \cosh y \pm \cosh x \sinh y \qquad (17.2.13)$$

$$\cosh (x \pm y) = \cosh x \cosh y \pm \sinh x \sinh y \qquad (17.2.14)$$

The values of the primary trig functions of certain angles are listed in Table 17.1.

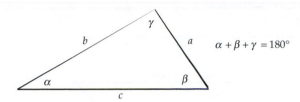

$$\alpha + \beta + \gamma = 180°$$

Figure 17.3 A general triangle.

TABLE 17.1 Functions of Certain Angles

	0	30°	45°	60°	90°	135°	180°	270°	360°
$\sin q$	0	1/2	$\sqrt{2}/2$	$\sqrt{3}/2$	1	$\sqrt{2}/2$	0	−1	0
$\cos q$	1	$\sqrt{3}/2$	$\sqrt{2}/2$	1/2	0	$\sqrt{2}/2$	−1	0	1
$\tan q$	0	$1/\sqrt{3}$	1	$\sqrt{3}$	∞	−1	0	$-\infty$	0

Example 17.4

Express $\cos^2 \theta$ as a function of $\cos 2\theta$.

Solution: Substitute Eq. 17.2.3 into Eq. 17.2.5 and obtain

$$\cos 2\theta = \cos^2 \theta - (1 - \cos^2 \theta)$$

$$= 2\cos^2 \theta - 1$$

There results

$$\cos^2 \theta = \frac{1}{2}(1 + \cos 2\theta)$$

Example 17.5

If $\sin\theta = x$, what is $\tan\theta$?

Solution: Think of $x = x/1$. Thus, the hypotenuse of an imaginary right triangle is of length unity and the leg opposite θ is of length x. The other leg is of length $\sqrt{1 - x^2}$. Hence,

$$\tan \theta = \frac{x}{\sqrt{1 - x^2}}$$

Example 17.6

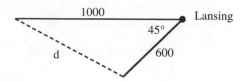

$$d^2 = 1000^2 + 600^2 - 2 \times 1000 \times 600 \cos 45°$$

$$= 511470$$

$$\therefore d = 715.2 \text{ km}$$

An airplane leaves Lansing flying due southwest at 300 km/hr, and a second plane leaves Lansing at the same time flying due west at 500 km/hr. How far apart are the airplanes after 2 hours?

Solution: After 2 hours, the respective distances from Lansing are 600 km and 1000 km. A sketch is quite helpful. The distance d that the two airplanes are apart is found using the law of cosines:

17.3 GEOMETRY

A regular polygon with n sides has a vertex angle (the central angle subtended by one side) of $2\pi/n$. The included angle between two successive sides is given by $\pi(n-2)/n$.

Some common geometric shapes are displayed in Fig. 17.4.

The equation of a straight line can be written in the general form

$$Ax + By + C = 0 \tag{17.3.1}$$

There are three particular forms that this equation can take. They are:

$$\text{Point – slope:} \quad y - y_1 = m(x - x_1) \tag{17.3.2}$$

$$\text{Slope – intercept:} \quad y = m(x - x_1) \tag{17.3.3}$$

$$\text{Two – intercept:} \quad \frac{x}{a} + \frac{y}{b} = 1 \tag{17.3.4}$$

In the above equations m is the slope, (x_1, y_1) a point on the line, "a" the x-intercept, and "b" the y-intercept (see Fig. 17.5). The perpendicular distance d from the point (x_3, y_3) to the line $Ax + By + C = 0$ is given by (see Fig. 17.5)

$$d = \frac{|Ax_3 + By_3 + C|}{\sqrt{A^2 + B^2}} \tag{17.3.5}$$

The equation of a plane surface is given as

$$Ax + By + Cz + D = 0 \qquad (17.3.6)$$

The general equation of second degree

$$Ax^2 + 2Bxy + Cy^2 + 2Dx + 2Ey + F = 0 \qquad (17.3.7)$$

represents a set of geometric shapes called *conic sections.* They are classified as follows:

Ellipse: $B^2 - AC < 0$ (circle: $B = 0$, $A = C$)

Parabola: $B^2 - AC = 0$ $\qquad\qquad\qquad\qquad$ (17.3.8)

Hyperbola: $B^2 - AC > 0$

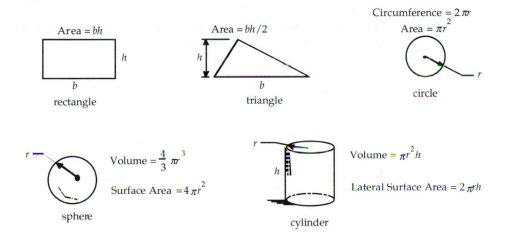

Figure 17.4 Common geometric shapes.

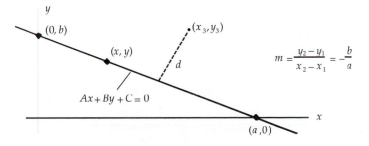

Figure 17.5 A straight line.

If $A = B = C = 0$, the equation represents a line in the xy-plane, not a parabola. Let's consider each in detail.

Circle: The circle is a special case of an ellipse with $A = C$. Its general form can be expressed as

$$(x - a)^2 + (y - b)^2 = r^2 \qquad (17.3.9)$$

where its center is at (a, b) and r is the radius.

Ellipse: The sum of the distances from the two foci, F, to any point on an ellipse is a constant. For an ellipse centered at the origin

$$\frac{x^2}{a^2} + \frac{y^2}{b^2} = 1 \qquad (17.3.10)$$

where a and b are the semi-major and semi-minor axes. The foci are at $(\pm c, 0)$ where $c^2 = a^2 - b^2$. See Fig. 17.6a.

Parabola: The locus of points on a parabola are equidistant from the focus and a line (the directrix). If the vertex is at the origin and the parabola opens to the right, it is written as

$$y^2 = 2px \qquad (17.3.11)$$

where the focus is at $(p/2, 0)$ and the directrix is at $x = -p/2$. See Fig. 17.6b. For a parabola opening to the left, simply change the sign of p. For a parabola opening upward or downward, interchange x and y.

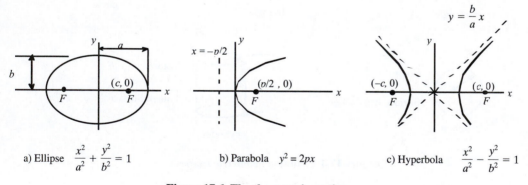

a) Ellipse $\dfrac{x^2}{a^2} + \dfrac{y^2}{b^2} = 1$ b) Parabola $y^2 = 2px$ c) Hyperbola $\dfrac{x^2}{a^2} - \dfrac{y^2}{b^2} = 1$

Figure 17.6   The three conic sections.

Hyperbola: The difference of the distances from the foci to any point on a hyperbola is a constant. For a hyperbola centered at the origin opening left and right, the equation can be written as

$$\frac{x^2}{a^2} - \frac{y^2}{b^2} = 1 \qquad (17.3.12)$$

The lines to which the hyperbola is asymptotic are asymptotes:

$$y = \pm \frac{b}{a} x \qquad (17.3.13)$$

If the asymptotes are perpendicular, a rectangular hyperbola results. If the asymptotes are the x and y axes, the equation can be written as

$$xy = \pm k^2 \tag{17.3.14}$$

Finally, in our review of geometry, we will present three other coordinate systems often used in analysis. They are the polar (r, θ) coordinate system, the cylindrical (r, θ, z) coordinate system, and the spherical (r, θ, ϕ) coordinate system. The polar coordinate system is restricted to a plane:

$$x = r \cos \theta, \quad y = r \sin \theta \tag{17.3.15}$$

For the cylindrical coordinate system

$$x = r \cos \theta, \quad y = r \sin \theta, \quad z = z \tag{17.3.16}$$

And, for the spherical coordinate system

$$x = r \, \sin \phi \, \cos \theta$$

$$y = r \, \sin \phi \, \sin \theta \tag{17.3.17}$$

$$z = r \, \cos \theta$$

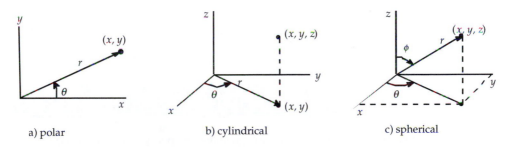

a) polar   b) cylindrical   c) spherical

Figure 17.7 The polar, cylindrical and spherical coordinate systems.

Example 17.7

What conic section is represented by $2x^2 - 4xy + 5x = 10$?

Solution: Comparing this with the general form Eq. 17.3.7, we see that

$$A = 2, \quad B = -2, \quad C = 0$$

Thus, $B^2 - AC = 4$, which is greater than zero. Hence, the conic section is a hyperbola.

Example 17.8

Calculate the radius of the circle given by $x^2 + y^2 - 4x + 6y = 12$.

Solution: Write the equation in standard form (see Eq. 17.3.9):

$$(x - 2)^2 + (y + 3)^2 = 25$$

The radius is $r = \sqrt{25} = 5$.
Note: the terms $4x$ and $6y$ demand that we write $(x - 2)^2$ and $(y + 3)^2$.

Example 17.9

Write the general form of the equation of a parabola, vertex at (2, 4), opening upward, with directrix at $y = 2$.

Solution: The equation of the parabola (see Eq. 17.3.11) can be written as

$$(x - x_1)^2 = 2p(y - y_1)$$

where we have interchanged x and y so that the parabola opens upward. For this example, $x_1 = 2$, $y_1 = 4$, and $p = 4$ ($p/2$ is the distance from the vertex to the directrix). Hence, the equation is

$$(x - 2)^2 = 2(4)(y - 4)$$

or, in general form,

$$x^2 - 4x - 8y + 36 = 0$$

17.4 COMPLEX NUMBERS

A complex number consists of a real part x and an imaginary part y, written as $x + iy$, where $i = \sqrt{-1}$. (In electrical engineering, however, it is common to let $j = \sqrt{-1}$ since i represents current.) In real number theory, the square root of a negative number does not exist; in complex number theory, we would write $\sqrt{-4} = \sqrt{4(-1)} = 2i$ The complex number may be plotted using the real x-axis and the imaginary y-axis, as shown in Fig. 17.8.

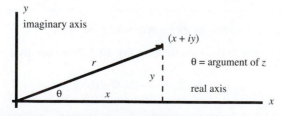

Figure 17.8 The complex number.

It is often useful to express a complex number in polar form as

$$x + iy = re^{i\theta} \tag{17.4.1}$$

where we use *Euler's equation*

$$e^{i\theta} = \cos\theta + i\sin\theta \tag{17.4.2}$$

to verify the relations

$$x = r\cos\theta, \quad y = r\sin\theta \tag{17.4.3}$$

Note that $e^{i\theta} = e^{i(\theta+2n\pi)}$ where n is an integer. This simply adds 360° (2π radians) to θ and hence in Fig. 17.8 $re^{i\theta}$ and $re^{i(\theta+2n\pi)}$ represent the identical point.

Multiplication and division are accomplished with either form:

$$(a + ib)(c + id) = ac - bd + i(ad + bc)$$

$$= r_1 e^{i\theta_1} r_2 e^{i\theta_2} = r_1 r_2 e^{i(\theta_1 + \theta_2)} \tag{17.4.4}$$

$$\frac{a + ib}{c + id} = \frac{a + ib}{c + id}\frac{c - id}{c - id} = \frac{(a + ib)(c - id)}{c^2 + d^2}$$

$$= \frac{r_1}{r_2} e^{i(\theta_1 - \theta_2)} \tag{17.4.5}$$

It is usually easier to find powers and roots of complex numbers using the polar form:

$$(x + iy)^k = r^k e^{ik\theta} \ , (x + iy)^{1/k} = r^{1/k} e^{i\theta/k} \tag{17.4.6}$$

When finding roots, more than one root results by using $e^{i\theta}$ and $e^{i(\theta+2n\pi)}$. An example illustrates. Remember, in mathematical equations we usually express θ in radians.

Using Euler's equation we can show that

$$\sin\theta = \frac{e^{i\theta} - e^{-i\theta}}{2i}, \quad \cos\theta = \frac{e^{i\theta} + e^{-i\theta}}{2} \tag{17.4.7}$$

Example 17.10

Divide $(3 + 4i)$ by $(4 + 3i)$.

Solution: We perform the division as follows:

$$\frac{3 + 4i}{4 + 3i} = \frac{3 + 4i}{4 + 3i} \cdot \frac{4 - 3i}{4 - 3i} = \frac{12 + 16i - 9i + 12}{16 + 9} = \frac{24 + 7i}{25} = 0.96 + 0.28i$$

Note that we multiplied the numerator and the denominator by the *complex conjugate* of the denominator. A complex conjugate is formed simply by changing the sign of the imaginary part.

Example 17.11

Find $(3 + 4i)^6$.

Solution: This can be done multiplying six times or using the polar form. Polar form is

$$r = \sqrt{3^2 + 4^2} = 5, \; \theta = \tan^{-1} 4/3 = 0.9273 \text{ rad}$$

We normally express θ in radians. The complex number, in polar form, is

$$3 + 4i = 5\, e^{0.9273i}$$

Thus,

$$(3 + 4i)^6 = 5^6\, e^{6(0.9273)i} = 5^6\, e^{5.564i}$$

Converting back to rectangular form we have

$$5^6\, \theta^{5.564i} = 15625(\cos 5.564 + i \sin 5.564)$$

$$= 11755 - 10293i$$

Example 17.12

Find the three roots of 1.

Solution: We express the complex number in polar form as

$$1 = 1e^{0i}$$

Since the trig functions are periodic, we know that

$$\sin \theta = \sin(\theta + 2\pi) = \sin(\theta + 4\pi)$$

$$\cos \theta = \cos(\theta + 2\pi) = \cos(\theta + 4\pi)$$

Thus, in addition to the first form, we have

$$1 = e^{2\pi i} = e^{4\pi i}$$

Taking the one-third root of each form, we find the three roots to be

$$1^{1/3} = 1e^{0i/3} = 1$$

$$1^{1/3} = 1e^{2\pi i/3}$$

$$= \cos 2\pi/3 + i \sin 2\pi/3 = -0.5 + 0.866i$$

$$1^{1/3} = 1e^{4\pi i/3}$$

$$= \cos 4\pi/3 + i \sin 4\pi/3 = -0.5 + 0.866i$$

If we added 6π to the angle we would be repeating the first root, so obviously this is not done.

17.5 LINEAR ALGEBRA

The primary objective in linear algebra is to find the solution to a set of n linear algebraic equations for n unknowns. To do this we must learn how to manipulate a matrix, a rectangular array of quantities arranged into rows and columns.

An $m \times n$ matrix has m rows (the horizontal lines) and n columns (the vertical lines). An $m \times n$ matrix multiplied by an $n \times s$ matrix produces an $m \times s$ matrix. When multiplying two matrices the columns of the first matrix must equal the rows of the second. Their product is a third matrix:

$$[c_{ij}] = \sum_{k=1}^{n} [a_{ik}][b_{kj}] \tag{17.5.1}$$

We are primarily interested in square matrices since we usually have the same number of equations as unknowns, such as

$$a_{11}x_1 + a_{12}x_2 + a_{13}x_3 + a_{14}x_4 = r_1$$

$$a_{21}x_1 + a_{22}x_2 + a_{23}x_3 + a_{24}x_4 = r_2 \tag{17.5.2}$$

$$a_{31}x_1 + a_{32}x_2 + a_{33}x_3 + a_{34}x_4 = r_3$$

$$a_{41}x_1 + a_{42}x_2 + a_{43}x_3 + a_{44}x_4 = r_4$$

In matrix form this can be written as

$$[a_{ij}][x_j] = [r_i] \quad \text{or} \quad \mathbf{Ax} = \mathbf{r} \tag{17.5.3}$$

where $[x_j]$ and $[r_i]$ are column matrices. (A column matrix is often referred to as a *vector*.) The coefficient matrix $[a_{ij}]$ and the column matrix $[r_i]$ are assumed to be known quantities. The solution $[x_j]$ is expressed as

$$[x_j] = [a_{ij}]^{-1}[r_i] \quad \text{or} \quad \mathbf{x} = \mathbf{A}^{-1}\mathbf{r} \tag{17.5.4}$$

where $[a_{ij}]^{-1}$ is the *inverse* matrix of $[a_{ij}]$. It is defined as

$$[a_{ij}]^{-1} = \frac{[a_{ij}]^{+}}{|a_{ij}|} \quad \text{or} \quad \mathbf{A}^{-1} = \frac{\mathbf{A}^{+}}{|\mathbf{A}|} \tag{17.5.5}$$

where $[a_{ij}]^{+}$ is the *adjoint* matrix and $|a_{ij}|$ is the *determinant* of $[a_{ij}]$. Let us review how the determinant and the adjoint are evaluated.

In general, the determinant may be found using the *cofactor* A_{ij} of the element a_{ij}. The cofactor is defined to be $(-1)^{i+j}$ times the *minor*, the determinant obtained by deleting the ith row and the jth column. The determinant is then

$$|a_{ij}| = \sum_{j=1}^{n} a_{ij} A_{ij} \tag{17.5.6}$$

where i is any value from 1 to n. Recall that the third-order determinant can be evaluated by writing the first two columns after the determinant and then summing the products of the elements of the diagonals, using negative signs with the diagonals sloping upward.

The elements of the adjoint $[a_{ij}]^+$ are the cofactors A_{ij} of the elements a_{ij}; for the matrix $[a_{ij}]$ of Eq. 17.5.2 we have

$$[a_{ij}]^+ = \begin{bmatrix} A_{11} & A_{21} & A_{31} & A_{41} \\ A_{12} & A_{22} & A_{32} & A_{42} \\ A_{13} & A_{23} & A_{33} & A_{43} \\ A_{14} & A_{24} & A_{34} & A_{44} \end{bmatrix}$$

(17.5.7)

Note that A_{ij} takes the position of a_{ji}.

Finally, the solution $[x_j]$ of Eq. 17.5.4 results if we multiply the square matrix $[a_{ij}]^{-1}$ by the column matrix $[r_i]$. In general, we multiply the elements in each left-hand matrix row by the elements in each right-hand matrix column, add the products, and place the sum at the location where the row and column intersect. The following examples will illustrate.

Before we work some examples though, we should point out that the above matrix presentation can also be presented as *Cramer's rule,* which states that the solution element x_n can be expressed as

$$x_n = \frac{|b_{ij}|}{|a_{ij}|}$$

(17.5.8)

where $|b_{ij}|$ is formed by replacing the nth column of $|a_{ij}|$ with the elements of the column matrix $|r_i|$.

Notes: If the system of equations is homogeneous, i.e., $r_i = 0$, a solution may exist if $|a_{ij}| = 0$. If the determinant of a matrix is zero, that matrix is *singular* and its inverse does not exist.

In the solution of a system of first-order differential equations, we encounter the matrix equation

$$(A - \lambda I)x = O$$

(17.5.9)

where O represents a matrix with all zero elements. The scalar λ is the *eigenvalue* and the vector x is the *eigenvector* associated with the eigenvalue. The matrix I is called the *unit matrix*, which for a 2 × 2 matrix is $\begin{bmatrix} 1 & 0 \\ 0 & 1 \end{bmatrix}$. If A is 2 × 2, λ has two distinct values, and if A is 3 × 3 it has three distinct values. Since $r_i = 0$ in Eq. 17.5.9, the scalar equation

$$|A - \lambda I| = 0$$

(17.5.10)

provides the equation that yields the eigenvalues. Then Eq. 17.5.9 is solved with each eigenvalue to give the eigenvectors. We often say that λ represents the eigenvalues of the matrix A.

Example 17.13

Calculate the determinants of $\begin{bmatrix} 2 & -3 \\ 1 & 4 \end{bmatrix}$ and $\begin{bmatrix} 2 & 3 & 0 \\ 1 & 4 & -2 \\ 0 & 3 & 5 \end{bmatrix}$.

Solution: For the first matrix we have

$$\begin{vmatrix} 2 & -3 \\ 1 & 4 \end{vmatrix} = 2 \times 4 - 1(-3) = 11$$

The second matrix is set up as follows:

$$= 40 + 0 + 0 - 0 - (-12) - 15 = 37$$

Example 17.14

Multiply the two matrices $\begin{bmatrix} 1 & 1 \\ -1 & -1 \end{bmatrix}$ and $\begin{bmatrix} 1 & 1 \\ -1 & -1 \end{bmatrix}$

Solution: Multiply the two matrices using Eq. 17.5.1. If we desire c_{12}, we use the first row of the first matrix and the second column of the second matrix so that

$$c_{12} = a_{11}b_{21} + a_{12}b_{22} = 1(1) + 1(-1) = 0$$

Doing this for all elements we find

$$\begin{bmatrix} 1 & 1 \\ -1 & -1 \end{bmatrix} \begin{bmatrix} 1 & 1 \\ -1 & -1 \end{bmatrix} = \begin{bmatrix} 0 & 0 \\ 0 & 0 \end{bmatrix}$$

Note: Even though a matrix has no zero elements, its square may be zero. Matrix multiplication is not like other forms of multiplication.

Example 17.15

Find the adjoint of the matrix

$$[a_{ij}] = \begin{bmatrix} 1 & 0 & -2 \\ -1 & 2 & 0 \\ 1 & 2 & 1 \end{bmatrix}$$

Solution: The cofactor of each element of $[a_{ij}]$ must be determined. The cofactor is found by multiplying $(-1)^{i+j}$ times the determinant formed by deleting the ith row and the jth column. They are found to be

$$
\begin{array}{lll}
A_{11} = 2, & A_{12} = 1, & A_{13} = -4 \\
A_{21} = -4, & A_{22} = 3, & A_{23} = -2 \\
A_{31} = 4, & A_{32} = 2, & A_{33} = 2
\end{array}
$$

The adjoint is then

$$
[a_{ij}]^+ = [A_{ji}] = \begin{bmatrix} 2 & -4 & 4 \\ 1 & 3 & 2 \\ -4 & -2 & 2 \end{bmatrix}
$$

Note: The matrix $[A_{ji}]$ is called the *transpose* of $[A_{ij}]$, i.e., $[A_{ji}] = [A_{ij}]^T$.

Example 17.16

Find the inverse of the matrix $\quad a_{ij} = \begin{bmatrix} 1 & 0 & -2 \\ -1 & 2 & 0 \\ 1 & 2 & 1 \end{bmatrix}$

Solution: The inverse is defined to be the adjoint matrix divided by the determinant $|a_{ij}|$. Hence, the inverse is (see Examples 17.13 and 17.15)

$$
[a_{ij}]^{-1} = \frac{1}{10} \begin{bmatrix} 2 & -4 & 4 \\ 1 & 3 & 2 \\ -4 & -2 & 2 \end{bmatrix} = \begin{bmatrix} 0.2 & -0.4 & 0.4 \\ 0.1 & 0.3 & 0.2 \\ -0.4 & -0.2 & 0.2 \end{bmatrix}
$$

Example 17.17

Find the solution to

$$
\begin{array}{llll}
x_1 & & - 2x_3 & = 2 \\
-x_1 & + 2x_2 & & = 0 \\
x_1 & + 2x_2 & + x_3 & = -4
\end{array}
$$

Solution: The solution matrix is (see Example 17.13 and 17.15 or use $[a_{ij}]^{-1}$ from Example 17.16)

$$
[x_j] = [a_{ij}]^{-1}[r_i] = \frac{[a_{ij}]^+}{|a_{ij}|}[r_i]
$$

$$
= \frac{1}{10} \begin{bmatrix} 2 & -4 & 4 \\ 1 & 3 & 2 \\ -4 & -2 & 2 \end{bmatrix} \begin{bmatrix} 2 \\ 0 \\ -4 \end{bmatrix}
$$

First, let's multiply the two matrices; they are multiplied row by column as follows:

$$2 \cdot 2 + (-4) \cdot 0 + 4 \cdot (-4) = -12$$

$$1 \cdot 2 + 3 \cdot 0 + 2 \cdot (-4) = -6$$

$$-4 \cdot 2 - 2 \cdot 0 + 2 \cdot (-4) = -16$$

The solution vector is then

$$[x_i] = \frac{1}{10} \begin{bmatrix} -12 \\ -6 \\ -16 \end{bmatrix} = \begin{bmatrix} -1.2 \\ -0.6 \\ -1.6 \end{bmatrix}$$

In component form, the solution is

$$x_1 = -1.2, \quad x_2 = -0.6, \quad x_3 = -1.6$$

Example 17.18

Use Cramer's rule and solve

$$
\begin{array}{rrrrr}
x_1 & & - 2x_3 & = & 2 \\
-x_1 & + 2x_2 & & = & 0 \\
x_1 & + 2x_2 & + x_3 & = & -4
\end{array}
$$

Solution: The solution is found (see Example 17.13) by evaluating the ratios as follows:

$$x_1 = \frac{\begin{vmatrix} 2 & 0 & -2 \\ 0 & 2 & 0 \\ -4 & 2 & 1 \end{vmatrix}}{D} = \frac{-12}{10} = -1.2 \qquad x_2 = \frac{\begin{vmatrix} 1 & 2 & -2 \\ -1 & 0 & 0 \\ 1 & -4 & 1 \end{vmatrix}}{D} = \frac{-6}{10} = -0.6$$

$$x_3 = \frac{\begin{vmatrix} 1 & 0 & 2 \\ -1 & 2 & 0 \\ 1 & 2 & -4 \end{vmatrix}}{D} = \frac{-16}{10} = -1.6$$

where

$$D = \begin{vmatrix} 1 & 0 & -2 \\ -1 & 2 & 0 \\ 1 & 2 & 1 \end{vmatrix} = 10$$

Note that the numerator is the determinant formed by replacing the ith column with right-hand side elements r_i when solving for x_j.

Example 17.19

Find the eigenvalues of $A = \begin{bmatrix} 4 & 0 & 2 \\ 0 & 8 & 0 \\ 3 & 0 & 5 \end{bmatrix}$.

Solution: To find the eigenvalues of the matrix, we form the equation

$$|A - \lambda I| = 0 \quad \text{or} \quad \begin{vmatrix} 4 - \lambda & 0 & 2 \\ 0 & 8 - \lambda & 0 \\ 3 & 0 & 5 - \lambda \end{vmatrix} = 0$$

Expanding the determinant using cofactors we have

$$(8 - \lambda) \begin{vmatrix} 4 - \lambda & 2 \\ 3 & 5 - \lambda \end{vmatrix} = (8 - \lambda)\left[(4 - \lambda)(5 - \lambda) - 6\right] = (8 - \lambda)\left[\lambda^2 - 9\lambda + 14\right] = 0$$

or

$$(8 - \lambda)(\lambda - 7)(\lambda - 2) = 0$$

The eigenvalues are then $\lambda = 8, 7, 2$.

17.6 CALCULUS

Differentiation

The slope of a curve $y = f(x)$ is the ratio of the change in y to the change in x as the change in x becomes infinitesimally small. This is the first derivative, written as

$$\frac{dy}{dx} = \lim_{\Delta x \to 0} \frac{\Delta y}{\Delta x} \qquad (17.6.1)$$

This may be written using abbreviated notation as

$$\frac{dy}{dx} = Dy = y' = \dot{y} \qquad (17.6.2)$$

The second derivative is written as

$$\frac{d^2 y}{dx^2} = D^2 y = y'' = \ddot{y} \qquad (17.6.3)$$

and is defined by

$$\frac{d^2 y}{dx^2} = \lim_{\Delta x \to 0} \frac{\Delta y'}{\Delta x} \qquad (17.6.4)$$

Some derivative formulas, where *f* and *g* are functions of *x*, and *k* is constant, follow.

$$\frac{dk}{dx} = 0$$

$$\frac{d(kx^n)}{dx} = k\,n\,x^{n-1}$$

$$\frac{d}{dx}(f + g) = f' + g'$$

$$\frac{df^n}{dx} = nf^{n-1}f'$$

$$\frac{d}{dx}(fg) = fg' + gf'$$ (17.6.5)

$$\frac{d}{dx}(\ln x) = \frac{1}{x}$$

$$\frac{d}{dx}(e^{kx}) = ke^{kx}$$

$$\frac{d}{dx}(\sin x) = \cos x$$

$$\frac{d}{dx}(\cos x) = -\sin x$$

Maxima and Minima

Derivatives are used to locate points of inflection, maxima, and minima. Note the following:

$f'(x) = 0$ at a maximum or a minimum.
$f''(x) = 0$ at an inflection point.
$f''(x) > 0$ at a minimum.
$f''(x) < 0$ at a maximum.

An inflection point always exists between a maximum and a minimum.

L'Hospital's Rule

Differentiation is also useful in establishing the limit of $f(x)/g(x)$ as $x \to a$ if $f(a)$ and $g(a)$ are both zero or $\pm\infty$. *L'Hospital's rule* is used in such cases and is as follows:

$$\lim_{x \to a}\frac{f(x)}{g(x)} = \lim_{x \to a}\frac{f'(x)}{g'(x)} = \lim_{x \to a}\frac{f''(x)}{g''(x)}$$ (17.6.6)

Integration

The inverse of differentiation is the process called *integration*. If a curve is given by $y = f(x)$, then the area under the curve from $x = a$ to $x = b$ is given by

$$A = \int_a^b y\,dx \tag{17.6.7}$$

The length of the curve between the two points is expressed as

$$L = \int_a^b (1 + y'^2)^{1/2}\,dx \tag{17.6.8}$$

Volumes of various objects are also found by an appropriate integration.

If the integral has limits, it is a *definite integral;* if it does not have limits, it is an *indefinite integral* and a constant is always added. Some common indefinite integrals follow:

$$\int dx = x + C$$

$$\int cy\,dx = c\int y\,dx$$

$$\int x^n dx = \frac{x^{n+1}}{n+1} + C \quad n \neq -1$$

$$\int x^{-1}dx = \ln x + C$$

$$\int e^{ax}dx = \frac{1}{a}e^{ax} + C \tag{17.6.9}$$

$$\int \sin x\,dx = -\cos x + C$$

$$\int \cos x\,dx = \sin x + C$$

$$\int \cos^2 x\,dx = \frac{x}{2} + \frac{1}{4}\sin 2x + C$$

$$\int u\,dv = uv - \int v\,du$$

This last integral is often referred to as "integration by parts." If the integrand (the coefficients of the differential) is not one of the above, then in the last integral, $\int v\,du$ may in fact be integrable. An example will illustrate.

Example 17.20

Find the slope of $y = x^2 + \sin x$ at $x = 0.5$.

Solution: The derivative is the slope:

$$y'(x) = 2x + \cos x$$

At $x = 0.5$ the slope is

$$y'(0.5) = 2 \cdot 0.5 + \cos 0.5 = 1.878$$

Note: Use 0.5 radians when evaluating cos 0.5.

Example 17.21

Find $\dfrac{d}{dx}(\tan x)$.

Solution: Writing $\tan x = \sin x / \cos x = f(x) \cdot g(x)$ we find

$$\frac{d}{dx}(\tan x) = \frac{1}{\cos x}\frac{d}{dx}(\sin x) + \sin x \frac{d}{dx}(\cos x)^{-1}$$

$$= \frac{\cos x}{\cos x} + \frac{\sin^2 x}{\cos^2 x} = 1 + \tan^2 x$$

$$= \frac{\cos^2 x + \sin^2 x}{\cos^2 x} = \frac{1}{\cos^2 x} = \sec^2 x$$

Either expression is acceptable.

Example 17.22

Locate the maximum and minimum points of the function $y(x) = x^3 - 12x - 9$ and evaluate y at those points.

Solution: The derivative is

$$y'(x) = 3x^2 - 12$$

The points at which $y'(x) = 0$ are at

$$x = 2, -2$$

At these two points the extrema are:

$$y_{\min} = (2)^3 - 12 \cdot 2 - 9 = -25$$

$$y_{\max} = (-2)^3 - 12(-2) - 9 = 7$$

Let us check the second derivative. At the two points we have:

$$y''(2) = 6 \cdot 2 = 12$$

$$y''(-2) = 6 \cdot (-2) = -12$$

Obviously, the point $x = 2$ is a minimum since its second derivative is positive there.

Example 17.23

Find the limit as of $x \rightarrow 0$ of sin $x \,/\, x$.

Solution: If we let $x = 0$ we are faced with the ratio of 0/0, an indeterminate quantity. Hence, we use L'Hospital's rule and differentiate both numerator and denominator to obtain

$$\lim_{x \to 0} \frac{\sin x}{x} = \lim_{x \to 0} \frac{\cos x}{1}$$

Now, we let $x = 0$ and find

$$\lim_{x \to 0} \frac{\sin x}{x} = \frac{1}{1} = 1$$

Example 17.24

Find the area of the shaded area in the figure.

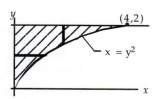

Solution: We can find this area by using either a horizontal strip or a vertical strip. We will use both. First, for a horizontal strip:

$$A = \int_0^2 x\,dy$$

$$= \int_0^2 y^2\,dy = \frac{y^3}{3}\bigg|_0^2 = 8/3$$

Using a vertical strip we have

$$A = \int_0^4 (2 - y)\, dx$$

$$= \int_0^4 (2 - x^{1/2})\, dx = \left[2x - \frac{2}{3} x^{3/2} \right]_0^4 = 8 - \frac{2}{3} 8 = 8/3$$

Either technique is acceptable. The first is the simpler one.

Example 17.25

Find the volume enclosed by rotating the shaded area of Example 17.24 about the *y*-axis.

Solution: If we rotate the horizontal strip about the *y*-axis we will obtain a disc with volume

$$dV = \pi x^2 dy$$

This can be integrated to give the volume, which is

$$V = \int_0^2 \pi x^2\, dy$$

$$= \pi \int_0^2 y^4 dy = \frac{\pi y^5}{5} \bigg|_0^2 = \frac{32\pi}{5}$$

Now, let us rotate the vertical strip about the *y*-axis to form a cylinder with volume

$$dV = 2\pi x (2 - y)dx$$

This can be integrated to yield

$$V = \int_0^4 2\pi x (2 - y)dx$$

$$= \int_0^4 2\pi x (2 - x^{1/2})dx = 2\pi \left[x^2 - \frac{2x^{5/2}}{5} \right]_0^4 = \frac{32\pi}{5}$$

Again, using the horizontal strip results in a simpler solution.

Example 17.26

Show that $\int xe^x\,dx = (x-1)e^x + C$.

Solution: Let's attempt the last integral of (17.6.9). Define the following:

$$u = x, \quad dv = e^x dx$$

Then,

$$du = dx, \quad v = \int e^x dx = e^x$$

and we find that

$$\int xe^x dx = x\,e^x - \int e^x dx$$

$$= x\,e^x - e^x + C = (x-1)e^x + C$$

17.7 PROBABILITY AND STATISTICS

Events are independent if the probability of occurrence of one event does not influence the probability of occurrence of other events. The number of permutations (a particular sequence) of n things taken r at a time is

$$P(n,r) = \frac{n!}{(n-r)!} \tag{17.7.1}$$

If the starting point is unknown, as in a ring, the *ring permutation* is

$$P(n,r) = \frac{(n-1)!}{(n-r)!} \tag{17.7.2}$$

The number of *combinations* (no order-conscious arrangement) of n things taken r at a time is given by

$$C(n,r) = \frac{n!}{r!(n-r)!} \tag{17.7.3}$$

For independent events of two sample groups A and B the following rules are necessary:

1. The probability of A or B occurring equals the sum of the probability of occurrence of A and the probability of occurrence of B; that is,

$$P(A \text{ or } B) = P(A) + P(B) \tag{17.7.4}$$

2. The probability of both *A* and *B* occurring is given by the product

$$P(A \text{ and } B) = P(A)P(B) \qquad (17.7.5)$$

3. The probability of *A* not occurring is given as

$$P(\text{not } A) = 1 - P(A) \qquad (17.7.6)$$

4. The probability of either *A* or *B* occurring is given by

$$P(A \text{ or } B) = P(A) + P(B) - P(A)P(B) \qquad (17.7.7)$$

The probability of an event occurring is in the range of 0 to 1. An impossible event has a probability of 0 and an event that is certain to occur has a probability of 1.

The data gathered during an experiment can be analyzed using quantities defined by the following:

1. The arithmetic mean $\overline{x}$ is the average of the observations; that is,

$$\overline{x} = \frac{x_1 + x_2 + x_3 + \cdots + x_n}{n}$$

2. The median is the middle observation when all the data is ordered by magnitude; half the values are below the median. The median for an even number of data is the average of the two middle values.
3. The mode is the observed value that occurs most frequently.
4. The standard deviation σ of the sample is a measure of variability. It is defined as

$$\sigma = \left[\frac{(x_1 - \overline{x})^2 + x_2 - \overline{x})^2 + \cdots + (x_n - \overline{x})^2}{n - 1} \right]^{1/2} \qquad (17.7.8)$$

$$= \left[\frac{x_1^2 + x_2^2 + \cdots + x_n^2 - n\overline{x}^2}{n - 1} \right]^{1/2} \qquad (17.7.9)$$

For large observations (over 50), it is customary to simply use *n*, rather that $(n - 1)$, in the denominators of the above. In fact, if *n* is used in the above equations, σ is often referred to as the *standard deviation of the population*.

5. The *variance* is defined to be σ^2.

Example 17.27

How many different ways can seven people be arranged in a lineup? In a circle?

Solution: This is the number of permutations of seven things taken seven at a time. The answer is

$$P(7, 7) = \frac{n!}{(n - r)!}$$

$$= \frac{7!}{(7 - 7)!} = 5040$$

In a circle we use the ring permutation:

$$P(7,7) = \frac{(n-1)!}{(n-r)!}$$

$$= \frac{(7-1)!}{(7-7)!} = 720$$

Note that $0! = 1$.

Example 17.28

How many different collections of eight people can fit into a six-passenger vehicle? (Only six will fit at a time.)

Solution: The answer does not depend on the seating arrangement. If it did, it would be a permutation. Hence, we use the combination relationship and find

$$C(8,6) = \frac{n!}{(n-r)!\,r!}$$

$$= \frac{8!}{(8-6)!6!} = 28$$

Example 17.29

A carnival booth offers $10 if you pick a red ball and then a white ball (the first ball is re-inserted) from a bin containing 60 red balls, 15 white balls, and 25 blue balls. If $1 is charged for an attempt, will the operator make money?

Solution: The probability of drawing a red ball on the first try is 0.6. If it is then re-inserted, the probability of drawing a white ball is 0.15. The probability of accomplishing both is then given by

$$P(\text{red and white}) = P(\text{red})\,P(\text{white})$$
$$= 0.6 \times 0.15 = 0.09$$

or 9 chances out of 100 attempts. Hence, the entrepreneur will pay out $90 for every $100 taken in and will thus make money.

Example 17.30

If the operator of the bin of balls in Example 17.29 offers a $1.00 prize to contestants who pick either a red ball or a white ball from the bin on the first attempt, and charges $0.75 per attempt, will the operator make money?

Solution: The probability of selecting either a red ball or a white ball on the first attempt is

$$P(\text{red and white}) = P(\text{red})\, P(\text{white})$$
$$= 0.6 \times 0.15 = 0.75$$

Consequently, 75 out of 100 gamblers will win and the operator must pay out $75 for every $75 taken in. The operator would not make any money.

Example 17.31

If the operator of Example 17.30 had two identical bins and offered a $1.00 prize for successfully withdrawing a red ball from the first bin or a white ball from the second bin, will the operator make money if he charges $0.75 per attempt?

Solution: The probability of selecting a red ball from the first bin (sample group A_i) or a white ball from the second bin (sample group B_i) is

$$P(\text{red and white}) = P(\text{red}) + P(\text{white}) - P(\text{red})\, P(\text{white})$$
$$= 0.6 + 0.15 - 0.6 \times 0.15 = 0.66$$

For this situation the owner must pay $66 to every 100 gamblers who pay $75 to participate. The profit is $9.

Example 17.32

The temperature at a given location in the south at 2 p.m. each August 10 for 25 consecutive years was measured, in degrees Celsius, to be 33, 38, 34, 26, 32, 31, 28, 39, 29, 36, 32, 29, 31, 24, 35, 34, 32, 30, 31, 32, 26, 40, 27, 33, 39. Calculate the arithmetic mean, the median, the mode and the sample standard deviation.

Solution: Using the appropriate equations, we calculate the arithmetic mean:

$$\overline{T} = \frac{T_1 + T_2 + \cdots + T_{25}}{25}$$

$$= \frac{33 + 38 + 34 + \cdots + 39}{25} = \frac{801}{25} = 32.04°\text{C}$$

The median is found by first arranging the values in order. We have 24, 26, 26, 27, 28, 29, 29, 30, 31, 31, 31, 32, 32, 32, 32, 33, 33, 34, 34, 35, 36, 38, 39, 39, 40. Counting 12 values in from either end, the median is found to be 32°C.

The mode is the observation that occurs most often; it is 32°C.

The sample standard deviation is found to be

$$\sigma = \left[\frac{T_1^2 + T_2^2 + \cdots + T_{25}^2 - n\overline{T}^2}{n - 1} \right]^{1/2}$$

$$= \left[\frac{33^2 + 38^2 + \cdots + 39^2 - 25 \times 32.04^2}{25 - 1} \right]^{1/2} = \sqrt{\frac{26{,}099 - 25{,}664}{24}} = 4.26$$

EXERCISES AND ACTIVITIES

Algebra

17.1   A growth curve is given by $A = 10e^{2t}$. At what value of t is $A = 100$?
   a) 5.261   b) 3.070   c) 1.151   d) 0.726

17.2   If $\ln x = 3.2$, what is x?
   a) 18.65   b) 24.53   c) 31.83   d) 64.58

17.3   If $\log_s x = -1.8$, find x.
   a) 0.00483   b) 0.0169   c) 0.0552   d) 0.0783

17.4   One root of the equation $3x^2 - 2x - 2 = 0$ is
   a) 1.215   b) 1.064   c) 0.937   d) 0.826

17.5 $\sqrt{4 + x}$ can be written as the series

   a) $2 - x/4 + x^2/64 + \cdots$ c) $2 - x^2/4 - x^4/64 + \cdots$

   b) $2 + x/8 - x^2/128 + \cdots$ d) $2 + x/4 - x^2/64 + \cdots$

17.6   Resolve $\dfrac{2}{x(x^2 - 3x + 2)}$ into partial fractions.

   a) $\dfrac{1}{x} + \dfrac{1}{x - 2} - \dfrac{2}{x - 1}$ c) $\dfrac{2}{x} - \dfrac{1}{x - 2} - \dfrac{2}{x - 1}$

   b) $\dfrac{1}{x} - \dfrac{2}{x - 2} + \dfrac{1}{x - 1}$ d) $-\dfrac{1}{x} + \dfrac{2}{x - 2} + \dfrac{1}{x - 1}$

17.7   Express $\dfrac{4}{x^2(x^2 - 4x + 4)}$ as the sum of fractions.

   a) $\dfrac{1}{x} - \dfrac{1}{x - 2} + \dfrac{1}{(x - 2)^2}$ c) $\dfrac{1}{x^2} + \dfrac{1}{(x - 2)^2}$

   b) $\dfrac{1}{x} + \dfrac{1}{x^2} - \dfrac{1}{x - 2} + \dfrac{1}{(x - 2)^2}$ d) $\dfrac{1}{x} + \dfrac{1}{x^2} + \dfrac{1}{x - 2} + \dfrac{1}{(x - 2)^2}$

17.8   A germ population has a growth curve of $Ae^{0.4t}$. At what value of t does its original value double?
   a) 9.682   b) 7.733   c) 4.672   d) 1.733

Trigonometry

17.9   If $\sin \theta = 0.7$, what is $\tan \theta$?
   a) 0.98   b) 0.94   c) 0.88   d) 0.85

17.10   If the short leg of a right triangle is 5 units long and the long leg is 7 units long, find the angle opposite the short leg, in degrees.
   a) 26.3   b) 28.9   c) 31.2   d) 35.5

17.11   The expression $\tan\theta \sec\theta (1 - \sin^2\theta) / \cos \theta$ simplifies to
   a) $\sin \theta$ b) $\cos \theta$ c) $\tan \theta$ d) $\sec \theta$

17.12   A triangle has sides of length 2, 3 and 4. What angle, in radians, is opposite the side of length 3?
   a) 0.55   b) 0.61   c) 0.76   d) 0.81

17.13   The length of a lake is to be determined. A distance of 850 m is measured from one end to a point x on the shore. A distance of 732 m is measured from x to the other end. If an angle of 154° is measured between the two lines connecting x, what is the length of the lake?
   a) 1542   b) 1421   c) 1368   d) 1261

17.14   Express $2\sin^2 \theta$ as a function of cos 2θ.
   a) cos 2θ − 1   b) cos 2θ + 1   c) cos 2θ + 2   d) 1 − cos 2θ

Geometry

17.15   The included angle between two successive sides of a regular eight-sided polygon is
   a) 150°   b) 135°   c) 120°   d) 75°

17.16   A large 15-m-dia cylindrical tank that sits on the ground is to be painted. If one liter of paint covers 10 m², how many liters are required if it is 10 m high? (Include the top.)
   a) 65   b) 53   c) 47   d) 38

17.17   The equation of a line that has a slope of −2 and intercepts the x-axis at $x = 2$ is
   a) $y + 2x = 4$ c) $y + 2x = -4$
   b) $y - 2x = 4$ d) $2y + x = 2$

17.18   The equation of a line that intercepts the x-axis at $x = 4$ and the y-axis at $y = -6$ is
   a) $2x - 3y = 12$ c) $2x + 3y = 12$
   b) $3x - 2y = 12$ d) $3x + 2y = 12$

17.19   The shortest distance from the line $3x - 4y = 3$ to the point (6, 8) is
   a) 4.8   b) 4.2   c) 3.8   d) 3.4

17.20   The equation $x^2 + 4xy + 4y^2 + 2x = 10$ represents which conic section?
   a) circle   b) ellipse   c) parabola   d) hyperbola

17.21   The x- and y-axes are the asymptotes of a hyperbola that passes through the point (2, 2). Its equation is
   a) $x^2 - y^2 = 0$ b) $xy = 4$ c) $y^2 - x^2 = 0$ d) $x^2 + y^2 = 4$

17.22   A 100-m-long track is to be built 50 m wide. If it is to be elliptical, what equation could describe it if the 100-m length is along the x-axis?
   a) $50x^2 + 100y^2 = 1000$ c) $4x^2 + y^2 = 2500$
   b) $2x^2 + y^2 = 250$ d) $x^2 + 2y^2 = 250$

17.23   The cylindrical coordinates (5, 30°, 12) are expressed in spherical coordinates as
   a) (13, 30°, 67.4°)   c) (15, 52.6°, 22.6°)
   b) (13, 30°, 22.6°)   d) (15, 52.6°, −22.6°)

17.24   The equation of a 4-m-radius sphere using cylindrical coordinates is
   a) $x^2 + y^2 + z^2 = 16$ c) $r^2 + z^2 = 16$
   b) $r^2 = 16$ d) $x^2 + y^2 = 16$

Complex Numbers

17.25   Divide $3 - i$ by $1 + i$.
    a) $1 - 2i$ b) $1 + 2i$ c) $2 - i$ d) $2 + i$

17.26   Find $(1 + i)^6$.
    a) $1 + i$ b) $1 - i$ c) $8i$ d) $-8i$

17.27   Find the root of $(1 + i)^{1/5}$ with the smallest argument.
    a) $0.168 + 1.06i$ c) $1.06 - 0.168i$
    b) $1.06 + 0.168i$ d) $0.168 - 1.06i$

17.28   Express $(3 + 2i)\, e^{2it} + (3 - 2i)\, e^{-2it}$ in terms of trigonometric functions.
    a) $3 \cos 2t - 4 \sin 2t$ c) $6 \cos 2t - 4 \sin 2t$
    b) $3 \cos 2t - 2 \sin 2t$ d) $3 \sin 2t + 2 \sin 2t$

17.29   Subtract $5e^{0.2i}$ from $6e^{2.3i}$.
    a) $-0.903 + 3.481i$ c) $-8.898 - 5.468i$
    b) $-8.898 + 3.481i$ d) $-0.903 - 5.468i$

Linear Algebra

17.30   Find the value of the determinant $\begin{vmatrix} 3 & 2 & 1 \\ 0 & -1 & -1 \\ 2 & 0 & 2 \end{vmatrix}$.

    a) 8   b) 4   c) -8 d) -4

17.31   Evaluate the determinant $\begin{vmatrix} 1 & 0 & 1 & 1 \\ 2 & -1 & 0 & 1 \\ 0 & 0 & 2 & 0 \\ 3 & 2 & 1 & 1 \end{vmatrix}$.

    a) 8   b) 4   c) 0   d) -4

17.32   The cofactor A_{21} of the determinant of Prob. 17.30 is
    a) -5 b) -4 c) 3   d) 4

17.33   The cofactor A_{34} of the determinant of Prob. 17.31 is
    a) 4   b) 6   c) -6 d) -4

17.34   Find the adjoint matrix of $\begin{vmatrix} 1 & -4 \\ 0 & 2 \end{vmatrix}$.

    a) $\begin{bmatrix} 4 & 2 \\ 0 & 1 \end{bmatrix}$ b) $\begin{bmatrix} 1 & 0 \\ 4 & 2 \end{bmatrix}$ c) $\begin{bmatrix} 2 & 4 \\ 1 & 0 \end{bmatrix}$ d) $\begin{bmatrix} 2 & 4 \\ 0 & 1 \end{bmatrix}$

17.35   The inverse matrix of $\begin{bmatrix} 2 & 3 \\ 1 & 1 \end{bmatrix}$ is

    a) $\begin{bmatrix} -1 & 3 \\ 1 & -2 \end{bmatrix}$ b) $\begin{bmatrix} 1 & -1 \\ -3 & 2 \end{bmatrix}$ c) $\begin{bmatrix} -1 & 1 \\ -3 & 2 \end{bmatrix}$ d) $\begin{bmatrix} -2 & 3 \\ 1 & -1 \end{bmatrix}$

17.36 Calculate $\begin{bmatrix} 2 & -1 \\ 3 & 2 \end{bmatrix} \begin{bmatrix} 2 \\ 1 \end{bmatrix}$.

a) $\begin{bmatrix} 8 \\ 3 \end{bmatrix}$ b) $\begin{bmatrix} 3 \\ 8 \end{bmatrix}$ c) $\begin{bmatrix} -3 \\ -8 \end{bmatrix}$ d) $[3, 8]$

17.37 Determine $\begin{bmatrix} 1 & 2 \\ 2 & 1 \end{bmatrix} \begin{bmatrix} -1 & 0 \\ 1 & 2 \end{bmatrix}$.

a) $\begin{bmatrix} 1 & 4 \\ -1 & 2 \end{bmatrix}$ b) $\begin{bmatrix} 1 & -1 \\ 4 & 2 \end{bmatrix}$ c) $\begin{bmatrix} 1 \\ -1 \end{bmatrix}$ d) $\begin{bmatrix} 4 \\ 2 \end{bmatrix}$

17.38 Solve for $[x_i]$.

$$3x_1 + 2x_2 = -2$$
$$x_1 - x_2 + x_3 = 0$$
$$4x_1 + 2x_3 = 4$$

a) $\begin{bmatrix} 2 \\ 4 \\ -6 \end{bmatrix}$ b) $\begin{bmatrix} -2 \\ 4 \\ 12 \end{bmatrix}$ c) $\begin{bmatrix} 2 \\ 4 \\ 8 \end{bmatrix}$ d) $\begin{bmatrix} -6 \\ 8 \\ 14 \end{bmatrix}$

17.39 Find the eigenvalues of $\begin{bmatrix} 1 & 2 \\ 3 & 2 \end{bmatrix}$.

a) 4, −1 b) 4, 1 c) 1, −4 d) 3, 2

Calculus

17.40 The slope of the curve $y = 2x^3 - 3x$ at $x = 1$ is
a) 3 b) 5 c) 6 d) 8

17.41 If $y = \ln x + e^x \sin x$, find dy/dx at $x = 1$.
a) 1.23 b) 3.68 c) 4.76 d) 6.12

17.42 At what value of x does a maximum of $y = x^3 - 3x$ occur?
a) 2 b) 1 c) 0 d) −1

17.43 Where does an inflection point occur for $y = x^3 - 3x$?
a) 2 b) 1 c) 0 d) −1

17.44 Evaluate $\lim\limits_{x \to \infty} \dfrac{2x^2 - x}{x^2 + x}$.

a) 2 b) 1 c) 0 d) −1

17.45 Find the area between the y-axis and $y = x^2$ from $y = 4$ to $y = 9$.
a) 29/3 b) 32/3 c) 34/3 d) 38/3

17.46 The area contained between $4x = y^2$ and $4y = x^2$ is
a) 10/3 b) 11/3 c) 13/3 d) 16/3

17.47 Rotate the shaded area of Example 17.24 about the x-axis. What volume is formed?
a) 4π b) 6π c) 8π d) 10π

17.48   Evaluate $\int_0^2 (e^x + \sin x)\, dx$.

    a) 7.81   b) 6.21   c) 5.92   d) 5.61

17.49   Evaluate $2 \int_0^1 e^x \sin x\, dx$.

    a) 1.82   b) 1.94   c) 2.05   d) 2.16

17.50   Derive an expression $\int x \cos x\, dx$.

    a) $x \cos x - \sin x + C$ c) $x \sin x - \cos x + C$
    b) $x \sin x + \cos x + C$ d) $x \cos x + \sin x + C$

Probability and Statistics

17.51   You reach into a jelly bean bag and grab one bean. If the bag contains 30 red, 25 orange, 15 pink, 10 green, and 5 black beans, the probability that you will get a black bean or a red bean is nearest to
    a) 0.6   b) 0.5   c) 0.4   d) 0.3

17.52   Two jelly bean bags are identical to that of Prob. 17.51. One bean is to be selected from each bag. The probability of selecting a black bean from the first bag and a red bean from the second bag is nearest to
    a) 1/100   b) 2/100   c) 3/100   d) 4/100

17.53   From the original bag of Prob. 17.52, the probability of selecting 5 beans, the first three of which are red and the next two of which are orange, is nearest to
    a) 3/1000   b) 4/1000   c) 5/1000   d) 6/1000

17.54   Two bags each contain 2 black balls, 1 white ball, and 1 red ball. One ball is to be selected from each bag. What is the probability of selecting the white ball from the first bag or the red ball from the second bag?
    a) 1/2   b) 7/16   c) 1/4   d) 3/8

17.55   A professor gives the following scores to her students. What is the mode?

frequency:	1	3	6	11	13	10	2
score:	35	45	55	65	75	85	95

    a) 65   b) 75   c) 85   d) 11

17.56   For the data of Prob. 17.55, what is the arithmetic mean?
    a) 68.5   b) 68.9   c) 69.3   d) 70.2

17.57   Calculate the sample standard deviation for the data of Prob. 17.55.
    a) 9.27   b) 10.11   c) 11.56   d) 13.78

Chapter 18

Engineering Fundamentals

INTRODUCTION

We have selected several subjects to be reviewed in this chapter that you have undoubtedly studied in your Physics courses, along with a short presentation of Economics. They are the introductory topics in several technical and scientific programs. You should be somewhat familiar with these subjects (with the possible exception of Economics), but there may be some material here that was either omitted in your Physics courses, or that was not included in the texts that you used. It is hoped that this short review may better prepare you for your future courses.

18.1 STATICS

Statics is concerned primarily with the equilibrium of bodies subjected to force systems. It has been introduced in Physics courses and is reviewed here to refresh your memories of the major components that make up the subject. Forces and moments are the two entities that are of most interest in Statics. Let's review them.

Forces, Moments, and Resultants

A force is the manifestation of the action of one body upon another. Forces arise from the direct action of two bodies in contact with one another, or from the "action at a distance" of one body upon another, as occurs with gravitational and magnetic forces. We classify forces as either *body forces*, which act (and are distributed) throughout the volume of the body, or as *surface forces*, which act over a surface portion of the body. If the surface over which the force system acts is very small, we usually assume localization at a specific point in the surface and speak of a *concentrated force* at that point.

Mathematically, forces are represented by *vectors*. Geometrically, a vector is a directed line segment having a head and a tail, i.e., an arrow. Its orientation defines the line of action, and the direction of the arrow (tail to head) gives the sense of the force.

Systems of concentrated forces are listed as *concurrent* when all of the forces act, or could act, at a single point; otherwise they are termed *non-concurrent* systems. Parallel force systems are in this second group. Also, force systems are often described as two-dimensional (acting in a single plane), or three-dimensional (spatial systems).

In addition to the "push or pull" effect on the point at which it acts, a force creates a *moment* about axes passing through the body. Conceptually, a moment may be thought of as a tendency to rotate the body upon which it acts about a certain axis. The moment of a force R, acting at a point A sketched in Fig. 18.1, about the y-axis is found by passing a plane normal to the y-axis through point A. The component of R in this plane, which we will call F, multiplied by the perpendicular distance d from the line of action of F to the y-axis, equals the moment of R about the y-axis, i.e.,

$$M_y = F \times d \qquad (18.1.1)$$

The *resultant* force of a system of forces is the equivalent force of the total system. It is the vector sum of the individual forces.

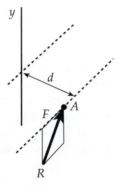

Figure 18.1 Moment of force R about the y-axis.

Example 18.1

Determine the magnitude of the resultant force R for the (a) plane, and (b) space concurrent systems shown (The force B is in the yz-plane).

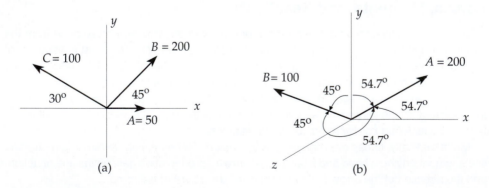

(a)   (b)

Solution: a) Sum components in the x-direction:

$$R_x = -100 \cos 30° + 200 \cos 45° + 50 = 104.8$$

Sum components in the y-direction:

$$R_y = 100 \sin 30° + 200 \sin 45° = 191.4$$

$$\therefore R = \sqrt{104.8^2 + 191.42^2} = 218.2$$

b) Sum components in the x-, y-, and z-directions:

$$R_x = 200 \cos 54.7° + 0 = 115.5$$

$$R_y = 200 \cos 54.7° + 100 \cos 45° = 186.3$$

$$R_z = 200 \cos 54.7° + 100 \cos 45° = 186.3$$

$$\therefore R = \sqrt{115.5^2 + 186.3^2 + 186.3^2} = 287.6$$

Example 18.2

Determine the moment of force F:

(a) about the x-axis.
(b) about the y-axis.

Distances are measured in meters.

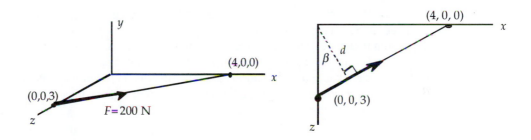

Solution: a) The line of action of F passes through the x-axis. Therefore, $d = 0$ and

$$M_x = 200 × 0 = 0$$

b) The force F acts in a plane normal to the y-axis so we must determine the perpendicular distance d, sketched above:

$$d = 3 \cos β = 3 × 0.8 = 2.4 \text{ m}$$

$$\therefore M_y = F × d = 200 × 2.4 = 480 \text{ N} \cdot \text{m}$$

Equilibrium

If the system of forces acting on a body is one whose resultant is absolutely zero (vector sum of all forces is zero, and the resultant moment of the forces about each of the axes is zero), the body is in *equilibrium*. Mathematically, equilibrium requires the six equations

$$\sum F_x = 0 \qquad \sum M_x = 0$$

$$\sum F_y = 0 \qquad \sum M_y = 0 \qquad \qquad (18.1.2)$$

$$\sum F_z = 0 \qquad \sum M_z = 0$$

which must hold for any orientation of the *xyz*-system. If the forces are concurrent and their vector sum is zero, the sum of moments will be satisfied automatically.

If all the forces act in a single plane, say the *xy*-plane, one of the above force equations and two of the moment equations are satisfied identically, so that equilibrium requires only

$$\sum F_x = 0, \qquad \sum F_y = 0 \qquad \sum M_z = 0 \qquad \qquad (18.1.3)$$

In this case we can solve for only three unknowns instead of six as when Eqs. 18.1.2 are required.

The solution for unknown forces and moments in equilibrium problems rests firmly upon the construction of a good *free body diagram*, abbreviated FBD, from which the detailed Eqs 18.1.2 or 18.1.3 may be obtained. A free body diagram is a neat sketch of the body (or of any appropriate portion of it) showing all forces and moments acting on the body, together with all important linear and angular dimensions.

A body in equilibrium under the action of two forces only is called a *two-force member*, and the two forces must be equal in magnitude and oppositely directed along the line joining their points of application. If a body is in equilibrium under the action of three forces (a *three-force member*) those forces must be coplanar, and concurrent (unless they form a parallel system). Examples are shown in Fig. 18.2.

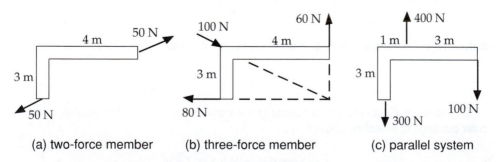

(a) two-force member   (b) three-force member   (c) parallel system

Figure 18.2 Plane force systems

A knowledge of the possible reaction forces and moments at various supports is essential in preparing a correct free body diagram. Several of the basic reactions are illustrated in

Fig. 18.3 showing a block of concrete subjected to a horizontal pull *P*, Fig. 18.3a, and a cantilever beam carrying both a distributed and concentrated load, Fig. 18.3b. The correct FBDs are on the right.

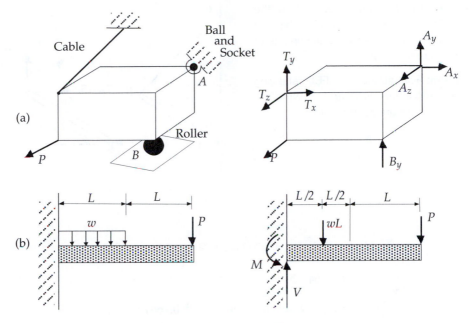

Figure 18.3 Free body diagrams.

Example 18.3

Determine the tension in the two cables supporting the 700 N block.

Solution: Construct the FBD of junction *A* of the cables. Sum forces in *x* and *y* directions:

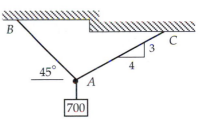

$$\sum F_x = -0.707T_{AB} + 0.8T_{AC} = 0$$

$$\sum F_y = 0.707T_{AB} + 0.6T_{AC} - 700 = 0$$

Solve, simultaneously, and find

$$T_{AC} = 700/1.4 = 500 \text{ N}$$

$$T_{AB} = 0.8(500)/0.707 = 565.8 \text{ N}$$

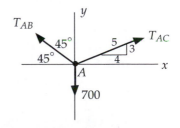

Example 18.4

A 12-m bar weighing 140 N is hinged to a vertical wall at *A*, and supported by the cable *BC*. Find the tension in the cable and the horizontal and vertical components of the force reaction at *A*.

Solution: Construct the FBD showing force components at *A* and *B*. Write the equilibrium equations and solve:

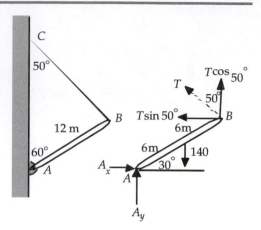

$$\sum M_A = 6T \sin 50° + 6\sqrt{3}T \cos 50° - 140(6)\sqrt{3}/2 = 0$$

$$\therefore T = 64.5 \text{ N}$$

$$\sum F_x = A_x - T \sin 50° = A_x - 64.5(0.766) = 0$$

$$\therefore A_x = 49.4 \text{ N}$$

$$\sum F_y = A_y + T \cos 50° - 140 = A_y + 64.5(0.643) - 140 = 0$$

$$\therefore A_y = 98.5 \text{ N}$$

EXERCISES—STATICS

18.1   Find the component of *A* = 200 in the direction of *B* = 100.
    a) 158
    b) 143
    c) 129
    d) 115

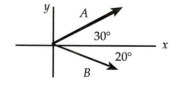

18.2   Find the magnitude of the resultant of *A* = 100, *B* = 50, and *C* = 120.
    a) 194
    b) 202
    c) 226
    d) 275

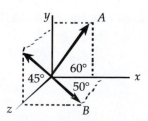

18.3   Determine the magnitude of the moment about the *y*-axis of the force $F = 500$
($F_x = 300$, $F_y = 200$, $F_z = ?$) acting at $(4, -6, 4)$.
a) 186   b) 1385   c) 2580   d) 3185

18.4   Find the magnitude of the moment of the two forces F_1 ($F_x = 50$, $F_y = 0$, $F_z = 40$) and
F_2 ($F_x = 0$, $F_y = 60$, $F_z = 80$) acting at $(2, 0, -4)$ and $(-4, 2, 0)$, respectively, about the
x-axis.
a) 0   b) 80   c) 160   d) 240

18.5   The force system shown may be referred to as being
a) non-concurrent, non-coplanar
b) coplanar
c) parallel
d) two-dimensional

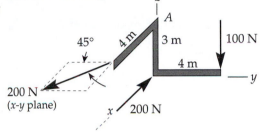

18.6   Equilibrium exists due to a rigid support at *A* in the figure of Prob. 18.5. Find the
magnitude of the force that exists at *A*.
a) 153 N   b) 173 N   c) 257 N   d) 382 N

18.7   Equilibrium exists on the object in Prob. 18.5. Find the magnitude of the reactive
moment that exists at support *A*.
a) 915 N · m   b) 691 N · m   c) 862 N · m   d) 721 N · m

18.8   If three nonparallel forces hold a rigid body in equilibrium, they must
a) be equal in magnitude.
b) be concurrent.
c) be non-concurrent.
d) form an equilateral triangle.

18.9   Find the magnitude of the reaction
at support *B*.
a) 400 N   c) 600 N
b) 500 N   d) 700 N

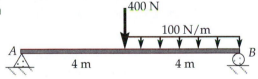

18.10   What moment *M* exists at suport *A*?
a) 5600 N · m   c) 4400 N · m
b) 5000 N · m   d) 4000 N · m

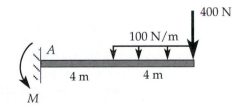

18.11   Calculate the reactive force at support *A*.
a) 250 N   c) 450 N
b) 350 N   d) 550 N

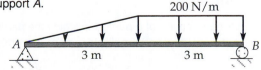

18.12   Find the support moment at A.
a) 66 N · m   c) 88 N · m
b) 77 N · m   d) 99 N · m

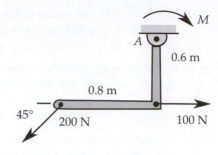

18.13   Calculate the magnitude of the equilibrating force at A for the three-force body shown.
a) 217 N   c) 343 N
b) 287 N   d) 385 N

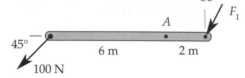

18.14   Find the magnitude of the equilibrating force at point A.
a) 187 N   c) 114 N
b) 142 N   d) 84 N

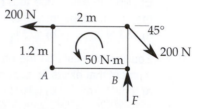

18.2 DYNAMICS

Dynamics is separated into two major divisions: *kinematics*, which is a study of motion without reference to the forces causing the motion, and *kinetics*, which relates the forces on bodies to their resulting motions. Newton's laws of motion are necessary in relating forces to motions; they are:

1st law: A particle remains at rest or continues to move in a straight line with a constant velocity if no unbalanced force acts on it.

2nd law: The acceleration of a particle is proportional to the force acting on it and inversely proportional to the particle mass; the direction of acceleration is the same as the force direction.

3rd law: The forces of action and reaction between contacting bodies are equal in magnitude, opposite in direction, and colinear.

Law of gravitation: The force of attraction between two bodies is proportional to the product of their masses and inversely proportional to the square of the distance between their centers.

Kinematics

In rectilinear motion of a particle in which the particle moves in a straight line, the acceleration a, the velocity, and the displacement s are related by

$$a = \frac{dv}{dt}, \quad v = \frac{ds}{dt}, \quad a = \frac{d^2s}{dt^2} = v\frac{dv}{ds} \qquad (18.2.1)$$

If the acceleration is a known function of time, the above can be integrated to give $v(t)$ and $s(t)$. For the important case of constant acceleration, integration yields

$$v = v_o + at$$

$$s = v_o t + at^2/2 \tag{18.2.2}$$

$$v^2 = v_o^2 + 2as$$

where at $t = 0$, $v = v_o$ and $s_o = 0$.

Angular displacement is the angle θ that a line makes with a fixed axis, usually the positive x-axis. Counterclockwise motion is assumed to be positive, as shown in Fig. 18.4. The angular acceleration α, the angular velocity ω, and θ are related by

$$\alpha = \frac{d\omega}{dt}, \quad \omega = \frac{d\theta}{dt}, \quad \alpha = \omega\frac{d\omega}{d\theta} = \frac{d^2\theta}{dt^2} \tag{18.2.3}$$

If α is a constant, integration of these equations gives

$$\omega = \omega_o + \alpha t \quad \theta = \omega_o t + \alpha t^2/2 \quad \omega^2 = \omega_o^2 + 2\alpha\theta \tag{18.2.4}$$

where we have assumed that $\omega = \omega_o$ and $\theta_o = 0$ at $t = 0$.

When a particle moves on a plane curve as shown in Fig. 18.5, the motion may be described in terms of coordinates along the normal n and the tangent t to the curve at the instantaneous position of the particle.

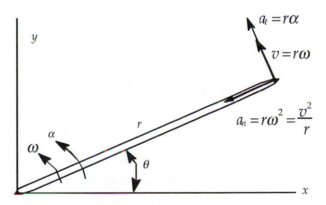

Figure 18.4 Angular motion.

The acceleration is the vector sum of the normal acceleration a_n and the tangential acceleration a_t. These components are given as

$$a_n = \frac{v^2}{r}, \quad a_t = \frac{dv}{dt} \tag{18.2.5}$$

where r is the radius of curvature and v is the magnitude of the velocity. The velocity is always tangential to the curve, so no subscript is necessary to identify the velocity.

It should be noted that a rigid body traveling without rotation can be treated as particle motion.

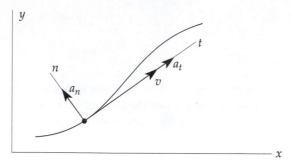

Figure 18.5 Motion on a plane curve.

Example 18.5

The velocity of a particle is $v(t) = 5 + 10t$ m/s. Find the acceleration and the displacement at $t = 10$ s, if $s_o = 0$ at $t = 0$.

Solution: The acceleration is found to be

$$a = \frac{dv}{dt} = 10 \text{ m/s}^2$$

The displacement is found using $v_o = 5$ m/s:

$$s = v_o t + at^2/2$$

$$= 5 \times 10 + 10 \times 10^2/2 = 550 \text{ m}$$

Example 18.6

An automobile skids to a stop 60 m after its brakes are applied while traveling 25 m/s. What is its acceleration?

Solution: Since speed and distance are given we use the relationship

$$v^2 = v_o^2 + 2as$$

Letting $v = 0$, we find,

$$a = -\frac{v_o^2}{2s} = -\frac{25^2}{2 \times 60} = -5.21 \text{ m/s}^2$$

Example 18.7

A wheel, rotating at 100 rad/s ccw (counterclockwise), is subjected to an angular accelera-tion of 20 rad/s^2 cw. Find the total number of revolutions (cw plus ccw) through which the wheel rotates in 8 seconds.

Solution: The time at which the angular velocity is zero is found as follows:

$$\cancel{\omega}^{0} = \omega_o + \alpha t$$

$$\therefore t = -\frac{\omega_o}{\alpha} = -\frac{100}{-20} = 5 \text{ s.}$$

After three additional seconds the angular velocity is determined by

$$\omega = \cancel{\omega_o}^{0} + \alpha t$$

$$= -20 \times 3 = -60 \text{ rad / s.}$$

The angular displacement from 0 to 5 s is

$$\theta = \omega_o t + \alpha t^2/2$$

$$= 100 \times 5 - 20 \times 5^2/2 = 250 \text{ rad}$$

During the next 3 s, the angular displacement is

$$\theta = \alpha t^2/2$$

$$= -20 \times 3^2/2 = -90 \text{ rad}$$

The total number of revolutions rotated is

$$\theta = (250 + 90)/2\pi = 54.1 \text{ rev}$$

Example 18.8

Consider idealized projectile motion (no air drag) in which $a_x = 0$ and $a_y = -g$. Find expres-sions for the range R and the maximum height H in terms of v_0 and θ.

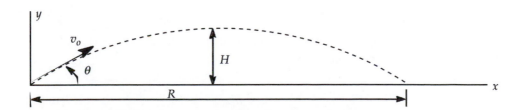

Solution: Using Eq. 18.2.2 for constant acceleration in the *y*-direction we have the point $(R, 0)$:

$$\cancel{y}^{\,0} = (v_o \sin \theta)t - gt^2/2.$$

$$\therefore t = 2v_o \sin \theta/g.$$

where $(v_y) = v_0 \sin \theta$. From the *x*-component equation recognizing that $a_x = 0$:

$$x = (v_o \cos \theta)t + \cancel{a_x}^{\,0} t^2/2.$$

$$\therefore R = (v_o \cos \theta)2v_o \sin \theta / g = v_o^2 \sin 2\theta / g.$$

using $2\sin \theta \cos \theta = \sin 2\theta$. Obviously, the maximum height occurs when the time is one-half that which yields the range *R*. Hence, with $y_{max} = H$,

$$H = (v_o \sin \theta) \frac{v_o \sin \theta}{g} - \frac{g}{2} \left(\frac{v_o \sin \theta}{g} \right)^2$$

$$= \frac{v_o^2}{2g} \sin^2 \theta$$

Note: The maximum *R* for a given v_o occurs when $\sin 2\theta = 1$, which means $\theta = 45°$ for R_{max}.

Example 18.9

It is desired that the normal acceleration of a satellite be 9.6m/s² at an elevation of 200 km. What should be the velocity be for a circular orbit? The radius of the earth is 6400 km.

Solution: The normal acceleration, which points toward the center of the Earth, is

$$a_n = \frac{v^2}{r}$$

$$\therefore v = \sqrt{a_n r} = \sqrt{9.6 \times (6400 + 200) \times 1000} = 7960 \text{ m/s}$$

The normal acceleration is essentially the value of gravity near the Earth's surface. Gravity varies only slightly if the elevation is small with respect to the Earth's radius.

Kinetics

To relate the force acting on a body to the motion of that body we use Newton's laws of motion. Newton's 2nd law is used in the form

$$\sum = F = m a \tag{18.2.6}$$

where the mass of the body is assumed to be constant and the vector is the acceleration of the center of mass (center of gravity) if the body is rotating.

The gravitational attractive force between one body and another is given by

$$F = K\frac{m_1 m_2}{r^2}$$ (18.2.7)

where $K = 6.67 \times 10^{-11}$ N · m²/kg². Note: Since metric units are used in the above relations, mass must be measured in kilograms. The weight is related to the mass by

$$W = mg$$ (18.2.8)

where we use $g = 9.8$ m/s², unless otherwise stated.

Example 18.10

Find the tension in the rope and the distance the 300 kg mass moves in 3 seconds. The mass starts from rest and the mass of the pulley is negligible. The force due to friction is μN where N is the normal force and μ is the coefficient of friction. The pully is frictionless.

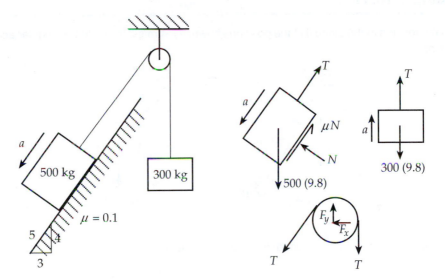

Solution: First, sketch the free bodies of the two masses and the pulley, as shown. The pulley does not change tension T in the rope. Next, sum forces normal to the surface:

$$\sum F_n = 0$$

$$N - 500 \times 9.8 \times 0.6 = 0. \quad \therefore N = 2940 \text{ N}. \quad \therefore F = \mu N = 294 \text{ N}$$

Applying Newton's 2nd law to the 500 kg mass parallel to the surface gives

$$\sum F = ma$$

$$0.8 \times 500 \times 9.8 - 294 - T = 500\,a$$

By studying the pulley we observe the 300 kg mass to also be accelerating at *a*. Hence, we have

$$\sum F = ma$$

$$T - 300 \times 9.8 = 300 \times a$$

Solving the above equations simultaneously results in

$$a = 0.8575 \text{ m/s}^2, \qquad T = 3197 \text{ N}$$

The distance the 300 kg mass moves is

$$s = \frac{1}{2}at^2 = \frac{1}{2} \times 0.8575 \times 3^2 = 3.31 \text{ m}$$

Example 18.11

Find the tension in the string if at the position shown, $v = 4$ m/s. Calculate the angular acceleration.

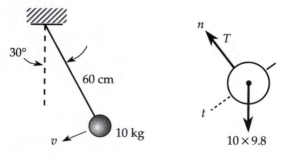

Solution: Sum forces in the normal direction and obtain

$$\sum F_n = ma_n = mv^2 / r$$

$$T - 10 \times 9.8 \cos 30° = 10 \times 4^2 / 0.6$$

$$\therefore T = 352 \text{ N}$$

Sum forces in the tangential direction and find

$$\sum F_t = ma_t = mr\alpha$$

$$10 \times 9.8 \sin 30° = 10 \times 0.6\alpha$$

$$\therefore \alpha = 8.17 \text{ rad/s}^2$$

EXERCISES—DYNAMICS

18.15 An object is moving with an initial velocity of 20 m/s. If it is decelerating at 5 m/s^2 how far does it travel before it stops?
a) 10 m   b) 20 m   c) 30 m   d) 40 m

18.16 A projectile is shot straight up with v_o = 40 m/s. After how many seconds will it return if drag is neglected?
a) 4 s   b) 6 s   c) 8 s   d) 10 s

18.17 An automobile is traveling at 25 m/s^2. It takes 0.3 s to apply the brakes after which the deceleration is 6.0 m/s^2. How far does the automobile travel before it stops?
a) 40 m   b) 45 m   c) 50 m   d) 60 m

18.18 A wheel accelerates from rest with α = 6 rad/s^2. How many revolutions are experienced in 4 s?
a) 7.64   b) 9.82   c) 12.36   d) 25.6

18.19 A 2-m-long shaft rotates about one end at 20 rad/s. It begins to accelerate with α = 10 rad/s^2. After how long will the velocity of the free end reach 100 m/s?
a) 3 s   b) 4 s   c) 5 s   d) 6 s

18.20 A roller-coaster reaches a velocity of 20 m/s at a location where the radius of curvature is 40 m. Calculate the acceleration, in m/s^2.
a) 8   b) 9   c) 10   d) 12

18.21 A bucket full of water is to be rotated in the vertical plane. What minimum angular velocity, in rad/s, is necessary to keep the water inside if the rotating arm is 120 cm?
a) 2.86   b) 3.15   c) 3.86   d) 4.26

18.22 An automobile is accelerating at 5 m/s^2 on a straight road on a hill where the radius of curvature of the hill is 200 m. What is the magnitude of the total acceleration when the car's speed is 30 m/s?
a) 5 m/s^2 b) 5.46 m/s^2 c) 6.04 m/s^2 d) 6.73 m/s^2

18.23 A particle experiences the displacement shown. What is its velocity at t = 1 s?

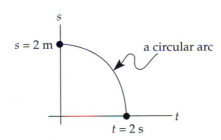

a) −0.58 m/s   b) −0.76 m/s   c) −0.92 m/s   d) −1.0 m/s

18.24 Neglecting the change of gravity with elevation, estimate the speed a satellite must have to orbit the earth at an elevation of 100 km. Earth's radius = 6400 km.
a) 4000 m/s   b) 6000 m/s   c) 8000 m/s   d) 10 000 m/s

18.25 Find an expression for the maximum range of a projectile with initial velocity v_o at angle θ with the horizontal.

a) $\dfrac{v_o^2}{g}\sin 2\theta$ b) $\dfrac{v_o^2}{2g}\sin^2\theta$ c) $\dfrac{v_o^2}{2g}\sin\theta$ d) $\dfrac{v_o^2}{g}\cos\theta$

18.26   Find the maximum height, in meters.

a) 295   b) 275   c) 255   d) 235

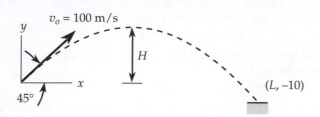

18.27   Calculate the time, in seconds, it takes the projectile of Problem 18.26 to reach the low point.

a) 14.6   b) 12.2   c) 11.0   d) 10.2

18.28   What is the distance L, in meters, in Problem 18.26?

a) 530   b) 730   c) 930   d) 1030

18.29   What is a_A?

a) 2.09 m/s^2

b) 1.85   m/s^2

c) 1.63   m/s^2

d) 1.47 m/s^2

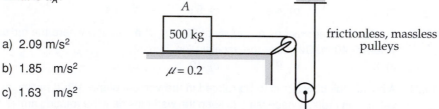

18.30   How far, in meters, will the weight move in 10 seconds, if released from rest? The friction force is $F_{friction} = \mu N$ where N is the normal force.

a) 350   c) 250

b) 300   d) 200

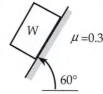

18.31   At what angle, in degrees, should a road be slanted to prevent an automobile traveling at 25 m/s from tending to slip? The radius of curvature is 200 m.

a) 22   b) 20   c) 18   d) 16

18.32   A satellite orbits the Earth 200 km above the surface. What speed, in m/s, is necessary for a circular orbit? The radius of the Earth is 6400 km and $g = 9.2$ m/s^2

a) 7800   b) 7200   c) 6600   d) 6000

18.33   Determine the mass of the Earth, in kg, if the radius of the Earth is 6400 km.

a) 6×10^{22} b) 6×10^{23} c) 6×10^{24} d) 6×10^{25}

18.34   The coefficient of sliding friction between rubber and asphalt is about 0.6. What minimum distance, in meters, can an automobile slide on a horizontal surface if it is traveling at 25 m/s?

a) 38   b) 43   c) 48   d) 53

18.3 THERMODYNAMICS

Thermodynamics involves the storage, transformation and transfer of energy. It is stored as internal energy, kinetic energy, and potential energy; it is transformed between these various forms; and it is transferred as work or heat transfer. We will present some introductory concepts as they apply primarily to ideal gases. These same concepts can be applied to substances that change phase, such as water and a refrigerant, however, tabulated properties, which are not included here, are needed.

Overview, Definitions, and Laws

Both a *system*, a fixed quantity of matter, and a *control volume*, a volume into which and/or from which a substance flows, can be used in thermodynamics. (A control volume may also be referred to as an *open system*.) A system and its surroundings make up the *universe*. Some useful definitions follow:

phase —	matter that has the same composition throughout; it is homogeneous
mixture —	a quantity of matter that has more than one phase
property —	a quantity which serves to describe a system
simple system —	a system composed of a single phase, free of magnetic, electrical, and surface effects. Only two properties are needed to fix a simple system
state —	the condition of a system described by giving values to its properties at the given instant
intensive property —	a property that does not depend on the mass
extensive property —	a property that depends on the mass
specific property —	an extensive property divided by the mass
thermodynamic equilibrium —	when the properties do not vary from point to point in a system and there is no tendency for additional change
process —	the path of successive states through which a system passes
quasi-equilibrium —	if, in passing from one state to the next, the deviation from equilibrium is infinitesimal
reversible process —	a process which, when reversed, leaves no change in either the system or surroundings
isothermal —	temperature is constant
isobaric —	pressure is constant
isometric —	volume is constant
isentropic —	entropy is constant
adiabatic —	no heat transfer (an insulated surface)

Experimental observations are organized into mathematical statements or *laws*. Some of those used in thermodynamics follow:

zeroith law of thermodynamics —	If two bodies are equal in temperature to a third, they are equal in temperature to each other.
first law of thermodynamics —	During a given process, the net heat transfer minus the net work output equals the change in energy.

second law of thermodynamics — During a process, the net entropy of the universe cannot decrease.

Boyle's law — The volume varies inversely with pressure for an ideal gas.

Charles' law — The volume varies directly with temperature for an ideal gas.

Avogadro's law — Equal volumes of different ideal gases with the same temperature and pressure contain an equal number of molecules.

Density, Pressure, and Temperature

The density ρ is the mass divided by the volume,

$$\rho = \frac{m}{V} \tag{18.3.1}$$

The specific volume is the reciprocal of the density,

$$v = \frac{1}{\rho} = \frac{V}{m} \tag{18.3.2}$$

The pressure P is the normal force divided by the area upon which it acts. In thermodynamics, it is important to use *absolute pressure,* defined by

$$P_{abs} = P_{gauge} + P_{atmospheric} \tag{18.3.3}$$

where the atmospheric pressure is taken as 100 kPa (14.7 psi), unless otherwise stated. If the gauge pressure is negative, it is a *vacuum.*

The temperature scale is established by choosing a specified number of divisions, called degrees, between the ice point and the steam point, each at 101 kPa absolute. In the Celsius scale, the ice point is set at 0°C and the steam point at 100°C. **In thermodynamics, pressures are always assumed to be given as absolute pressures.** Expressions for the absolute temperature in kelvins and degrees Rankine are, respectively,

$$T = T_{celsius} + 273; \quad T = T_{fahrenheit} + 460 \tag{18.3.4}$$

The temperature, pressure and specific volume for an ideal (perfect) gas are related by the *ideal gas law*

$$Pv = RT, \quad P = \rho RT, \quad \text{or} \quad PV = mRT \quad R = \overline{R} / M \tag{18.3.5}$$

where the *universal gas constant* $\overline{R}$ = 8.314 kj/kmol · K, M is the molar mass, and R is the gas constant which for air is 0.287 kJ/kg · K.

Example 18.12

What mass of air is contained in a room 20 m × 40 m × 3 m at standard conditions?

Solution: Standard conditions are $T = 25°C$ and $P = 100$ kPa. Using $M = 28.97$,

$$\rho = \frac{P}{RT}$$

$$= \frac{100}{0.287 \times 298} = 1.17 \text{ kg/m}^3, \quad \text{where } R = \frac{8.314}{28.97} = 0.287 \text{ kJ/kg} \cdot \text{K}$$

$$\therefore m = \rho V$$

$$= 1.17 \times 20 \times 40 \times 3 = 2810 \text{ kg}$$

The First Law of Thermodynamics for a System

The first law of thermodynamics, referred to as the "first law" or the "energy equation," is expressed for a cycle as

$$Q_{net} = W_{net} \tag{18.3.6}$$

and for a process as

$$Q - W = \Delta E \tag{18.3.7}$$

where Q is the heat transfer, W is the work, and E represents the energy (kinetic, potential, and internal) of the system. Heat transfer to the system is positive and work done by the system is positive. (Some authors define work done on the system as positive so that $Q + W = \Delta U$.) In thermodynamics attention is focused on internal energy with kinetic and potential energy changes neglected (unless otherwise stated) so that we have

$$Q - W = \Delta U \quad \text{or} \quad q - w = \Delta u \tag{18.3.8}$$

where the specific internal energy is

$$u = \frac{U}{M} \quad \text{and} \quad q = \frac{Q}{m}, \quad w = \frac{W}{m} \tag{18.3.9}$$

Heat transfer may occur during any of the three following modes :

•**Conduction**—heat transfer due to molecular activity. For steady-state heat transfer through a constant wall area *Fourier's law* states

$$\dot{Q} = kA\Delta T/L = A\Delta T/R \tag{18.3.10}$$

where k is the *conductivity* (dependent on the material), R is the *resistance factor*, and the length, L, is normal to the heat flow. A dot signifies a rate, so that $\dot{Q}$ has units of J/s.

•**Convection**—heat transfer due to fluid motion. The mathematical expression used is

$$\dot{Q} = hA\Delta T = A\Delta T/R \tag{18.3.11}$$

where *h* is the *convective heat transfer coefficient* (dependent on the surface geometry, the fluid velocity, the fluid viscosity and density, and the temperature difference).

•*Radiation*—heat transfer due to the transmission of waves. The heat transfer from body 1 is

$$\dot{Q} = \sigma \epsilon A(T_1^4 - T_2^4)F_{1-2} \tag{18.3.12}$$

in which the *Stefan-Boltzmann constant* is, $\sigma = 5.67 \times 10^{-11} \text{kJ/s} \cdot \text{m}^2 \cdot \text{K}^4$, ε is the *emissivity* ($\varepsilon = 1$ for a black body), and F_{1-2} is the *shape factor* ($F_{1-2} = 1$ if body 2 encloses body 1).

For a two layer composite wall with an inner and outer convection layer we use the resistance factors and obtain

$$\dot{Q} = A\Delta T/(R_i + R_1 + R_2 + R_o) = U A \Delta T \tag{18.3.13}$$

where *U* is the *overall heat transfer coefficient*. It is not to be confused with internal energy of Eq. 18.3.8.

Work can be accomplished mechanically by moving a boundary, resulting in a quasi-equilibrium work mode

$$W = \int PdV \tag{18.3.14}$$

It can also be accomplished in non-quasi-equilibrium modes such as with paddle wheel or by electrical resistance. But then Eq. 18.3.14 cannot be used.

We introduce *enthalpy* for convenience, and define it to be

$$H = U + PV \tag{18.3.15}$$

$$h = u + Pv$$

For ideal gases we assume constant specific heats and use

$$\Delta u = c_v \Delta T \tag{18.3.16}$$

$$\Delta u = c_p \Delta T \tag{18.3.17}$$

where c_v is the *constant volume specific heat,* and c_p is the *constant pressure specific heat.* From the above we can find

$$c_p = c_v + R \tag{18.3.18}$$

We also define the *ratio of specific heats k* to be

$$k = c_p/c_v \tag{18.3.19}$$

For air $c_v = 0.716$ kJ/kg · K, $c_p = 1.00$ kJ/kg · K, and $k = 1.4$. For most solids and liquids we can find the heat transfer using

$$Q = mc_p \Delta T \tag{18.3.20}$$

For water $c_p = 4.18$ kJ/kg · °K, and for ice $c_p \cong 2.1$ kJ/kg · °K.

When a substance changes phase, *latent heat* is involved. The energy necessary to melt a unit mass of a solid is the *heat of fusion*; the energy necessary to vaporize a unit mass of liquid is the *heat of vaporization*; the energy necessary to vaporize a unit mass of solid is the *heat of sublimation*. For ice, the heat of fusion is approximately 320 kJ/kg and the heat of sublimation is about 2040 kJ/kg; the heat of vaporization of water varies from 2050 kJ/kg at 0°C to zero at the point where no vaporization occurs (at very high pressures).

We consider the preceding paragraphs and summarize as follows for quasi-equilibrium processes:

Constant Temperature (Isothermal)

$$1\text{st law:} \qquad Q - W = m\Delta u \quad \text{or} \quad q - w = \Delta u \qquad (18.3.21)$$

$$\text{ideal gas:} \qquad Q = W = mRT \ln \frac{v_2}{v_1} = mRT \ln \frac{P_1}{P_2} \qquad (18.3.22)$$

$$P_2 = P_1 v_1 / v_2 \qquad (18.3.23)$$

Constant Pressure (Isobaric)

$$1\text{st law:} \qquad Q = m\Delta u \quad \text{or} \quad q = \Delta u \qquad (18.3.24)$$

$$W = 0 \qquad (18.3.25)$$

$$\text{ideal gas:} \qquad Q = mc_p \Delta T \qquad (18.3.26)$$

$$T_2 = T_1 \, v_2 / v_1 \qquad (18.3.27)$$

Constant Volume (Isometric)

$$1\text{st law:} \qquad Q = m\Delta u \quad \text{or} \quad q = \Delta u \qquad (18.3.28)$$

$$W = 0 \qquad (18.3.29)$$

$$\text{ideal gas:} \qquad Q = mc_v \Delta T \qquad (18.3.30)$$

$$T_2 = T_1 P_2 / P_1 \qquad (18.3.31)$$

Adiabatic Process (Isentropic)

1st law: $-W = m\Delta u$ or $-w = \Delta u$ (18.3.32)

$$Q = 0$$ (18.3.33)

ideal gas: $-W = mc_v\,\Delta T$ (18.3.34)

$$T_2 = T_1(v_1/v_2)^{k-1} = T_1(P_2/P_1)^{(k-1)/k}$$ (18.3.35)

$$P_2 = P_1(v_1/v_2)^k$$ (18.3.36)

Example 18.12

How much heat is needed to raise the temperature of 100 kg of ice at T_1 $-10°C$ to $80°C$?

Solution: The heat transfer is related to the enthalpy by

$$Q = m\Delta h$$

$$= m(c\Delta T_{ice} + \text{heat of fusion} + c\Delta T_{water})$$

Using the values given in this article,

$$Q = 100(2.1 \times 10 + 320 + 4.18 \times 80)$$

$$= 67\,540 \text{ kJ or } 67.54 \text{ MJ}$$

Example 18.13

Calculate the work done by a piston if the 2 m³ volume of air is tripled while the temperature is maintained at 40°C. The initial pressure is 400 kPa.

Solution: The mass is needed in order to use Eq. 18.3.22 to find the work; it is

$$m = \frac{PV}{RT} = \frac{400 \times 2}{0.287 \times 313} = 8.91 \text{ kg}$$

The work is then found to be

$$W = mRT \ln v_2/v_1$$

$$= 8.91 \times 0.287 \times 313 \ln 3 = 879 \text{ kJ}$$

Note: The temperature is expressed as 40 + 273 = 313 K.

Example 18.14

How much work is necessary to compress air in an insulated cylinder from 0.2 m³ to 0.01 m³? Use $T_1 = 20°C$ and $P_1 = 100$ kPa.

Solution: For an adiabatic process $Q = 0$ so that the first law is

$$-W = m(u_2 - u_1)$$

$$= mc_v(T_2 - T_1)$$

To find the mass m we use the ideal gas equation:

$$m = \frac{PV}{RT} = \frac{100 \times 0.2}{0.287 \times 293} = 0.2378 \text{ kg}$$

The temperature T_2 is found to be

$$T_2 = T_1(v_1/v_2)^{k-1}$$

$$= 293(0.2/0.01)^{0.4} = 971.1 \text{ K}$$

The work is then

$$W = 0.2378 \times 0.716(971.1 - 293) = 115.5 \text{ kJ}$$

Example 18.15

A 10-cm-thick wall made of pine wood is 3 m high and 10 m long. Calculate the heat transfer rate if the temperature is 25°C on the inside and –20°C on the outside. Neglect convection. Use $k = 0.15$ J/(s · m·°C).

Solution: The heat transfer occurs due to conduction. Using the given k Eq. 18.3.10 provides

$$\dot{Q} = kA\Delta T/L$$

$$= 0.15 \times (3 \times 10) \times [25 - (-20)]/0.1 = 2025 \text{ J/s}$$

Example 18.16

The surface of the glass in a 1.2m $\times$ 0.8m skylight is maintained at 20°C. If the air temperature is –20°C, estimate the rate of heat loss from the window. Use $h = 12$ J/(s $\cdot$ m^2 $\cdot$ °C).

Solution: The convective heat transfer coefficient depends on several parameters so, as usual, it is specified. Using Eq. 18.3.11 the rate of heat loss is

$$\dot{Q} = hA\Delta T$$

$$= 12 \times (1.2 \times 0.8) \times [20 - (-20)] = 461 \text{ J/s}$$

Example 18.17

A 2-cm-diameter heating oven is maintained at 1000°C and the oven walls are at 500°C. If the emissivity of the element is 0.85, estimate the rate of heat loss from the 2-m-long element.

Solution: The heat loss will be primarily due to radiation. Neglecting any convection loss, using $F_{1-2} = 1$ since the oven encloses the element, and Eq. 18.3.12 provides us with

$$\dot{Q} = \sigma\epsilon A(T_1^4 - T_2^4)$$

$$= 5.67 \times 10^{-11} \times 0.85 \times (\pi \times 0.02 \times 2)(1273^4 - 773^4) = 13.7 \text{ kJ/s}$$

Note that absolute temperature must be used. Also, the area A is the surface area of the cylinder, i.e., πDL.

The First Law of Thermodynamics for a Control Volume

The continuity equation, which accounts for the conservation of mass, may be used in a certain situation involving a *control volume* (a volume into which and/or from which a fluid flows). It is stated as

$$\dot{m} = \rho_1 A_1 V_1 = \rho_2 A_2 V_2 \tag{18.3.37}$$

where, in control volume formulations, V is the velocity and $\dot{m}$ is called the *mass flux*. In the above continuity equation, we assume *steady flow*, that is, the variables are independent of time, and one inlet and one outlet. For such steady state flow situations, the first law takes the form

$$\frac{\dot{Q} - \dot{W}_s}{\dot{m}} = \frac{V_2^2 - V_1^2}{2} + h_2 - h_1 + g(z_2 - z_1) \tag{18.3.38}$$

where the dot signifies a rate so that $\dot{Q}$ and $\dot{W}_s$ have the units of kJ/s. In most devices the potential energy change is negligible. Also, the kinetic energy change can often be ignored (but if sufficient information is given, it should be included, as in a nozzle) so that the first law is most often used in the simplified form

$$\dot{Q} - \dot{W}_s = \dot{m}(h_2 - h_1)$$ (18.3.39)

Particular devices are of special interest. The energy equation for a *valve* or a *throttle plate* is simply

$$h_2 - h_1$$ (18.3.40)

neglecting kinetic energy change.

For a *turbine* expanding a gas, the heat transfer is negligible so that

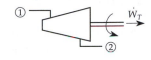

$$\dot{W}_T = \dot{m}(h_1 - h_2)$$ (18.3.41)

The work input to a *gas compressor* with negligible heat transfer is

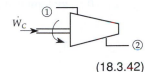

$$\dot{W}_C = \dot{m}(h_2 - h_1)$$ (18.3.42)

A *boiler* and a *condenser* are simply heat transfer devices. The first law then simplifies to

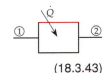

$$\dot{Q} = \dot{m}(h_2 - h_1)$$ (18.3.43)

For a *nozzle* or a *diffuser* there is no work or heat transfer; we must include, however, the kinetic energy change, resulting in

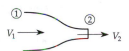

$$0 = \frac{V_2^2 - V_1^2}{2} + h_2 - h_1$$ (18.3.44)

For a *pump* or a *hydroturbine* we take a slightly different approach. We return to Eq. 18.3.38 and write it using $v = 1/\rho$, and with Eq. 18.3.15, as

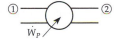

$$\frac{\dot{Q} - \dot{W}_S}{\dot{m}} = \frac{V_2^2 - V_1^2}{2} + u_2 - u_1 + \frac{P_2 - P_1}{\rho} + g(z_2 - z_1)$$ (18.3.45)

For most ideal situations we do not transfer heat and assume constant temperature so that $u_2 = u_1$. Often the kinetic and potential energy changes are negligible so

$$-\dot{W}_S = \dot{m}\frac{P_2 - P_1}{\rho}$$ (18.3.46)

This would provide the minimum pump power requirement or the maximum turbine power output. The inclusion of an efficiency would increase the pump power requirement or decrease the turbine output.

A gas turbine or compressor efficiency is based on an isentropic process as the ideal process. For a gas turbine or a compressor we have

$$\eta_T = \frac{\dot{W}_a}{\dot{W}_s} = \frac{w_a}{w_s} \qquad \eta_C = \frac{\dot{W}_s}{\dot{W}_a} = \frac{w_s}{w_a} \qquad (18.3.47)$$

where $\dot{W}_a$ is the actual power and $\dot{W}_s$ is the power assuming an isentropic process.

Example 18.18

What is the minimum power requirement of a pump that is to increase the pressure from 2 kPa to 6 MPa for a mass flux of 10 kg/s of water?

Solution: With a liquid we let $h_2 - h_1 = u_2 - u_1 + (P_2 - P_1)v$ since v = const. We let $u_2 - u_1 = 0$ so that the first law simplifies to

$$\dot{W}_P = \dot{m}\frac{P_2 - P_1}{\rho}$$

$$= 10\frac{6000 - 2}{1000} = 59.98 \text{ kW}$$

Example 18.19

A nozzle accelerates air from 100 m/s, 400°C and 400 kPa to a receiver where $P = 20$ kPa. Assuming an isentropic process, find V_2.

Solution: The energy equation takes the form

$$0 = \frac{V_2^2 - V_1^2}{2} + h_2 - h_1$$

Assuming air to be an ideal gas with constant c_p we have

$$c_p(T_1 - T_2) = \frac{V_2^2 - V_1^2}{2}$$

We can find T_2 from Eq. 18.3.35 to be

$$T_2 = T_1(P_2/P_1)^{k-1/k}$$

$$= 673(20/400)^{0.4/1.4} = 286 \text{ K}$$

The exiting velocity is found as follows:

$$1000(673 - 286) = \frac{V_2^2 - 100^2}{2}$$

$$\therefore V_2 = 885 \text{ m/s}$$

Note: c_p must be used as 1000 J/kg $\cdot$ K so that the units are consistent.

EXERCISES—THERMODYNAMICS

18.35   Which would be considered a system rather than a control volume?
a) a pump   b) a tire   c) a pressure cooker   d) a turbine

18.36   Which of the following is an extensive property?
a) temperature   b) velocity   c) pressure   d) mass

18.37   An automobile heats up while sitting in a parking lot on a sunny day. The process can be assumed to be:
a) isothermal   b) isobaric   c) isometric   d) isentropic

18.38   A scientific law is a statement that
a) we postulate to be true.
b) is generally observed to be true.
c) is a summary of experimental observation.
d) is agreed upon by the scientific community.

18.39   In a quasiequilibrium process involving a system
a) the pressure remains constant.
b) the pressure varies with temperature.
c) the pressure is constant at an instant at each point in the system.
d) the pressure force results in a positive work input.

18.40   The density of air at vacuum of 40 kPa and –40°C is, in kg/m^3,
a) 0.598   b) 0.638   c) 0.897   d) 0.753

18.41   A cold tire has a volume of 0.03 m^3 at –10°C and 180 kPa gauge. If the pressure and temperature increase to 210 kPa gauge and 30°C, find the final volume in m^3.
a) 0.0304   b) 0.0308   c) 0.0312   d) 0.0316

18.42   A 300-watt light bulb provides energy in a 10-m-diameter spherical space. If the outside temperature is 20°C, find the inside steady-state temperature, in °C, if $R = 1.5$ hr $\cdot$ m^2 $\cdot$ °C/kj for the wall.
a) 40.5   b) 35.5   c) 32.6   d) 25.2

18.43   Select a correct statement of the first law.
a) Heat transfer equals the work done for a process.
b) Net heat transfer equals the net work for a cycle.
c) Net heat transfer equals net work plus internal energy change for a cycle.
d) Heat transfer minus work equals change in enthalpy.

18.44   A cycle undergoes the following processes. All units are kJ. Find E_{after} for the process 1→2.

	Q	W	ΔE	E_{before}	E_{after}
1→2	20	5		10	
2→3		−5	5		
3→1	30			30	

a) 10   b) 15   c) 20   d) 25

18.45   For the cycle of Problem 18.44 find W_{3-1}.
a) 50   b) 60   c) 70   d) 80

18.46   Ten kilograms of −10°C ice is added to 100 kg of 20°C water. What is the eventual temperature, in °C, of the water? Assume an insulated container.
a) 9.2   b) 10.8   c) 11.4   d) 12.6

18.47   Select the incorrect statement. $(Q - W = \Delta U)$
a) Work and heat transfer represent energy crossing a boundary.
b) The differentials of work and heat transfer are exact.
c) Work and heat transfer are path integrals.
d) Net work and net heat transfer are equal for a cycle.

18.48   Determine the work, in kJ, necessary to compress 2 kg of air from 100 kPa to 4000 kPa if the temperature is held constant 300°C.
a) −1210   b) −1105   c) −932   d) −812

18.49   There are 200 people in a 2000 m² room, lighted with 30 W/m². Estimate the maximum temperature increase, in °C, if the ventilation system fails for 20 min. Each person generates 400 kJ/h. The room is 3 m high.
a) 5.6   b) 6.8   c) 8.6   d) 13.4

18.50   Estimate the average c_p value, in kJ/kg · K, of a gas if 522 kJ/kg of heat are necessary to raise the temperature from 300 K to 800 K holding the pressure constant.
a) 1.000   b) 1.026   c) 1.038   d) 1.044

18.51   How much heat, in kJ, must be transferred to 10 kg of air to increase the temperature from 10°C to 230°C if the pressure is maintained constant?
a) 2200   b) 2090   c) 1890   d) 1620

18.52   A tire is pressurized to 100 kPa gauge in Michigan where $T = 0$°C. In Arizona the tire is at 70°C. Assuming a rigid tire, estimate the pressure in kPa gauge.
a) 120   b) 130   c) 140   d) 150

18.53   Air expands in an insulated cylinder from 200°C and 400 kPa to 20 kPa. Find T_2 in °C.
a) −24   b) −28   c) −51   d) −72

18.54   During an isentropic expansion of air, the volume triples. If the initial temperature is 200°C, find T_2 in °C.
a) 32   b) 28   c) 16   d) 8

18.55   Find the work, in kJ/kg, needed to compress air isentropically from 20°C and 100 kPa to 6 MPa.
a) −523   b) −466   c) −423   d) −392

18.56   What is the energy requirement, in kW, for a pump that is 75% efficient if it increases the pressure of 10 kg/s of water from 10 kPa to 6 MPa?
    a) 60   b) 70   c) 80   d) 90

18.57   A river 60 m wide and 2 m deep flows at 2 m/s. A hydro plant develops a pressure of 300 kPa gage just before the turbine. What maximum power, in MW, is possible?
    a) 72   b) 64   c) 56   d) 48

18.4 ELECTRICAL CIRCUITS

Circuits

Electric circuits are an interconnection of electrical components for the purpose of either generating and distributing electrical power; converting electrical power to some other useful form such as light, heat, or mechanical torque; or processing information contained in an electrical form (i.e., electrical signals). Most electrical circuits contain a source (or sources) of electrical power, passive components which store or dissipate energy, and possibly active components which change the electrical form of the energy or information being processed by the circuit.

Circuits may be classified as *Direct Current* (DC) circuits when the currents and voltages do not vary with time or as *Alternating Current* (AC) circuits when the currents and voltage vary sinusoidally with time. Both DC and AC circuits are said to be operating in the *steady state* when their respective current/voltage time variation is purely constant or purely sinusoidal with time. A *transient circuit* condition occurs when a switch is thrown that turns a source either on or off. This review will cover the DC steady-state. The primary quantities of interest in making circuit calculations are presented in Table 18.1.

TABLE 18.1   Quantities Used in Electric Circuits

Quantity	Symbol	Unit	Defining Equation	Definition
Charge	Q	coulomb	$Q = \int I dt$	
Current	I	ampere	$I = \dfrac{dQ}{dt}$	Time rate of flow of charge past a point in the circuit.
Voltage	V	volt	$V = \dfrac{dW}{dQ}$	Energy per unit charge either gained or lost through a circuit element.
Energy	W	joule	$W = \int V dQ = \int P dt$	
Power	P	watt	$P = \dfrac{dW}{dt} = IV$	Power is the time rate of energy flow.

Circuit Components

The circuits reviewed in this section will contain one or more sources interconnected with passive components. These passive circuit components include resistors, inductors and capacitors.

a) *Resistors* absorb energy and have a resistance value *R* measured in ohms

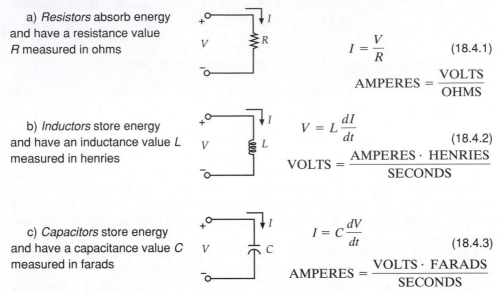

$$I = \frac{V}{R} \qquad (18.4.1)$$

$$\text{AMPERES} = \frac{\text{VOLTS}}{\text{OHMS}}$$

b) *Inductors* store energy and have an inductance value *L* measured in henries

$$V = L\frac{dI}{dt}$$

$$\qquad (18.4.2)$$

$$\text{VOLTS} = \frac{\text{AMPERES} \cdot \text{HENRIES}}{\text{SECONDS}}$$

c) *Capacitors* store energy and have a capacitance value *C* measured in farads

$$I = C\frac{dV}{dt}$$

$$\qquad (18.4.3)$$

$$\text{AMPERES} = \frac{\text{VOLTS} \cdot \text{FARADS}}{\text{SECONDS}}$$

Sources of Electrical Energy

Sources in electric circuits can be either independent of current and/or voltage values elsewhere in the circuit, or they can be dependent upon them. In this section only independent sources will be considered. Fig. 18.6 shows both ideal and linear models for current and voltage sources.

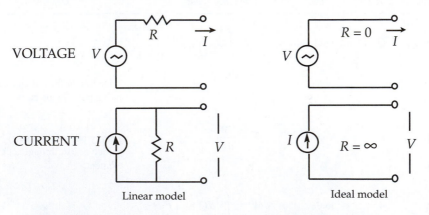

Figure 18.6   Ideal and linear models of current and voltage sources.

Kirchhoff's Laws

Two laws of conservation govern the behavior of all electrical circuits:

a) *Kirchhoff's Voltage Law* (KVL), for the conservation of energy, states that the sum of voltage rises or drops around any closed path in an electrical circuit must be zero:

$$\sum V_{DROPS} = 0 \quad \sum V_{RISES} = 0 \ (\text{around closedpath})$$
(18.4.4)

b) *Kirchhoff's Current Law* (KCL), for the conservation of charge, states that the flow of charges either into (positive) or out of (negative) any node in a circuit must add to zero:

$$\sum I_{IN} = 0 \quad \sum I_{OUT} = 0 \quad (\text{at node})$$
(18.4.5)

Ohm's Law

Ohm's Law is a statement of the relationship between the voltage across an electrical component and the current through the component. For DC circuits, where the components are resistors, Ohm's law is

$$V = IR \quad \text{or} \quad I = V/R$$
(18.4.6)

For AC circuits, with resistors, capacitors and inductors, Ohm's law, stated in terms of the component impedance Z, is

$$V = IZ \quad \text{or} \quad I = V/Z$$
(18.4.7)

where all variables can be complex (real and imaginary parts).

Reference Voltage Polarity and Current Direction

Circuit analysis requires defining first a reference current direction with an arrow placed next to the circuit component. For each of the components a reference current direction is arbitrarily defined. Once the current reference direction is defined, the voltage reference polarity marks can be placed on each component. The polarity marks on passive components are always placed so that the current flows from the plus (+) mark to the minus (−) mark, the passive sign convention.

Current values can be either positive or negative. A positive current value shows that the current does in fact flow in the reference direction. A negative current value shows that the current flows opposite to the reference direction. Voltage values can be either positive or negative. A positive voltage value indicates a loss of energy or reduction in voltage when moving through the circuit from the plus polarity mark to the minus polarity mark. A negative voltage value indicates a gain of energy when moving through the circuit from the plus polarity mark to the minus polarity mark.

It is proper electrical terminology to talk about voltage drops and voltage rises. A voltage drop is experienced when moving through the circuit from the plus (+) polarity mark to the minus (−) polarity mark. A voltage rise is experienced when moving through the circuit from the minus (−) polarity mark to the plus (+) polarity mark.

Circuit Equations

When writing circuit equations, the current is assumed to have a positive value in the reference direction and the voltage is assumed to have a positive value as indicated by the polarity marks. To write the KVL circuit equation one must move around a closed path in the circuit and sum all the voltage rises or all the voltage drops. For example, for the circuit in Fig. 18.7 begin at point a and move to b, then c, then d, and back to a. For $\Sigma V_{RISES} = 0$ obtain

$$V_s - IR_1 - IR_2 - IR_3 = 0 \qquad (18.4.8)$$

For ΣV_{DROPS} obtain

$$-V_s + IR_1 + IR_2 + IR_3 = 0 \qquad (18.4.9)$$

Either of these equations can now be solved for the one unknown current

$$I = \frac{V_s}{R_1 + R_2 + R_3} = \frac{V_s}{R_{eq}} \qquad (18.4.10)$$

where R_{eq} is the equivalent resistance for the circuit.

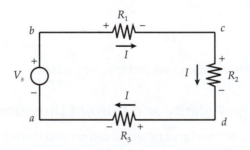

Figure 18.7 A simple circuit.

Circuit Equations Using Branch Currents

The circuit in Fig. 18.8 has meshes around which two voltage equations (KVL) can be written. There are, however, three branches in the circuit. An unknown current with a reference direction is assumed in each branch. The polarity marks are indicated for each resistor so the KVL can be written. Write two KVL equations, one around each mesh. Using $\Sigma V_{DROPS} = 0$ obtain:

$$-V_s + I_1 R_1 + I_3 R_2 + I_1 R_3 = 0$$

$$-I_3 R_2 + I_2 R_4 + I_2 R_5 + I_2 R_6 = 0$$

$$(18.4.11)$$

Write one KCL equation at circuit node *a*. This additional equation is necessary since there are three unknown currents:

$$I_1 - I_2 - I_3 = 0 \qquad (18.4.12)$$

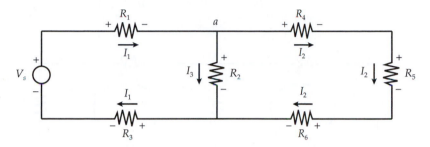

Figure 18.8 A circuit with two meshes and three branches.

These three equations can be solved for I_1, I_2, and I_3. The current I_1 is (see Eq. 16.5.8)

$$I_1 = \frac{\begin{vmatrix} V_s & 0 & R_2 \\ 0 & R_4 + R_5 + R_6 & -R_2 \\ 0 & -1 & -1 \end{vmatrix}}{\begin{vmatrix} R_1 + R_3 & 0 & R_2 \\ 0 & R_4 + R_5 + R_6 & -R_2 \\ 1 & -1 & -1 \end{vmatrix}} \qquad (18.4.13)$$

Circuit Equations Using Mesh Currents

A simplification in writing the circuit equations for Fig. 18.8 occurs if mesh currents are used. Note that

$$I_3 = I_1 - I_2 \qquad (18.4.14)$$

Redefine the reference currents in the network of Fig. 18.8 as shown in Fig. 18.9. Now there are only two unknown currents to solve for instead of the three in Fig. 18.8. The current through R_1 and R_3 is I_1. The current through R_4, R_5 and R_6 is I_2. The current through R_2 is $I_1 - I_2$ which is consistent with the KCL equation written for the network of Fig. 18.8. Write two KVL equations, one around each mesh:

$$-V_s + I_1(R_1 + R_2 + R_3) - I_2 R_2 = 0$$

$$\qquad (18.4.15)$$

$$-I_1 R_2 + I_2(R_2 + R_4 + R_5 + R_6) = 0$$

These two equations can be solved for I_1 and I_2. The current I_1 is equivalent to that of Eq. 18.4.13:

$$I_1 = \frac{\begin{vmatrix} V_s & -R_2 \\ 0 & R_2 + R_4 + R_5 + R_6 \end{vmatrix}}{\begin{vmatrix} R_1 + R_2 + R_3 & -R_2 \\ -R_2 & R_2 + R_4 + R_5 + R_6 \end{vmatrix}} \qquad (18.4.16)$$

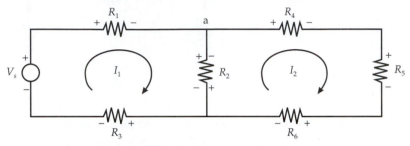

Figure 18.9 A circuit with two meshes.

Circuit Simplification

It is possible to simplify a circuit by combining components of the same kind that are grouped together in the circuit. The formulas for combining several R's to a single R, several L's to a single L and several C's to a single C are found using Kirchhoff's laws (see Eq.18.4.10). Consider two inductors in series as shown in Fig. 18.11. We see that $L_{eq} = L_1 + L_2$. Combinations of circuit components are summarized in Table 18.2.

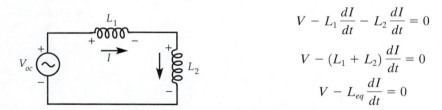

$$V - L_1 \frac{dI}{dt} - L_2 \frac{dI}{dt} = 0$$

$$V - (L_1 + L_2) \frac{dI}{dt} = 0$$

$$V - L_{eq} \frac{dI}{dt} = 0$$

Figure 18.11 Circuit Simplification.

TABLE 18.2 Circuit Components in Series and Parallel

Component	Series	Parallel
R	$R_{eq} = R_1 + R_2 + \cdots + R_N$	$\dfrac{1}{R_{eq}} = \dfrac{1}{R_1} + \dfrac{1}{R_2} + \cdots + \dfrac{1}{R_N}$
L	$L_{eq} = L_1 + L_2 + \cdots + L_N$	$\dfrac{1}{L_{eq}} = \dfrac{1}{L_1} + \dfrac{1}{L_2} + \cdots + \dfrac{1}{L_N}$
C	$\dfrac{1}{C_{eq}} = \dfrac{1}{C_1} + \dfrac{1}{C_2} + \cdots + \dfrac{1}{C_N}$	$C_{eq} = C_1 + C_2 + \cdots + C_N$

DC Circuits

In a DC circuit the only crucial components are resistors. Another component, the inductor, appears as a zero resistance connection (or short circuit) and a third component, a capacitor, appears as an infinite resistance, or open circuit. The three circuit components are summarized in Table 18.3.

TABLE 18.3 DC Circuit Components

Component		Impedance	Current	Power	Energy Stored
Resistor		R	$I = V/R$	$P = I^2 R = V^2/R$	None stored
Inductor		Zero (Short Circuit)	Unconstrained	None dissipated	$W_L = \dfrac{1}{2} L I^2$
Capacitor		Infinite (Open Circuit)	Zero	None dissipated	$W_C = \dfrac{1}{2} C V^2$

Example 18.20

Compute the current in the 10 Ω resistor.

Solution: Assume loop currents I_1 and I_2. Write KVL around both meshes:

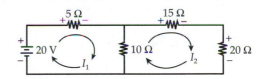

$$\sum V_{DROPS} = 0$$

Mesh 1: $-20 + 5I_1 + 10I_1 - 10I_2 = 0$

Mesh 2: $-10I_1 + 10I_2 + 15I_2 + 20I_2 = 0$

These are arranged as

$$15I_1 - 10I_2 = 20$$

$$-10I_1 + 45I_2 = 0$$

The solution is

$$I_1 = 1.57 \text{ A}, \quad I_2 = 0.35 \text{ A}$$

The current in the 10Ω resistor is

$$I = I_1 - I_2$$

$$= 1.57 - 0.35 = 1.22 \text{ A}$$

Example 18.21

Compute the power delivered to the 6Ω resistor.

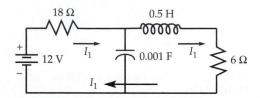

Solution: Only one current path exists since no DC current flows through a capacitor; the voltage drop across the inductor is zero. Write KVL:

$$\sum V_{DROPS} = 0$$

$$-12 + 18I_1 + 6I_1 = 0$$

$$I_1 = 0.5 \text{ A}$$

The power is then

$$P = I^2 R$$

$$= 0.5^2 \times 6 = 1.5 \text{ W}$$

EXERCISES—ELECTRICAL CIRCUITS

18.58   For the circuit below, with voltages' polarities as shown, KVL in equation form is

a) $v_1 + v_2 + v_3 - v_4 + v_5 = 0$

b) $-v_1 + v_2 + v_3 - v_4 + v_5 = 0$

c) $v_1 + v_2 - v_3 - v_4 + v_5 = 0$

d) $-v_1 - v_2 - v_3 + v_4 + v_5 = 0$

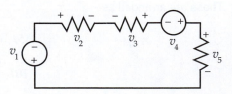

18.59   Find I_1 in amps.
   a) 12
   b) 15
   c) 18
   d) 21

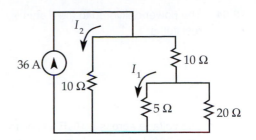

18.60   Find the magnitude and sign of the power, in watts, absorbed by the circuit element
   in the box.
   a) −20
   b) −8
   c) 8
   d) 12

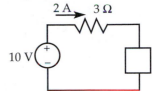

18.61   For the circuit shown, the voltage across the 4 ohm resistor is, with $v = 1$ V
   a) 1/4
   b) 1/2
   c) 2/3
   d) 2

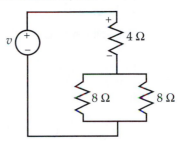

18.62   A 100μF capacitor has $I_c(t)$. The capacitor voltage $V_c(t)$ at $t = 2.5$ seconds ($V(0) = 1.0$V) is most nearly
   a) −24
   b) −25
   c) 25
   d) 26

18.63   The voltage across the 5 ohm resistor in the circuit shown is
   a) 1.0
   b) 2.5
   c) 3.0
   d) 5.83

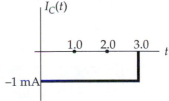

18.64   The power delivered to the 5 ohm resistor is
a) 1.5
b) 2.15
c) 2.85
d) 3.2

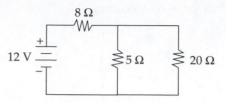

18.65   The voltage across a 10μF capacitor is $50t^2$ V. The time, in seconds, it will take to store 200 J of energy is most nearly
a) 0.15   b) 0.21   c) 1.38   d) 11.25

18.66   The equivalent resistance, in ohms, between points *a* and *b* in the circuit below is
a) 3
b) 5
c) 7
d) 8

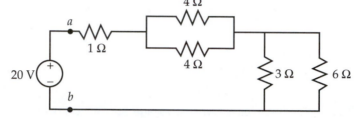

18.67   The voltage V_2 is
a) 6.4
b) 4.0
c) 2.0
d) 5.6

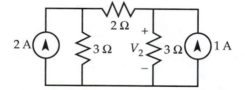

18.68   Find I_1 in amperes.
a) 4.0
b) 2.0
c) 4.11
d) 2.11

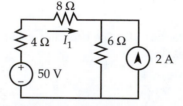

18.5 ECONOMICS

All designs are intended to produce good results. In general, the good results are accompanied by undesirable effects including the costs of manufacturing or construction. Selecting the best design requires the designer to anticipate and compare the good and bad outcomes. If outcomes are evaluated in dollars, and if "good" is defined as positive monetary value, then design decisions may be guided by an economic analysis.

Value and Interest

"Value" is not synonymous with "amount." The value of an amount of money depends on when the amount is received or spent. For example, the promise that you will be given a dollar one year from now is of less value to you than a dollar received today. The difference between the anticipated amount and its current value is called *interest* and is frequently expressed as a time rate. If an interest rate of 10% per year is used, the expectation of receiving $1.00 one year hence has a value now of about $0.91. In economics, interest usually is stated in percent per year. If no time unit is given, "per year" is assumed.

Example 18.22

What amount must be paid in two years to settle a current debt of $1,000 if the interest rate is 6%?

Solution: Value after one year $= 1000 + 1000 \times 0.06$
$= 1000(1 + 0.06)$
$= \$1060$

Value after two years $= 1060 + 1060 \times 0.06$
$= 1000(1 + 0.06)^2$
$= \$1124$

Hence, $1,124 must be paid in two years to settle the debt.

Cash Flow Diagrams

As an aid to analysis and communication, an economics problem may be represented graphically by a horizontal time axis and vertical vectors representing dollar amounts. The cash flow diagram for Example 18.22 is sketched in Fig. 18.11. Income is up and expenditures are down. It is important to pick a point of view and stick with it. For example, the vectors in Fig. 18.11 would have been reversed if the point of view of the lender had been adopted. It is a good idea to draw a cash flow diagram for every economics problem that involves amounts occurring at different times.

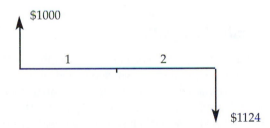

Figure 18.11 Cash flow diagram for Example 18.22.

In economics, amounts are almost always assumed to occur at the ends of years. Consider, for example, the value today of the future operating expenses of a truck. The costs probably will be paid in varied amounts scattered throughout each year of operation, but for computational ease the expenses in each year are represented by their sum (computed without consideration of interest) occurring at the end of the year. The error introduced by neglecting interest for partial years is usually insignificant compared to uncertainties in the estimates of future amounts.

Cash Flow Patterns

Economics problems involve the following four patterns of cash flow, both separately and in combination. They are illustrated in Fig. 18.12.

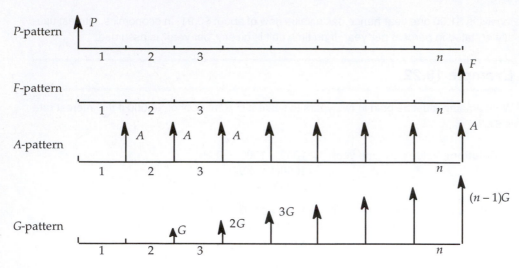

Figure 18.12 Four cash flow patterns.

P-pattern: A single amount *P* occurring at the beginning of *n* years. *P* frequently represents "present" amounts.

F-pattern: A single amount *F* occurring at the end of *n* years. *F* frequently represents "future" amounts.

A-pattern: Equal amounts *A* occurring at the ends of each of *n* years. The *A*-pattern frequently is used to represent "annual" amounts.

G-pattern: End-of-year amounts increasing by an equal annual gradient *G*. Note that the first amount occurs at the end of the second year. *G* is the abbreviation of "gradient."

Equivalence of Cash Flow Patterns

Two cash flow patterns are said to be equivalent if they have the same value. Most of the computational effort in economics problems is directed at finding a cash flow pattern that is equivalent to a combination of other patterns. Example 18.22 can be thought of as finding the amount in an *F*-pattern that is equivalent to $1,000 in a *P*-pattern. The two amounts are proportional, and the factor of proportionality is a function of interest rate *i* and number of periods *n*. There is a different factor of proportionality for each possible pair of the cash flow patterns defined in the next section. To minimize the possibility of selecting the wrong factor, mnemonic symbols are assigned to the factors. For Example 18.23, the proportionality factor is written and solution is achieved by evaluating

$$F = (F/P)_n^i\, P.$$

To analysts familiar with the canceling operation of algebra, it is apparent that the correct factor has been chosen. However, the letters in the parentheses together with the sub- and super-scripts constitute a single symbol; therefore, the canceling operation is not actually performed. Table 18.4 lists symbols and formulas for commonly used factors. Table 18.5, located at the end of this chapter, presents a convenient way to find numerical values of interest factors. Those values are tabulated for selected interest rates *i* and number of interest periods *n*; linear interpolation for intermediate values of *i* and *n* is acceptable for most situations.

TABLE 18.4 Formulas for Interest Factors

Symbol	To Find	Given	Formula
$(F/P)_n^i$	F	P	$(1 + i)^n$
$(P/F)_n^i$	P	F	$\dfrac{1}{(1 + i)^n}$
$(A/P)_n^i$	A	P	$\dfrac{i(1 + i)^n}{(1 + i)^n - 1}$
$(P/A)_n^i$	P	A	$\dfrac{(1 + i)^n - 1}{i(1 + i)^n}$
$(A/F)_n^i$	A	F	$\dfrac{i}{(1 + i)^n - 1}$
$(F/A)_n^i$	F	A	$\dfrac{(1 + i)^n - 1}{i}$
$(A/G)_n^i$	A	G	$\dfrac{1}{i} \dfrac{n}{(1 + i)^n - 1}$
$(F/G)_n^i$	F	G	$\dfrac{1}{i}\left[\dfrac{(1 + i)^n - 1}{i} - n\right]$
$(P/G)_n^i$	P	G	$\dfrac{1}{i}\left[\dfrac{(1 + i)^n - 1}{i(1 + i)^n} - \dfrac{n}{(1 + i)^n}\right]$

Example 18.23

A new widget twister, with a life of six years, would save $2,000 in production costs each year. Using a 12% interest rate, determine the highest price that could be justified for the machine. Although the savings occur continuously throughout each year, follow the usual practice of lumping all amounts at the ends of years.

Solution: First, sketch the cash flow diagram.

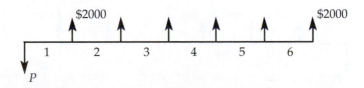

The cash flow diagram indicates that an amount in a *P*-pattern must be found that is equivalent to $2,000 in an *A*-pattern. The corresponding equation is

$$P = (P/A)_n^i \, A$$

$$= (P/A)_6^{12\%} \, 2000$$

Table 18.5 is used to evaluate the interest factor for $i = 12\%$ and $n = 6$:

$$P = 4.1114 \times 2000$$

$$= \$8223$$

Example 18.24

How soon does money double if it is invested at 8% interest?

Solution: Obviously, this is stated as

$$F = 2P.$$

Therefore,

$$(F/P)_n^{8\%} = 2$$

In the 8% interest table, the tabulated value for (F/P) that is closest to 2 corresponds to $n = 9$ years.

Example 18.25

Find the value in 2002 of a bond described as "Acme 8% of 2015" if the rate of return set by the market for similar bonds is 10%.

Solution: The bond description means that the Acme company has an outstanding debt that it will repay in the year 2015. Until then, the company will pay out interest on that debt at the 8% rate. Unless otherwise stated, the principal amount of a single bond is $1,000. If it is assumed that the debt is due December 31, 2015, interest is paid every December 31, and the bond is purchased January 1, 2002, then the cash flow diagram, with unknown purchase price *P*, is:

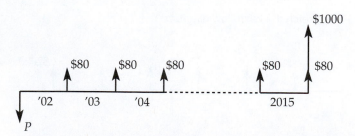

The corresponding equation is

$$P = (P/A)_{14}^{10\%}\, 80 + (P/F)_{14}^{10\%}\, 1000$$

$$= 7.3667 \times 80 + 0.2633 \times 1000$$

$$= \$853$$

That is, to earn 10% the investor must buy the 8% bond for $853, a "discount" of $147. Conversely, if the market interest rate is less than the nominal rate of the bond, the buyer will pay a "premium" over $1,000.

The solution is approximate because bonds usually pay interest semiannually, and $80 at the end of the year is not equivalent to $40 at the end of each half year. But the error is small and is neglected.

Example 18.26

You are buying a new appliance. From past experience you estimate future repair costs as:

First Year	$ 5
Second Year	15
Third Year	25
Fourth Year	35

The dealer offers to sell you a four-year warranty for $60. You require at least a 6% interest rate on your investments. Should you invest in the warranty?

Solution: Sketch the cash flow diagram.

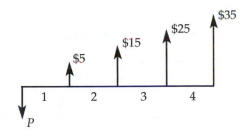

The known cash flows can be represented by superposition of a $5 A-pattern and a $10 G-pattern. Verify that statement by drawing the two patterns. Now it is clear why the standard G-*pattern* is defined to have the first cash flow at the end of the second year. Next, the equivalent amount P is computed:

$$P = (P/A)_{4}^{6\%}\, A + (P/G)_{4}^{6\%}\, G$$

$$= 3.4651 \times 5 + 4.9455 \times 10$$

$$= \$67$$

Since the warranty can be purchased for less then $67, the investment will earn a rate of return greater than the required 6%. Therefore, you should purchase the warranty.

If the required interest rate had been 12%, the decision would be reversed. This demonstrates the effect of a required interest rate on decision-making. Increasing the required rate reduces the number of acceptable investments.

Example 18.27

Compute the annual equivalent maintenance costs over a 5-year life of a laser printer that is warranted for two years and has estimated maintenance costs of $100 annually. Use $i = 10\%$.

Solution: The cash flow diagram appears as:

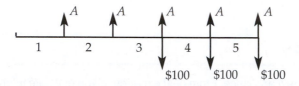

There are several ways to find the 5-year A-pattern equivalent to the given cash flow. One of the more efficient methods is to convert the given 3-year A-pattern to an F-pattern, and then find the 5-year A-pattern that is equivalent to that F-pattern. That is,

$$A = (A/F)_5^{10\%} \, (F/A)_3^{10\%} \, 100$$

$$= \$54$$

Unusual Cash Flows and Interest Periods

Occasionally an economics problem will deviate from the year-end cash flow and annual compounding norm. The examples in this section demonstrate how to handle these situations.

Example 18.28

PAYMENTS AT BEGINNINGS OF YEARS

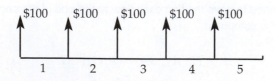

Using a 10% interest rate, find the future equivalent of:
Solution: Shift each payment forward one year. Therefore,

$$A = (F/P)_1^{10\%} \, 100 = \$110$$

This converts the series to the equivalent *A*-pattern:

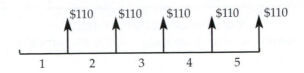

and the future equivalent is found to be

$$F = (F/A)_5^{10\%} \, 110 = \$672$$

Alternative Solution: Convert to a six-year series:

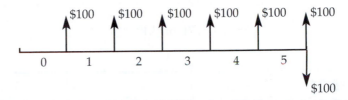

The future equivalent is

$$F = (F/A)_6^{10\%} \, 100 - 100 = \$672$$

Example 18.29

ANNUAL PAYMENTS WITH INTEREST COMPOUNDED *m* TIMES PER YEAR
Compute the effective annual interest rate equivalent to 5% nominal annual interest compounded daily. (There are 365 days in a year.)

Solution: The legal definition of nominal annual interest is

$$i_n = mi$$

where *i* is the interest rate per compounding period. For the example,

$$i = i_n/m$$

$$= 0.05/365 = 0.000137 \text{ or } 0.0137\% \text{ per day}$$

Because of compounding, the effective annual rate is greater than the nominal rate. By equating (*F*/*P*)-factors for one year and *m* periods, the effective annual rate i_e may be computed as follows:

$$(1 + i_e)^1 = (1 + i)^m$$

$$i_e = (1 + i)^m - 1$$

$$= (1.000137)^{365} - 1 = 0.05127 \text{ or } 5.127\%$$

Example 18.30

CONTINUOUS COMPOUNDING

Compute the effective annual interest rate i_e equivalent to 5% nominal annual interest compounded continuously.

Solution: As m approaches infinity, the value for i_e is found as follows:

$$i_e = e^i - 1$$

$$= e^{0.05} - 1$$

$$= 0.051271 \text{ or } 5.1271\%$$

Example 18.31

ANNUAL COMPOUNDING WITH m PAYMENTS PER YEAR

Compute the year-end amount equivalent to twelve end-of-month payments of $10 each. Annual interest rate is 6%.

Solution: The usual simplification in economics is to assume that all payments occur at the end of the year, giving an answer of $120. This approximation may not be acceptable for a precise analysis of a financial agreement. In such cases, the agreement's policy on interest for partial periods must be investigated.

Example 18.32

ANNUAL COMPOUNDING WITH PAYMENT EVERY m YEARS

With interest at 10% compute the present equivalent of

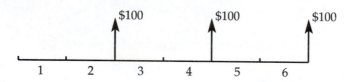

Solution: First convert each payment to an A-pattern for the m preceding years. That is,

$$A = (A/F)_2^{10\%} \, 100$$

$$= \$47.62$$

Then, convert the A-pattern to a P-pattern:

$$P = (P/A)_6^{10\%} \, 47.62$$

$$= \$207$$

EXERCISES—ECONOMICS

18.69   Which of the following would be most difficult to monetize?
a) maintenance cost   b) selling price c) fuel cost   d) prestige

18.70   If $1,000 is deposited in a savings account that pays 6% annual interest and all the interest is left in the account, what is the account balance after three years?
a) $840   b) $1,000   c) $1,180   d) $1,191

18.71   Your perfectly reliable friend, Frank, asks for a loan and promises to pay back $150 two years from now. If the minimum interest rate you will accept is 8%, what is the maximum amount you will loan him?
a) $119   b) $126   c) $129   d) $139

18.72 $12,000 is borrowed now at 12% interest. The first payment is $4000 and is made 3 years from now. The balance of the debt immediately after the payment is
a) $4000   b) $8000   c) $12,000   d) $12,860

18.73   An alumnus establishes a perpetual endowment fund to help Saint Louis University. What amount must be invested now to produce income of $100,000 one year from now and at one-year intervals forever? Interest rate is 8%.
a) $8000   b) $100,000   c) $1,250,000 d) $10,000,000

18.74   The annual amount of a series of payments to be made at the end of each of the next twelve years is $500. What is the present worth of the payments at 8% interest compounded annually?
a) $500   b) $3,768   c) $6,000   d) $6,480

18.75   Consider a prospective investment in a project having a first cost of $300,000, operating and maintenance costs of $35,000 per year, and an estimated net disposal value of $50,000 at the end of thirty years. Assume an interest rate of 8%.

What is the present equivalent cost of the investment if the planning horizon is thirty years?
a) $670,000   b) $689,000   c) $720,000   d) $791,000

If the project replacement will have the same first cost, life, salvage value, and operating and maintenance costs as the original, what is the capitalized cost of perpetual service?
a) $670,000   b) $689,000   c) $720,000   d) $765,000

18.76   Maintenance expenditures for a structure with a twenty-year life will come as periodic outlays of $1,000 at the end of the fifth year, $2,000 at the end of the tenth year, and $3,500 at the end of the fifteenth year. With interest at 10%, what is the equivalent uniform annual cost of maintenance for the twenty-year period?
a) $200   b) $262   c) $300   d) $325

18.77   An alumnus has given Michigan State University ten million dollars to build and operate a laboratory. Annual operating cost is estimated to be one hundred thousand dollars. The endowment will earn 6% interest. Assume an infinite life for the laboratory and determine how much money may be used for its construction.
a) 5.00×10^6
b) 8.33×10^6
c) 8.72×10^6
d) 9.90×10^6

18.78   An investment pays $6000 at the end of the first year, $4000 at the end of the second year, and $2000 at the end of the third year. Compute the present value of the investment if a 10% rate-of-return is required.
a) $8333   b) $9667   c) $10,300   d) $12,000

18.79   An amount F is accumulated by investing a single amount P for n compounding periods with interest rate of i. Select the formula that relates P to F.
a) $P = F(1 + i)^{-n}$
b) $P = F(1 + i)^{-n}$
c) $P = F(1 + n)^{-i}$
d) $P = F(1 + ni)^{-1}$

18.80   At the end of each of the next ten years, a payment of $200 is due. At an interest rate of 6%, what is the present worth of the payments?
a) $27   b) $200   c) $1472   d) $2000

18.81   The purchase price of an instrument is $12,000 and its estimated maintenance costs are $500 for the first year, $1500 for the second and $2500 for the third year. After three years of use the instrument is replaced; it has no salvage value. Compute the present equivalent cost of the instrument using 10% interest.
a) $14,070   b) $15,570   c) $15,730   d) $16,500

18.82   If an amount invested five years ago has doubled, what is the annual interest rate?
a) 15%   b) 12%   c) 10%   d) 6%

18.83   After a factory has been built near a stream, it is learned that the stream occasionally overflows its banks. A hydrologic study indicates that the probability of flooding is about 1 in 8 in any one year. A flood would cause about $20,000 in damage to the factory. A levee can be constructed to prevent flood damage. Its cost will be $54,000 and its useful life is thirty years. Money can be borrowed at 8% interest. If the annual equivalent cost of the levee is less than the annual expectation of flood damage, the levee should be built. The annual expectation of flood damage is $(1/8) \times 20,000 =$ $2,500. Compute the annual equivalent cost of the levee.
a) $1,261   b) $1,800   c) $4,320   d) $4,800

18.84   If $10,000 is borrowed now at 6% interest, how much will remain to be paid after a $3,000 payment is made four years from now?
a) $7,000   b) $9,400   c) $9,625   d) $9,725

18.85   The maintenance costs associated with a machine are $2,000 per year for the first ten years, and $1,000 per year thereafter. The machine has an infinite life. If interest is 10%, what is the present worth of the annual disbursements?
a) $16,145   b) $19,678   c) $21,300   d) $92,136

18.86   A bank currently charges 10% interest compounded annually on business loans. If the bank were to change to continuous compounding, what would be the effective annual interest rate?
a) 10%   b) 10.517%   c) 12.5%   d) 12.649%

TABLE 18.5 — Interest Tables (Compound Interest Factors)

$i = 6.00\%$

n	(P/F)	(P/A)	(P/G)	(F/P)	(F/A)	(A/P)	(A/F)	(A/G)	n
1	.9434	0.9434	−0.0000	1.0600	1.0000	1.0600	1.0000	−0.0000	1
2	.8900	1.8334	0.8900	1.1236	2.0600	0.5454	0.4854	0.4854	2
3	.8396	2.6730	2.5692	1.1910	3.1836	0.3741	0.3141	0.9612	3
4	.7921	3.4651	4.9455	1.2625	4.3746	0.2886	0.2286	1.4272	4
5	.7473	4.2124	7.9345	1.3382	5.6371	0.2374	0.1774	1.8836	5
6	.7050	4.9173	11.4594	1.4185	6.9753	0.2034	0.1434	2.3304	6
7	.6651	5.5824	15.4497	1.5036	8.3938	0.1791	0.1191	2.7676	7
8	.6274	6.2098	19.8416	1.5938	9.8975	0.1610	0.1010	3.1952	8
9	.5919	6.8017	24.5768	1.6895	11.4913	0.1470	0.0870	3.6133	9
10	.5584	7.3601	29.6023	1.7908	13.1808	0.1359	0.0759	4.0220	10
11	.5268	7.8869	34.8702	1.8983	14.9716	0.1268	0.0668	4.4213	11
12	.4970	8.3838	40.3369	2.0122	16.8699	0.1193	0.0593	4.8113	12
13	.4688	8.8527	45.9629	2.1329	18.8821	0.1130	0.0530	5.1920	13
14	.4423	9.2950	51.7128	2.2609	21.0151	0.1076	0.0476	5.5635	14
15	.4173	9.7122	57.5546	2.3966	23.2760	0.1030	0.0430	5.9260	15
16	.3936	10.1059	63.4592	2.5404	25.6725	0.0990	0.0390	6.2794	16
17	.3714	10.4773	69.4011	2.6928	28.2129	0.0954	0.0354	6.6240	17
18	.3503	10.8276	75.3569	2.8543	30.9057	0.0924	0.0324	6.9597	18
19	.3305	11.1581	81.3062	3.0256	33.7600	0.0896	0.0296	7.2867	19
20	.3118	11.4699	87.2304	3.2071	36.7856	0.0872	0.0272	7.6051	20
21	.2942	11.7641	93.1136	3.3996	39.9927	0.0850	0.0250	7.9151	21
22	.2775	12.0416	98.9412	3.6035	43.3923	0.0830	0.0230	8.2166	22
23	.2618	12.3034	104.7007	3.8197	46.9958	0.0813	0.0213	8.5099	23
24	.2470	12.5504	110.3812	4.0489	50.8156	0.0797	0.0197	1.87951	24
25	.2330	12.7834	115.9732	4.2919	54.8645	0.0782	0.0182	9.0722	25
26	.2198	13.0032	121.4684	4.5494	59.1564	0.0769	0.0169	9.3414	26
28	.1956	13.4062	132.1420	5.1117	68.5281	0.0746	0.0146	9.8568	28
30	.1741	13.7648	142.3588	5.7435	79.0582	0.0726	0.0126	10.3422	30
∞	.0000	16.6667	277.7778	∞	∞	0.0600	0.0000	16.667	∞

$i = 8.00\%$

n	(P/F)	(P/A)	(P/G)	(F/P)	(F/A)	(A/P)	(A/F)	(A/G)	n
1	.9259	0.9259	−0.0000	1.0800	1.0000	1.0800	1.0000	−0.0000	1
2	.8573	1.7833	0.8573	1.1664	2.0800	0.5608	0.4808	0.4808	2
3	.7938	2.5771	2.4450	1.2597	3.2464	0.3880	0.3080	0.9487	3
4	.7350	3.3121	4.6501	1.3605	4.5061	0.3019	0.2219	1.4040	4
5	.6806	3.9927	7.3724	1.4693	5.8666	0.2505	0.1705	1.8465	5
6	.6302	4.6229	10.5233	1.5869	7.3359	0.2163	0.1363	2.2763	6
7	.5835	5.2064	14.0242	1.7138	8.9228	0.1921	0.1121	2.6937	7
8	.5403	5.7466	17.8061	1.8509	10.6366	0.1740	0.0940	3.0985	8
9	.5002	6.2469	21.8081	1.9990	12.4876	0.1601	0.0801	3.4910	9
10	.4632	6.7101	25.9768	2.1589	14.4866	0.1490	0.0690	3.8713	10
11	.4289	7.1390	30.2657	2.3316	16.6455	0.1401	0.0601	4.2395	11
12	.3971	7.5361	34.6339	2.5182	18.9771	0.1327	0.0527	4.5957	12
13	.3677	7.9038	39.0463	2.7196	21.4953	0.1265	0.0465	4.9402	13
14	.3405	8.2442	43.4723	2.9372	24.2149	0.1213	0.0413	5.2731	14
15	.3152	8.5595	47.8857	3.1722	27.1521	0.1168	0.0368	5.5945	15
16	.2919	8.8514	52.2640	3.4259	30.3243	0.1130	0.0330	5.9046	16
17	.2703	9.1216	56.5883	3.7000	33.7502	0.1096	0.0296	6.2037	17
18	.2502	9.3719	60.8426	3.9960	37.4502	0.1067	0.0267	6.4920	18
19	.2317	9.6036	65.0134	4.3157	41.4463	0.1041	0.0241	6.7697	19
20	.2145	9.8181	69.0898	4.6610	45.7620	0.1019	0.0219	7.0369	20
21	.1987	10.0168	73.0629	5.0338	50.4229	0.0998	0.0198	7.2940	21
22	.1839	10.2007	76.9257	5.4365	55.4568	0.0980	0.0180	7.5412	22
23	.1703	10.3711	80.6726	5.8715	60.8933	0.0964	0.0164	7.7786	23
24	.1577	10.5288	84.2997	6.3412	66.7648	0.0950	0.0150	8.0066	24
25	.1460	10.6748	87.8041	6.8485	73.1059	0.0937	0.0137	8.2254	25
26	.1352	10.8100	91.1842	7.3964	79.9544	0.0925	0.0125	8.4352	26
28	.1159	11.0511	97.5687	8.6271	95.3388	0.0905	0.0105	8.8289	28
30	.0994	11.2578	103.4558	10.0627	113.2832	0.0888	0.0088	9.1897	30
∞	.0000	12.500	156.2500	∞	∞	0.0800	0.0000	12.5000	∞

Compound Interest Tables (*continued*)

$i = 10.00\%$

n	(P/F)	(P/A)	(P/G)	(F/P)	(F/A)	(A/P)	(A/F)	(A/G)	n
1	.9091	0.9091	-0.0000	1.1000	1.0000	1.1000	1.0000	-0.0000	1
2	.8264	1.7355	0.8264	1.2100	2.1000	0.5762	0.4762	0.4762	2
3	.7513	2.4869	2.3291	1.3310	3.3100	0.4021	0.3021	0.9366	3
4	.6830	3.1699	4.3781	1.4641	4.6410	0.3155	0.2155	1.3812	4
5	.6209	3.7908	6.8618	1.6105	6.1051	0.2638	0.1638	1.8101	5
6	.5645	4.3553	9.6842	1.7716	7.7156	0.2296	0.1296	2.2236	6
7	.5132	4.8684	12.7631	1.9487	9.4872	0.2054	0.1054	2.6216	7
8	.4665	5.3349	16.0287	2.1436	11.4359	0.1874	0.0874	3.0045	8
9	.4241	5.7590	19.4215	2.3579	13.5795	0.1736	0.0736	3.3724	9
10	.3855	6.1446	22.8913	2.5937	15.9374	0.1627	0.0627	3.7255	10
11	.3505	6.4951	26.3963	2.8531	18.5312	0.1540	0.0540	4.0641	11
I2	.3186	6.8137	29.9012	3.1384	21.3843	0.1468	0.0468	4.3884	12
13	.2897	7.1034	33.3772	3.4523	24.5227	0.1408	0.0408	4.6988	13
14	.2633	7.3667	36.8005	3.7975	27.9750	0.1357	0.0357	4.9955	14
15	.2394	7.6061	40.1520	4.1772	31.7725	0.1315	0.0315	5.2789	15
16	.2176	7.8237	43.4164	4.5950	35.9497	0.1278	0.0278	5.5493	16
17	.1978	8.0216	46.5819	5.0545	40.5447	0.1247	0.0247	5.8071	17
18	.1799	8.2014	49.6395	5.5599	45.5992	0.1219	0.0219	6.0526	18
19	.1635	8.3649	52.5827	6.1159	51.1591	0.1195	0.0195	6.2861	19
20	.1486	8.5136	55.4069	6.7275	57.2750	0.1175	0.0175	6.5081	20
21	.1351	8.6487	58.1095	7.4002	64.0025	0.1156	0.0156	6.7189	21
22	.1228	8.7715	60.6893	8.1403	71.4027	0.1140	0.0140	6.9189	22
23	.1117	8.8832	63.1462	8.9543	79.5430	0.1126	0.0126	7.1085	23
24	.1015	8.9847	65.4813	9.8497	88.4973	0.1113	0.0113	7.2881	24
25	.0923	9.0770	67.6964	10.8347	98.3471	0.1102	0.0102	7.4580	25
26	.0839	9.1609	69.7940	11.9182	109.1818	0.1092	0.0092	7.6186	26
28	.0693	9.3066	73.6495	14.4210	134.2099	0.1075	0.0075	7.9137	28
30	.0573	9.4269	77.0766	17.4494	164.4940	0.1061	0.0061	8.1762	30
∞	.0000	10.0000	100.0000	∞	∞	0.1000	0.0000	10.0000	∞

$i = 12.00\%$

n	(P/F)	(P/A)	(P/G)	(F/P)	(F/A)	(A/P)	(A/F)	(A/G)	n
1	.8929	0.8929	-0.0000	1.1200	1.0000	1.1200	1.0000	-0.0000	1
2	.7972	1.6901	0.7972	1.2544	2.1200	0.5917	0.4717	0.4717	2
3	.7118	2.4018	2.2208	1.4049	3.3744	0.4163	0.2963	0.9246	3
4	.6355	3.073	4.1273	1.5735	4.7793	0.3292	0.2092	1.3589	4
5	.5674	3.6048	6.3970	1.7623	6.3528	0.2774	0.1574	1.7746	5
6	.5066	4.1114	8.9302	1.9738	8.1152	0.2432	0.1232	2.1720	6
7	.4523	4.5638	11.6443	2.2107	10.0890	0.2191	0.0991	2.5515	7
8	.4039	4.9676	14.4714	2.4760	12.2997	0.2013	0.0813	2.9131	8
9	.3606	5.3282	17.3563	2.7731	14.7757	0.1877	0.0677	3.2574	9
10	.3220	5.6502	20.2541	3.1058	17.5487	0.1770	0.0570	3.5847	10
11	.2875	5.9377	23.1288	3.4785	20.6546	0.1684	0.0484	3.8953	11
12	.2567	6.1944	25.9523	3.8960	24.1331	0.1614	0.0414	4.1897	12
13	.2292	6.4235	28.7024	4.3635	28.0291	0.1557	0.0357	4.4683	13
14	.2046	6.6282	31.3624	4.8871	32.3926	0.1509	0.0309	4.7317	14
15	.1827	6.8109	33.9202	5.4736	37.2797	0.1468	0.0268	4.9803	15
16	.1631	6.9740	36.3670	6.1304	42.7533	0.1434	0.0234	5.2147	16
17	.1456	7.1196	38.6973	6.8660	48.8837	0.1405	0.0205	5.4353	17
18	.1300	7.2497	40.9080	7.6900	55.7497	0.1379	0.0179	5.6427	18
19	.1161	7.3658	42.9979	8.6128	63.4397	0.1358	0.0158	5.8375	19
20	.1037	7.4694	44.9676	9.6463	72.0524	0.1339	0.0139	6.0202	20
21	.0926	7.5620	46.8188	10.8038	81.6987	0.1322	0.0122	6.1913	21
22	.0826	7.6446	48.5543	12.1003	92.5026	0.1308	0.0108	6.3514	22
23	.0738	7.7184	50.1776	13.5523	104.6029	0.1296	0.0096	6.5010	23
24	.0659	7.7843	51.6929	15.1786	118.1552	0.1285	0.0085	6.6406	24
25	.0588	7.8431	53.1046	17.0001	133.3339	0.1275	0.0075	6.7708	25
26	.0525	7.8957	54.4177	19.0401	150.3339	0.1267	0.0067	6.8921	26
28	.0419	7.9844	56.7674	23.8839	190.6989	0.1252	0.0052	7.1098	28
30	.0334	8.0552	58.7821	29.9599	241.3327	0.1241	0.0041	7.2974	30
∞	.0000	8.333	69.4444	∞	∞	0.1200	0.0000	8.3333	∞

Chapter 19

The Campus Experience

Webster's New World Dictionary defines *orienting* as "to face in any specific direction" or "to adjust to a situation." Both of these definitions describe the moment of truth when you enter the college campus of your choice for the first time. You have set your sights in a specific direction, and in most cases you will have to adjust to many situations that you have never before experienced. The road ahead can be fairly bumpy, but you can develop some very workable tools that will make the experience enjoyable and valuable.

This chapter introduces some of the things you can do to adjust to the college/university experience and, more specifically, to your life as an engineer on campus. This is a transition period in your life and much will be asked of you. You will need to make numerous decisions relating to your future career. In this chapter you will be provided with a number of exercises that will encourage you to explore your surroundings. Hopefully, you will be creative and find many other ways to acclimate yourself to your new campus. So let's get started.

19.2 EXPLORING YOUR NEW HOME AWAY FROM HOME

The Art of Map Reading

At first, you might be reminded of your Scouting days when you had to learn how to read a map of some unknown territory. Well, come to think of it, that might be a good analogy. When a member of the Scouts is asked to perform an orienteering activity, it involves following a map with some particular set of instructions. They must quickly acclimate themselves to their surroundings. How many seniors, or even campus employees, know all that much about their surroundings. If you make the effort to familiarize yourself with various locations on campus from the beginning, you will feel much more at home. Locating the proper offices for needed assistance will then be effortless. On the next page is a sample map of the campus at Michigan State University.

Sample Campus Map

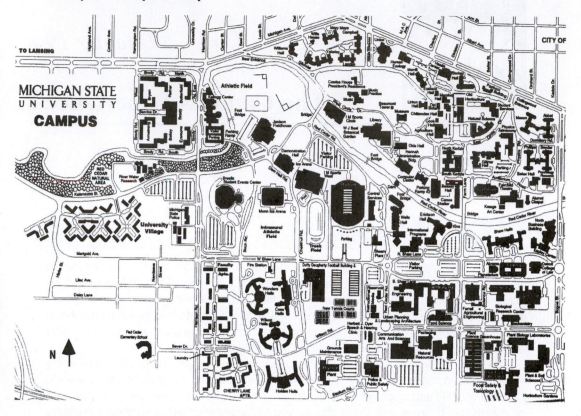

19.3 DETERMINING AND PLANNING YOUR MAJOR

There are many sources on campus to help you with the big decisions ahead of you. We are concerned here with determining your major, so we will limit our discussion to those related sources. You will need to begin making it a practice to ask as many questions as you can of those who can provide you with insightful answers. This is vital. Learn all you can from everyone about the fields that you think you might find worthwhile. (Chapters 2 and 3 will be a big help here.)

19.4 GET INTO THE HABIT OF ASKING QUESTIONS

Asking Questions

Asking a multitude of questions will make life a lot easier. In life the active questioners are the ones who learn the most, who understand the course in a useful, permanent way, and always seem to get assignments done correctly. It's not just luck! As the saying goes: "The harder I work, the luckier I get." Questioning is good work! These students make sure they understand everything, and then they move on to complete the task.

One of the most important things you can do as an engineer is to clearly convey information. You can practice this by verifying information you get from others.

19.5 THE 'PEOPLE ISSUE'

Meeting People

It is sometimes very difficult to strike up a conversation with a perfect stranger, but this can be vitally important to your present course work and to your career. You will find that working with others is one of the most important elements of the learning process. People are necessary for your success. It is only the very few who survive without constant collaboration with others. Make every effort to talk to other students in your classes. Find out what their interests are, where they're from, their abilities, and where they might be struggling. By uncovering such information you will discover opportunities for further conversation, opportunities where you can learn from them, and opportunities in which you can help them. A little pleasant conversation at the beginning of the term may lead to opportunities that spell the difference between getting a passing grade or acing the course. Making friends can also help you refine your perspective on engineering (and life) as you converse with peers, and with those who've already been through what you're just beginning (professors included).

Your Academic Advisor

When you enrolled in college you most likely found that you were assigned an academic advisor. In the early orientation period you might have had contact with this individual, but many students fail to speak with their advisor again until they prepare for graduation. Some students even find that their graduation is held up because they failed to meet one or two requirements. Advisors can help guide you in the right direction—to take specific courses and to avoid others that you're not qualified to take. They make sure that your prerequisites are fulfilled and that you take the necessary courses to adequately prepare for your career. Advisors also know a tremendous amount about the university itself. They can guide you to the right people to discuss financial aid or administrative problems. Advisors know what each course requires, and they know which instructors are good for you and which may not be.

The problem that many people encounter when entering college is a fear of asking questions—simple questions whose answers, in the long run, could greatly simplify the educational process. Questioning requires constant practice to fine-tune. Start by practicing on your advisor!

Investigating Instructors

Many students go through their academic years without engaging their instructors. Courses are taken, tests and assignments are graded, and that's it. With a little effort you can discover a great deal about your instructors and their methods of teaching.

Before blindly signing up for courses, take the time to ask questions of students who have taken the course with those instructors. Ask pertinent questions that will help you make decisions about selecting a particular instructor or opting for someone else.

- Does the instructor challenge students to really learn the material?
- Is the material covered logically?
- Can students ask questions?
- Is there an evident organization to the course?

Another method you can try is to ask to sit in on a class, or even simply sit outside a classroom and listen to what is going on. It's quite easy in such ways to get valuable info about the approach of the instructor and the set-up of the course.

Networking

One of the most important things you can do early on is networking—building contacts that may be useful to you for the rest of your engineering career. Friendships can help you grow as a person, help you keep perspective in hard times, and avoid pitfalls. Friendships can also be helpful in the classroom. They can help with projects and homework, and can give access to a multitude of activities and opportunities that you would have never known about otherwise. This group includes the instructors in your courses, your classmates, and your neighbors in dormitories and apartments. By keeping contact with them, you will always have a group of people who can offer assistance and support, and with whom you can practice your skills by helping in return!

19.6 SEARCHING FOR CAMPUS RESOURCES

Finding Out What's Available on Campus

Every school should have a document of some sort that lists the activities and opportunities on campus. It is your responsibility to find this document and use it. Nowadays, there may be a website that ties you into services available to you as a student. Michigan State University's website (http://www.msu.edu/msuinfo/), for example, lists the following items to help a student start a search for helpful resources.

Things to Do, Places to Go

Things to do on Campus
Abrams Planetarium
W.J. Beal Botanical Garden
Kellogg Hotel and Conference Center
Kresge Art Museum
MSU Museum
Wharton Center for Performing Arts
Breslin Center
Horticultural Gardens
MSU Butterfly House and Bug House
MSU Union
Michigan State University Celebrity Lecture Series
Pavilion for Agriculture and Livestock Education

What's Happening

Academic Calendar
The Academic Calendar (kept by the Office of the Registrar)
Calendar of Events (MSU News Bulletin)

Diversity Resources

News and Information

The State News
MSU News Bulletin

WKAR Radio and TV
MSU Press
Common Phone Numbers
Mailing to Campus Addresses
Campus Security and Crime Awareness

Collections

W.J. Beal Botanical Garden
Michigan State University Herbarium
Botany and Plant Pathology Live Plant Collection
Campus Wood Plant Collection
Entomology Museum
Human Environment and Design Collections
MSU Museum
Kresge Art Museum
MSU Libraries Special Collections
University Archives and Historical Collections
G. Robert Vincent Voice Library

The website also informs you about:

- Library location and hours—a good place to study
- Services that will benefit you—legal aid, counseling, financial aid, advisors
- Extracurricular activities—
  - sports programs
  - leisure time activities
  - focused clubs: Formula Car, Hybrid Vehicle, Bridge Building, and more

19.7 OTHER IMPORTANT ISSUES

"If you haven't got the time to do it right, when will you have the time to do it over?"

—Jeffrey Mayer

It is important that you take the initiative to look into areas where you may be slightly uncomfortable and seek information that will help you make your career preparation worthwhile. Asking questions and seeking help is truly something some people never seem to learn. For the list of topics below, we suggest some resources of which you may want to take advantage. They are simply some of the sources that you can check. The list is endless when it comes to books that you can access for help. Always keep your eyes open for resources that will help, and save any information that you find especially valuable.

Managing Time

If you have no idea where your time goes, you need to discover ways to get that time element under control.

Recommended reading: *Making Time, Making Money*—Rita Davenport; *Seven Habits of Highly Successful People*—Steven Covey

The Usefulness of Reading

Engineering involves a lot of reading. You will be evaluating reports and consulting a wide range of technical texts in order to prepare projects and documents. This may be the first time you are required to digest large amounts of technical information. To prepare you for this, there are entire books devoted to teaching the art of reading. If you feel that you are a slow reader or that you have difficulty with reading comprehension, Student Services can steer you toward offices on campus that can provide information to improve your abilities.

A few things to help get you started:

- Is there a short synopsis at the beginning of the chapter? If there is, read it carefully.
- Are there questions at the end of the chapter? If so, read them first. This will focus your mind on what is important in the reading.
- Are there headings at the beginning of each section? If so, use these as the main points of your outline for taking notes.
- Each paragraph should have a central point and the rest of the paragraph should support that topic. Make sure that you understand each paragraph's central point.

Recommended reading: *College Reading and Study Skills*—Kathleen McWhorter

Fulfilling Duties

Be proud of your abilities and aware of your obligations. It is important that you realize that you have much to offer your school. It is also vital that you realize that the years of training and the financial breaks (underwritten by tax dollars) which our society offers students, oblige you to hold up your end of the deal in every way. You ought to fulfill the civic duty of your training just as soldiers fulfill the military duty of their training. It's not just a matter of money. Other members of our society look up to and expect to rely on those whom it frees up to gain high level training. So if you start to feel cynical about college, realize that in truth it is a high calling. It truly is an occasion to rise to. Use this awareness as a motivation that lasts longer than just getting an "A"! By giving, you receive—your abilities greatly improve and become clear to others. Work hard to provide quality feedback when it is asked for. You'll find your life to be more worthwhile when you create worth. To develop such helpfulness and creativity, be assertive. You have to be willing to work against inertia, to raise your hand, to volunteer, if you want to be your best. When you contribute, that's when you get what you need. So don't ever feel that you have nothing to offer. Contributing brings its own reward. When things are not going the way you think they should be, ask questions. One answer may solve a vast number of problems.

Using the Web

The World Wide Web is an invaluable tool that allows you to travel the world to discover information in seconds. Spend some time looking for specific information, not just wandering haphazardly. Just log onto a browser and travel. You'll soon see how this tool can be vital to your success, both academically and professionally.

Sending E-mail

Much of what you discover from your contacts will come in the form of e-mail. Get used to the fact that you will be required to converse with people through this electronic medium.

Learn to type, if you have not already done so, and familiarize yourself with standard e-mail software.

Test-Taking Skills

Preparing outlines of the information that you have been presented in class or in books that you have been required to read for your courses will help you to focus on the course content. Team up with other members of your classes to discuss the information. Listen carefully to what the group is saying. Ask questions. The questions that you ask may clarify points that would have hindered you from doing the best job you could on a test.

Check with Student Services on campus and find out when they provide seminars or meetings on test-taking techniques. Most campuses offer these kinds of activities, and you will find them beneficial.

Recommended reading:

Proven Strategies for Successful Test Taking—Thomas Sherman
Effective Study Skills—James Semone

Taking Notes

You will be required to take a mountain of notes in your four- or five-year undergraduate career. You will need to know how to organize the information and sort out what is essential from what is not. This requires specialized skill, effort, and good help.

Recommended reading:

Note-Taking Made Easy—Kesselman-Turkel/Peterson
Tips for College Test Taking—Learning Skills Center
Achieving Academic Success—Kendall Hunt Publishing

Study Skills

We all need to study efficiently. Time is valuable, so you have to evaluate the effectiveness with which you study. Do you take initiative and use your time well, or do you rush to get the studying in at the last minute? Do you prepare an outline and notes to study with, or do you just pile up your books and hope to hit the right points as your fingers pass over the pages? Do you set yourself up for interruptions—or do you carefully protect your set-aside study time? Do you find studying to be tiring and stressful?

You should learn to make studying a calm, structured, routine part of your everyday life as soon as possible because you have plenty of it ahead of you! It's essential to learning, so if you find that you begrudge the time for it, try to look at it a new way. Have you thought about rewarding yourself for studying? Don't study in a way that gets you tense. Take periodic breaks for a change of pace. It's recommended to stand up, move around, and stretch a few times an hour. In fact, it is necessary for optimal study health. Find some fun things to do as a reward after you have devoted a proper amount of time to getting that studying done.

Recommended reading:

Improving Study Skills —Conrad Lashley
Orientation to College Learning —Dianna Van Blerkow

Teaching Styles

You will encounter a variety of instructors in college—from graduate assistants to distinguished full-professors. Your experience will run the gamut. Many will be excellent teachers and communicators, while some will only know how to monotonously present stale facts. Some will have accents that are difficult to understand, and some will downright dislike being in the classroom. It is your job to deal with these varying personalities and gain as much information as you can. You may have to rely more on background information for one professor; you may have to actively participate in another professor's course to draw out the information you need most through aggressive questioning; and you may be pushed to your limits just to keep up to gain the full benefits of another professor's lectures. You must fully engage in the learning process, not by just sitting there but by learning how to adjust to the teaching style of each instructor. By knowing how they differ, you can adjust the way in which you react to their particular style.

Learning Styles

We all learn in different ways. You have your own particular style for assimilating information, so it is necessary for you to make sure that you understand your learning style and incorporate effective methods of learning. Do you memorize everything you read or hear the first time? Do you have to write out facts several times in order to remember them? Do you study only at the last hour, or only when others force you to? Do you have particular places you go when you need to study? Do you have any idea how you study best? If not, you will need to learn.

Perspectives of Others

You will find that being able to keep an open mind about the thoughts, feelings, and beliefs of others will make your college career more valuable. There is no need to go along with everything that goes on around campus, but there is a need to be able to listen and rationally discuss your beliefs and the beliefs of others, and to work together to put them all to equal tests. Your view might pass your own tests just fine, but maybe you're missing some angles. Utilizing other people is the best way to get help to do this. If you find it hard to be empathetic in certain discussions, then it is better to seek help from campus resources that might at least help you widen your ability to listen, or not block the rights of others to speak. You also have a right to be talked to at a level that shows respect to you. No one should badger or assault you verbally with their ideas.

Managing Diversity —Norma Carr-Rufino

Listening Skills

You must listen with greater intensity in college than you have previously. Courses in subjects that you are unfamiliar with will require that you concentrate harder than you ever have before. You will need to focus your attention on what instructors are saying and how it relates to other sources of information presented in the course. You will find that your new environment consists of students, faculty, and staff from myriad backgrounds. This does not require that you agree with everything you hear, but it does require you to allow the other individual to speak so that you can in turn discuss your beliefs and explain your conclusions. Dialogue does not have to be confrontational if both parties are willing to listen respectfully.

Building Active Listening Skills—Judi Brownell

Handling Stress

Develop study skills that are as agreeable to you as possible, but always take some time to relax. Do all things in moderation. This doesn't mean you should overdo one thing, then overdo another to offset the first and supposedly "balance" it. Such a way of handling stress creates more stress.

A common way to create unhealthy stress is to take a class for the wrong reason. You should study to learn and for no other reason. Your grade will, in the end, reflect what you learn. And the way you arrange your daily life will be set up around study, for the most part. But you shouldn't simply study for a grade and you shouldn't take a class just because you have to. Some required classes might seem pointless to you, but instead of getting a bad attitude, you should rise to the occasion, trying to see why the class was required so you can get more out of the class than you thought possible. If you resent your studies, you're in for trouble.

You shouldn't even necessarily resent your weakpoints, since everyone has plenty of features that require improvement. Inevitable failures will roll off you like water off a duck if you approach them positively. After all, as the saying goes, "Success is but a series of failures handled properly." You must go beyond what you know, and are already good at, to learn and grow.

"Grades, sports, and partying" is an approach to school that some students *devolve* to, even as they think they're being clever. A results-oriented bias that over-emphasizes a balance of study, exercise, and fun can put students in this mind-set even without them hardly noticing a change. To whatever degree we take any type of performance more seriously than actual learning, we might tend to over-do one or more aspects, then try to balance it out against the others, so that more study requires more fun, on and on. . . . This can start a stressful trend, even a downward spiral. If you find yourself flopping between extremes, work to develop a better initial understanding of what your real purpose is in life and in school. This should defuse the stress, even if it does seem to make you a bit more "boring" to others or make your study seem to take more effort.

If you notice these things, that's good! Paying attention is a great way to beat stress. Notice the results of your increased effort. See how steady study really doesn't hurt you. Notice how, even when things aren't exciting or you can't seem to see any obvious improvement, if you have a good attitude—including patience!—things work out better in the end.

A "heads up" approach which lets you see the *benefits* of stress beyond its sometimes gloomy appearance will give you better endurance, toleration, and perspective.

You can't avoid stress. But you can *grow* through it if you use it right. To do so, you have to assert yourself. You have to go against the grain as you work to fulfill your obligations. If some of your fellow students seem to be negative about school sometimes, it doesn't mean you have to be. This sort of approach can protect you from the harmful effects of stress.

Of course, even with a good attitude and lifestyle you will have many challenges and problems as you grow through your college career. Even the best study system can break down. You will have unpleasant surprises and consequences. But if you handle them right, you can turn them to your favor even as things look bad outwardly. If handled improperly, even success can cause damage to your system and to your progress in learning.

Talking it out can help, so develop a varied group of people to whom you can direct your concerns over stressful situations (and who know they can let you know if they see you getting into trouble). This group should include friends and peers, but also professors, older students, mentors, role models, clergy, and anyone you look up to. Even your parents! (Oh yes, they'll soon start to seem wiser and more intelligent.)

There are also a number of agencies that provide listeners who can offer a sympathetic shoulder for you to unburden your anxieties. Operations like The Listening Ear let you talk to anonymous people on the phone, giving you a chance to start the venting process with no embarrassment. Never let the pressure build. Release the steam before it creates a damaging situation. But be aware that the build-up itself is a sign that you might need to change daily habits and expectations.

Recommended reading:

Career Success/Personal Stress—Christine Ann Leatz
Why Zebras Don't Get Ulcers—Robert Spolsky

19.8 FINAL THOUGHTS

The ideas expressed in this chapter should help you settle in with greater confidence at college. Efforts will be made on the part of many to make you a part of that life, but it still will depend on your curiosity and steadfastness to investigate every element that makes up the total package we call the college experience. You can do that if you get involved.

EXERCISES AND ACTIVITIES

19.1   With a map of the university in front of you, find the following:
a) a research unit on your campus
b) a place you can go if you are feeling ill
c) the offices dealing with financial aid
d) a dormitory with an odd or interesting name

After locating these sites, plan the shortest route to visit each of the buildings and then do so. Write a brief paragraph that describes the journey and the buildings.

19.2   Find a building on campus that seems relatively mysterious to you. Visit the building and ask questions about the activities within it. Report your findings.

19.3   On your way to your classes, walk through buildings that you have never been in. What goes on in those buildings?

19.4   Begin a walking tour and enter every building in a chosen section of campus. Take time to look at what is in the buildings—departments, faculty, services—and make notes of these findings.

19.5   Stop in one of the offices along your way, identify yourself, and ask for information concerning what goes on in the office or building. Write a memo to your instructor concerning the information that you discovered.

19.6   Find at least five buildings on campus that you have never entered. Investigate what goes on in them, investigate their history, and find out what is planned for them in the future. Write a brief report on your experience.

19.7   Find the oldest building on campus. What is its history?

19.8   Are there any buildings on campus that have served more than one purpose in their existence?

19.9   How much office space is allotted to each faculty member in the engineering department?

19.10   Do the athletic facilities have areas that only certain students can use?

19.11   Find a bridge on campus and document when it was built. What was happening on campus at that time?

19.12   Find someone who graduated from your school at least 10 years ago. What seems different to them about today's campus?

19.13   When high school students visit your campus, what do they want to see?

19.14   After a chosen lecture, pose a set of questions that you could ask in order to clarify information from that lecture. Make a point of asking those questions during the next class period, or send an e-mail to the instructor focusing on those questions.

19.15   Suggest to your instructor that he/she have the class turn in a set of questions that relate to the previous lecture or lectures. Ask that the common questions be addressed during the next class period or placed on a Web site for easy viewing by the class.

19.16   Find out the following from your classmates:
a) who has traveled to the most countries of the world
b) who has gone to a recent concert
c) who wants to follow your same career path
d) who thinks that communication is critical to success in engineering
e) has any home-state student never traveled out of the state within which the college is located

19.17   As the semester goes on, find out who is getting a 4.0 in your courses. Ask them if they incorporate any unique methods in their study that help them obtain those high marks.

19.18   Make an effort to help someone who is having problems in one of your courses. Through helping them, you may well come to understand the material more fully.

19.19   Make an appointment with your academic advisor to discuss your future courses and the requirements for entering the engineering profession. Report on your conversation.

19.20   Collect information from other students about instructors in your major department.

19.21   When you have free time with no classes, ask if you can sit in on courses that you might take in the future. Write up your evaluations of the experience.

19.22   Make an effort to get to know at least two of your instructors. Interview them and ask about their educational experience. What suggestions can they give you about your own education?

19.23   In one of your courses, create a diagram of those areas where you feel you will need help in understanding the material. Find students who can provide you with the information that will help you succeed.

Chapter 20

Financial Aid

20.1 INTRODUCTION

Since you're just beginning your academic journey, you're probably concerned about the amount of money it takes to complete a college degree. According to a 1998 U.S. General Accounting Office report, tuition at four-year public colleges and universities rose 234% from 1980 to 1994—nearly triple the 82% growth in median household income during that time. Average student loan amounts rose 467%, from $518 in 1980 to more than $2,400 in 1995, the report states. The nationwide average for total school-year costs in 2010–2011, as listed in catalogs and on websites, was $35,000 for private colleges and $14,000 for public universities.

Observe the list in Table 20.1 and note that it represents only the tuition and fees of the institutions in question. This list does not reflect cost-of-living expenses and those extras that everyone needs. With these costs in mind, it is critical that you plan well financially for your education.

It might be worth noting here that some countries provide college education free of charge to students, basing admittance to even the elite schools on test scores alone.

It is important to remember, though you might sometimes be frustrated with your financial planning, that many schools do receive large tax subsidies based on enrollment. Taxpayers support you with their hard-won earnings. The true costs of college are far higher than the tuition you pay at such schools. Furthermore, the "cost" of college goes beyond money in a way that might affect how you view the whole college experience.

Our society frees up young people for these years not cynically to make money off them as they struggle to pay for their education, but so these students can develop into productive contributors to society and "pay back" society in that way. It is your responsibility to use both your society-supported time and your tax-supported resources wisely. Even though at times it can seem like you're on your own, you're not: your culture and heritage are backing you up and expecting great things from you. As a student, you fulfill your most serious duty only when you study your best and then do the best you can with your degree afterwards. The pride that comes from a good attitude here easily makes worthwhile the effort that it takes to deal with your financial obligations.

So not only do you have to think about leaving home and all the new responsibilities that come with being on your own, but you also have to solve the problem of securing enough money to pay the out-of-pocket cash costs of the entire college experience.

TABLE 20.1 Sample Tuition for Various Colleges 2010-2011*

Massachusetts Institute of Technology – Massachusetts	$38,940
Rensselaer Polytechnic Institute – New York	40,680
Tufts University – Massachusetts	41,598
Notre Dame University – Indiana	39,920
California Institute of Technology – California	36,282
St. Louis University – Missouri	32,180
Seattle University – Washington	30,825
John Brown University – Arkansas	19,730
Pennsylvania State University – Pennsylvania	15,250
Purdue University – Indiana	9,070
University of Missouri at Rolla – Missouri	9,460
Wright State University – Ohio	7,059
Michigan State University – Michigan	11,434
State University of New York at Buffalo – New York	7,014
New Mexico State University – New Mexico	5,364
University of New Orleans – Louisiana	4,811
Florida State University – Florida	5,238
California State University at Los Angeles – California	1,616

*Tuition and fees in most cases - in state

The bottom line here is that there are five main areas from which you can draw money: parents, scholarships and grants, loans, work-study programs, and work. This chapter briefly addresses a number of issues. You'll have to take the initiative to investigate the details involved with many of these areas.

The chapter is divided into the following sections:

- Parental Assistance
- Financial Assistance Info
- Scholarships/Grants
- Loans
- Work-Study and Just Plain Work
- Scams to Beware

20.2 PARENTAL ASSISTANCE

If you are fortunate enough to have parents who can cover all your college expenses, then the rest of this chapter may have little significance for you. If you have no concerns about expenses, then the issue is finished, except for how you handle your relations with your family. On the other hand, if you are like the majority of students planning a college career, you have a real need to be money-conscious. On average, nine million students must find ways to fund their college education every fall. You will be in competition with those students for available financial resources. You've got your work cut out for you, but the effort you make will probably pay off.

Don't get discouraged if at first you feel that there is no money available to you from any source. This chapter contains some very simple forms for you to complete. By submitting these forms to the proper offices, you may find financial assistance you hadn't expected.

20.3 IS FINANCIAL ASSISTANCE FOR YOU?

Applying for Financial Aid

Financial aid programs are categorized into three areas:

- grants and scholarships
- loans
- work

Some institutions are so expensive that they assume you will need some financial assistance and will automatically offer money to help defray costs. This outright gift will cover only a portion of your expenses, leaving you on your own to pay the rest.

It is never too early to start planning for college expenses. Your parents may have already put money aside for your college education. If they have, your worries have been lessened somewhat. If they haven't been able to, you need to figure out where the money is going to come from.

Several factors must be evaluated when estimating your actual financial needs. One element is need versus non-need. Sometimes students feel that their parents make too much money and that they will not be eligible for financial aid. You must be very careful not to let money slip through your fingers by misinterpreting the conditions for obtaining certain funds.

Many scholarships and grants are provided solely upon your need based upon family income. The limits of this income vary between scholarships. Other scholarships are based on your academic qualifications and have nothing to do with the income of your family. These scholarships are based on your scholastic performance in high school and/or your performance on certain exams.

Why Apply?

After all is said and done, almost every student qualifies for some form of financial aid. But institutions cannot determine your eligibility without the right data. Applications are free and you can, of course, decline any aid you are offered that you don't want.

Budgeting

Advisors are usually available to assist you with personal budgeting. Most institutions want you to start off with everything you need to succeed. They can help you estimate your costs and income and develop a plan to manage your school-related finances. Often, you can find this information on their Web sites.

How to Apply

To determine your eligibility for federal aid, state aid, and other types of aid administered through your institution, you will need to fill out the Free Application for Federal Student Aid (FAFSA), which is described below and pictured later in this chapter. You must file this form *annually* to maintain your eligibility. You have your choice of paper or electronic application forms. The paper form is available early in January in the Office of Financial Aid at most institutions (including high schools). If you prefer, you may apply on-line through FAFSA on the Web as described below. By submitting this form, you will be able to determine your eligibility for the following Federal Financial Student Aid Programs:

- Federal Pell Grants
- Federal Supplemental Educational Opportunity Grants (FSEOG)
- Federal Subsidized and Unsubsidized Stafford Loans
- Federal Direct Subsidized and Unsubsidized Stafford Loans
- Federal Perkins Loans
- Federal Work Study (FWS)
- Title VII and Public Health Act Programs

The Free Application for Federal Student Aid (FAFSA)

The *Free Application for Federal Student Aid* (FAFSA) online registration system is found on the Web at http://www.fafsa.ed.gov. It offers easy step-by-step instructions and can be used from any computer operating system. It is now the most common way of registering. We present the printer version of the form in its entirety following this explanatory section.

The FAFSA form is the first thing you should complete to become eligible for aid. It is best to apply for aid for the next fall semester as soon after January 1 as possible, since funds for some programs may become depleted during the year. However, the application can be filed at any time.

The lists that follow will give you an idea of the information requested. Keep in mind that the reason for answering all these questions is to give everyone submitting a form a fair chance of receiving financial aid. There are five sections of the application. They ask questions about:

- you (the student)
- your school plans
- your household and finances
- your parents' household and finances (if applicable)
- your assets and other information

Remember that when you are going through these questions, "you" and "your" always refer to you, the student, who is applying for aid, not your parents.

The following questions provide you with the core information needed to complete the FAFSA. If you can provide an answer for each of these, the completion of the actual form will be relatively easy.

- What is your Social Security number?
- What is your date of birth?
- Are you a U.S. citizen?
- What is your permanent mailing address?
- What is your state of legal residence?
- What was the month and year you began living in your state of legal residence? If you have always lived there, give the month and year you were born.
- What is your permanent telephone number?
- Do you have a driver's license?
- What is your driver's license number?
- What was the state that issued your driver's license?

- As of today, what is your marital status?
- Are you an orphan or ward of the court, or were you a ward of the court until age 18?
- Do you have legal dependents (other than a husband or wife)? (Dependents are people who live with you and for whom you provide more than half the financial support.)
- Are you a veteran of the U.S. armed forces?
- Are you an "early analysis" student? (You are an "early analysis" student if you are using this application to estimate your financial need, and you have applied, or will be applying for, "early action" or "early decision" at a college before your freshman year.)
- Will you have a high school diploma or GED before you enroll in college?
- What was the highest education level completed by your father?
- What was the highest education level completed by your mother? (This question is for state scholarship programs only. Your answer does not affect your eligibility for any federal student aid.)
- What most likely will your enrollment status be in the next school year?
- What will be your major course of study in college?
- What degree or certificate will you earn from your currently planned major or course of study?
- What will be the month and year you will earn your degree or certificate?
- What college(s) will you be attending?
- In addition to grants, what kinds of federal student aid are you interested in? (Choose all that you think you might want.)
  - — I am interested in student employment.
  - — I am interested in parent loans.
  - — I am interested in student loans.
- How much money did you (the student) earn by working during the year prior to your application?
- If you are married, how much money did your spouse earn by working during the year prior to your application?
- What federal IRS tax form did you file for that year?
- What was your adjusted gross income?
- How many exemptions did you claim?
- If you received any Aid to Families with Dependent Children (AFDC), ADC, or Temporary Assistance for Needy Families (TANF), what was the total amount of benefits you received during that year?
- What was the total amount of child support you received for all children?
- Did you receive any of the following during that year:

  1. welfare benefits
  2. veterans' education benefits or veterans' noneducation benefits
  3. credit for federal tax on special fuels
  4. a foreign income exclusion
  5. workers compensation
  6. housing, food, or other allowances paid to military members
  7. other untaxed income or benefits
  8. earnings from Federal Work-Study or other need-based programs
  9. allowances or benefits from AmeriCorps
  10. student grants and scholarships in excess of tuition, and supplies

- What is the total amount you currently have in cash, savings, and checking accounts? If you are married, include the cash, savings, and checking accounts of your spouse.

- Do you have any investments or own real estate, a business, an investment farm? Do not include your home.
- What is the current total amount your parents have in cash, savings and checking accounts?
- Do your parents have any investments or own real estate, a business, or an investment farm? Do not include their home.
- What is the current value of your parents' real estate? Remember, don't include their home. "Real estate" means rental property, land, and second or summer homes.
- What is your parents' current real estate debt?
- What is the current value of your parents' investments?
- What is your parents' current investments debt?
- What is the current value of your parents' business? Include the value of land, buildings, machinery, equipment, and inventories.
- What is your parents' current business debt? Include only the present mortgage and related debts for which the business was used as collateral.

This may seem like a great number of questions, and possibly you are thinking about how much detailed information these questions request, and that you might not be able to determine the answers very readily. You might even worry that they invade your privacy. However, you should realize that they are necessary in order to make reasonable decisions on who will be granted financial assistance.

You will also need to have available the following specific materials that will help you to complete the forms:

- Your Social Security card and driver's license
- W-2 forms or other records of income earned
- Your (and your spouse's, if married) federal income tax return
- Records of other untaxed income received such as welfare benefits, Social Security benefits, AFDC, TANF, ADC, veteran's benefits, or military or clergy allowances
- Current bank statements, records of stocks, bonds, etc.
- Business or farm records, if applicable
- Your alien registration card (if you are not a U.S. citizen)
- A flash drive, if you want to save your data from the on-line form

Sample Form and Deadlines

Make sure that you also check with the financial aid administrator at your university for information about school and state deadline dates. The 2010-2011 application that follows is presented primarily so that you can see what is required. It's best to use the official form or online registration for your actual submission. The latest version may have changes from the one presented. A current paper form should be filled out and mailed to: Federal Student Air Programs, P. O. Box 4001, Mt. Vernon, IL 62864-8601.

20.4 SCHOLARSHIPS/GRANTS

Grants and scholarships are educational funds that do not need to be repaid. Some programs require that you have "need" as determined by the FAFSA. Others may require a certain level of academic achievement or other qualifications. Grant and scholarship funding may originate from public, private or university sources. You are automatically considered for many scholarships and grants by submitting your admissions and financial aid applications.

2010–2011
FAFSA ON THE WEB WORKSHEET
www.fafsa.gov

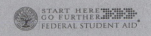
START HERE
GO FURTHER
FEDERAL STUDENT AID

DO NOT MAIL THIS WORKSHEET.

You must complete and submit a *Free Application for Federal Student Aid* (FAFSA) to apply for federal student aid and to apply for most state and college aid. Applying online with *FAFSA on the Web* at **www.fafsa.gov** is faster and easier than using a paper FAFSA.

This worksheet has been designed to provide a preview of the questions that you may be asked on *FAFSA on the Web*. Write down notes to help you easily complete your FAFSA anytime after January 1, 2010.

See the table to the right for state deadlines. Check with your high school counselor or your college's financial aid administrator about other deadlines.

- **This Worksheet is optional and should only be completed if you plan to use *FAFSA on the Web*.**

- Sections in purple are for parent information.

- **This Worksheet does not include all the questions from the FAFSA. The questions that are included are ordered as they appear on *FAFSA on the Web*. When you are online you may be able to skip some questions based on your answers to earlier questions.**

Apply Faster—Sign your FAFSA with a Federal Student Aid PIN.

If you do not have a PIN, you can apply for one at **www.pin.ed.gov**. Your PIN allows you to electronically sign when you submit your FAFSA. If you are providing parent information, one parent must also sign your FAFSA. To sign electronically, your parent should also apply for a PIN.

> You do not have to pay to get help or submit your FAFSA. Submit your FAFSA for **free** online at **www.fafsa.gov**. Federal Student Aid provides **free** help online at **www.fafsa.gov** or you can call 1-800-4-FED-AID. TTY users (hearing impaired) may call 1-800-730-8913.

NOTES:

STATE AID DEADLINES

Check with your financial aid administrator for these states and territories:

AL, AS *, AZ, CO, FM *, GA, GU *, HI *, MH *, MP *, NC, NE, NM, NV *, PR, PW *, SD *, TX *, UT, VA *, VI *, VT *, WA, WI and WY *.

Pay attention to the symbols that may be listed after your state deadline.

AK	April 15, 2010 *(date received)*
AR	Academic Challenge - June 1, 2010 *(date received)*
	Workforce Grant - Contact the financial aid office.
	Higher Education Opportunity Grant
	- June 1, 2010 (fall term) *(date received)*
	- November 1, 2010 (spring term) *(date received)*
CA	Initial awards - March 2, 2010 + *
	Additional community college awards
	- September 2, 2010 *(date postmarked)* + *
CT	February 15, 2010 *(date received)* # *
DC	June 30, 2010 *(date received by state)* # *
DE	April 15, 2010 *(date received)*
FL	May 15, 2010 *(date processed)*
IA	July 1, 2010 *(date received)*
ID	Opportunity Grant - March 1, 2010 *(date received)* # *
IL	As soon as possible after 1/1/2010. Awards made until funds are depleted.
IN	March 10, 2010 *(date received)*
KS	April 1, 2010 *(date received)* # *
KY	March 15, 2010 *(date received)* #
LA	July 1, 2010 *(date received)*
MA	May 1, 2010 *(date received)* #
MD	March 1, 2010 *(date received)*
ME	May 1, 2010 *(date received)*
MI	March 1, 2010 *(date received)*
MN	30 days after term starts *(date received)*
MO	April 1, 2010 *(date received)*
MS	MTAG and MESG Grants - September 15, 2010 *(date received)* #
	HELP Scholarship - March 31, 2010 *(date received)* #
MT	March 1, 2010 *(date received)* #
ND	March 15, 2010 *(date received)*
NH	May 1, 2010 *(date received)*
NJ	2009-2010 Tuition Aid Grant recipients - June 1, 2010 *(date received)*
	All other applicants
	- October 1, 2010, fall & spring terms *(date received)*
	- March 1, 2011, spring term only *(date received)*
NY	May 1, 2011 *(date received)* + *
OH	October 1, 2010 *(date received)*
OK	April 15, 2010 *(date received)* #
OR	OSAC Scholarship - March 1, 2010
	Oregon Opportunity Grant - Contact the financial aid office.
PA	All 2009-2010 State Grant recipients & all non-2009-2010 State Grant recipients in degree program - May 1, 2010 *(date received)* *
	All other applicants - August 1, 2010 *(date received)* *
RI	March 1, 2010 *(date received)* #
SC	Tuition Grants - June 30, 2010 *(date received)*
	SC Commission on Higher Education - no deadline
TN	State Grant - February 15, 2010 *(date received)* #
	State Lottery - September 1, 2010 *(date received)* #
WV	April 15, 2010 *(date received)* # *

For priority consideration, submit application by date specified.
+ Applicants encouraged to obtain proof of mailing.
* Additional form may be required.

STATE AID DEADLINES

Federal Student Aid logo and FAFSA are service marks or registered service marks of Federal Student Aid, U.S. Department of Education.

WWW.FAFSA.GOV   2010-2011 FAFSA ON THE WEB WORKSHEET   PAGE 1

SECTION 1 - STUDENT INFORMATION

After you are online, you can add up to ten colleges on your FAFSA. The colleges will receive the information from your processed FAFSA.

Student's Last Name	First Name	Social Security Number

Student Citizenship Status (check one of the following)

❑ U.S. citizen (U.S. national)   ❑ **Neither citizen nor eligible noncitizen**

❑ **Eligible noncitizen** (Enter your Alien Registration Number in the box to the right.)

Your Alien Registration Number

Generally, you are an eligible noncitizen if you are:
- A permanent U.S. resident with a Permanent Resident Card (I-551);
- A conditional permanent resident (I-551C); or
- The holder of an Arrival-Departure Record (I-94) from the Department of Homeland Security showing any of the following designations: "Refugee," "Asylum Granted," "Parolee" (I-94 confirms paroled for a minimum of one year and status has not expired), "Victim of human trafficking," T-Visa holder (T-1, T-2, T-3, etc.) or "Cuban-Haitian Entrant."

`A` ☐☐☐☐☐☐☐☐☐

Student Marital Status (check one of the following)

❑ Single   ❑ Married or remarried   ❑ Separated   ❑ Divorced or widowed

You will be asked to provide information about your spouse if you are married or remarried.

Selective Service Registration

If you are male and 25 or younger, you can use the FAFSA to register with Selective Service.

Student Aid Eligibility Drug Convictions

❑ I have never attended college   ❑ I have never received federal student aid   ❑ I have never had a drug conviction

If you did not check any of these boxes, you will be asked more questions online.

Highest school your father completed	❑ Middle school/Jr. high ❑ High school	❑ College or beyond ❑ Other/unknown
Highest school your mother completed	❑ Middle school/Jr. high ❑ High school	❑ College or beyond ❑ Other/unknown

SECTION 2 - STUDENT DEPENDENCY STATUS

If you can check ANY of the following boxes, you will not have to provide parental information. Skip to page 4.
If you check NONE of the following boxes, you will be asked to provide parental information. Go to the next page.

❑ I was born before January 1, 1987	❑ I am married	❑ I will be working on a master's or doctorate program (e.g., MA, MBA, MD, JD, PhD, EdD, graduate certificate)	
❑ I am serving on active duty in the U.S. Armed Forces	❑ I am a veteran of the U.S. Armed Forces	❑ I have children and I provide more than half of their support	
❑ Since I turned age 13, both of my parents were deceased	❑ I was in foster care since turning age 13	❑ I have dependents (other than children or my spouse) who live with me and I provide more than half of their support	
❑ I was a dependent or ward of the court since turning age 13	❑ I am currently or I was an emancipated minor	❑ I am currently or I was in legal guardianship	❑ I am homeless or I am at risk of being homeless

NOTES:

SECTION 3 - PARENT INFORMATION

Who is considered a parent? "Parent" refers to a biological or adoptive parent. Grandparents, foster parents, legal guardians, older siblings, and uncles or aunts are **not** considered parents on this form unless they have legally adopted you. In case of divorce or separation, give information about the parent you lived with most in the last 12 months. If you did not live with one parent more than the other, give information about the parent who provided you the most financial support during the last 12 months or during the most recent year you received support. If your divorced or widowed parent has remarried, also provide information about your stepparent.

Providing your father's information? You will need:	**Providing your mother's information? You will need:**
Father's/Stepfather's Social Security Number	Mother's/Stepmother's Social Security Number
Father's/Stepfather's name	Mother's/Stepmother's name
Father's/Stepfather's date of birth	Mother's/Stepmother's date of birth
❑ Check here if your father/stepfather is a dislocated worker	❑ Check here if your mother/stepmother is a dislocated worker

Did your parents file or will they file a 2009 income tax return?

❑ My parents have already completed a tax return

❑ My parents will file, but have not yet completed a tax return

❑ My parents are not going to file an income tax return

Your parents will need their tax returns and/or W-2 forms to complete the FAFSA.

What was your parents' adjusted gross income for 2009?

Skip this question if your parents did not file taxes. Adjusted gross income is on IRS Form 1040—Line 37; 1040A—line 21; or 1040EZ—line 4.

$ _____

The following questions ask about earnings (wages, salaries, tips, etc.) in 2009. Answer the questions whether or not a tax return was filed. This information may be on the W-2 forms, or on the IRS Form 1040—Line 7 + 12 + 18 + Box 14 of IRS Schedule K-1 (Form 1065); 1040A—line 7; or 1040EZ—line 1

How much did your father/stepfather earn from working in 2009? $ _____

How much did your mother/stepmother earn from working in 2009? $ _____

In 2008 or 2009, did anyone in your parents' household receive:

❑ Supplemental Security Income   ❑ Temporary Assistance for Needy Families (TANF)

❑ Food Stamps   ❑ Special Supplemental Nutrition Program for Women, Infants and Children (WIC)

❑ Free or Reduced Price School Lunch

Note: Food Stamps and TANF may have a different name in your state. Call 1-800-4-FED-AID to find out the name of the state's program.

Did your parents have any of the following items in 2009?
Check all that apply. Once online, you may be asked to report amounts paid or received by your parents.

Additional Financial Information	**Untaxed Income**	
❑ Hope and Lifetime Learning tax credits	❑ Payments to tax-deferred pension and savings plans	❑ Untaxed portions of pensions
❑ Child support paid	❑ Child support received	❑ Housing, food and other living allowances paid to members of the military, clergy and others
❑ Taxable earnings from work-study, assistantships or fellowships	❑ IRA deductions and payments to self-employed SEP, SIMPLE and Keogh	❑ Veterans noneducation benefits
❑ Grant and scholarship aid reported to the IRS	❑ Tax exempt interest income	❑ Other untaxed income not reported, such as workers' compensation or disability
❑ Combat pay or special combat pay	❑ Untaxed portions of IRA distributions	
❑ Cooperative education program earnings		

Your parents may be asked to provide more information about their assets.
Your parents may need to report the net worth of their current businesses and/or investment farms.

NOTES:

SECTION 4 - STUDENT INFORMATION

Did you file or will you file a 2009 income tax return?
❏ I have already completed my tax return
❏ I will file, but I have not completed my tax return
❏ I'm not going to file an income tax return

You will need your tax returns and/or W-2 forms to complete the FAFSA.

What was your (and spouse's) adjusted gross income for 2009?
Skip this question if you or your spouse did not file taxes. Adjusted gross income is on IRS Form 1040—Line 37; 1040A—line 21; or 1040EZ—line 4.

$ _____

The following questions ask about earnings (wages, salaries, tips, etc.) in 2009. Answer the questions whether or not a tax return was filed. This information may be on the W-2 forms, or on the IRS Form 1040—Line 7 + 12 +18 + Box 14 of IRS Schedule K-1 (Form 1065); 1040A—line 7; or 1040EZ—line 1.

How much did you earn from working in 2009?
❏ Check here if you are a dislocated worker

$ _____

How much did your spouse earn from working in 2009?
❏ Check here if your spouse is a dislocated worker

$ _____

In 2008 or 2009, did anyone in your household receive:
❏ Supplemental Security Income   ❏ Temporary Assistance for Needy Families (TANF)
❏ Food Stamps   ❏ Special Supplemental Nutrition Program for Women, Infants and Children (WIC)
❏ Free or Reduced Price School Lunch

Note: Food Stamps and TANF may have a different name in your state. Call 1-800-4-FED-AID to find out the name of the state's program.

Did you or your spouse have any of the following items in 2009?
Check all that apply. Once online you may be asked to report amounts paid or received.

Additional Financial Information	**Untaxed Income**	❏ Untaxed portions of pensions
❏ Hope and Lifetime Learning tax credits	❏ Payments to tax-deferred pension and savings plans	❏ Housing, food and other living allowances paid to members of the military, clergy and others
❏ Child support paid	❏ Child support received	❏ Veterans noneducation benefits
❏ Taxable earnings from work-study, assistantships or fellowships	❏ IRA deductions and payments to self-employed SEP, SIMPLE and Keogh	❏ Other untaxed income not reported, such as workers' compensation or disability
❏ Grant and scholarship aid reported to the IRS	❏ Tax exempt interest income	❏ Money received or paid on your behalf
❏ Combat pay or special combat pay	❏ Untaxed portions of IRA distributions	
❏ Cooperative education program earnings		

You may be asked to provide more information about your (and spouse's) assets.
You may need to report the net worth of current businesses and/or investment farms.

NOTES:

Do not mail this Worksheet. Go to www.fafsa.gov to complete and submit your application.

For more information on federal student aid, visit **www.FederalStudentAid.ed.gov**.
You can also talk with your college's financial aid office about other types of student aid that may be available.

You must seek private scholarships on your own. There are books available in the library to help you search for private funding, or you may conduct an on-line scholarship search.

Do not hesitate to pursue every possible avenue for funds for your education. These funds may come from your local high school, professional groups, corporations, service organizations, the government (both state and federal) or from the college that you will attend. Consider every aspect of public, private, or government sectors in your effort to find available money.

It is your responsibility to seek out private scholarships/grants. These may include awards given by you or your parents' employers; grants given to members of certain ethnic groups; or funds earmarked for specific majors, locations of colleges, or future career choices (i.e., receiving your degree in mechanical engineering and then planning to work in Indonesia). There are many books available that can provide you with a vast number of places from which you can search for scholarships that pertain to both you and your needs. There is also a free national computerized scholarship search service called "fastWEB" at http://www.fastweb.com/.

20.5 LOANS

Loans may be secured from lending institutions and state and federal loan programs. These sources can be located by asking questions of the admissions officers on the campuses you consider attending. They can provide you with the needed paperwork to apply for these loans.

Students who apply for financial aid will be notified of their eligibility for both student and parent federal loans. Parents of dependent students may apply for the PLUS (Parent Loan for Undergraduate Students) without filing the FAFSA. Other loans, often referred to as alternative loans, are available to parents or students who qualify. For more information on these alternative loans, call the Office of Financial Aid at the school of your choice and request information.

Loans can also be obtained from parents or relatives who feel that you should repay the money that is required to "put you through school." There may be very good reasons for requesting repayment—you may have brothers and sisters who will also need to attend college and there may not be enough money to go around. Repayment will ensure that everyone gets the chance to obtain a college degree. It is also a sign of maturity when you are able to stand on your own and pay your own way. Never feel that repaying a loan to family members is an imposition. It creates in you a strong dedication for achieving the degree and teaches you responsibility.

Activity 20.1

Contact your school's loan office and find out what monetary aid is available to you. Is there a work/study program? Can you apply for loans immediately or are there particular conditions that must be met first? Do your parents have to fill out forms in order for you to be given the loan? Have you discussed this with your parents?

20.6 WORK-STUDY OR JUST PLAIN WORK

Scholarships are wonderful because you don't have to repay them; loans might be next in order of preference because you don't have to repay them until after you have finished your degree; but there is another option that you may want to consider: a job. "Earning money the old-fashioned way" will probably be one of the principal ways for you to help pay for your education. A job can also help pay incidental costs immediately. This shouldn't be under appreciated. It can also help you pay off more quickly any loan you do take out. The sooner you can pay for even minor expenses, the better!

Work can be approached in a number of different ways:

- on-campus or off-campus employment during school
- summer jobs
- internships
- co-ops
- a combination of the above

All of these jobs require that you put in a certain amount of time, and for that you receive pay. This is not a situation where you receive money and have to pay it back later. You must make time for work on top of your normal schedule. This requires that you manage your time carefully, so that in addition to work, you can go to class, eat, sleep, study and have adequate free time. Working is definitely something that you have to approach carefully and with a healthy respect for how much you can do and what impact it will have upon your studies. In some cases, it can prolong your school years. That might not sound so good right now, but in the end you may be much better off financially. It is therefore important to investigate the opportunities that are available to you on campus at the Student Services office.

Work-Study

After you have completed your *Free Application for Federal Student Aid*, you will be given information as to whether you will be allowed to participate in work-study programs at your institution. Work-study employment is subsidized by the federal or state government (the government reimburses the employer for a portion of wages paid to recipients). Job listings in the Student Employment Offices of colleges and universities and on their Web sites are categorized by work-study and non-work-study. If you are eligible for work-study employment it will be listed on your financial aid award letter. If you are, it will be your responsibility to find a position that you would like to apply for that accepts only work-study students.

Sample Work-Study Student Jobs

The following examples are just a few of the many jobs that are available on campus for work-study students. Take a close look at the job descriptions, requirements for specific time commitments, and the rate of pay. These could be jobs that you could hold for the entire time you are on campus.

ANIMAL CARETAKER I

CARCINOGENESIS LABORATORY/COM/DEAN. No experience necessary. Pre-vet majors preferred. Will be responsible for routine care and maintenance of an athymic mouse colony. Assist with procedures such as breeding, weaning, injections, and tumor removal. Hours: 10-30/wk. Rate of pay: $7.40/hr.

AUDIO VISUAL AIDE I

INSTRUCTIONAL MEDIA CENTER. No experience necessary. Will be responsible for providing media support in classroom. Will be responsible for delivery, set-up, and operation of audio-visual equipment at various on-campus locations, and other duties as assigned. Hours: To be arranged between 7:30am and 10:00pm.

AUDIO VISUAL AIDE II

STUDENT LIFE. Must have good verbal communication skills and have the ability to speak in front of groups. Must have a good technical aptitude and a valid driver's license. Will be responsible for delivering, setting up, and operating a karaoke system (training provided); working with D.J.; maintaining records; and related tasks as assigned. Hours: Late afternoons and evenings. Rate of pay: $7.40/hr.

CASHIER I

MSU MUSEUM. Experience with the public, good math skills and/or past retail work desired. Natural or cultural history knowledge a plus. Will be responsible for opening and closing store, assisting customers with purchases, using register and credit card reader, and answering general public questions about museums and exhibits. Hours: Saturday 10am-5pm, Sunday 1pm-5pm. Rate of pay: $7.40/hr.

CUSTODIAL WORKER I

DORM COMPLEX. No experience necessary. Will be responsible for painting, cleaning, conference set-up, and general cleaning of facilities. Hours: Mon-Fri 7:00am-3:30pm. Rate of pay: $7.40/hr.

CLERK I

FINANCIAL AID. Will be responsible for filing, mailings, data entry. Hours: To be arranged. Rate of pay: $7.40.

ENGLISH. Will train. Will be responsible for processing mail, running errands, operating office equipment including photocopy machine. Hours: To be arranged. Rate of pay: $7.40/hr.

CLERK/TYPIST II

OFFICE FOR INTERNATIONAL STUDENTS AND SCHOLARS. Must have an interest in international students and cross-cultural awareness. Must be flexible and have the ability to work for multiple staff and with frequent interruptions. Will be responsible for greeting students at front counter and answering phones. Will be responsible for filing, copying, running errands, and some typing. May compile reports and assist with general office projects. Hours: 10-15/wk. Rate of pay: $7.40/hr.

UNIVERSITY OUTREACH. Must be a self-starter requiring little supervision, with previous office experience, typing/computer skills, and a valid driver's license. Will be responsible for filing, maintaining records, copying, giving information, correspondence, use of university vehicle, typing reports, answering telephones, data entry, word processing, and related tasks as assigned. Hours: To be arranged. Rate of pay: $7.40/hr.

COMPUTER ASSISTANT III

H-NET HUMANITIES & SOCIAL SCIENCES ONLINE. Must be fluent in Japanese, familiar with HTML and various e-mail programs. Knowledge of Japanese word processing programs a plus. Will be responsible for editing Japanese language listserv logs and Web pages. Hours: 10/wk. Rate of pay: $7.40/hr.

DEPARTMENTAL AIDE I

RESIDENCE LIFE. Will be responsible for supervising campus during evening hours as a camp assistant. Hrs: M-F/5 pm–7 am.

"Just Plain Work"

If you are not interested in work-study positions, or if you fail to qualify for aid, you may want to look both on campus and off campus for positions that will provide you an income and a job that is meaningful to you. So, where should you go? Most schools have Student Employment offices, usually within some form of Student Services operation. Career Services and Placement offices on campus also may have opportunities for students who do not fit into the work-study system.

Some of the jobs that these offices may suggest relate to:

- Volunteer offices
- School newspaper
- Departments on campus
- Public safety
- Dormitory operations
- Theater/auditorium
- Athletic operations
- Food courts
- Union operations
- Administration buildings
- Golf course
- Swimming pool
- Intramural facilities

There also may be opportunities for you in special skill areas:

- Machine shops
- Physical plants
- Grounds
- Museum
- Art gallery

Volunteering

Not all jobs pay wages. You may want to consider positions that will be purely voluntary. Volunteering will allow you to perform a needed non-economic civil service. One often takes a paid job because of need. With volunteering, it's a special chance to do work purely of your own volition. You'll notice the difference! Appreciating the benefits of such initiative can help you in many other ways as well (e.g., when you fill out that application for medical school or law school). Also, if you are so inclined, it may open up opportunities for paid positions. Here is a list of some of the various programs for which you could become a volunteer:

- Elementary school science enrichment; high school science tutoring
- The Listening Ear (counseling); youth mentoring; senior programs
- Volunteer youth probation officer
- Child development, linguistics, pre-school/day care, mental health
- Women's Resource Center; Council Against Domestic Assault
- Tutoring (any level); gifted enrichment program; special ed
- Volunteer income tax assistance
- Animal hospitals and clinics
- Hospitals

- High school language tutor; international student assistance
- Educational enrichment programs

Full Semester Off-Campus Employment

There are many jobs that you can seek out yourself in the community neighboring the college or university that you attend. The student services operation at your school will have some listings of these opportunities. Local newspapers also are a good resource. You also may try the old-fashioned way of just going from business to business asking for employment.

If you are trying to make your employment activities match your academic studies, then you should consider Cooperative Engineering Education. Although it is described in detail in Chapter 21 of this textbook, the following brief information will give you an indication of its worth to you.

Cooperative Education

Cooperative Education (co-op) is an academic program in which college students are employed in positions directly related to their major field of study. *(Note: A more complete detailing of the co-op experience is given in the next chapter.)* Co-op is a three-way partnership among a student, an employer and a college. National studies show that approximately 80% of co-op students who work for such organizations eventually receive an offer for permanent employment. Not only is this a great way for you to obtain money for your education, but it also gives you a possible avenue into a future employment opportunity with the company.

There are many types of co-op education:

- *Alternating*—a full semester of employment alternated with full semesters of school
- *Parallel*—part-time employment and school on a concurrent basis
- *Back-to-back semesters*—two consecutive semesters of employment, then back to school for two semesters.

To provide you with basic information concerning cooperative engineering education, here are just a few facts from the Michigan State University Co-operative Engineering Education Division of the College of Engineering:

- During 2009–2010, over 400 engineering and computer science students participated in the MSU co-op program.
- Co-op participants include students from sophomore through senior years.
- Currently, several graduate students are also involved. Freshmen are eligible to complete an application at any time.
- MSU co-op participants earned an estimated $4,700,000 last year, in total. Estimated salaries range from $13.17/hr to $15.63/hr. Salaries vary depending on geographical location, class level, employer, major and experience.
- 17% of MSU co-ops are women and 16% are minorities.
- The grade point averages of co-op participants are significantly higher than the averages of those who do not participate in the co-op program.
- 76% of co-ops and interns are employed within the Michigan region (Michigan, Indiana, Illinois, Ohio, Wisconsin, and Minnesota) by industries and governmental units. 24% work with out-of-state organizations.

If this makes you wonder about the possibility of your participation in co-op, stop by the office in your college that handles this form of employment. Remember, this opportunity is useful not only for obtaining money, but also for the experience.

20.7 SCAMS TO BEWARE

While it is important to search for sources of money, it is equally important that you remember the old saying "There is no free lunch!" Very often you may hear of a chance to pick up an easy buck, where you can get this or that scholarship with little or no effort. Maybe you only have to pay a third party $50 or $100 and it's all refundable if you don't secure at least $1000 in scholarships. Beware! It may be that the only thing you get is an empty wallet.

If you think you may have already been caught in such a scam, read on and refer to the contact people at the end of the section.

The section that follows is from a public bulletin the federal government released concerning scholarship fraud, located online at:

http://www.ftc.gov/bcp/menu-jobs.htm

FTC Cautions: Do Your Own Homework to Avoid Scholarship Service Rip-offs

As more than nine million students nationwide start college classes again this semester, tens of thousands of them will have been victimized by fraudulent companies that pose as legitimate foundations, scholarship sponsors and scholarship search services, the Federal Trade Commission (FTC) reports. And the number of victims is growing as the rise in college costs outpaces families' ability to pay. Capitalizing on this gap, scam artists guarantee "free money for college" in campus newspapers, flyers, postcards and on the Internet, and charge from $10 to $400 for their services. According to the FTC, however, students who rely on a fraudulent search service instead of doing their homework when applying for scholarships or grants will face the upcoming school year bills with nothing to show for this effort but a hard-earned lesson.

"Bogus scholarship search services are just a variation on the prize-promotion scam, targeted to a particular audience—students and their parents who are anxious about paying for college," says Jodie Bernstein, director of the FTC's Bureau of Consumer Protection. "They guarantee students and their families free scholarship money and all they have to do to claim it is pay an up-front fee. In comparison to the scholarship, the fee seems paltry. But even $50 is a significant loss to a young person who is trying to pull together college funding, and many schemes charge much more than that. You don't need college-level math to know that multiplying those fees by tens of thousands of victims adds up to big money for the scam artists. The potential loss to students is even greater when you add in the lost time and money that could have been spent working with legitimate sources."

Fraudulent scholarship search services go through databases of public information and provide a list of purported scholarships and grants for which a student supposedly is eligible to apply. Some of the scholarships listed are for specific disciplines and many students are ineligible. In other situations, the deadlines have expired. In addition, many of these outfits actually turn out to be either loan programs or contests. The bottom line is, a scholarship search service cannot truthfully guarantee that a student will receive a scholarship.

"Filling out scholarship applications is like doing your homework: If you're going to get anything out of it, you have to do your own work," Bernstein says. "These fraudulent search firms are not going to fill out those applications, gather recommendations and write those essays for you, even for $300."

According to Mark Kantrowitz, author of the Financial Aid Information Page on the World Wide Web, many fraudulent search services use some variation on a theme that billions in unclaimed scholarship aid is available, deceptively capitalizing on a two-decade-old report about employer-provided tuition benefits. In fact, he says, the benefits touted in that report were available only to the companies' own employees who had children enrolled in college—not to the general public. About 95 percent of college aid comes from state or federal government agencies, according to authorities.

Other techniques employed by fraudulent companies include using official-sounding names that are slight variations of the names of legitimate government and private organizations, and Washington, D.C., addresses that turn out to be post office boxes rather than business locations, according to the FTC. Some companies claim to be holding scholarships for consumers, and ask them to provide their checking account numbers to confirm their eligibility.

"Don't do it," the FTC's Bernstein cautions. "A con artist can use the number to drain your checking account. They don't need your signature on a check . . . only the account number."

These and other tips are included in flyers, bookmarks, brochures and on-line public service announcements that are part of the massive Project $cholar$cam consumer education campaign to help students, parents, educators and financial aid administrators identify and avoid scholarship scams. As part of the campaign, the following Web sites include information on both scholarships and scholarship scams, and how to avoid the scams:

- The FTC's Web site   www.ftc.gov
- The Student Marketing Association   www.salliemae.com
- The Financial Aid Information Page   www.financialaid.com
- The National Association of Student Financial Aid Administrators
                                                        www.nasfaa.org
- The American Counseling Association   www.counseling.org

20.8 THE ROAD AHEAD AWAITS

It is now time for you to examine the many different sources available to you for obtaining the funds needed for your college expenses. It will require effort and thought, but you can do it. You will have to determine how much you actually need to survive in college; you will have to complete the correct forms in order to be considered for scholarships, loans, and work/study; and you will have to meet deadlines. This is a major test on the road to achieving that college degree. Can you pass? Sure you can!

EXERCISES AND ACTIVITIES

20.1   Get on the Internet and search for scholarships and grants for which you may apply. Research the scholarships and grants offered by companies, foundations, and institutions. What pertinent information do these applications request?

20.2   Copy the federal student aid application presented in this chapter and fill in the necessary information.

Chapter 21

Engineering Work Experience

"How do you get experience without a job, and how do you get a job without experience?"

This is the job-seeker's first, natural question. It was used in a national publicity campaign by the National Commission for Cooperative Education a number of years ago. It's one that all engineering students have considered or will consider at some point during their college career. Career-related experience, commonly called "experiential learning," has become an important supplement to a student's engineering education. Throughout the years, and despite the changing times of up-and-down job markets, numerous employers of engineering graduates, and engineering graduate schools have continued to stress the fact that they seek to fill their annual openings by selecting only from those candidates who have done outstanding things during their college years to distinguish themselves from their peers. In most cases, those students who have participated in some form of experiential learning have gained the types of skills and competencies that prospective employers and graduate schools are seeking.

A college education has evolved over the past century to be quite a different experience from that of our predecessors. For most engineering students during the first half of the 20th century, college was completed in four years. Since studying engineering has always been a rigorous experience, most students chose to devote their time and efforts to experiences in the classroom, library, or laboratory. Those who chose to work while going to school were viewed by the conventional student population as "non-traditional," representing the "working class" who were forced to work for pay solely as a means of affording their schooling. However, in the latter part of the 20th century the costs of a college education soared. It is now the rule, rather than the exception, that most students are forced to work for pay in order to keep up with steadily increasing tuition, room and board, and associated expenses such as books, computers, and transportation. (Again, some countries fully subsidize college education—with one of the benefits being that students can concentrate fully on their studies, with less need for extracurricular employment during school.)

For many students, the idea of combining their academic studies with a career-related employment experience is something they had not previously considered. The thought of stretching their college education to five or even six years can seem to be an unnecessary burden. To their parents who grew up with the notion that a college education is a four-year plan with a direct path into the world of employment, lengthening the time spent in school is also an uncomfortable concept.

Consider the change in employment trends in recent years. The current generation of college students was born, for the most part, in the mid-1980s. During the relatively short span of their lifetime, this group has witnessed firsthand several major shifts within corporate America. Prior to the years when many of these students were born, a major recession was experienced in some of the largest manufacturing sectors. Especially hard hit was the automobile industry and the surrounding Midwest region of the country. Dubbed the "Rust-Belt," this area saw a significant reduction in the work force, especially in the traditional "blue-collar" jobs. Jobs that had traditionally been viewed as secure for a worker's lifetime were now eroding and in many cases disappearing. During this period, the employment of professionals remained fairly stable, and new engineering graduates were having only minor problems gaining employment at graduation time.

The period 1983–1986 saw a revival in the U.S. economy and this was reflected by an increase in hiring of new engineers, with salaries well above the rate of inflation.

The period 1988–1994 was one of major restructuring within corporate America. Many major firms instituted a series of layoffs and experienced periods of downsizing. This period was notable for how the "white-collar" work force was affected. Professionals either were being laid off or provided with incentives to retire early. The net effect was a significant restructuring, reflected by one of the weakest labor markets ever for engineers.

From 1994 to early 2001, the economy rebounded vigorously. Increases in productivity, the general vitality of the global marketplace, mergers and acquisitions, and a lower birth rate that produced one of the lowest supplies of graduating engineers in decades all contributed to one of the strongest job markets in many years. That group of graduating engineers was receiving multiple job offers—some with extra incentive packages including tuition reimbursement, extra benefits, and signing bonuses.

Since mid-2001, the job market for graduating engineers has declined. According to surveys of employers conducted by the Collegiate Employment Research Institute at Michigan State University, overall hiring contracted nearly 50% in 2002–2003. However, the survey of 450 employers for 2003–2004 shows an improving economy and a more stabilized labor market. As this trend continues, opportunities for engineers should also increase.

Truly, the current generation of college students has witnessed a roller-coaster type of economy. Many realize that the rules of the game have changed and that lifelong job security with a single employer is a thing of the past.

However, one fact remains clear from employers in good times and bad: the need for college graduates with experience is a constant. Many employers expect today's college graduates to have significant work experience. Why? Because engineering graduates with career-related work experience require less training and can produce results more quickly than the typical recent college graduate. To an employer, this is sound economic policy.

There are many forms of experiential learning available to engineering students. This chapter will discuss several of the most popular options, including on- and off-campus jobs, summer work, volunteer experiences, academic internships, research assistantships, and perhaps the most popular and beneficial program for students, cooperative education. Each of these has its own advantages and disadvantages, depending on a student's particular situation and background. These issues will be examined as part of this chapter.

21.2 SUMMER JOBS AND ON- AND OFF-CAMPUS WORK EXPERIENCES

Most engineering students arrive at college with some type of work experience. Some have mowed lawns, done baby-sitting, or worked in fast-food franchises, while others may have had the opportunity to actually work in a business or industry that was involved in some type of engineering activity. Whatever the experience, it can be helpful in preparing you for your future. Career-related work experience is clearly preferred by engineering employers, but one has to start someplace in developing a work history.

There is value in working in the residence hall cafeteria, or delivering pizzas, or helping out in one of the campus offices. These experiences begin to build a foundation which your classroom learning and other career-related experiences will supplement. On- and off-campus jobs and summer work experiences may not seem highly relevant, but they can help develop important skills. Through such experiences, many students have strengthened their communication skills, learned to work as part of a team, and developed problem-solving abilities and an organizational style that can be applied to engineering courses and future employment positions. Jobs that you hold as a student will influence your perceptions about the things you like to do, the conditions and environments in which you like to do them, and the types of people and things with which you like to work. These jobs can be useful in helping you identify your personal strengths and weaknesses, interests, personal priorities, and the degree of challenge you are willing to accept in a future position.

"Real world" experience can come in many forms. While career-related experience is important, do not underestimate the importance of learning some basic "on the job" skills. If you are seeking to develop some basic job skills or merely trying to earn some extra money to assist with your college expenses, you should not be afraid to explore some of the many employment options that are currently available on and off campus during the school year and in the summer.

21.3 VOLUNTEER OR COMMUNITY SERVICE EXPERIENCES

Another popular form of experiential learning is found in the many opportunities associated with volunteer or community service experiences. While these programs are available to engineering students of all class levels, they can be particularly beneficial to freshman or sophomore students who are trying to gain some practical experience. Generally, volunteer or community service projects can be short- or long-term arrangements. Typically, these positions are with nonprofit organizations such as human service groups, educational institutions at all levels, a variety of local, state, and federal government agencies, or small businesses. These groups are particularly interested in engineering students for their strong backgrounds in math, science, and computer skills, which can be applied to a variety of projects and activities.

Even though these positions usually are not paid, students can gain significant experience and develop important skills. Students have found that working with different groups can enhance their communication abilities, as they are often interacting with people much different from those in their collegiate environment. Through the various projects and activities, students can develop organizational skills, initiative, and the desire for independent learning that can be useful in many settings including the classroom and other employment situations. Students involved in volunteer and community service experiences find these opportunities helpful with career decisions, while building confidence with the satisfaction of helping others and providing needed services.

For those students interested in developing a basic set of skills which can be useful throughout their professional engineering career, volunteer and community service work is worth exploring. For more information, contact your campus placement center or a particular firm or agency in your area. A faculty member or academic advisor may be able to provide assistance in this very valuable endeavor.

21.4 SUPERVISED INDEPENDENT STUDY OR RESEARCH ASSISTANTSHIP

The *Supervised Independent Study* or *Research Assistantship* is a form of experiential learning which is designed primarily for the advanced undergraduate engineering student who is considering graduate school and/or a career in research and development. Generally, a supervised independent study or research assistantship is a planned program of study or research that involves careful advanced planning between the instructor and student, with the goals, the scope of the project, and evaluation method specified in writing. Some of these opportunities are paid, while others involve the award of credit, and possibly both.

For students considering graduate school or a career in research and development, these opportunities can provide experience working in an environment which can be quite different from other engineering functions. Graduate school and research work involve an application of technical skills at a very theoretical level. Students should enjoy the challenges of advanced level problem solving which will require a very in-depth use of their engineering, math, and science background.

To learn more about supervised independent study programs or research assistantships, talk with some of your professors whom you enjoy working with and with whom you share similar engineering interests.

21.5 INTERNSHIPS

What is an internship? Of all forms of experiential learning, some think that this is the most difficult to describe, as it can have different meanings, uses, and applications depending on a variety of circumstances. Generally, *internships* can be either paid or unpaid work experience that is arranged for a set period of time. This usually occurs during the summer so there is the least disruption to a student's academic schedule. For most students and employers, an internship is a one-time arrangement with no obligations by either party for future employment. Objectives and practices will vary greatly depending on the employer and the student's class level in school. Some internships involve observing practicing engineers and professionals to see what working in the field is "really like." The intern's actual work often involves menial tasks designed to support the needs of the office. However, other internships can be structured as a "capstone experience" which permits students to apply the principles and theory taught on campus to some highly technical and challenging real-world engineering problems.

At most engineering schools, internships are treated as an informal arrangement between the student and the employer. Therefore, this activity is usually under the general supervision of an experienced professional in the field in a job situation which places a high degree of responsibility on the student for success or achievement of desired outcomes. While many campus placement centers provide facilities and assist with arrangements for internship interviews, there generally is not a formal evaluation of the internship job description by any engineering faculty member. As a result, it is often difficult for students to understand how, and whether, this internship will be related to their engineering field of study. Since the typical engineering internship does not involve faculty input, there is usually not any formal evaluation or monitoring of the experience by any college officials. For these reasons, college credit is usually not awarded for these experiences.

For many students, however, an internship can be a very valuable part of their college education. Under the best of circumstances, the internship will be structured by the employer so it will be directly related to the student's field of study. It is advantageous for the student to try to obtain, in advance, a description of the duties and responsibilities of the position. The student may wish to review this material with an academic advisor or faculty member to gain an opinion on the relevance to the student's selected field. These individuals may be able to make suggestions and comments on ways that the student can maximize the benefits of the experience. The most productive internships are obtained by upper-level undergraduates who have completed a significant amount of engineering course work. It is often easier for employers to match these students and their qualifications to specific projects and activities which will take advantage of the student's academic experiences. Students gain some practical real-world experience which will provide them with a better perspective of the engineering field that has been selected. For the employer, internships provide a mechanism to supplement their workforce in order to complete many short-term engineering projects. It also enables the employer to evaluate students in a variety of work-related situations. Many employers will use internship programs to screen candidates for possible full-time employment after graduation.

For students with restrictions related to time constraints, curriculum flexibility, scheduling difficulties, location or geographical preferences, the one-time internship program can serve as a valuable supplement to their engineering education. While internships do not provide the depth of experience found in other forms of experiential learning, they can help students gain an appreciation for the engineering profession and perhaps provide additional opportunities for full-time employment after graduation.

21.6 COOPERATIVE EDUCATION

At most engineering schools, cooperative education programs are viewed and promoted as the preferred form of experiential learning.

> *Engineering cooperative education (co-op) programs integrate theory and practice by combining academic study with work experiences related to the student's academic program. Employing organizations are invited to participate as partners in the learning process and should provide experiences that are an extension of and complement to classroom learning.*

This background statement, endorsed by the Cooperative and Experiential Education Division of the American Society of Engineering Education, describes the basic foundation of *cooperative education* and some of the critical components of a quality engineering co-op program. Perhaps the most important part of any cooperative education program is its linkage to a college's engineering programs. Co-op is considered to be an academic program, and in fact at many colleges of engineering it is administered centrally, by the college itself.

As an academic program, co-op is structured so student assignments are directly related to their major field of study. Typically, participating employers, who have been previously approved by the program administrators will forward detailed job descriptions, including salary information of available positions, to co-op faculty and staff. These descriptions are reviewed to evaluate their correlation to the academic content and objectives of their educational program. Employers also provide a set of qualifications for the candidates they are seeking to provide appropriate matching of students with the positions they have available.

The integration of school and work is another important component of cooperative education programs. This is typically achieved by alternating several periods of increasingly complex co-op assignments with periods of study at school, almost always with the same employer. This program is called the "*alternating-term co-op*" and is the most common form of co-op scheduling. For example, a student might be at the XYZ Company during the fall semester, back at school for the spring semester, then return to XYZ for the summer semester, back to school for the fall, etc. This type of schedule allows students to apply classroom learning to real-world problems in an engineering setting. Upon returning to school, students are then able to relate the learning in the classroom to the experience gained on the job. After another semester of schoolwork, students return to the employers with an even greater knowledge of engineering and scientific principles, which can be used for even more complex assignments in the workplace. As students mature academically and professionally they become stronger students and valuable employees. It is truly a winning combination for the employers, the students, and the schools.

Another form of cooperative education scheduling which is popular at some colleges and universities is called "parallel co-op." Under this arrangement, students usually enroll for a partial load of class work while engaging in co-op assignments for 20 to 25 hours per week. This type of co-op program allows students to continue to make progress toward their degrees while gaining valuable work experience. The disadvantage of this arrangement is that students may encounter difficulties if the demands of the co-op position begin to interfere with the time needed for school assignments, or if the time devoted to schoolwork begins to hamper the performance on the co-op assignments. Great care and judgment should be exercised when students and employers consider undertaking this co-op option.

Under another arrangement, co-op assignments are scheduled consecutively, during back-to-back semesters. This option is particularly useful for co-op projects that require a longer commitment than a single semester, or where the activity is seasonal in nature. For example, many civil engineering students are employed by highway construction firms for assignments that typically last for a summer and fall semester. A student involved in this type of co-op may work two consecutive semesters, followed by a semester or two in which they attend school and then a return to the employer for one or two additional co-op assignments. This pattern can provide many of the same benefits of the alternating plan if it is scheduled in a way which ensures that the academic course work is still a major component of the overall plan.

Some co-op schedules are actually a combination of these different patterns. Due to the sequential scheduling of a typical engineering curriculum, it can be difficult to adhere to a true alternating co-op program. Therefore, some students and employers will develop a series of assignments that may begin as an alternating plan for the first two or three rotations, but which may then require a back-to-back schedule or a parallel semester to complete the full schedule plan due to a student's particular academic schedule.

Cooperative education is rarely a summer-only program. Interrupting work for two semesters every year makes it difficult to maintain continuity in the integration of academics and the employment experience. In addition, most participating employers have needs which require them to have co-op students available for these positions throughout the year. In fact, most co-op employers plan their hiring so that while one student is away at school, another student is on-site working on a given co-op assignment.

In order for both an employer and a student to gain maximum benefit from the co-op experience, most employers and schools require that a student make a commitment for three to four semesters. This guarantees stability for the employers while ensuring the student at least a full year of job-related experience. Most engineering schools typically will recognize the student's participation in co-op after completion of the equivalent of one year of work experience by placing an official designation on the student's transcript and/or diploma.

Advantages of Cooperative Education Programs

. . . Advantages for Students:

Cooperative education programs provide students with an advantage for employment or graduate school consideration. Many employers seek to fill permanent positions with students who have had significant work experience. These employers use the cooperative education program to evaluate candidates on the job. For many students this usually leads to an offer of permanent employment upon graduation. Students are under no obligation to accept full-time employment from their co-op employer, but studies have demonstrated that many will accept these offers. Students who choose to work for their co-op employer often begin with higher seniority and benefits, including larger starting salaries, increased vacation time, and other fringe benefits. Those students who elect to interview with other firms usually find that their co-op experience places them far ahead of their peers, as they are sought more often for interviews and plant trips, and receive more offers, and earn higher starting salaries than their classmates.

Students who have participated in co-op programs usually comment on how they have been able to develop their technical skills. The nature of co-op assignments allows students to use the skills gained in the classroom and apply them to actual engineering problems. In the classroom, students' assignments and laboratory experiments are graded based on how close they come to achieving the correct answer. As they quickly learn on the job, the real-life assignments and projects often are not directed toward a specific answer, but a variety of potential solutions. This is where the student's technical background is developed and applied. Much of the actual technical learning occurs when students are forced to apply various principles to these unique situations. The learning that is achieved as part of the co-op experience provides students with new knowledge that often will make them better prepared for classroom challenges. The opportunity to "learn by doing" has been an important component of human development throughout time.

Many students who participate in cooperative education programs also find that the experiences can help solidify career decisions. Many engineering students are not fully aware of the spectrum of options and opportunities available to them. By engaging in the daily responsibilities of engineers, interacting with both the technical and non-technical individuals that work with engineers, and applying their academic background to their positions, students begin to gain a better understanding of the engineering profession. As a result, many find that their selection of a particular discipline has been confirmed, reinforced, and even strengthened. Others find that they have been exposed to new fields and technologies of which they had not previously been aware. Some change majors to reflect their new interests, while others modify their course schedule to include new courses to incorporate some of the exciting new areas to which they have been exposed.

All cooperative education assignments are, by definition, paid positions. Therefore, students find that the salaries they earn as co-op participants can greatly assist them in meeting the high costs of education. Recent studies have demonstrated that average salaries paid to engineering co-op students are in the neighborhood of $2,400 per month. For students who are working between four and seven months per year, co-ops can provide significant resources to meet their education expenses. It should be noted that average salaries can vary greatly depending on such factors as the employer, geographical location (pay is often higher in high-cost-of-living areas), the students' majors, class levels, and the amount of experience they have had.

. . . Advantages for Employers:

Many students often wonder why employers become involved in cooperative education programs. Some of the most common reasons given by participating employers are:

- The cost of recruiting co-op students averages sixteen times less than recruiting college graduates.
- Almost 50% of co-op students eventually accept permanent positions with their co-op employers. The retention of college graduates after five years of employment is 30% greater for co-op graduates.
- Typically, co-op students receive lower salaries and fewer fringe benefits than permanent employees. Total wages average 40% less for co-op students. In addition, employers are not required to pay unemployment compensation taxes on wages of co-op students if they are enrolled in a qualified program.
- The percentage of minority members hired is twice as high among co-op students as among other college graduates, thus assisting co-op employers in meeting EEO objectives.
- Co-op programs provide an opportunity to evaluate employees prior to offering them full-time employment.
- The co-op graduate's work performance is often superior to that of a college graduate without co-op experience. Co-op students are typically more flexible, and easily adapt to a professional environment.
- Regular staff members are freed up from more rudimentary aspects of their jobs to focus on more complex or profitable assignments.
- Co-op programs often supply students who have fresh ideas and approaches and who bring state-of-the-art technical knowledge to their work assignments.
- Co-op graduates are capable of being promoted sooner (and farther) than other graduates.
- Co-op programs build positive relationships between businesses and schools, which in turn helps employers with their recruiting.

. . . Advantages for Schools:

There also are many benefits to educational institutions that offer cooperative education programs to their engineering students.

- Cooperative work experiences provide for an extension of classroom experience, thus integrating theory and practice.
- Cooperative education keeps faculty members better informed of current trends in business and industry.
- Co-op programs build positive relationships between schools and businesses, and provide faculty members with access to knowledgeable people working in their fields.
- Co-op programs enhance the institution's reputation and attract students interested in the co-op plan to their school.

- Cooperative education provides schools with additional business and industry training facilities that would be unaffordable otherwise.
- Cooperative education lowers placement costs for graduates.

More Benefits of Co-op

Co-op students have found that many professional development skills so critical for success in today's workplace can be enhanced through their co-op experiences.

1) **Written and oral communication skills:** The ability to communicate ideas, both in written form and orally, is a skill that is critical. Most co-op assignments require students to document their findings and report them to others. This may take the form of an e-mail to a supervisor, a technical memo, a presentation to a group of engineers and technicians, or even a formal presentation to a group of directors or other executives. While communications and presentations may not be a favorite activity for many engineering students, co-ops provide students the opportunity to improve and develop these critical skills. For those who find this communication a liability, cooperative education assignments can help turn it into an asset.

2) **Networking:** Most co-op students find the opportunity to develop new contacts and work with people from a variety of backgrounds a great asset to their short- and long-term career objectives. Co-op students come in contact with many individuals who are able to provide advice, lend support with projects, and help get things done more efficiently. These same people can recommend new assignments to co-op students, and maybe even write an important letter of recommendation.

3) **Self-discipline:** Most co-op students quickly discover that the transition from school to work requires that they develop an organizational style which may be superior to that used in college. They find that within the co-op structure they are given a great amount of responsibility and freedom to succeed (or fail) with their projects. Time management, punctuality, adequate preparation, organization, and specific protocol and procedures are learned on the job. As they acquire and develop these skills, they find that they are able to apply them back at school, which in turn makes them more productive and successful students.

4) **Interactions with a variety of people and groups:** Co-op students deal with a wide range of individuals as they complete their tasks. Typical individuals may include engineers, technicians, scientists, production workers, labor union representatives, clerical staff, managers, directors, and perhaps even CEOs. Each of these will contribute to the students' learning as they experience the successes, challenges, frustrations, and friendships that will evolve from these interactions. These individuals can help the co-op student evolve from a student to a successful engineer.

5) **Supervisory and management experience:** Many cooperative education assignments require that the student take responsibility for other individuals or groups in order to accomplish their tasks. These opportunities to manage the work of others can provide invaluable experience to students, especially those who aspire to eventual management or leadership positions.

A Note of Caution

As discussed above, cooperative education programs can provide many positive benefits to students, employers, and schools. However, a co-op program may not be the appropriate form of experiential education for every engineering student. There are many aspects to the co-op structure that may not be consistent with a particular student's overall goals and objectives.

The biggest drawback to co-op participation is the intermingling of work and study. During each of the semesters that students are off campus working on their co-op assignments, they are not making progress toward their degree requirements. This means that typical co-op students add an extra year to the duration of their academic programs. For many, this means that co-op participation will extend their engineering degree program to five or even six years. The overwhelming majority of co-op students, schools, and employers feel that this sacrifice is minor compared to the outstanding benefits received from this type of education. However, students and parents should carefully weigh this factor.

There are other factors which also should be considered. The continual relocation between school and the work is a minor inconvenience, but one which takes time and effort—especially if the two sites are a substantial distance apart.

The engineering curricula also can present a difficulty due to the nature of the sequential offering of certain classes and their prerequisites. Students should work out a long-term schedule with their academic advisor and co-op coordinator so that any class scheduling problems can be handled.

To find out more about engineering cooperative education programs at your school, contact your co-op office or placement center, or talk with your academic advisor or a faculty member. If your school doesn't have a co-op program, you may want to write or talk with a corporation of interest to you. You may be able to work out a co-op assignment with the help of an advisor, or faculty member, or the campus career center.

21.7 WHICH IS BEST FOR YOU?

Many forms of experiential learning have been discussed in this chapter. Each of these opportunities has advantages and disadvantages associated with it. Only you can determine which program is best suited to your particular situation. Begin by asking yourself these types of questions:

- Is obtaining career-related work experience a high priority in my educational planning?
- Am I willing to sacrifice personal convenience to gain the best possible experience?
- Will I be flexible in considering all available work opportunities, or do I have special personal circumstances (class schedules, time constraints, geographic or location restrictions) which limit my choices?
- Do I have the drive and commitment necessary to succeed in a co-op program?

Your answers to such questions will be important as you move forward in your engineering career. Work with your academic advisor, a professor, or your campus placement office to get as much information as possible to make an informed choice in determining which is right for you. The resources are available to you—it is up to you to take advantage of them.

EXERCISES AND ACTIVITIES

21.1   Develop a list of your skills and abilities. Determine which could be improved with work experience and which will be developed in the classroom.

21.2   Interview one of your professors who is involved with a research project. Discuss the advantages and disadvantages of a career in research.

21.3   Talk with other students who have done volunteer and community service projects. List the benefits they have gained which have helped them in their engineering classes.

21.4   Visit your engineering co-op office or campus career planning or placement center and read the recruiting brochures of some of the employers in which you have interest. Make a list of the types of qualifications these firms are seeking for full-time and co-op or intern positions.

21.5   Attend a co-op orientation session. List some of the advantages and disadvantages as they relate to your personal priorities.

21.6   Attend your campus career/job fair. Talk with at least three recruiters from different firms. Write a report comparing their views of cooperative education experiences, internships, and other forms of experiential learning.

21.7   Visit with some upper-class engineering students who have been co-ops or interns. Write an essay that compares and contrasts their experiences. Which students do you feel gained the most from their experience, and why?

21.8   Visit the campus placement center. Gather information concerning the starting salaries and benefits of those students who have participated in cooperative education compared to those who have done internships or some other form of experiential learning.

21.9   Visit the engineering co-op office and read some of the job descriptions of current openings. Select one of these descriptions and write an essay discussing the benefits you feel would be gained from this position.

21.10   Meet with your academic advisor and develop a long-range course schedule. Modify this plan so it shows how a co-op and an internship would affect this schedule. Write a short paper discussing the advantages and disadvantages of extending your education to include extra time for work experience.

21.11   Develop a list of your personal career priorities. How can this list be enhanced by some form of experiential learning?

21.12   Develop a list of your personal and educational abilities that you think need to be strengthened. Write an essay that discusses how work experience could improve these skills for you.

21.13   Interview an engineering professor or graduate student who had been a co-op student as an undergraduate. Discuss how the co-op experience influenced them to pursue an advanced degree in engineering.

21.14   Visit <http://epics.ecn.purdue.edu> and review the volunteer student team projects that have been developed at this university. Prepare a short paper that discusses the benefits gained by these students and how these experiences can be an advantage in academic studies.

21.15   Attend a campus career fair. Talk with at least three recruiters about their opinions concerning career-related work experience. Develop a report that discusses the reasons that employers prefer to hire students for permanent positions who have had significant work experience as part of their undergraduate education.

21.16   Talk with an advanced engineering student who changed their engineering field of study as a result of a co-op or intern work experience. What factors in their experience caused them to change their major? Do they view this work experience as a disadvantage or as an advantage to their overall career planning?

Chapter 22

Connections: Liberal Arts and Engineering

22.1 WHAT ARE "CONNECTIONS"?

The connections that we will discuss in this chapter refer to the connections that exist between engineering and all those areas that are commonly referred to as liberal arts. Yes, liberal arts! You remember those courses you took in high school and may have referred to as the "soft courses"—courses that were not based on what engineers typically think of as science. Can you name a few? How about:

Literature	History
Music	Art
Social Studies	Philosophy

Did you, along with many of your science-oriented friends, feel that such courses would be of no significance to you as an engineer? Can you hear yourself saying, "I will never read another book, write another paper, or ever think about why some characters on stage are doing what they're doing"? Have you heard those sentiments echoed by fellow engineers? Well, if so, the purpose of this chapter is to get you to re-think those opinions.

It is important that you look closely at what engineers really are and what they really do. The root skills needed for art are similar to those needed for science. In the end, they're both about the quest for truth, as well as for what works. Art is amazingly rational and rigorous. It involves thinking about all aspects of physical objects, just like engineering, but it extends thinking past the usual limits of engineering. In creating their work, artists often have to think like engineers. But the flip side is also true: to do their best work engineers have to think like artists. Working to extend your mental horizons makes you a better engineer—and a better person. And it lets you see clearly that our human faculties do not stand alone, nor is one more important than another. Thought, emotion, will-power—all these need full development if we're to be the best we can be, on the job or anywhere else.

For a more specific definition of liberal arts, the dictionary says "liberal" comes from *liberty*, so that liberal arts means "works befitting a free man." Its goal is "a quest for truth, goodness, and beauty." Those are qualities that are difficult to put into equations, but all good equations do reflect them! Such a quest has everything, in the end, to do with engineering, but it's probably necessary to investigate it in more detail to show why this is so.

The Need for a General Education

As for "liberal education," it is half of the focus of many university degrees. It signifies a general "big picture" education as distinct from technical training. The liberal aspect of even technical degrees was developed because people have a need for a strong, open mind in addition to a specialty in order to be well-rounded. So they can have true liberty, and not be trapped by cultural blind spots. Technical training needs to be grounded in a general liberal education if it's going to have a healthy perspective.

It is often comforting to engineering students when they realize that a liberal education is not a vague project. There is a set of skills to learn and hone, and a modest amount of knowledge to acquire. There are definite ways to develop those skills. You learn them like you learn science skills. In fact, your talent in science will serve you well throughout your liberal education! There are also straightforward road maps to guide you so you get the most from the arts. You need to be taught how to use them like anything else worth learning. When we say the arts are natural, we mean they're what all humans can acquire, or draw out of themselves, through education. (In fact, the word *education* comes from the Latin "to draw out.") Let's look further into why this is so.

22.2 WHY STUDY LIBERAL ARTS?

As an engineer, you will be considered an educated person. Your knowledge of science and mathematics will lead you to be held in esteem simply because of the fact that those two areas require a rigorous amount of study. As an engineer, you won't be sitting at a desk putting square pegs into square holes and round pegs into round holes. You will be investigating the world around you and creatively imagining how you will attack the problems with which you are confronted. You will not be a machine. You will be a living, breathing problem-solver. Problem-solving requires a creative and active mind. What do the liberal arts have to do with this?

> *Liberal arts help you improve your intellectual competence and expand your creative powers.*

The Liberal Arts Help Improve . . . Your Broadness

Liberal arts give you practice at looking in many directions at once. They force you to ask questions about areas that do not have pre-set answers, and to evaluate why you have chosen the answers you have. This vibrant activity will keep your mind from becoming stagnant. It will allow you to grow and become more qualified to look at the world around you—a world of people who may not always act in obviously predictable ways.

The liberal arts also give you skills that are required in the real world, beyond school. As engineers in the 21st century you will be expected to be leaders. You will not be confined to laboratories. You must communicate with a wide variety of individuals from corporate executives to the custodian in your plant, from the stockholder to the individual bringing a liability suit against your company. All of these people will expect you to listen to their concerns, evaluate the problems they raise, and speak to each of them in a way that tells them that you care and understand. Can the study of science and mathematics provide you with that training? It is possible, but improbable.

> *The liberal arts help you to develop the qualities of character and personality that will truly make you a leader in your field.*

The Arts Improve . . . Your Perspective

It is important to engineers that they see the "big picture" of life so they can learn to use their skills in a wholesome way for everyday living. Otherwise, it's all too easy to let our lifestyles become fragmented, where what you do at work has little bearing on the rest of your life. The effort toward integrity, on the other hand, will always let you keep a good perspective and will give new insights into your engineering.

There's a strong tendency to become one-sided in modern life and in college. We then balance this one-sidedness with equally intense recreation of some type. This creates dangerous blind spots and stresses. Today, many students turn their university education into a vocational degree, like a shop mechanic would get at a trade school. We tend to focus on just getting that good job after we graduate. Our difficult studies make it even easier to become one-sided. A few years after we graduate, though, we often wish we'd taken the time for broader study. We finally see that money or a job isn't the goal of life, that they're just *means* to an end. We wish we'd studied the *ends* more seriously. They're a tougher nut to crack than even the toughest job. We easily overlook the fact that college is the one time in life—the best time in life—to develop the judgment that adult life requires. Thankfully, this is why engineers have graduation requirements that include the liberal arts! Even though they might seem annoying and beside the point at first, if we are to grow past immaturity we need to be fully-integrated humans, with no part overlooked. High school didn't do it, work life doesn't allow the time—college is just right!

Here are some examples of how it works. We need Philosophy to teach us a wide range of reasoning, and to appreciate the seriousness of our duty to use it right. Psychology, Social Science, and Anthropology help us to see that there are two sides to everything, that we don't always see our own motivations clearly, and that imbalances create disease in society and ourselves. Engineers in particular need to learn that technology is a double-edged sword, the use of which requires great judgment. If we study History, we learn a perspective that keeps us humble despite the powerful techniques that we master in science. History provides us with many role models, and shows how risky life can be. In general, the integrating, freeing benefits of liberal studies help us cover all the bases for a disciplined life, where we have the strength, skill and insight to "do the right thing," while keeping to the Golden Mean—avoiding excess in anything.

These skills give us the ability to do and to be our best in ways that transcend fads, trends, or temporary excitements. They give us integrity for a lifetime.

Perhaps this is a good time to make your own practical investigation into liberal arts and education, to help you see what we mean.

Activities 22.1

a) Look carefully at your world. What part does engineering play in the lives of the people around you? Collect six examples that show the ways that engineering is an integral part of the everyday lives of people.

b) Now think about those items from the standpoint of the engineers who created them. What did those engineers have to think about as they were working on perfecting the item? Did they only consider the science and mathematics of the object, or was there more?

c) Look at the latest models of automobiles. Are these automobiles designed only for gas efficiency and aerodynamics? Do we really need "rich Corinthian leather"? Heated seats? Thermometers in the rear view mirror? Why is there so much chrome? Why don't we just build an automobile that gets 80 miles to the gallon and gets us from point A to point B efficiently? Why is it necessary to try to produce beautiful automobiles? Why are they appealing?

The Arts Improve . . . Your Balance

The cultivation of an appreciation of art and its importance to the engineering field will allow you to see how artists work within their media and how the public is affected by their efforts. Because artists reflect their culture's view of the world, an artistic education will help you to see more clearly how the general public views the world. The insights that you gain will tell you that people do not always look at life with a scientific eye; many times they look with their hearts. They look with their whole person. And they need to keep this perspective, even when they work at specialized jobs, including engineering. This outlook perhaps can be thought of as a balance of views—energy, thought and emotion, all working together, in their proper ways. There are many ways to express it, but if one is one-sided, it is hard to create work that has integrity or lasting value. A study of art can be beneficial in explaining this process. You likely will find a medium or style which best speaks to you, but all the disciplines and styles have something to contribute. Altogether, they've brought us to where we are now. Don't worry if some forms seem obscure: art and its meanings need to be taught as much as any other skill. With a good teacher, you'll soon understand. In this way, we introduce another benefit of liberal studies:

> *Liberal arts enrich our personal lives with new knowledge, insights, and a keener appreciation of beauty.*

A well-rounded education that gives you a sense of what is happening within other areas of study cannot help but increase your ability to develop and present engineering ideas to the world.

A liberal education also provides you with practice in dealing with a variety of diverse ideas. Philosophy courses ask you to compare and contrast the varied interpretations of experience of individuals who are sometimes pitted in violent struggles against each other. Consider the difficulties in Northern Ireland or the Middle East or in the inner cities of the United States. How, as an engineer, do you explain the problems that have existed there for hundreds of years? Do you simply say, "$x + y = 3/z$"? Can mathematical thinking explain pain and suffering? Since math strives to realize harmony, in a certain way the answer is Yes. But you need to develop the skill to rigorously extend reasoning in new directions. Philosophy is not a "soft" science, after all. You need practice in deciphering and evaluating the misunderstandings of experience, laziness of reason, and fear of truth which lie at the source of so many troubles of life. . . . And which can easily derail engineering projects.

The Arts Improve . . . Your People Skills

The "people factors" are always involved in the answers to these questions, and engineers need to know about people and how they act and react. Liberal studies of these factors will help us grow as humans, and as such can't help but have ancillary benefits for our careers. For instance, we are living in a global society where the introduction of a car called the "Nova" in a Spanish-speaking country raises the eyebrows of customers. The automobile may be the most fantastic vehicle ever built, but since in Spanish "nova" means "it does not go," those customers may never spend their money on it. Learning to see all sides of a situation, including factors which are not at all obvious, will help you avoid such trouble!

The liberal education that is received in college will provide you with the tools to understand these "people" issues, issues that may not have been considered noteworthy 50 years ago. Additionally, it will help you be aware of things that modern tendencies avoid or neglect. It may be that some colleges don't have many liberal requirements any more, so you might find that you have to go against the grain to get the education you need. Be sure to let your advisor know your goals! He or she can help you avoid wasting your time or spinning your wheels. Not all liberal arts courses are equal. Some may not serve your needs. Some may be as technically specialized as any engineering course, which is not what you want. This is why it's best that a school plans a basic liberal curriculum to help fulfill your electives. Fifty years ago college was a place where even if you were getting a technical degree much of your effort went to developing yourself and your moral sense, as well as your awareness of the world. Today your liberal education could be neglected as you pursue a degree. To quote a commercial, "This isn't your father's Oldsmobile!" Times have changed. You, the engineer, will have to make sure you are a well-rounded, integrated individual. You'll be working in a global marketplace and you'll need to be ready to understand all sorts of cultural factors that may positively or negatively affect you and your company. If you have had training in the liberal arts, you can meet those circumstances head-on. If not, your short-sightedness will hurt you.

The Arts Improve . . . Your Sense of Duty and Responsibility

People the world over look to engineers to provide them with clean water, safe homes and transportation systems, warm places to sleep and work environments that will not injure or kill them, with neighborhoods that are free of toxic fumes, and lakes uncongested with algae. —And with technology experts who will not mislead them into putting too much focus on material goods in the first place. As you work to achieve these goals, the engineering profession will improve; but it will only improve with the human element as its core concern. Thus:

> *The liberal arts elevate, integrate, and unify the standards of the profession, focus them on the human, and create greater awareness and esteem for the engineer.*

A "people orientation" provides you with important tools to use for society:

> *You are better able to fulfill your duty in life, so society respects you more. Thus you are able to convince the public of issues that need to be resolved through effective engineering.*

> *By understanding people and their overall needs, you can apply your engineering expertise properly to any problem.*

> *By demonstrating the submission of science to the basic needs of mankind, you secure the dignity of the engineering profession.*

You are on the threshold of a great, yet humble, undertaking. You will be able to investigate the laws of science, examine data that will prove your assumptions, learn the ways of engineering, *and* contribute to making a better world. As a complete engineer and as a whole person, you will not only understand scientific parameters, but you will understand the people and the world in which you will provide your services.

EXERCISES AND ACTIVITIES

22.1   Assume that you have been requested by the department chairman within your major to explain to incoming freshmen why non-engineering courses are important to engineers. Discuss the role non-engineering courses can play in one's engineering career.

22.2   Evaluate a textbook you are using in a non-engineering course. Compare it to an engineering textbook that you are using. Do you see similarities and differences? Discuss them in a brief report.

Appendix A:

The Basics of PowerPoint

A.1 INTRODUCTION

It is assumed that the user is familiar with general Windows methods of menu use, selecting of files, text entry and editing, drag-and-drop, etc. The Copy, Paste, and Cut features of the Edit Menu of Windows work within PowerPoint.

The information presented here is very basic but should be enough to prepare a simple PowerPoint presentation. This is in keeping with the "20-80 Rule for Software"; namely, that if you learn to do 20 key procedures within a software package, you will be able to do 80% of anything you will ever want to do with the software. Here you learn about:

> . . . choosing a first slide,
> . . . adding a background to that slide or all slides of the presentation,
> . . . adding and arranging pictures or clipart on a slide,
> . . . adding and editing basic text on a slide,
> . . . getting another blank slide to work on,
> . . . arranging slides that have been made, and
> . . . displaying slides as a show.

The basics are all here but many bells and whistles that are available are not discussed. This includes adding of notes, charts, tables, geometric shapes, movies, and sounds to slides. Once the basic procedure is learned, users are urged to experiment with the many Menu Bar items and Toolbar buttons (particularly those in the Draw toolbar) that are available to see what can be discovered.

For additional information about using PowerPoint, the following sources might be useful:

1. Cliffsnotes at www.cliffsnotes.com offers publications about several versions of PowerPoint.
2. Enter "PowerPoint tutorials" or "PowerPoint books" into a search engine such as Google. This will yield a huge number of internet sites as potential resources. Some are selling related items but many have free online information.

A.2 THE BASICS OF POWERPOINT

1. Open the program and choose the "**Blank Presentation**" option.
   *Note: For convenience, activate the following toolbars: **Standard, Formatting, Drawing, Picture, WordArt.** To do so, right click anywhere on the Toolbar area and click on these items if a check mark does not already appear beside them. To identify what an icon button represents, move the mouse to it and wait. Only a few of the buttons that appear are referenced in the following. Experiment to see what others will do.*

2. Select a slide type and display your first slide. [It is advisable to begin saving your work at this point. Use **File-Save As** and give the new file (which will have a .ppt extension) a name and a location for storing it. Then, save periodically as you create new slides.]

3. If you would like a background for your slides, choose from the Menu bar the sequence **Format-Background.** A dialog box will appear from which you can choose a solid color and various fill patterns. The background will always remain behind any text or pictures you might add. The background can be uniform for all slides or individualized.

4. If a text block is included in the slide type chosen, click on it and enter the text you want. Click outside of the block to remove the block's border. To reactivate for editing, click on the text.

   . . . Edit as you would any word processor text.
   . . . You can drag and drop the text block to reposition it.
   . . . You can use the Free Rotate button on the WordArt toolbar to rotate the text block.
   . . . You can change the shape of the block by dragging any of its 8 handles (the small squares).
   . . . You can change the text characteristics (font, size, bold, center, etc.) using toolbar options. You must highlight the text to which the changes are to apply.
   . . . Right click on a text block for a menu of manipulations pertaining to it.

5. If additional text is desired, click on the "**Text Box**" button, or from the Menu Bar choose **Insert-Text Box**. Drag-and-drop the mouse to form a text block, which can be used as above.

6. Text in WordArt format can be added. Click on the "**Insert WordArt**" button on the Toolbar, select the desired format, enter the text, resize it, and reshape it if you wish. Explore the Toolbar buttons for manipulations that can be applied to WordArt text as well as regular text.

7. To import a picture file, click on the "**Insert Picture From File**" button on the toolbar or choose from the Menu bar the sequence **Insert-Picture-From File**. The small squares are the "handles."

   . . . To resize the picture, move the mouse to any of the 8 handles and drag to the desired size.
   . . . To move the picture, place the mouse inside the picture; when a 4-headed arrow set appears, drag and drop the picture to the desired location.
   . . . To resize or reposition a picture if the handles are not showing, first click on the picture to reveal the handles and then proceed as above.
   . . . Buttons on the Picture Toolbar allow you to crop the picture (choose the **Crop** button and move any handle) and to change the contrast and brightness of it.
   . . . Right click on a picture for a menu of manipulations pertaining to it.

8. Clip art can be added by clicking on the "**Insert Clip Art**" toolbar icon, or using the Menu bar sequence **Insert-Picture-ClipArt**. Procedures for resizing and repositioning are the same as for pictures.

9. When a new item (text, picture, clip art, etc.) is added, it will be on top of anything already there. To change the visibility of an item, select the item to be moved by clicking on any visible part of it. Then, right click on the selected item, click on **Order**, and choose the destination of the selected item.

10. To delete any item at any time, display its handles (click on it) and press **Delete** or **Backspace.**

11. To begin working on a new slide, click on the "**New Slide**" button of the Toolbar or use **Insert-New Slide** from the Menu bar. Select a slide format and proceed as with the first slide.

12. After you have created several slides, you can view them all in thumbnail form by clicking on the small button "**Slide Sorter View**," which is one of the buttons in the lower left corner just above any toolbars that might be there.

   . . . You can change the arrangement of the slides by dragging and dropping any slide into any other position. A vertical line indicates the location of the slide as you drag it.

   . . . To bring a slide back to the full screen for editing or additional work, double click on it.

   . . . To delete a slide, select the slide and press **Delete**.

   . . . To copy a slide, select it and use **Edit-Duplicate**.

13. Be sure you have saved your work.

14. To display your slide show:

   . . . If there are slides in the slide sorter view you do not want to include in the slide show, click on the slide to select it. Then, use the Menu bar sequence **Slide Show-Hide Slide** to mark it as hidden.

   . . . Use **Slide Show-Set Up Show** to choose options on how the show will be presented.

   . . . Use **Slide Show-Slide Transition** to choose a visual and/or audio transition pattern from slide to slide, if desired. A drop-down menu offers numerous transitions that can be applied individually or universally.

   . . . If your slide contains text, pictures, or clip art, you can assign an animation to each specifying the manner in which they appear on the screen. Select the item to be animated by clicking on it, then use **Slide Show-Preset Animation,** and then the available type of animation to be used.

   . . . If you assign multiple animations to a slide, you can change the order of the animations by using **Slide Show-Custom Animation**. Highlight a line in the "**Animation Order**" box and use arrows at the right to change the order of the animations.

   . . . To view the show, use **Slide Show-View Show**. Click the mouse to progress from slide to slide if automatic advancement is not in effect.

   . . . If you press **Ctrl** and the letter **P** at the same time, you will have a "pen" that can be used to draw on the current slide by clicking and moving the mouse. (Marks are not permanent; the slide itself is unaffected.)

*Note: PowerPoint makes an excellent layout for printing of a single picture or a simple collage of pictures. Prepare one slide with the format and content that you want to print, and then print that slide using **File-Print**.*

Appendix B:

An Introduction to MATLAB

B.1 INTRODUCTION

The purpose of this chapter is to provide a brief introduction to the MATLAB® computing environment. (MATLAB is a registered trademark of The MathWorks, Inc.) There are numerous texts written about MATLAB and its use. Our intent in this single chapter is not to provide a thorough understanding of the MATLAB environment, as there are entire texts that do this well. (Our favorite is "Introduction to Technical Problem Solving with MATLAB" by Sticklen & Eskil). Rather it is to begin exploring the possibilities of this powerful computer tool that you will be using throughout your engineering studies and career. As a way of introducing MATLAB, we will use examples of calculations that may be required in your freshman-engineering courses.

The MATLAB program was originally written as a matrix manipulator, and named MATRIX LABORATORY. It has proven to be a very powerful tool in performing engineering computations and has grown beyond its original intent. Some of the first applications of MATLAB were in the areas of image and signal processing which are part of electrical and computer engineering. It has since become widely accepted in a broad range of other engineering areas and programs. Many engineering curricula have moved to making MATLAB the primary computing tool in its undergraduate program, replacing traditional computer programming. (For MATLAB product information, please contact: The MathWorks, Inc., 3 Apple Hill Dr., Natick, MA 01760-2098, ph: 508-647-7000, fax: 508-647-7101, email: info@mathworks.com, web: www.mathworks.com.)

One reason for its popularity is that it is relatively easy to use and has the power of a computer language. In a traditional computer language, such as Fortran, C, or C++, a program is created. It then must be compiled, which means that it is translated into a language that the computer can understand. This compiled version of the program is then run by the computer. MATLAB provides the flexibility of programming in the traditional sense of a computer language but also has the functionality of a calculator. MATLAB removes the compiling step for the user.

MATLAB can be run on many different platforms including UNIX, PC, and Macintosh. The programs, data, and results are the same across platforms making it easy to use and easy to work collaboratively on larger projects. The examples for this chapter have been created on a PC using the student version of MATLAB that is widely available from campus bookstores. This student edition contains many of the capabilities of the full professional version but with the flexibility of being able to install the version on your own computer. MATLAB has numerous functions that can be added through extra software packages called toolboxes. These toolboxes contain built-in functions that provide special capabilities in areas such as financial analysis, image and signal processing, fuzzy logic, control systems design, and symbolic calculations.

B.2 MATLAB ENVIRONMENT

Starting MATLAB is very simple. On a UNIX platform, simply type MATLAB at the UNIX prompt. On a PC or Mac, double click on the MATLAB icon. A window will appear. This window consists of three components, your Command window, the Command History window and the Launch Pad window. The Command window is what you will use to run your programs and to see the results as shown in figure B.1. A prompt should appear that looks like >> as in the figure. This indicates that MATLAB is ready for a command.

The Launch Pad window allows you to start applications and demonstrations by clicking the icons in the window. This window can also show the variables that are active in the current session. To view these, click on the tab that is labeled "workspace." The Command History window shows a history of the commands that have been entered into the command window.

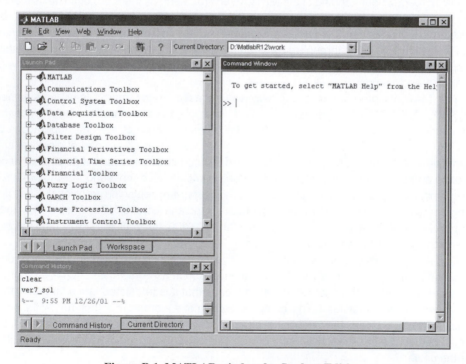

Figure B.1 MATLAB window for Student Edition

This window is one of three types of MATLAB windows. There are also graphics windows that will be discussed later in this chapter in Section B.9 on plotting, and editor windows which will be discussed with writing programs in Section B.10. A great way to begin to explore the possibilities of MATLAB is to run the demonstration programs or demos. This is done by simply typing **demo** at the MATLAB prompt and then **enter**. Alternatively, the demo can be run by clicking on the "+" symbol next to MATLAB in the Launch Pad and then double-clicking on the Demos icon. This will result in a graphical window shown in Fig. B.2. Select from the MATLAB topics on the left column to explore the capabilities of MATLAB. It is suggested that new MATLAB users spend time exploring the demos. They also provide a great way to brush up on MATLAB commands if you have not used them for a while.

MATLAB has built-in help functions that are very useful. As with any software, understanding how to find and use the help functions will allow the users to expand their abilities to use this powerful tool. The two main help commands are **help** and **lookfor**. To use **help**, you need to specify a MATLAB command (e.g., help command name). The help function can be accessed in the workspace using help command followed by the MATLAB command name or in a separate help window using **helpwin**. The **helpwin** command will spawn a separate window in which the functions can be described. The advantage of this separate window is that the workspace window is not taken up with the help topic descriptions. This saves you from having to scroll past the help descriptions to find your earlier work.

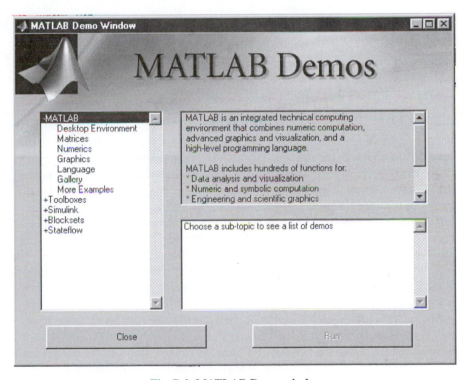

Fig. B.2 MATLAB Demo window

The **lookfor** function searches keywords with MATLAB functions and commands. This is very helpful if you don't know the command you are looking for but know the topic. MATLAB uses keywords contained in its functions, so if a word does not yield the desired result using **lookfor**, try synonyms for that word.

Example B.1

Use the ***lookfor*** command to find the MATLAB commands associated with logarithms. In the MATLAB workspace type the following:

        » lookfor logarithm

The result is that MATLAB displays the following:

        LOGSPACE Logarithmically spaced vector.
        LOG Natural logarithm.
        LOG10 Common (base 10) logarithm.
        LOG2 Base 2 logarithm and dissect floating point number.
        BETALN Logarithm of beta function.
        GAMMALN Logarithm of gamma function.
        LOGM Matrix logarithm.
    »

Notice that the LOG function in MATLAB is for natural logarithms and that base 10 logarithms are calculated by log10.

MATLAB can be used to perform simple operations. The following example shows operations that could be performed on a calculator. In Example B.2, we can see that MATLAB performs simple arithmetic functions that are on most calculators. MATLAB also has features that calculators do not have. For example, MATLAB stores the variables and the results that you have produced in memory in the workspace. Values can be assigned to variables as is done in Example B.2. At any time, the user can see the variables currently in memory by typing the command ***who***. To erase the memory, type the command ***clear*** at the MATLAB prompt. To erase a variable, type ***clear*** followed by the variable name. For example, in Example B.2, we could remove the variable **apples** by entering ***clear* apples** at the MATLAB prompt.

Variable names in MATLAB must begin with a letter and can be comprised of up to 19 characters that include letters, numbers and underscores (_). MATLAB is case sensitive so the variable **Apple** is different from *APPLE*, which is different from **apple**, which is different from **aPPLe**, which is different from **ApPIE**, and so on. It is advisable to use a single case when naming variables; having two variables with the same name but different cases can easily lead to confusion and errors.

Names that MATLAB already uses should be avoided when naming variables. If you use a name that MATLAB already uses for a function, you can disable that MATLAB function and create errors. Some of the names to avoid are **ans**, **Inf**, **NaN**, **i**, **j**, **eps**, **flops**, **nargin**, **narout**, **realmin**, and **realmax**, along with names of MATLAB functions. To determine if a name is in use by MATLAB, simply type that name at the prompt. If it is not being used by MATLAB, an error message will be returned "**??? Undefined function or variable.**" Alternatively, you can type ***help*** followed by the name you wish to use. If it is not being used as a MATLAB function, the response "**not found**" will be returned. A suggestion is to use underscores in the variable names as no MATLAB function contains an underscore. For example, **good_name**.

Example B.2

This example shows some simple arithmetic using MATLAB. (Note that the workspace tab in the Launchpad window has been selected and shows the current variables for this session.)

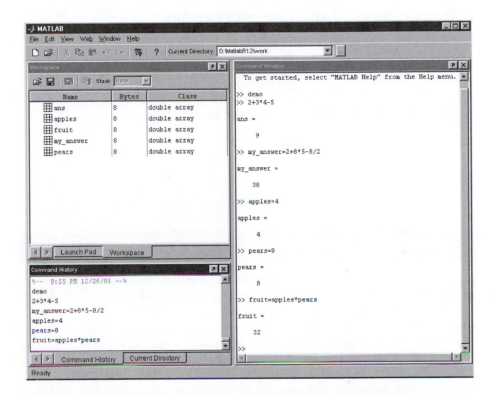

B.3 SYMBOLIC MANIPULATIONS

MATLAB can also be used as a symbolic solver. To allow MATLAB to do symbolic manipulations, variables must be declared symbols. This can be done using the **syms** command. Variables listed after the **syms** command are designated as symbols, as shown in Example B.3. Symbolic manipulations can be done using algebra, integral, or differential calculus as shown in the example.

Example B.3

MATLAB can be used to perform symbolic manipulations. First, the variables x and y are declared to be symbols:

```
» syms x y
```

Algebraic expressions can be solved using the ***solve*** command

```
» solve(x^2–4)
ans =
[ 2]
[–2]
```

Symbolic derivatives can be taken using the ***diff*** command.

```
» diff(y^3)
ans =
3*y^2
```

Symbolic integrals can also be taken using the ***int*** command.

```
» int(sin(x))

ans =
–cos(x)
```

B.4 SAVING AND LOADING FILES

It is possible to save your workspace in MATLAB for future use. This is done using the ***save*** command at the MATLAB prompt. On a PC or Mac, this can also be done under the file menu using the option for saving the workspace.

When you save a file, you need to know where the file is saved. As a default, MATLAB will put files in your present working directory. To save files to another directory, you must so specify. To find out the identity of your working directory, type ***pwd*** (print working directory) at the MATLAB prompt. This will tell you the address of your current directory. To change the working directory, you may use the ***cd*** (change directory) command (e.g., **cd c:\matlab\mystuff** would change the directory on a PC to the folder **mystuff**), o ruse the directory window above the command window.

To save a file to a different directory, include the path name as part of the file name. On a PC or Mac, a window will appear that lets you select the directory in which you save the MATLAB files. Remember to use a .mat extension when saving your workspace. If you use **save** and **return** with no file name, MATLAB will save the file under the name **matlab.mat**. When choosing a name, similar restrictions apply as with variables. Remember to avoid file names that MATLAB already uses for functions.

To retrieve your file and return to your workspace, use the command ***load*** followed by the file name. You do not have to use the **.mat** extension with the load command. For example, if the workspace is saved under the name "**my_workspace.mat**," it can be reloaded by typing ***load my_workspace*** at the MATLAB prompt.

When loading files in MATLAB, make sure that MATLAB can find the files. MATLAB will look for files in your current directory. Then, it will use the paths that it has saved. To see a listing of the paths that MATLAB is currently using, type ***path*** at the MATLAB prompt. What will follow is a list of the directories that MATLAB will search for files. If files are stored in a directory other than your current working directory, you can add this location to the path list using the command ***addpath*** followed by the location. To check to see that the path has been added, type **path** again. The new path will appear at the beginning of the new list so you may need to scroll back to the beginning.

B.5 VECTORS

MATLAB treats all numbers and variables as matrices. Recall the original name of Matrix Laboratory. Earlier, we did simple arithmetic with scalars, or single numbers. In MATLAB, we can also use vectors and matrices.

A vector is simply a row or column of numbers. To create a vector, enter a variable name along with an equals sign just as was done with a single variable. The difference for a vector is that the numbers are enclosed in square brackets. To make a column vector, the numbers are separated by semicolons. To make a row vector, the numbers are separated by spaces or comas.

Example B.4

Create a row vector and a column vector.

```
» row_vector=[1 3 6 9 12]

row_vector =
    1    3    6    9   12

» col_vector=[2;4;6;8;10]

col_vector =

     2
     4
     6
     8
    10
```

Vectors are commonly encountered in engineering applications. For vectors to be added and subtracted, the vectors must be of the same type (row or column). The MATLAB command **transpose** can change a column vector into a row vector and vice versa. The vectors also must be the same size (or length) as MATLAB will add/subtract the first numbers of each, then the second, and so on. If the vectors have different sizes, there will be numbers left over which will result in an error.

If we want to multiply or divide the numbers of one vector with the other, similarly to how we added and subtracted, there are special MATLAB symbols we must use. To multiply the elements of one vector by the elements of the other, use the command ".*". For division use "./". These symbols are referred to as element by element multiplication and division, respectively.

Vectors can be multiplied using the * operation and divided using the / operation. However, these are matrix algebra operations and follow matrix algebra rules. They are described along with matrices in the following section.

Example B.5

Create two vectors and multiply and divide the elements.

```
» clear
» vec_1=[1 2 3 4 5]

vec_1 =
   1   2   3   4   5

» vec_2=[2 4 6 8 10]

vec_2 =
   2   4   6   8  10

» X=vec_1 .* vec_2

X =
   2   8  18  32  50

» Y=vec_2 ./ vec_1

Y =
   2   2   2   2   2
```

B.6 MATRICES

Matrices, like vectors, are very common to engineering applications, and many engineering curricula require entire courses on matrices (or "linear algebra," as it might be called). For this brief introduction, we will provide some simple examples to show some of the uses of MATLAB for manipulating matrices.

A matrix is simply a group of numbers arranged in columns and rows. A vector is a special type of a matrix with a single row or column. An example of a matrix is

Every matrix has a size. The size tells the numbers of rows and columns, respectively.

$$A = \begin{bmatrix} 1 & 2 & 3 \\ 4 & 5 & 6 \\ 7 & 8 & 9 \end{bmatrix}$$

The size of **A** is 3×3 (three by three), three rows and three columns. The vectors in Example B.5 have sizes of 1×5 (one row by five columns). The *size* command in MATLAB gives the size of a matrix. For example, **size(A)** in MATLAB would return **3 3** for three rows and three columns.

Each element of a matrix is identified by the use of a numbering system called indexing. Each element has two numbers or indices associated with it. The first index is the row number and the second is the column number. For matrix A, each element and its index is shown:

$$A = \begin{bmatrix} a_{1,1} & a_{1,2} & a_{1,3} \\ a_{2,1} & a_{2,2} & a_{2,3} \\ a_{3,1} & a_{3,2} & a_{3,3} \end{bmatrix}$$

So, element $a_{1,1}$ (first row and first column of matrix **A**) = 1, $a_{3,2}$ = 8, etc. In this way, individual elements or sections of matrices can be identified and used in calculations or graphing. Often in engineering applications, data is stored as a matrix, and pieces of the data are extracted to be analyzed or plotted. MATLAB can extract an entire row, say the first row, by entering **A(1,:)** which tells MATLAB to take the first row and all columns. Similarly, we can ask for all rows and the first column of **A** by indicating **A(:,1)**. If we wanted to take elements $a_{1,1}$, $a_{1,2}$, $a_{2,1}$ and $a_{2,2}$, we could do so by entering **A(1:2,1:2)**. By separating the row and column numbers, we have required MATLAB to take rows one through two and columns one through two.

Example B.6

Create a matrix and extract some elements.

```
» A=[1 2 3;4 5 6;7 8 9]

A =

   1  2  3
   4  5  6
   7  8  9

» matrix_size = size(A)

matrix_size =
   3  3

» ele_21 = A(2,1)

ele_21 =
   4

» ele_32 = A(3,2)

ele_32 =
   8

» vec_row1= A(1,:)

vec_row1 =
   1  2  3

» vec_col2 = A(:, 2)

vec_col2 =
   2
   5
   8

» upper_square = A(1:2,1:2)

upper_square =
   1  2
   4  5
```

A matrix can be any size, and the number of rows does not have to equal the number of columns as in the previous example. Arithmetic operations with matrices follow special rules. As with vectors, we can add, subtract, multiply, and divide two matrices, element by element, using the +, −, .* and ./ symbols, respectively.

Example B.7

Element by element operations can be performed using MATLAB. Several examples follow. For addition:

$$\begin{bmatrix} a_{1,1} & a_{1,2} \\ a_{2,1} & a_{2,2} \end{bmatrix} + \begin{bmatrix} b_{1,1} & b_{1,2} \\ b_{2,1} & b_{2,2} \end{bmatrix} = \begin{bmatrix} a_{1,1} + b_{1,1} & a_{1,2} + b_{1,2} \\ a_{2,1} + b_{2,1} & a_{2,2} + b_{2,2} \end{bmatrix}$$

For subtraction:

$$\begin{bmatrix} a_{1,1} & a_{1,2} \\ a_{2,1} & a_{2,2} \end{bmatrix} - \begin{bmatrix} b_{1,1} & b_{1,2} \\ b_{2,1} & b_{2,2} \end{bmatrix} = \begin{bmatrix} a_{1,1} - b_{1,1} & a_{1,2} - b_{1,2} \\ a_{2,1} - b_{2,1} & a_{2,2} - b_{2,2} \end{bmatrix}$$

For multiplication:

$$\begin{bmatrix} a_{1,1} & a_{1,2} \\ a_{2,1} & a_{2,2} \end{bmatrix} * \begin{bmatrix} b_{1,1} & b_{1,2} \\ b_{2,1} & b_{2,2} \end{bmatrix} = \begin{bmatrix} a_{1,1} \times b_{1,1} & a_{1,2} \times b_{1,2} \\ a_{2,1} \times b_{2,1} & a_{2,2} \times b_{2,2} \end{bmatrix}$$

For division:

$$\begin{bmatrix} a_{1,1} & a_{1,2} \\ a_{2,1} & a_{2,2} \end{bmatrix} / \begin{bmatrix} b_{1,1} & b_{1,2} \\ b_{2,1} & b_{2,2} \end{bmatrix} = \begin{bmatrix} a_{1,1} \div b_{1,1} & a_{1,2} \div b_{1,2} \\ a_{2,1} \div b_{2,1} & a_{2,2} \div b_{2,2} \end{bmatrix}$$

Matrices can be multiplied using the * command (e.g., **A** * **B**), but the operation follows linear algebra rules. Under these rules, the respective elements of the first row of matrix **A** are multiplied by the respective elements of the first column of matrix **B** and added. The sum becomes the element in the upper left corner (1,1) of the product matrix. The second row of matrix **A** is multiplied by the first row of matrix **B**, and so on. To illustrate, use the example of two, 2×2 matrices called **A** and **B** where

$$A = \begin{bmatrix} a_{1,1} & a_{1,2} \\ a_{2,1} & a_{2,2} \end{bmatrix} \qquad B = \begin{bmatrix} b_{1,1} & b_{1,2} \\ b_{2,1} & b_{2,2} \end{bmatrix}$$

A * **B** becomes

$$A * B \begin{bmatrix} a_{1,1} & a_{1,2} \\ a_{2,1} & a_{2,2} \end{bmatrix} * \begin{bmatrix} b_{1,1} & b_{1,2} \\ b_{2,1} & b_{2,2} \end{bmatrix} = \begin{bmatrix} a_{1,1} \times b_{1,1} + a_{1,2} \times b_{2,1} & a_{1,1} \times b_{1,2} + a_{1,2} \times b_{2,2} \\ a_{2,1} \times b_{1,1} + a_{2,2} \times b_{2,1} & a_{2,1} \times b_{1,2} + a_{2,1} \times b_{2,2} \end{bmatrix}$$

Note that, in general, these rules do not have the same commutative property as scalar multiplication: **A*B** ≠ **B*A**. For two matrices to be multiplied together, the number of columns of the first matrix must equal the number of rows in the second. For example, for **A*B**, the number of columns in **A** must equal the number of rows in **B**. This is illustrated in Example B.8.

Example B.8

Multiply the matrices of different sizes: **A*x,** where **A** and **x** are defined as

$$A = \begin{bmatrix} 1 & 2 \\ 3 & 4 \end{bmatrix} \text{ and } x = \begin{bmatrix} 5 \\ 6 \end{bmatrix}$$

A is a 2 × 2 and **x** is a 2 × 1 matrix. So **A*x** becomes (2 × 2)*(2 × 1). For the operation to be possible, the underlined numbers (middle numbers) must be the same. The resultant matrix becomes a 2 × 1, which are the remaining or outside numbers. The result of the multiplication is

$$A * x = \begin{bmatrix} 1 \times 5 + 2 \times 6 \\ 3 \times 5 + 4 \times 6 \end{bmatrix} = \begin{bmatrix} 17 \\ 39 \end{bmatrix}$$

B.7 SIMULTANEOUS EQUATIONS

One of the most helpful applications of using MATLAB to manipulate matrices is the ability to quickly solve simultaneous equations. To illustrate, consider three equations:

$$8Z + 3X + 6Y = 25$$
$$5Y + 7Z - 36 = 0$$
$$4X - 6Y - 8Z = -18$$

While we could solve these equations using algebra, it would take a lot of time and effort. It would take even longer if there were more equations and unknowns. With MATLAB, as the number of equations increases, the only time that significantly increases is the time it takes to input the elements. MATLAB uses matrix algebra to solve the equations, so the equations must be written in matrix form. There will be three matrices. One is a vector of the variables (*x*, *y* and *z*) that we will call **x**. The second matrix contains the coefficients of the variables and is called **B**. The third vector **b** contains the constant terms of the equations.

The first step in writing the equations in matrix form it to put them in standard form:

$$3X + 6Y + 8Z = 25$$
$$5Y + 7Z = 36$$
$$4X - 6Y - 8Z = -18$$

Now the matrix form of **Ax = b** can be written as

$$\begin{bmatrix} 3 & 6 & 8 \\ 0 & 5 & 7 \\ 4 & -6 & -8 \end{bmatrix} \begin{bmatrix} X \\ Y \\ Z \end{bmatrix} = \begin{bmatrix} 25 \\ 36 \\ -18 \end{bmatrix}$$

If we follow the rules of matrix multiplication, we get our original three equations back. If the equation of **Ax = b** were an algebraic one, we would simply divide both sides by **A** and solve for **x** as **x = b/A.** With MATLAB, we use the left division operator as **x = A\b.**

Example B.9

Solve the following simultaneous equations using MATLAB:

$$3X + 6Y + 8Z = 25$$
$$5Y + 7Z = 36$$
$$4X - 6Y - 8Z = -18$$

In MATLAB, enter:

```
» A=[3 6 8;0 5 7;4 -6 -8]
A =
    3   6   8
    0   5   7
    4  -6  -8
» b=[25; 36; -18]
b =
    25
    36
   -18
» x=A\ b

x =
     1.0000
   -67.0000
    53.0000
```

B.9 PLOTTING

MATLAB is very useful for plotting data. For the purposes of this chapter, we will discuss two-dimensional plotting on linear, semi-log and log scales. However, there are other types of plots including 3-D plotting available in MATLAB. By using the **demo** command, the demonstrations, which were referred to earlier, will provide a good introduction.

The basic function in graphing is **plot** and is used to generate linear *xy* plots. The plot function requires:

plot(*x* axis values, *y* axis values, 'symbol or line type')

The *x*- and *y*-axis values must be stored in MATLAB as vectors and they must have the same number of elements. The symbol or line type allows the user to specify symbols, type of line, and color. Note that the line type and symbol are enclosed in single quotes. The MATLAB names for different symbols, line types, and colors are shown in Table B.1. Symbols and line types can be combined but they must be in the same color.

To plot multiple data sets, we can use the **plot** command by listing two different pairs of data. For example, if we want to plot data stored in vectors **x** and **y** as well as **w** and **z**, we can enter:

plot(x,y,'-k',w,z,':g')

The result will be a plot of **x** and **y** with a solid black line and **w** and **z** with a dotted green line.

A second way to plot multiple data sets is to use the **hold** command. By entering **hold on**, subsequent plotting commands will be plotted on the same plot. Entering **hold off** can turn off the hold command. Example B.10 shows how this can be used to plot data multiple times. In this example, the axes are labeled using the commands **xlabel**, **ylabel**, and **title**. Notice that the descriptive text is enclosed in parentheses and single quotes. The axes can be scaled to different values using the **axis** command.

If you wish to open multiple figure windows, MATLAB has the command **figure(n),** where *n* is the figure number. For instance, **figure(2)** would open a second figure window and label it "figure 2." This will make it active and any subsequent plotting will be done in that window. To switch back from **figure 2** to the first figure, simply type **figure(1)** or click on **figure 1** with the mouse.

The command **subplot** is used to generate multiple plots on a single figure. The command is used as **subplot(i,j,k)**, where *i* and *j* indicate the number of rows and columns of figures and *k* represents the graph number. The graph numbers begin in the upper left-hand corner and move left to right across a row and then down to the left side of the next row. For example, if we wish to put six plots on a single figure, we would enter **subplot(3,2,1)** to plot the figure in the upper left hand corner. To activate the plot on the right side of the second row, enter **subplot(3,2,4)**. Example B.11 demonstrates the use of the **subplot** command. In this example, four types of 2-D plots are demonstrated.

TABLE B.1 Line and Symbol Options

Color	Symbol	Line type
y yellow	. point	- solid
m magenta	o circle	: dotted
c cyan	x x-mark	-. dashdot
r red	+ plus	--- dashed
g green	* star	
b blue	s square	
w white	d diamond	
k black	v triangle (down)	
	^ triangle (up)	
	> triangle (right)	
	p pentagram	
	h hexagram	

Semi-log plots are also shown in Example B.11. In a semi-log plot, one axis has a log scale while the other has a linear scale. MATLAB has **semilogx** and **semilogy** commands to make the *x*- and *y*-axis the log scale, respectively. To make both scales log, use the plotting command **loglog**.

Example B.10

Create a plot of *x* and *y* values by entering the data into vectors named **x** and **y**. Then plot the data and label the graph as shown below.

```
» x=[1 2 3 4 5 6 7 8 9]

x =

   1  2  3  4  5  6  7  8  9
```

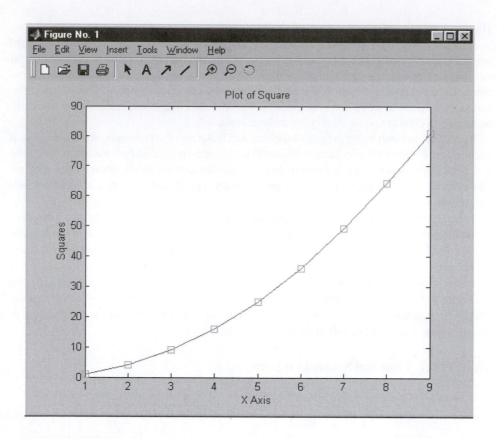

» y=[1 4 9 16 25 36 49 64 81]

y =
  1 4 9 16 25 36 49 64 81

» plot(x,y,'-b')
» hold on
» plot(x,y,'rs')
» title('Plot of Square')
» xlabel('X Axis')
» ylabel('Squares')

Example B.11

Plot the function $y = e^x$ on linear, semilog, and log-log plots. Place four plots on a single figure.
  MATLAB commands:

        »figure(2) % generates a new figure window

        » x=[1 2 3 4 5 6 7 8 9] % enter the x data
        x =
            1 2 3 4 5 6 7 8 9

        » y=exp(x) % generating y data using the x data

```
y =
Columns 1 through 6
    2.7183  7.3891  20.086  54.598  148.41  403.43
Columns 7 through 9
    1096.6   2981  8103.1

» subplot(2,2,1) % selects the upper left hand corner plot
» plot(x,y)
» subplot(2,2,2) % selects the upper right hand corner plot
» loglog(x,y) % plots log scales on both x and y axes
» subplot(2,2,3)
» semilogx(x,y)
» subplot(2,2,4)
» semilogy(x,y)
```

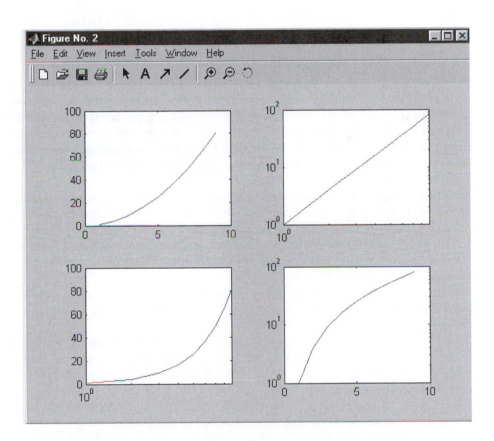

B.10 PROGRAMMING

In MATLAB we can write programs similar to a programming language. These programs, also called scripts, consist of a series of MATLAB commands that can be saved to run later.

To create a MATLAB program, we need to open an editor. In UNIX, open any editor of your choice. On a PC or Mac, there is a MATLAB editor that can be accessed through the file menu. Under **new**, select **M-file** to open the programming editor.

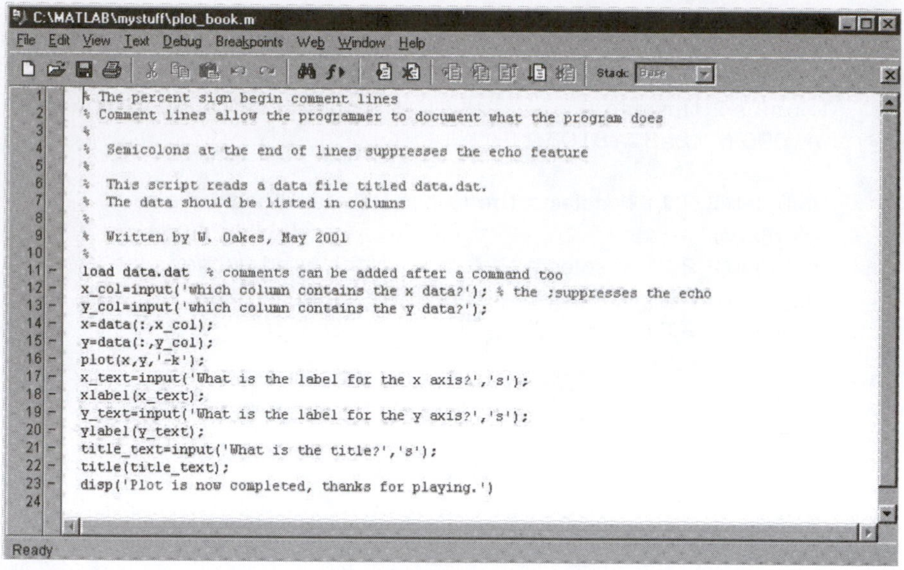

Figure B.3 Editor window.

In the editor, you can enter MATLAB commands just like you would type them into the workspace. You can also add comment statements by using the % symbol. Anything entered after the % will be ignored by MATLAB so this is where you can enter comments for your own use. It is good programming practice, regardless of the programming language, to use comment statements to document your program. They will help you understand what the program does when you come back to it later.

Once the commands and comment lines have been entered into the file, it must be saved in a file with a **.m** extension. The naming rules are the same as with the **.mat** (workspace) files described earlier in this chapter. Remember to avoid file names that MATLAB already uses. Also, similar to saving the **.mat** (workspace) files, you must also save the file in a directory that MATLAB will access. This may require using the ***path*** and ***addpath*** commands outlined earlier.

After the file is saved, it can be executed by typing the file name (without the **.m** extension) at the MATLAB prompt. For example, if the script name was "**example_script.m**," you would type "**example script**" at the MATLAB prompt.

There are times when you will require the user to enter data into a program. The ***input*** command is very useful for this function. This command allows the user to enter data in response to a MATLAB prompt. For example, if we want the user to enter a number that the program might use, we could write

W = input('Enter a number to be used by the program')

The result is that MATLAB will display the prompt

Enter a number to be used by the program

and will wait for the user to enter a number and then press return. Once the number is entered, it is stored as the value for the variable *W* and can be used in the program as *W*.

If a word or letters are required rather than numbers, we must tell MATLAB to expect letters, which are called strings. This requires adding a **,'s'** to the end of the input line. To ask the user for a string would require the following:

my_word = input('Enter a word: ','s')

Example B.12 uses string inputs to label the plots.

If data is to be displayed, the function ***disp*** can be used. This function will display the contents of a variable. ***Disp(W)*** would display the value of *W* while ***disp(my_word)*** would display the contents of the variable **my-word**. Text can be displayed as it is in Example B.12 using single quotes. To wish the user good luck, enter ***disp('good luck')*** and "**good luck**" will be displayed on the screen.

Example B.12

In an editor, enter and save the following code under the name '**plot_example.m**' in the directory **c:\matlab\mystuff**.

```
% The percent sign (%) begins comment lines
% Comment lines allow the programmer to document the program
%
% Semicolons at the end of lines suppresses the echo
% feature in MATLAB
%
% This script reads a data file titled data.dat.
% The data to be plotted is listed in columns
%
% Written by W. Oakes
%
load data.dat % comments can also be added after a command
x_col=input('which column contains the x data? ');
y_col=input('which column contains the y data? ');
x=data(:,x_col);
y=data(:,y_col);
plot(x,y,'-k');
x_text=input('What is the label for the x axis? ','s');
xlabel(x_text);
y_text=input('What is the label for the y axis? ','s');
ylabel(y_text);
title_text=input('What is the title? ','s');
title(title_text);
disp('Plot is now completed, thanks for playing.')
```

To run the program, the following was typed in Matlab

```
» addpath c:\matlab\mystuff
» plot_example
which column contains the x data? 1
which column contains the y data? 2
```

What is the label for the x axis? x axis
What is the label for the y axis? y axis
What is the title? Title of a Wonderful Graph
Plot is now completed, thanks for playing.
    »

The result is a plot with a title and *x*-and *y*-axes labeled.

For MATLAB product information, please contact: The MathWorks, Inc., 3 Apple Hill Dr., Natick, MA 01760-2098, ph: 508-647-7000, fax: 508-647-7101, email: info@mathworks .com, web: www.mathworks.com.

EXERCISES

B.1   Use the *lookfor* command to find the MATLAB commands that deal with:
a) sine
b) plot
c) logarithm
d) variable
e) path
f) exponential

B.2   Use the *help* function in MATLAB to find out what five functions do. Summarize these functions in a short report.

B.3   Run the MATLAB demo by typing *demo* at the MATLAB prompt. Explore at least two of the functions and prepare a one-page report on how MATLAB will be helpful to you as an engineering student.

B.4   Run the MATLAB demo by clicking on the demos icon in the Launchpad window to explore the graphics capabilities of MATLAB and prepare a one-page report on how MATLAB 's graphics capabilities could be used in engineering applications.

B.5   Use MATLAB to do simple arithmetic using variables. Print the screen you are working with to show the variables you created.

B.6   Solve $\mathbf{y} = 0.5\mathbf{x} - \mathbf{w}$, where $\mathbf{x} = $ [2 64 82 106 94] and $\mathbf{w} = $ [5 6 7 8 9] using MATLAB.

B.7   Write the following equations in matrix form ($\mathbf{Ax} = \mathbf{b}$):

a) $x - 14 = -2y - 3z$

b) $40 - 4x = 6y + 8z$

c) $x + 3y + 7z - 28 = 0$

B.8   Solve Exercise B.7 using MATLAB.

B.9   Write a brief description of the following MATLAB commands:

a) round

b) format

c) ceil

d) floor

e) fix

f) abs

g) rem

h) exp

B.10   Create a **y** vector such that $y = e^x$ where **x** = [1,2,3,4,5,6,7,8,9]. Plot the result on an *xy* graph. Label the *x*- and *y*-axes and place a title on the graph.

B.11   Create a graph of **x** = [2 4 6 8 10 12] and **y** = [8 32 72 128 200 288]. Label the graph using red squares for the symbols connected with a dashed blue line.

B.12   Replot the data from Exercise B.10. Is there a graph type (e.g., linear, semilog, or log-log) where the graph becomes a straight line?

B.13   Replot the data from Exercise B.11. Is there a graph type (e.g., linear, semilog, or log-log) where the graph becomes a straight line?

B.14   Create symbol *x* and find the roots of:

a) $x^2 - 4$

b) $x^2 - 2x - 195$

c) $x^3 + 2x^2 - 13x + 10$

B.15   Use the **simple** command to multiply $(x + 4)(x - 5)(x + 6)(x - 2)$.

B.16   Use MATLAB to integrate the following:

a) $3x^2 + 2x + 1$

b) $\sin(x)$

c) $\tan(x)$

d) $\sin^2(x)$

B.17   Use MATLAB to differentiate the following:

a) $3x^3 + 4x^2 - 2x - 8$

b) $\cos(x)$

c) $\tan^2(x)$

B.18   Create a workspace with variables and save the file in a directory named **mat-lab_stuff**. Clear the workspace and reload the file.

B.19   Write a MATLAB script that will plot data that the user inputs. Select some data and show a plot.

B.20   Write a MATLAB script that will ask the user their name and age and calculate the number of days the person has been living. Demonstrate the script with an example.

B.21   For x=2 and y=5, compute the following using MATLAB
a.) yx3/(x-y)

b.) 3x/2y

c.) 3xy/2

d.) x5/(x5-1)

B.22   For x=3 and y=4, use MATLAB to solve the following

a.) (1-(1/x5))-1

b.) 3ρx2

c.) 3y/(4x-8)

d.) 4(y-5)/(3x-6)

B.23   Suppose x takes on the values of x=1, 1.2, 1.4, . . . ,5. Use MATLAB to compute the array y that results from the function y=7sin(4x). Plot the results

B.24   Create, save a run a script that solves the following set of equations for given values of a,b and c. Check your script for the values of a=95, b=−50 and c=145.

7x+14y-6z=a

12x-5y+9z=b

−5x+7y+15z=c

B.25   The volume of a sphere is given by $V=4\pi r^3/3$, where r is the radius of the sphere. Make a plot of volume vs. radius for radii of 0.1 to 10.

B.26   Repeat the previous problem using four subplots, one with linear axes, one with a semilog on the x axix, one with a semilog on the y axis and one with both axes having a log scale (log-log). Label each plot properly.

B.27   Given the Matrices

$$A = \begin{bmatrix} -7 & 16 \\ 4 & 9 \end{bmatrix} \quad B = \begin{bmatrix} 6 & -5 \\ 12 & -2 \end{bmatrix} \quad C = \begin{bmatrix} -3 & -9 \\ 6 & 8 \end{bmatrix}$$

use MATLAB to find

a.) A+B+C

b.) A-B+C

c.) Verify the associative law (A+B)+C=A+(B+C)

d.) Verify the commutative law A+B+C=A+C+B=B+C+A

B.28   Use MATLAB to solve (for x=2)

$$(6x^3+4x^2-5)/(12x^3-7x^2+3x+9)$$

B.29   . . . for x=1,1.1,1.2, . . . , 2 and $y=3x^3+4x^2-5x-7$, plot x vs y using a blue dashed line with green circles for the data points. Provide a title and labels for the x and y axes.

B.30   . . . for x=1,1.1,1.2, . . . , 2 and $y=x^2$, plot x and y using:

a.) Red squares for the data points

b.) Red solid line and blue squares for the data points

c.) Yellow dotted line with green triangles for the data points

d.) Green dashed line with black circles for the data points.

Index